# Gmelin Handbook of Inorganic Chemistry

8th Edition

## Volumes published on "Rare Earth Elements" (Syst.-No. 39)

| | |
|---|---|
| ***A 1** | Introductory Review. History. Occurrence – 1938 |
| ***A 2** | Scandium: History. Occurrence – 1973 |
| ***A 3** | Y, La, and the Lanthanides: Cosmochemistry – 1980 |
| ***A 4** | Y, La, and the Lanthanides: Crystal Chemistry – 1979 |
| ***A 5** | Y, La, and the Lanthanides: Geochemistry – Total Earth. Magmatic Cycle – 1981 |
| **°B 1** | Sc, Y, La, and the Lanthanides: History, Position in the Periodic System. Separation from Raw Materials – 1976 |
| ***B 2** | Sc, Y, La, and the Lanthanides: Separation of the Rare Earth Elements from One Another. Preparation of the Pure Metals. Uses. Toxicology – 1976 |
| ***B 3** | Sc, Y, La and the Lanthanides: Physical Properties of the Metals – 1974 |
| **°B 4** | Sc, Y, La, and the Lanthanides: Nuclear, Atomic, and Molecular Properties – 1976 |
| ***B 5** | Sc, Y, La, and the Lanthanides: Preparation, Enrichment, and Separation of Isotopes (Sc, Y, La to Sm) – 1978 |
| ***B 6** | Sc, Y, La, and the Lanthanides: Preparation, Enrichment, and Separation of Isotopes (Eu to Lu). Detection and Determination of Isotopes. Chemical Reactions of the Elements – 1978 |
| ***B 7** | Sc, Y, La, and the Lanthanides: Reactions of the Ions in Solution. Electrochemical Behavior – 1979 |
| ***C 1** | Sc, Y, La, and the Lanthanides: Hydrides. Oxides – 1974 |
| ***C 2** | Sc, Y, La, and the Lanthanides: Compounds with H+O, Alkali Metals+O, N, as well as the Related Alkali Double Salts – 1974 |
| ***C 3** | Sc, Y, La, and the Lanthanides: Fluorides, Fluoride Oxides, and the Related Alkali Double Salts – 1976 |
| **C 4a** | Sc, Y, La, and the Lanthanides: Chlorides. Comparative Data – 1982 |
| ***C 5** | Sc, Y, La, and the Lanthanides: Oxide Chlorides, Hydroxide Chlorides, Salts of Oxoacids of Chlorine, and Alkali Chlorometallates – 1977 |
| ***C 6** | Sc, Y, La, and the Lanthanides: Bromides. Iodides, Corresponding Basic Halides. Salts of Halogen Oxoacids and Alkali Double Salts – 1978 |
| **°C 8** | Sc, Y, La, and the Lanthanides: Sulfide Halides. Sulfates, Salts of Other Sulfuric Acids and Corresponding Alkali Double Compounds – 1981 |
| **D 1** | Sc, Y, La, and the Lanthanides: Coordination Compounds 1 – 1980 |
| **D 2** | Sc, Y, La, and the Lanthanides: Coordination Compounds 2 – 1981 |
| **D 3** | Sc, Y, La, and the Lanthanides: Coordination Compounds 3 – 1981 |

* Completely or ° in part in German

# Gmelin Handbuch der Anorganischen Chemie

Achte völlig neu bearbeitete Auflage

BEGRÜNDET VON Leopold Gmelin

ACHTE AUFLAGE BEGONNEN im Auftrag der Deutschen Chemischen Gesellschaft von R.J. Meyer

FORTGEFÜHRT VON E.H.E. Pietsch und A. Kotowski
Margot Becke-Goehring

HERAUSGEGEBEN VOM Gmelin-Institut für Anorganische Chemie der Max-Planck-Gesellschaft zur Förderung der Wissenschaften
Direktor: Ekkehard Fluck

Springer-Verlag
Berlin · Heidelberg · New York 1982

# Gmelin Handbook of Inorganic Chemistry

8th Edition

---

# Sc, Y, La–Lu
# RARE EARTH ELEMENTS

**Part C 4a**

Chlorides. Comparative Data

With 51 illustrations


AUTHORS Inge Flachsbart, Hiltrud Hein, Gerhard Kirschstein, Ernst Koch, Irmgard Kreuzbichler, Peter Kuhn, Herbert Lehl †, Ursula Vetter

EDITORS Hartmut Bergmann, Hiltrud Hein, Ernst Koch, Peter Merlet, Ursula Vetter

CHIEF EDITOR Hartmut Bergmann


**System Number 39**


Springer Verlag
Berlin · Heidelberg · New York 1982

LITERATURE CLOSING DATA: END OF 1980
IN MANY CASES MORE RECENT DATA HAVE BEEN CONSIDERED



Library of Congress Catalog Card Number: Agr 25-1383

ISBN 3-540-93422-7 Springer-Verlag, Berlin · Heidelberg · New York
ISBN 0-387-93422-7 Springer-Verlag, New York · Heidelberg · Berlin



Gesamtherstellung Universitätsdruckerei H. Stürtz AG, Würzburg

## Foreword

The volumes "Rare Earth Elements" C 4 a and C 4 b deal with the rare earth chlorides and the rare earth chloride systems. Comparative and broadly valid data for these compounds and systems are presented in volume C4a, data on individual chlorides and chloride systems are given in volume C4b.

Molecules and ions in the gas phase or as matrix-isolated species are described in the first section of the present volume C 4 a. The subsequent sections treat the properties of chlorides in the solid state, melts, and solutions. Several interesting crystallographic properties have been found for the chlorides of rare earths in lower oxidation states. The most space in this volume is occupied by the description of the anhydrous trichlorides. Main topics of this section are preparation, crystallographic properties, resonance spectroscopy, thermal properties, and chemical reactions; many of the data are tabulated. Similar data are presented for the hydrated trichlorides. Another large section covers the physical properties of trichloride solutions.

The physical properties of rare earth ions in crystal lattice as well as the chemistry of complexes are only shortly described, since these topics are reserved for later volumes of the Gmelin Handbook. Values on physical properties are converted into SI units only when it seems necessary for uniformity. The most important conversion factors are compiled in a table on pp. 271/2.

Frankfurt am Main
December 1981

H. Bergmann

# Table of Contents

Page

# 11 Chlorides

General Literature:

J.M. Haschke in: K.A. Gschneidner, L.R. Eyring, Handbook on the Physics and Chemistry of Rare Earths, Vol. 4, North-Holland Publishing Company, Amsterdam-New York-Oxford 1979, pp. 89/151.

Ya.I. Gerasimov, A.N. Krestovnikov, A.F. Kuznetsov, V.I. Lavrent'ev, A.L. Lomov, Khimicheskaya Termodinamika v Tsvetnoi Metallurgii, Vol. 7, Termodinamika Skandiya Ittriya Redkozemel'nykh Elementov i Ikh Vazhneishikh Soedinenii, Metallurgiya, Moscow 1975, pp. 123/37.

D. Brown, Halides of the Lanthanides and Actinides, Wiley-Interscience, London-New York-Sydney 1968.

R.E. Thoma in: L. Eyring, Progress in the Science and Technology of the Rare Earths, Vol. 2, Pergamon Press, Oxford-London-Edinburgh-New York-Toronto-Sydney-Paris-Braunschweig 1966, pp. 90/122.

V.V. Serebrennikov, Khimiya Redkozemel'nykh Elementov, Vol. 1, Tomsk 1959, pp. 237/43.

## 11.1 Molecules and Ions

This section describes molecules and ions only in the gas phase or as matrix-isolated species (in noble gases, nitrogen, etc.). Ions and molecules in the crystal lattice, in solution, and in melts will be treated in Section 11.2, p. 19.

Review

MCl molecules (M=Sc, Y, Yb, and Lu) have been identified and studied in the gas phase of low pressure discharges. Molecular constants were determined by analyzing the electronic emission spectra.

Molecular constants for some $MCl_2$ molecules have been evaluated from the infrared spectra of the matrix-isolated species. $MCl_2$ molecules are probably bent.

Electron diffraction shows that the $MCl_3$ molecules for M=Y, La, Pr, Gd, Tb, Ho, and Lu are pyramidal. There are relatively few spectra of the vapor. Many of the molecular constants of the $MCl_3$ as well as for MCl and $MCl_2$ molecules have been estimated.

Both the dichlorides and trichlorides form volatile complexes with the trichlorides of aluminium.

### 11.1.1 Molecules of the Type MCl

#### MCl with M=Sc, Y, La

Molecular Constants and Spectra

ScCl is formed by electrical discharge in low pressure (2 to 4 Torr) $ScCl_3$-Ar or $SCl_3$-Ar-He mixtures (cold spot temperature ≈500 °C) and is identified by its emission spectrum. The spectrum between 900 and 200 nm consists of nine band systems, of which five correspond to transitions from singlet states to the $^1\Sigma$ ground state and three to transitions between triplet states. The ninth band system (368 to 360 nm) was not assigned. For the singlet states, analysis of the bands gives the term values

$T_e$ or $T_o$, the vibrational frequencies $\omega_e$, the anharmonicity constants $\omega_e x_e$, the rotational constants $B_e$ (or $B_o$) and $D_o$, the vibration-rotation interaction constants $\alpha$, and the interatomic distances $r_e$ or $r_o$ shown in the following table. The $r_e$ and $r_o$ values are in Å; all other values are in $cm^{-1}$.

| state | $T_e(T_o)$ | $\omega_e$ | $\omega_e x_e$ | $B_e(B_o)$ | $10^7 D_o$ | $10^4 \alpha$ | $r_e(r_o)$ |
|---|---|---|---|---|---|---|---|
| $F\,^1\Pi$ | 31249.9 | 364.7 | 1.0 | – | – | – | – |
| $E\,^1\Sigma$ | 27033.3 | 472.1 | 1.32 | – | – | – | – |
| $D\,^1\Pi$ | 21521.1 | 373.1 | 1.6 | 0.1569($B_o$) | 1.6 | – | 2.337($r_o$) |
| $B\,^1\Pi$ | 17576.6($T_o$) | 374.3 | 2.3 | 0.1551($B_o$) | 1.1 | – | 2.351($r_o$) |
| $A\,^1\Sigma(Sc^{35}Cl)$ | 12431.2 | 373.9 | 0.9 | 0.1574($B_o$) | 1.1 | – | 2.333($r_o$) |
| $A\,^1\Sigma(Sc^{37}Cl)$ | 12393.1($T_o$) | – | – | 0.1524($B_o$) | 0.9 | – | 2.335($r_o$) |
| $X\,^1\Sigma(Sc^{35}Cl)$ | 0 | 447.4 | 1.8 | 0.1725 | 1.1 | 10 | 2.229 |
| $X\,^1\Sigma(Sc^{37}Cl)$ | 0 | – | – | 0.1664($B_o$) | 0.8 | – | 2.234($r_o$) |

The radiative lifetimes of the (0,0) bands of three singlet systems are (in ns) 21±2 (B–X), 20±2 (D–X), 17±2 (E–X) [24].

Two of the triplet systems correspond to the transitions $c\,^3\Delta - a\,^3\Delta$ and $b\,^3\Phi - a\,^3\Delta$, where $c\,^3\Delta$ ($\omega_e$=355.9, $\omega_e x_e$=2.18 $cm^{-1}$) and $b\,^3\Phi$ are, respectively, 13113.8 and 12567.6 $cm^{-1}$ above $a\,^3\Delta$ ($\omega_e$=398.3, $\omega_e x_e$=1.36 $cm^{-1}$). The third triplet system transition involves a $^3\Sigma$ term ($\omega_e$=297.3, $\omega_e x_e$=0.61 $cm^{-1}$) and a $^3\Pi$ term ($\omega_e$=312.5, $\omega_e x_e$= 0.55 $cm^{-1}$) which is 22260.9 $cm^{-1}$ higher, Shenyavskaya et al. [1].

Mass-spectrometric measurements of the Y-$Cl_2$ surface reaction at temperatures up to 1750 K show $YCl^-$. Presumably $YCl_2$, and not YCl, is the initial product, McKinley [3]. YCl, like ScCl, is formed from the trihalide in a low pressure discharge; the emission spectrum was observed with the cold spot temperature at 700 to 800 °C, Janney [2]. Between 695 and 630 nm only one electronic transition of the $^1\Sigma - {}^1\Sigma$ type is identified. Analysis of the vibrational band systems and of the rotational structure of the (0,0), (0,1), and (1,0) bands gives the following values for the molecular constants (in $cm^{-1}$) and interatomic distances (in Å) [2]:

| state | $T_0$ | $\omega_e$ | $\omega_e x_e$ | $B_e$ | $10^8 D_0$ | $10^4 \alpha_e$ | $r_e$ |
|---|---|---|---|---|---|---|---|
| $Y^{35}Cl$ upper state | x+14877.60 | 324.5 | 1.14 | 0.1089 | 9.0 | 7 | 2.483 |
| $Y^{35}Cl$ lower state | x | 380.7 | 1.3 | 0.1162 | 9.3 | 3 | 2.404 |
| $Y^{37}Cl$ upper state | x′+14878.09 | 317.9 | 1.07 | 0.1048 | 8.5 | 8 | 2.488 |
| $Y^{37}Cl$ lower state | x′ | 372.9 | 1.2 | 0.1118 | 8.3 | 5 | 2.403 |

Two higher excited singlet states were found by Fischell et al. [24], who recorded the excitation spectrum of YCl between 460 and 360 nm with the laser-induced fluorescence method. The following molecular constants and interatomic distances were derived (presumably for $Y^{35}Cl$):

| $T_e$ | $\omega_e$ | $\omega_e x_e$ | $B_e$ | $r_e$ |
|---|---|---|---|---|
| 22787±2 | 345±2 | 1.0±0.3 | 0.1108 | 2.463±0.005 |
| 27116±2 | 335±2 | 1.0±0.3 | 0.1101 | 2.470±0.005 |

The radiative lifetimes of the (0,0) bands are 28±2 and 21±2 ns, respectively, as compared with 36±3 ns for the (0,0) band of the $^1\Sigma - {}^1\Sigma$ system [24].

The vibrational frequencies estimated by Krasnov [4] for ScCl (478 $cm^{-1}$) and YCl (410 $cm^{-1}$) are too high compared to the experimental values in the preceeding tables. Consequently, $\omega_e$ for LaCl must also be smaller than the estimated value (376 $cm^{-1}$ [4]). The interatomic distances of ScCl, YCl, and LaCl must also be shorter than the values estimated by Spiridonov, Tatevskii [5].

Neither ScCl nor YCl are the emitting species of the chemiluminescence spectra observed in the reactions of Sc or Y with $Cl_2$ as was claimed by Gole et al. [25]. These spectra, rather, are emitted by $ScCl_2$ or $YCl_2$, [26], see p. 6.

Thermodynamic Data

The following dissociation energies D were derived from band-spectra measurements: for ScCl, $D_e$ is 3.4 eV [1]; for YCl, $D_e$ is 3.5 eV in the ground state and 2.9 eV in the excited state A [2]. Since estimated D values for ScCl (4.0 eV, Krasnov [6]) and for YCl (5.1 eV, Krasnov, Timoshinin [7]) are distinctly higher, the D value estimated for LaCl (4.8 eV [6, 7]) may also be too high. For LaCl, Chervonnyi [8] estimated $D_{298} = 115 \pm 6$ kcal/mol (=5.0 eV) by a method similar to that used for estimating the dissociation energy of MCl, where M=Ce through Lu (see p. 4).

Recommended values for ScCl and YCl, respectively: enthalpies of formation from the elements at 0 K are $\Delta H_0^\circ = 27$ and 48 kcal/mol; $\Delta H_{298}^\circ = 26.9$ and 47.8 kcal/mol; $\Delta G_{298}^\circ =$ 20.6 and 41.5 kcal/mol, Wagman et al. [9]. For the formation of ScCl, YCl, and LaCl from the atoms, values of the change in enthalpy, $-\Delta H_f^\circ = 89$, 94, and 99 kcal/mol, were extrapolated by Novikov [10] from data for other transition metal monohalides.

Pre-1969 molecular data for ScCl (probably estimated), YCl (experimental data), and LaCl (estimated data) were used by Krasnov, Danilova [11] to calculate thermodynamic potentials $\Phi^\circ$, which in turn were used to calculate entropies S° and enthalpies $H^\circ - H_0^\circ$ at 298.15, 1000, 2000, and 3000 K. The following table shows the values at 298.15 K [11]; recommended values [9] are given in parentheses:

| MCl | ScCl | YCl | LaCl |
|---|---|---|---|
| $\Phi^\circ$ in $cal \cdot mol^{-1} \cdot K^{-1}$ | 48.450 | 50.644 | 52.010 |
| S° in $cal \cdot mol^{-1} \cdot K^{-1}$ | 55.966 (56.00) | 58.290 (58.33) | 59.703 |
| $H^\circ - H_0^\circ$ in cal/mol | 2241 (2246) | 2280 (2286) | 2294 |

The heat capacities of gaseous ScCl and YCl are $C_p^\circ = 8.40$ and 8.56 $cal \cdot mol^{-1} \cdot K^{-1}$ at 298 K [9].

## MCl with M=Ce to Lu

Formation

Heating elemental rare earths with $BaCl_2$ or $EuCl_2$ in a Knudsen cell gives MCl (as well as $MCl_2$ and $MCl_3$) as shown by mass-spectrometric measurement of the gas phase. Specifically, MCl was identified at temperatures greater than 1100 K over M-$BaCl_2$ mixtures (M=Ce, Pr, Nd, Sm, Eu, Tb, Dy, Ho, Er, Tm, Yb, or Lu) and above M-$EuCl_2$ mixtures (M=Nd, Gd, Dy, or Tm). Quantitative interpretation of the mass-spectrometric data gave an equilibrium constant of 0.259 for the reaction $Tb(g) + BaCl(g) \rightleftharpoons Ba(g) + TbCl(g)$ at 1523 K, Chervonnyi [12]. YbCl has been generated for spectroscopic studies either by the reaction of Yb with $ClO_2$ under molecular beam conditions, Lee, Zare [13], or from $HgCl_2$ and Yb in a microwave discharge, Kramer [14]. Spectroscopic evidence shows the formation of LuCl by microwave discharge in a mixture of Hg, Lu, $HgCl_2$, CsCl, and Ar, Kramer [15].

### Term Values, Vibrational Constants, Interatomic Distances

For YbCl (and YbF, cf. "Seltenerdelemente" C3, 1976, p. 9) there are three known excited states ($^2\Pi_{1/2}$, $^2\Pi_{3/2}$, and $^2\Sigma^+$), which give rise to three band systems ($A^2\Pi_{1/2}-X^2\Sigma$, $A^2\Pi_{3/2}-X^2\Sigma$, and $B^2\Sigma^+-X^2\Sigma$). The bands were first analyzed by Gatterer et al. [16]; results are also reported in Barrow, Wieland [17]. For YbCl the band heads of the three systems have the wave numbers $T_0=17892.4$, 19381.2, and 19940.6 $cm^{-1}$ [13]. More recent measurements give $T_0=17892.1$ and 19369.2 $cm^{-1}$ for the two $A^2\Pi-X^2\Sigma$ bands and give the following molecular constants (in $cm^{-1}$; prime and double prime denote upper and lower states, respectively) [14]:

| state | $T_e$ | $\omega_e'$ | $\omega_e'x_e'$ | $\omega_e''$ | $\omega_e''x_e''$ |
|---|---|---|---|---|---|
| $^2\Pi_{1/2}$ | 17882.0±1.3 | 313.0±0.6 | 0.89±0.03 | 292.3±0.7 | 0.89±0.04 |
| $^2\Pi_{3/2}$ | 19357.3±1.2 | 310.8±1.7 | 1.18±0.53 | 287.8±2.1 | 1.98±0.64 |

For the $B^2\Sigma^+$ state, the molecular constants are $T_e=19927.7\pm3.3$, $\omega_e'=313.60\pm0.28$, and $\omega_e'x_e'=0.844\pm0.005$ $cm^{-1}$ [13]. Earlier, Filippenko et al. [18] estimated $\omega_e''\approx370$ $cm^{-1}$, and the interatomic distance r=2.28 Å for YbCl, furthermore $\omega_e''\approx375$ $cm^{-1}$, and r≈2.43 Å for EuCl.

The emission spectrum of LuCl in the visible consists of two red-degraded band systems, each with a different lower state (possibly $^1\Sigma^+$ and $^1\Delta$). Values for the one system are $T_e'-T_e''=20844.9\pm2.1$, $\omega_e'=323.0\pm2.2$, $\omega_e'x_e'=1.05\pm0.39$, $\omega_e''=350.6\pm2.6$, and $\omega_e''x_e''=0.78\pm0.51$ $cm^{-1}$; for the other system are $T_e'-T_e''=21828.1\pm2.2$, $\omega_e'=369.3\pm2.4$, $\omega_e'x_e'=1.79\pm0.18$, $\omega_e''=386.4\pm2.5$, and $\omega_e''x_e''=1.76\pm0.21$ $cm^{-1}$, Kramer [15].

### Ionization

The ionization energies of NdCl and GdCl are $E_i=7.6$ and 7.4 eV, respectively, as determined from the appearance potentials of the fragment ions from $NdCl_3$ (17.3 eV) and $GdCl_3$ (17.5 eV), Ciach et al. [19]. The ionization energy of EuCl was estimated as $E_i=5.4$ eV by comparison with the measured ionization energies of the alkaline earth monochlorides, Hariharan, Eick [20]. Estimated values for YbCl are $E_i=6.05\pm0.20$ eV, Hariharan et al. [21], and $E_i\leqq6$ eV, Diebold et al. [22].

### Dissociation

For EuCl, the dissociation energy $D_0^\circ=98\pm5$ kcal/mol (=4.3±0.2 eV) was obtained from $D_0^\circ$ values of the alkaline earth monochlorides, Hariharan, Eick [20]. For YbCl, the values D≦3.2 eV (=73.8 kcal/mol) [12] or $D_0=2.96$ eV (=68.3 kcal/mol) [14] were derived from the band spectrum, and Polyachenok [23] used an empirical formula to calculate $D_0=90$ kcal/mol. For TbCl, $D_{298}=110\pm5$ kcal/mol was derived from the heat of reaction for BaCl+Tb→TbCl+Ba and from the dissociation energy of BaCl ($D_{298}=107\pm2$ kcal/mol). The difference between the dissociation energy of TbCl and the atomization energy of $TbCl_2$ is 114 kcal/mol. Assuming that the difference between the dissociation energy of the monohalide and the atomization energy of the dihalide for the other rare earth also equals 114 kcal/mol, Chervonnyi [8] obtained the following values for $D_{298}$ of MCl (in kcal/mol±6 kcal/mol):

| M . . . . | Ce | Pr | Nd | Pm | Sm | Eu | Gd | Tb | Dy | Ho | Er | Tm | Yb | Lu |
|---|---|---|---|---|---|---|---|---|---|---|---|---|---|---|
| $D_{298}$ . . . | 113 | 110 | 107 | 103 | 101 | 98 | 111 | 110 | 97 | 100 | 103 | 91 | 85 | 108 |

Values estimated earlier for EuCl (77 kcal/mol) and YbCl (76 kcal/mol) [18] are thus too low.

References:

[1] E.A. Shenyavskaya, A.A. Mal'tsev, D.I. Kataev, L.V. Gurvich (Opt. Spektroskopiya **26** [1969] 937/44; Opt. Spectrosc. [USSR] **26** [1969] 509/12), E.A. Shenyavskaya, A.A. Mal'tsev, L.V. Gurvich (Vestn. Mosk. Univ. Ser. II Khim. **22** No. 4 [1967] 104/5; C.A. **68** [1968] No. 100373). – [2] G.M. Janney (J. Opt. Soc. Am. **56** [1966] 1706/11; Diss. Georgetown Univ. 1965; Diss. Abstr. **26** [1965/66] 3416). – [3] J.D. McKinley (J. Chem. Phys. **42** [1965] 2245/6; J. Chem. Phys. **41** [1964] 2814/7). – [4] K.S. Krasnov (Izv. Vysshikh Uchebn. Zavedenii Khim. Khim. Tekhnol. **10** [1967] 594/6). – [5] V.P. Spiridonov, V.M. Tatevskii (Zh. Fiz. Khim. **36** [1962] 2024/9; Russ. J. Phys. Chem. **36** [1962] 1086/9).

[6] K.S. Krasnov (Izv. Vysshikh Uchebn. Zavedenii Khim. Khim. Tekhnol. **12** [1969] 578/82). – [7] K.S. Krasnov, V.S. Timoshinin (Teplofiz. Vysokikh Temp. **7** [1969] 365/7; High Temp. [USSR] **7** [1969] 333/4). – [8] A.D. Chervonnyi (Zh. Fiz. Khim. **51** [1977] 1308/12; Russ. J. Phys. Chem. **51** [1977] 772/5). – [9] D.D. Wagman, W.H. Evans, V.P. Parker et al. (NBS-TN-270-5 [1971] 25, 30). – [10] G.I. Novikov (Probl. Sovrem. Khim. Koord. Soedin. **1970** No. 3, pp. 33/48, 47; C.A. **75** [1971] No. 10210).

[11] K.S. Krasnov, T.G. Danilova (Teplofiz. Vysokikh Temp. **7** [1969] 1213/6; High Temp. [USSR] **7** [1969] 1131/3). – [12] A.D. Chervonnyi (Zh. Fiz. Khim. **51** [1977] 1144/9; Russ. J. Phys. Chem. **51** [1977] 672/5). – [13] H.U. Lee, R.N. Zare (J. Mol. Spectrosc. **64** [1977] 233/43), H.U. Lee (Diss. Columbia Univ. 1976, pp. 1/120; Diss. Abstr. Intern. B **37** [1976] 2879). – [14] J. Kramer (J. Chem. Phys. **68** [1978] 5370/7). – [15] J. Kramer (J. Chem. Phys. **69** [1978] 2601/8).

[16] A. Gatterer, G. Piccardi, F. Vincenzi (Ricerche Spettroscopiche Lab. Astrofis. Specola Vaticana **1** [1942] 181/200; C. **1943** I 486). – [17] R.F. Barrow, K. Wieland (in: B. Rosen, Données Spectroscopiques Relatives aux Molécules Diatomiques, Oxford 1970, p. 410). – [18] N.V. Filippenko, E.V. Morozov, N.I. Giricheva, K.S. Krasnov (Izv. Vysshikh Uchebn. Zavedenii Khim. Khim. Tekhnol. **15** [1972] 1416/8). – [19] S. Ciach, A.J.C. Nicholson, D.L. Swingler, P.J. Thistlethwaite (Inorg. Chem. **12** [1973] 2072/4). – [20] A.V. Hariharan, H.A. Eick (High Temp. Sci. **4** [1972] 91/8).

[21] A.V. Hariharan, N.A. Fishel, H.A. Eick (High Temp. Sci. **4** [1972] 405/10). – [22] G.J. Diebold, F. Engelke, H.U. Lee, J.C. Whitehead, R.N. Zare (Chem. Phys. **20** [1977] 265/9). – [23] O.G. Polyachenok (Zh. Fiz. Khim. **40** [1966] 2644/7; Russ. J. Phys. Chem. **40** [1966] 1420/2). – [24] D.R. Fischell, H.C. Brayman, T.A. Cool (J. Chem. Phys. **73** [1980] 4260/72), D.R. Fischell (Diss. Cornell Univ. 1980; Diss. Abstr. Intern. B **41** [1980] 243). – [25] J.L. Gole, D.R. Preuss, C.L. Chalek (J. Chem. Phys. **66** [1977] 548/58), J.L. Gole, C.L. Chalek (Proc. Electrochem. Soc. **1978**, pp. 278/309; C.A. **90** [1979] No. 130198).

[26] H.C. Brayman, D.R. Fischell, T.A. Cool (J. Chem. Phys. **73** [1980] 4247/59).

## 11.1.2 Molecules of the Type $MCl_2$

### Formation

The gaseous dichlorides are obtained by vaporizing the solid compounds. Association to form $(MCl_2)_2$ could not be observed in the case of $EuCl_2$, and was not expected in the case of the other rare earth dihalides, Hastie et al. [1], Hariharan, Eick [2].

Quantitative mass-spectrometric measurements of the gas above condensed M-$BaCl_2$ or M-$EuCl_2$ systems (M=rare earth metals) at high temperatures allow calculation of the equilibrium constants for the gas-phase reaction

$$M+BaCl_2(EuCl_2) \rightleftharpoons MCl_2+Ba(Eu).$$

The equilibrium constants are as follows; temperatures are in parentheses [3]:

Reaction with $BaCl_2$

| | | | | | |
|---|---|---|---|---|---|
| Ce (1398 K) | Pr (1418 K) | Nd (1333 K) | Sm (1102 K) | Eu (1316 K) | Tb (1476 K) |
| 1.96 | 0.998 | 0.820 | 0.0641 | 0.0327 | 0.215 |
| Dy (1292 K) | Ho (1342 K) | Er (1326 K) | Tm (1353 K) | Yb (1198 K) | Lu (1368 K) |
| $5.20\times10^{-3}$ | $1.77\times10^{-2}$ | $9.08\times10^{-2}$ | $1.12\times10^{-3}$ | $2.04\times10^{-4}$ | 0.428 |

Reaction with $EuCl_2$

| | | | |
|---|---|---|---|
| Nd (1370 K) | Gd (1414 K) | Dy (1430 K) | Tm (1220 K) |
| 5.47 | 21.39 | 0.322 | 0.0180 |

$YCl_2$ is also formed according to $Y+Cl_2 \rightarrow YCl+Cl$ and $YCl+Cl_2 \rightarrow YCl_2+Cl$; when the YCl is vibrationally excited to levels of $v \geqq 8$, electronically excited $YCl_2$ molecules are formed, which are the emitting species of the chemiluminescence of the Y-$Cl_2$ reaction system [23].

### Structure

The electron configuration of Eu in $EuCl_2$ is $4f^76s^n6p^n$ with $n=0.53\pm0.05$ as derived from the Mössbauer spectrum of $^{151}EuCl_2$ in Ar ($\delta=-13.4\pm0.1$ mm/s, relative to $Sm_2O_3$; $e^2qQ=-235\pm5$ MHz), Baggio-Saitovitch et al. [4]. From the electron configuration of the valence states, Charkin and Dyatkina [5] conclude that the dichlorides of Sc, Y, and La, like the difluorides, have a bent structure. This is probably true also of the other rare earth dichlorides. Nevertheless, the expected deformation vibration has only been observed for $EuCl_2$ in solid Kr, Hastie et al. [6]. The conclusion that the Cl-M-Cl angle $\alpha$ differs from 180° was also deduced indirectly from the isotope effect on the stretching vibrations, de Kock et al. [7]. Initial estimates indicated an interatomic distance in $EuCl_2$ of $r=2.67\pm0.05$ Å and $\alpha=135\pm8°$ [6]; later estimated values are $r=2.52$ Å [8] and $\alpha=130\pm15°$ [7]. The bond angle in $SmCl_2$ is also reported as $130\pm15°$, and for $YbCl_2$, $\alpha$ is estimated to be $140\pm15°$ [7]. These values were used to estimate distances and angles for all the rare earth dichlorides: r decreases monotonically from $2.72\pm0.05$ Å ($LaCl_2$) to $2.49\pm0.05$ Å ($LuCl_2$), while $\alpha$ increases monotonically from $125\pm10°$ to $142\pm10°$, Chervonnyi [9].

### Vibrational Spectra and Force Constants

The only IR spectra to have been studied are those of $EuCl_2$ in solid Ne, Ar, Kr, and $N_2$ [6] and of $SmCl_2$, $EuCl_2$, and $YbCl_2$ in solid Ar, Kr, and $N_2$ [7]. The band observed with $EuCl_2$ in Kr at $64\pm10\ cm^{-1}$ corresponds to the deformation vibration $\nu_2$ [6]. Measured wave numbers for $\nu_1$ and $\nu_3$ are as follows:

| $MCl_2$ . | $SmCl_2$ | | | $EuCl_2$ | | | | $YbCl_2$ | | |
|---|---|---|---|---|---|---|---|---|---|---|
| Matrix . | Ar | Kr | $N_2$ | Ne | Ar | Kr | $N_2$ | Ar | Kr | $N_2$ |
| $\nu_1$ . . | – | – | 247 | 276 | 270 | 262 | 248 | – | – | 250 |
| $\nu_3$ . . | 283 | 276 | 252.1 ($Sm^{37}Cl_2$) 255.5 ($Sm^{35}Cl^{37}Cl$) 257.2 ($Sm^{35}Cl_2$) | 287 | 282 | 272 | 252.5 ($Eu^{37}Cl_2$) 255.9 ($Eu^{35}Cl^{37}Cl$) 257.5 ($Eu^{35}Cl_2$) | 294 | 287 | 258.6 ($Yb^{37}Cl_2$) 262.2 ($Yb^{35}Cl^{37}Cl$) 263.9 ($Yb^{35}Cl_2$) |
| Ref. . . | [7] | [7] | [7] | [6] | [6] | [6] | [7] | [7] | [7] | [7] |

If the force constants are assumed to increase monotonically in the series $LaCl_2$ to $LuCl_2$ ($f_r$=122 to 146, $f_{rr}$=10.2 to 12.2, $f_\alpha$=6.1 to 7.3 N/m), then the following wave numbers are obtained for the $MCl_2$ vibrations [9]:

| M . . | La | Ce | Pr | Nd | Pm | Sm | Eu | Gd | Tb | Dy | Ho | Er | Tm | Yb | Lu |
|---|---|---|---|---|---|---|---|---|---|---|---|---|---|---|---|
| $\nu_1$ . . | 273 | 275 | 276 | 278 | 279 | 281 | 281 | 282 | 283 | 285 | 286 | 287 | 287 | 289 | 290 |
| $\nu_2$ . . | 61 | 62 | 62 | 63 | 63 | 63 | 63 | 64 | 64 | 65 | 65 | 66 | 66 | 66 | 67 |
| $\nu_3$ . . | 274 | 276 | 278 | 280 | 282 | 283 | 283 | 284 | 287 | 288 | 290 | 292 | 292 | 293 | 295 |

An earlier estimate gave $\nu_1$=303, $\nu_2$=71, and $\nu_3$=306 $cm^{-1}$ for $YbCl_2$ [8].

Measured wave numbers [6, 7] were used to obtain the following force constants (all in N/m):

| | $SmCl_2$ | $EuCl_2$ | $EuCl_2$ | $YbCl_2$ |
|---|---|---|---|---|
| $f_r$ . . . . . . . . | 132 | 132 | 140 | 144 |
| $f_{rr}$ . . . . . . . . | 11 | 11 | $14\pm2$ | 12 |
| $f_\alpha$ . . . . . . . . | – | – | $3.1\pm0.8$ | – |
| Ref. . . . . . . | [7] | [7] | [6] | [7] |

### Ionization

The ionization energies of $NdCl_2$, $E_i$=7.1 eV, and $GdCl_2$, $E_i$=6.9 eV, were estimated from mass-spectrometric measurements with $NdCl_3$ and $GdCl_3$, Ciach et al. [10]. The appearance potentials of $EuCl_2^+$ and $YbCl_2^+$ have been measured; $AP(EuCl_2^+/EuCl_2)=10.5\pm0.5$ eV, Hastie et al. [1], and $AP(YbCl_2^+/YbCl_2)=9.73\pm0.21$ eV, Hariharan et al. [11].

### Dissociation

Estimated bond dissociation energies D(ClM–Cl) for $ScCl_2$, $YCl_2$, and $LaCl_2$ are 105, 119, and 119 kcal/mol, Krasnov, Timoshinin [12].

The dissociation energy (complete dissociation) of $EuCl_2$ from vapor pressure measurements is $D_0^\circ=212.0\pm2.8$ ($D_{298}^\circ=213.0\pm2.8$) kcal/mol. The value $D_0^\circ=216\pm11$ kcal/mol was calculated from mass-spectrometric data, i.e., from $AP(Eu^+/EuCl_2)$ and $E_i(Eu)$. The bond dissociation energy $D_0^\circ$(ClEu–Cl) = 113 ± 5 kcal/mol was calculated from AP($EuCl^+$/$EuCl_2$) and $E_i$(EuCl), Hariharan, Eick [2]. In the same way, Hariharan et al. [11] used vapor pressure measurements to get $D_0^\circ$ = 193.5 ± 3.2 ($D_{298}^\circ$ = 194.6 ± 3.2) kcal/mol for $YbCl_2$ and used mass-spectrometric data to calculate $D_0^\circ$(ClYb–Cl) = 107 ± 7 kcal/mol.

Chervonnyi [14] used the dissociation energy of $BaCl_2$ ($D_{298}$ = 220.1 ± 4.0 kcal/mol) and the heat of reaction of $BaCl_2 + Eu \rightarrow EuCl_2 + Ba$ to calculate the dissociation energy of $EuCl_2$, $D_{298}$ = 211.5 ± 4.2 kcal/mol, and used this value and the heat of reaction of $EuCl_2 + M \rightarrow MCl_2 + Eu$ (where M = Nd, Dy, Tm) [13] to calculate $D_{298}$ for $NdCl_2$, $DyCl_2$, and $TmCl_2$. These values and estimated $D_{298}$ values for the other lanthanide dichlorides are as follows [14]:

| | | | | | |
|---|---|---|---|---|---|
| M . . . . | La | Ce | Pr | Nd | Pm |
| $D_{298}$ . . . | 229 ± 5 | 227.1 ± 4.2 | 224.4 ± 4.2 | 220.6 ± 4.2 | 217 ± 5 |
| M . . . . | Sm | Eu | Gd | Tb | Dy |
| $D_{298}$ . . . | 214.6 ± 4.2 | 211.5 ± 4.2 | 225.3 ± 4.2 | 223.5 ± 4.2 | 211.1 ± 4.2 |
| M . . . . | Ho | Er | Tm | Yb | Lu |
| $D_{298}$ . . . | 213.9 ± 4.2 | 217.0 ± 4.2 | 204.6 ± 4.2 | 199.2 ± 4.2 | 222.4 ± 4.2 |

For older estimated $D_{298}$ values see Feber [15].

The Gibbs free energy change in the dissociation reaction $MCl_2(g) \rightarrow M(l) + Cl_2(g)$ for M = Sm, Eu, and Yb is ΔG = 137, 147, and 118 kcal/mol at 1000 K and 142, 152, and 123 kcal/mol at 1900 K, Novikov, Baev [16].

## Thermodynamic Data of Formation

The heats of formation from the elements of $ScCl_2$, $YCl_2$, and $LaCl_2$ in the gas phase were estimated from measured data for the dichlorides of other transition metals as $-\Delta H_f^\circ$ = 95, 85, and 78 kcal/mol, respectively, Novikov [17]. From molecular data, partly estimated, Filippenko et al. [8] derived $-\Delta H_f$ for $EuCl_2$ and $YbCl_2$ as 102 ± 20 and 104 ± 20 kcal/mol. For $EuCl_2$, Schumm et al. [18] give $-\Delta H_{298}^\circ$ = 110 kcal/mol. Earlier values for $-\Delta H_f^\circ$ at 298 K in kcal/mol [15]: 84 for $ScCl_2$ and 91 for $YCl_2$ (both values derived from measured data for the solid compound) and 121.0 for $SmCl_2$, 129.3 for $EuCl_2$, and 112.1 for $YbCl_2$.

## Additional Thermodynamic Data

Estimated molecular parameters were used by Krasnov, Danilova [19] to derive the thermodynamic potential $\Phi_T^\circ$ (in cal·mol$^{-1}$·K$^{-1}$), the enthalpy $H_T^\circ - H_0^\circ$ (in cal/mol), and the entropy $S_T^\circ$ (in cal·mol$^{-1}$·K$^{-1}$) for $ScCl_2$, $YCl_2$, and $LaCl_2$. These values are shown below. Given in parentheses are the recommended values of Wagman et al. [20].

| T in K | . . . . . . . . . . | 298.15 | 1000 | 2000 | 3000 |
|---|---|---|---|---|---|
| $ScCl_2$ | $\Phi_T^\circ$ . . . . . . . | 59.576 | 73.645 | 82.602 | 88.006 |
| | $H_T^\circ - H_0^\circ$ . . . . | 3123 (3100) | 12577 | 26416 | 40302 |
| | $S_T^\circ$ . . . . . . . | 70.049 (72.5) | 86.222 | 95.810 | 101.440 |
| $YCl_2$ | $\Phi_T^\circ$ . . . . . . . | 61.452 | 75.835 | 84.884 | 90.321 |
| | $H_T^\circ - H_0^\circ$ . . . . | 3218 (3200) | 12752 | 26611 | 40504 |
| | $S_T^\circ$ . . . . . . . | 72.246 (75.0) | 88.587 | 98.190 | 103.823 |

| T in K | | 298.15 | 1000 | 2000 | 3000 |
|---|---|---|---|---|---|
| $LaCl_2$ | $\Phi^\circ_T$ | 62.942 | 77.433 | 86.513 | 91.961 |
| | $H^\circ_T-H^\circ_0$ | 3253 | 12810 | 26675 | 40570 |
| | $S^\circ_T$ | 73.853 | 90.243 | 99.850 | 105.484 |

The molar heat capacities $C^\circ_p$ at 298 K for gaseous $ScCl_2$ and $YCl_2$ are given as 12.6 and 12.9 $cal \cdot mol^{-1} \cdot K^{-1}$ [20]. Structural data and vibration frequencies were used to calculate $H^\circ_T-H^\circ_{298}$, $S^\circ_T$, and the heat capacity $C^\circ_p$ for $EuCl_2(g)$ and $YbCl_2(g)$ [2, 11] for temperatures from 298.15 to 1600 K. From these data and the estimated interatomic distances and vibration wave numbers, Chervonnyi [9] obtained values for the thermodynamic potential, the enthalpy, and the entropy of the dichlorides of La through Lu. Selected values are as follows [9]:

| T in K | | 298.15 | 500 | 700 | 1000 | 1500 | 2000 |
|---|---|---|---|---|---|---|---|
| $LaCl_2$ | $H^\circ_T-H^\circ_0$ | 3450 | 6190 | 8943 | 13094 | 20032 | 26979 |
| | $S^\circ_T$ | 78.40 | 85.50 | 90.18 | 95.17 | 100.86 | 104.90 |
| $NdCl_2$ | $H^\circ_T-H^\circ_0$ | 3438 | 6175 | 8927 | 13077 | 20014 | 26961 |
| | $S^\circ_T$ | 80.05 | 87.13 | 91.82 | 96.80 | 102.49 | 106.54 |
| $EuCl_2$ | $H^\circ_T-H^\circ_0$ | 3433 | 6169 | 8920 | 13070 | 20007 | 26953 |
| | $S^\circ_T$ | 80.97 | 88.06 | 92.74 | 97.72 | 103.41 | 107.46 |
| $DyCl_2$ | $H^\circ_T-H^\circ_0$ | 3423 | 6157 | 8907 | 13056 | 19992 | 26938 |
| | $S^\circ_T$ | 80.40 | 87.48 | 92.16 | 97.14 | 102.83 | 106.87 |
| $YbCl_2$ | $H^\circ_T-H^\circ_0$ | 3414 | 6146 | 8895 | 13043 | 19980 | 26926 |
| | $S^\circ_T$ | 78.46 | 85.54 | 90.21 | 95.20 | 100.89 | 104.93 |
| $LuCl_2$ | $H^\circ_T-H^\circ_0$ | 3412 | 6143 | 8892 | 13040 | 19975 | 26921 |
| | $S^\circ_T$ | 77.52 | 84.58 | 89.26 | 94.25 | 99.93 | 103.98 |

## Chemical Reactions in the Gas Phase

The following changes of Gibbs free energy (in kcal/mol) have been calculated from literature data for the reactions $LnCl_2(g) + 2M^I(g) \rightarrow 2M^ICl(g) + Ln(l)$ and $LnCl_2(g) + M^{II}(g) \rightarrow M^{II}Cl_2(g) + Ln(l)$ for Ln = Sm, Eu, and Yb at $T_1 = 1000$ and $T_2 = 1900$ K:

| | Ln = Sm | | Ln = Eu | | Ln = Yb | |
|---|---|---|---|---|---|---|
| $M^I$ | $T_1$ | $T_2$ | $T_1$ | $T_2$ | $T_1$ | $T_2$ |
| Li | −23 | −9 | −13 | 1 | −42 | −28 |
| Na | 5 | 17 | 15 | 27 | −14 | −2 |
| K | −2 | 7 | 8 | 17 | −21 | −12 |

| | Ln = Sm | | Ln = Eu | | Ln = Yb | |
|---|---|---|---|---|---|---|
| $M^{II}$ | $T_1$ | $T_2$ | $T_1$ | $T_2$ | $T_1$ | $T_2$ |
| Mg | 10 | 35 | 21 | 45 | −8 | −16 |
| Ca | −29 | −12 | −19 | −2 | −48 | −31 |
| Zn | 60 | 85 | 70 | 95 | 41 | 56 |

The corresponding values for the reactions $LnCl_2(g) + H_2(g) \rightarrow 2HCl(g) + Ln(l)$ are 79, 89, 60 kcal/mol at 1000 K, 71, 81, 52 kcal/mol at 1900 K [16].

$SmCl_2$, like some of the rare earth trichlorides (see p. 16), forms gaseous red-brown clusters with $Al_2Cl_6$. However, the composition of the complex with $SmCl_2$ is not well defined, Papatheodorou, Kucera [21]. With $EuCl_2$, gaseous complexes of composition $EuAl_3Cl_{11}$ and $EuAl_4Cl_{14}$ are formed at 570 and 755 K as indicated by spectrometric measurements in the near UV, Sørlie, Øye [22].

References:

[1] J.W. Hastie, P. Ficalora, J.L. Margrave (J. Less-Common Metals **14** [1968] 83/91). – [2] A.V. Hariharan, H.A. Eick (High Temp. Sci. **4** [1972] 91/8). – [3] A.D. Chervonnyi (Zh. Fiz. Khim. **51** [1977] 1144/9; Russ. J. Phys. Chem. **51** [1977] 672/5). – [4] E. Baggio-Saitovitch, F.J. Litterst, H. Micklitz (J. Phys. [Paris] **37** [1976] Suppl. C6-529/C6-530), F.J. Litterst, A. Schichl, E. Baggio-Saitovitch, H. Micklitz, J.M. Friedt (Ber. Bunsenges. Physik. Chem. **82** [1978] 73/5). – [5] O.P. Charkin, M.E. Dyatkina (Zh. Strukt. Khim. **6** [1965] 579/90; J. Struct. Chem. [USSR] **6** [1965] 550/60), O.P. Charkin (Izv. Akad. Nauk SSSR Ser. Khim. **1970** 41/6; Bull. Acad. Sci. USSR Div. Chem. Sci. **1970** 37/41).

[6] J.W. Hastie, R.H. Hauge, J.L. Margrave (High Temp. Sci. **3** [1971] 56/72). – [7] C.W. deKock, R.D. Wesley, D.D. Radtke (High. Temp. Sci. **4** [1972] 41/7). – [8] N.V. Filippenko, E.V. Morozov, N.I. Giricheva, K.S. Krasnov (Izv. Vysshikh Uchebn. Zavedenii Khim. Khim. Tekhnol. **15** [1972] 1416/8). – [9] A.D. Chervonnyi (VINITI No. 7455-73 [1973]; C.A. **87** [1977] No. 44994). – [10] S. Ciach, A.J.C. Nicholson, D.L. Swingler, P.J. Thistlethwaite (Inorg. Chem. **12** [1973] 2072/4).

[11] A.V. Hariharan, N.A. Fishel, H.A. Eick (High Temp. Sci. **4** [1972] 405/10). – [12] K.S. Krasnov, V.S. Timoshinin (Teplofiz. Vysokikh Temp. **7** [1969] 365/7; High Temp. [USSR] **7** [1969] 333/4). – [13] A.D. Chervonnyi, V.I. Il'in, O.P. Charkin, A.V. Baluev, V.I. Evdokimov (VINITI No. 1657-74 [1974]; C.A. **86** [1977] No. 128429). – [14] A.D. Chervonnyi (Zh. Fiz. Khim. **51** [1977] 1308/12; Russ. J. Phys. Chem. **51** [1977] 772/5). – [15] R.C. Feber (LA-3164 [1965] 49, 90, 175, 177/9, 180; N.S.A. **19** [1965] No. 30148).

[16] G.I. Novikov, A.K. Baev (Izv. Vysshikh Uchebn. Zavedenii Khim. Khim. Tekhnol. **9** [1966] 180/4). – [17] G.I. Novikov (Probl. Sovrem. Khim. Koord. Soedin. **1970** No. 3, pp. 33/48, 34; C.A. **75** [1971] No. 10210). – [18] R.H. Schumm, D.D. Wagman, S. Bailey et al. (NBS-TN-270-7 [1973] 34). – [19] K.S. Krasnov, T.G. Danilova (Teplofiz. Vysokikh Temp. **7** [1969] 1213/6; High Temp. [USSR] **7** [1969] 1131/3). – [20] D.D. Wagman, W.H. Evans, V.B. Parker et al. (NBS-TN-270-5 [1971] 25, 30).

[21] G.N. Papatheodorou, G.H. Kucera (Inorg. Chem. **18** [1979] 385/9). – [22] M. Sørlie, H.A. Øye (J. Inorg. Nucl. Chem. **40** [1978] 493/8). – [23] H.C. Brayman, D.R. Fischell, T.A. Cool (J. Chem. Phys. **73** [1980] 4247/59), H.C. Brayman (Diss. Cornell Univ. 1980; Diss. Abstr. Intern. B **41** [1980] 243).

### 11.1.3 Molecules of the Type $MCl_3$

#### Formation

The gaseous trichlorides are obtained by vaporization of the solid compounds. They are, like the trifluorides, predominantly monomeric, with small concentrations of the dimer and negligible concentrations of the trimer. According to Wagner, Schäfer [3], the saturated vapor over $ScCl_3$ at 850 K has an $ScCl_3:Sc_2Cl_6$ ratio of 11.3:1, and the fraction of $Sc_2Cl_6$ increases with increasing temperature. In contrast, Rat'kovskii et al. [1] report $ScCl_3:Sc_2Cl_6:Sc_3Cl_9=0.01:1:0.001$; however, their enthalpy of dimerization is in conflict with the results of other investigators [2, 3, 4]. For additional information on the composition of the vapor phase, see p. 113.

Mass-spectrometric investigations of $ScCl_3$ give a dimerization enthalpy of $-\Delta H=$ 46.9 kcal/mol at 772 K, which converts to $-\Delta H=48.8\pm2$ kcal/mol at 298 K, Schäfer, Binnewies [2]. Further measurements between 730 and 850 K lead to $-\Delta H=47.6$

$\pm1$ kcal/mol at 298 K [3]. This confirms the measurements of Patrikeev et al. [4] from which $-\Delta H=47.1$ can be deduced [3], while the results of Rat'kovskii et al. [1] are obviously incorrect [3]. The attachment of an additional $ScCl_3$ molecule to $Sc_2Cl_6$ has a $-\Delta H$ of $53.3\pm2$ kcal/mol at 298 K [3].

Schäfer, Binnewies [2] also investigated $YCl_3$ and found the enthalpy of dimerization to be $-\Delta H=49.6\pm2$ kcal/mol at 298 K. From McKinley's work [5] on the mass spectrum of $YCl_3$ vapor at 850 K, $\Delta H$ values of only $-28$ [2] or $-29\pm6$ [6] kcal/mol were derived, corresponding to $-20$ at 298 K [2].

Hastie et al. [6] determined the enthalpies of dimerization $\Delta H$ for $LaCl_3$, $EuCl_3$, and $LuCl_3$ by mass spectrometry and then estimated the $\Delta H$ values for the other rare earth trichlorides by interpolation. Their measured and estimated enthalpies of dimerization at 1000 K (in kcal/mol, with uncertainties of about $\pm4$ kcal/mol) are as follows:

| M . . . . . . | La | Ce | Pr | Nd | Pm | Sm | Eu | Gd | Tb | Dy | Ho | Er | Tm | Yb | Lu |
|---|---|---|---|---|---|---|---|---|---|---|---|---|---|---|---|
| $-\Delta H$ . . . . . | 48 | 45 | 43 | 41 | 38 | 35 | 32 | 41 | 43 | 40 | 37 | 34 | 30 | 28 | 47 |

The vapor over high-temperature condensed systems of elemental rare earths with $BaCl_2$ or $EuCl_2$ was studied by quantitative mass spectrometry. The results were used to calculate the equilibrium constants for the reaction in the gas phase:

$$2M+3BaCl_2(EuCl_2) \rightleftharpoons 2MCl_3+3Ba(Eu).$$

The following table shows the equilibrium constants; temperatures in parentheses [7]:

Reaction with $BaCl_2$

| Ce (1398 K) | Pr (1418 K) | Nd (1333 K) | Tb (1476 K) | Dy (1292 K) | Ho (1342 K) |
|---|---|---|---|---|---|
| 1.05 | 0.0867 | 0.00718 | 0.0113 | $7.86\times10^{-7}$ | $2.72\times10^{-5}$ |

| Reaction with $BaCl_2$ | | Reaction with $EuCl_2$ | | |
|---|---|---|---|---|
| Er (1326 K) | Lu (1368 K) | Nd (1370 K) | Gd (1414 K) | Dy (1430 K) |
| $3.11\times10^{-3}$ | $3.18\times10^{-3}$ | 4.75 | 0.0694 | 0.0532 |

**Structure**

From model calculations and comparisons of experimental thermodynamic data Myers et al. [8, 9, 10] assume that the trichlorides, like the other trihalides, have predominantly covalent bonds.

Electron diffraction measurements of $YCl_3$, $LaCl_3$, and $NdCl_3$ led Akishin et al. [11] to conclude that these molecules are planar with the following interatomic distances: r(Y–Cl) = $2.47\pm0.03$ Å, r(La–Cl) = $2.60\pm0.03$ Å, and r(Nd–Cl) = $2.58\pm0.03$ Å. Data for $ScF_3$ were used to calculate r(Sc–Cl) = 2.32 Å in $ScCl_3$ [12]. A later calculation with the use of ionic radii also gives r(Sc–Cl) = 2.32 Å [13]. However, recent measurements of $YCl_3$ and of several lanthanide trichlorides show the molecules to have pyramidal structures with the following interatomic distances (in Å) and bond angles:

| $MCl_3$ | $YCl_3$ | $LaCl_3$ | $PrCl_3$ | $GdCl_3$ |
|---|---|---|---|---|
| $r_g$(M–Cl) | 2.44±0.01 | 2.587±0.007 | 2.553±0.006 | 2.489±0.006 |
| $r_g$(Cl···Cl) | 4.08±0.03 | 4.300±0.071 | 4.204±0.071 | 4.150±0.043 |
| α(Cl–M–Cl) | 113.5°±2° | – | 110.8°±3.0° | 113°±1.5° |
| Ref. | [14] | [15] | [16] | [17] |

| $MCl_3$ | $TbCl_3$ | $HoCl_3$ | $LuCl_3$ |
|---|---|---|---|
| $r_g$(M–Cl) | 2.478±0.005 | 2.459±0.006 | 2.417±0.006 |
| $r_g$(Cl···Cl) | 4.070±0.048 | 4.058±0.038 | 3.996±0.037 |
| α(Cl–M–Cl) | 110.5°±1.7° | 111.2°±2.0° | 111.5°±2.0° |
| Ref. | [18] | [16] | [19, 20, 21] |

Using earlier measured data [11, 21], Krasnov et al. [20] interpolated the following r(M–Cl) interatomic distances:

| M | La | Ce | Pr | Nd | Pm | Sm | Eu |
|---|---|---|---|---|---|---|---|
| r in Å | 2.597 | 2.578 | 2.561 | 2.545 | 2.529 | 2.515 | 2.501 |

| M | Gd | Tb | Dy | Ho | Er | Tm | Yb |
|---|---|---|---|---|---|---|---|
| r in Å | 2.496 | 2.484 | 2.472 | 2.460 | 2.450 | 2.438 | 2.428 |

MO calculations (extended Hückel procedure) also indicate the pyramidal structure to be the more stable [10]. For $EuCl_3$ and $YbCl_3$, Filippenko et al. [22] estimated r(M–Cl) = 2.56 Å and 2.41 Å, respectively, and α(Cl–M–Cl) = 117° for both molecules.

### Vibration Spectra and Force Constants

The wave numbers of $ScCl_3$, $YCl_3$, and $LaCl_3$ were estimated by Krasnov [23] with the assumption that these molecules are planar. Of the four IR bands predicted by theory (cf. "Seltenerdelemente" C3, 1976, p. 19), only those corresponding to the $\nu_3$ vibration were found by Selivanov et al. [24] in measurements of $ScCl_3$, $YCl_3$, and lanthanide trichloride vapors (1300 to 1500 °C) in the region 200 to 700 $cm^{-1}$:

| $MCl_3$ | $ScCl_3$ | $YCl_3$ | $LaCl_3$ | $PrCl_3$ | $NdCl_3$ | $GdCl_3$ |
|---|---|---|---|---|---|---|
| $\nu_3$ in $cm^{-1}$ | 477 | 370 | 316 | 320 | 324 | 326 |

$LaCl_3$, $GdCl_3$, and $LuCl_3$ in an Xe matrix at 20 to 30 K show $\nu_2$, $\nu_3$, and $\nu_4$ bands. Wave numbers (in $cm^{-1}$) are shown in the following table; estimated values for $\nu_1$ bands and for the $YCl_3$ and $ErCl_3$ bands are in parentheses, Perov et al. [25].

| | $YCl_3$ | $LaCl_3$ | $GdCl_3$ | $ErCl_3$ | $LuCl_3$ |
|---|---|---|---|---|---|
| $\nu_1$ | (294) | (260) | (270) | (302) | (280) |
| $\nu_2$ | (62) | 52 | 53 | (55) | 60 |
| $\nu_3$ | (351) | 300 | 318 | (328) | 331 |
| $\nu_4$ | (98) | 74 | 82 | (84) | 88 |

The wave number values in the table for $LaCl_3$ supersede the estimated values of [26]. The estimated wave numbers of [23, 26] served as the basis for the force constants estimated initially by Krasnov [27] and more recently by Phongsatha, Cyvin [28] for $ScCl_3$, $YCl_3$, and $LaCl_3$.

$NdCl_3$ vapor (≈1000 °C) shows four band groups, from which $\nu_1=349$, $\nu_2=177$, $\nu_3=306$, and $\nu_4=120\ cm^{-1}$ were derived; these values were used to calculate the force constants: $f_r=172$, $f_{rr}=21$, $f_\alpha=17$, and $f_{\alpha\alpha}=5$ N/m [29]. For $LuCl_3$, the frequencies of Perov et al. [25] (but with $\nu_1=332\pm8\ cm^{-1}$ instead of $280\ cm^{-1}$, cf. above) and the mean-square amplitudes of vibration from electron diffraction measurements [20, 21] were used to determine the force constants: $F_{11}=231\pm3$, $F_{12}=7\pm11$, $F_{22}=7\pm1$, $F_{33}=179\pm5$, $F_{34}=-2\pm8$, and $F_{44}=5\pm1$ N/m [19]. Estimated wave numbers for $EuCl_3$: 322, 57, 330, and 80 $cm^{-1}$; for $YbCl_3$: 338, 62, 348, and 78 $cm^{-1}$, Filippenko et al. [22].

**Ionization**

Mass spectrometric studies of $MCl_3$ vapor (≈1000 K) gave the following appearance potentials (AP) for $MCl_3^+$:

| | | | | | |
|---|---|---|---|---|---|
| $MCl_3^+$ . . . | $YCl_3^+$ | $LaCl_3^+$ | $NdCl_3^+$ | $EuCl_3^+$ | $LuCl_3^+$ |
| AP in eV . | $12.8\pm0.5$ | $13.8\pm1$ | $<11.4$ | $13\pm1$ | $11.5\pm0.5$ |
| Ref. . . . | [5] | [6] | [30] | [6] | [6] |

**Dissociation**

At T=298 K, the bond dissociation energies (in kcal/mol) $D(Cl_2M-Cl)=118$ for $ScCl_3$, 128 for $YCl_3$, and 126 for $LaCl_3$ were obtained by subtracting the estimated atomization enthalpy of $MCl_2$ from that of $MCl_3$, Krasnov, Timoshinin [31].

The average bond energy E of $LaCl_3$ has been calculated by Allen [33] from thermodynamic data of Spedding, Flynn [32] as 123 kcal/mol. Using a method which gave E= 121 kcal/mol for $LaCl_3$ resulted in E=106 kcal/mol for $ScCl_3$ [33]. More recently, Dudchik et al. [34] have used vapor pressure measurements to derive the following E values (in kcal/mol) for the M–Cl bond in $MCl_3$:

| | | | | | | | | |
|---|---|---|---|---|---|---|---|---|
| M . . . . . | Sc | Y | La | Ce | Pr | Nd | Sm | Eu |
| E . . . . . | 111 | 116 | 122 | 124 | 116 | 111 | 101 | 92 |
| M . . . . . | Gd | Tb | Dy | Ho | Er | Tm | Yb | Lu |
| E . . . . . | 116 | 116 | 107 | 107 | 110 | 102 | 91 | 114 |

An estimate based on thermodynamic data gives E=92 to 104 kcal/mol for $YbCl_3$ [35].

The following values for the enthalpy of atomization $\Delta H_{at}$ at 298.15 K (i.e. the energy required for complete dissociation in the gas phase) were derived from thermodynamic data (in kcal/mol):

| | | | | | | | |
|---|---|---|---|---|---|---|---|
| $MCl_3$ . . . . | $LaCl_3$ | $CeCl_3$ | $PrCl_3$ | $NdCl_3$ | $SmCl_3$ | $EuCl_3$ | $GdCl_3$ |
| $\Delta H_{at}$ . . . . | 366 | 359 | 351 | 334 | 305 | 301 | 349 |
| $MCl_3$ . . . . | $TbCl_3$ | $DyCl_3$ | $HoCl_3$ | $ErCl_3$ | $TmCl_3$ | $YbCl_3$ | $LuCl_3$ |
| $\Delta H_{at}$ . . . . | 349 | 327 | 330 | 330 | 315 | 286 | 350 |

Myers [8].

Heats of reaction (some measured, others calculated) for $3\,BaCl_2+2\,M \rightleftharpoons 2\,MCl_3+3\,Ba$ were used by Chervonnyi [36] to calculate $\Delta H_{at}(T=298\ K)$ for all the lanthanide trichlorides; accuracy is 5 kcal/mol:

| | | | | | | | | |
|---|---|---|---|---|---|---|---|---|
| $MCl_3$ . . . . | $LaCl_3$ | $CeCl_3$ | $PrCl_3$ | $NdCl_3$ | $PmCl_3$ | $SmCl_3$ | $EuCl_3$ | $GdCl_3$ |
| $\Delta H_{at}$ . . . . | 351 | 348.5 | 346.6 | 339.6 | 339 | 336 | 333 | 347.8 |
| $MCl_3$ . . . . | $TbCl_3$ | $DyCl_3$ | $HoCl_3$ | $ErCl_3$ | $TmCl_3$ | $YbCl_3$ | $LuCl_3$ | |
| $\Delta H_{at}$ . . . . | 347.3 | 330.8 | 336.0 | 339.8 | 326 | 321 | 343.0 | |

Earlier $\Delta H_{at}$ values were based on the estimated enthalpies of formation of Feber [37], which are given in the subsequent section.

Novikov, Baev [38] calculated the change in free energy $\Delta G$ (in kcal/mol) at 1000 and 1900 K for the dissociation reaction $MCl_3(gas) \rightarrow M(liquid) + {}^3/_2 Cl_2(gas)$:

| | | | | | | | | |
|---|---|---|---|---|---|---|---|---|
| $MCl_3$ . . . . | $YCl_3$ | $LaCl_3$ | $CeCl_3$ | $PrCl_3$ | $NdCl_3$ | $PmCl_3$ | $SmCl_3$ | $EuCl_3$ |
| $\Delta G$ (1000 K) . | 118 | 135 | 134 | 134 | 130 | 130 | 128 | 126 |
| $\Delta G$ (1900 K) . | 93 | 111 | 110 | 110 | 106 | 106 | 104 | 101 |
| $MCl_3$ . . . . | $GdCl_3$ | $TbCl_3$ | $DyCl_3$ | $HoCl_3$ | $ErCl_3$ | $TmCl_3$ | $YbCl_3$ | $LuCl_3$ |
| $\Delta G$ (1000 K) . | 128 | 127 | 123 | 121 | 118 | 116 | 113 | 110 |
| $\Delta G$ (1900 K) . | 103 | 102 | 97 | 95 | 92 | 90 | 87 | 85 |

The energies of dissociation of a few of the trihalides into $M^{3+} + 3X^-$ at $T = 0$ K have been calculated from an equation, which involves the polarizability of $X^-$, the interatomic distance, and the repulsive coefficient. Such dissociation energies in kcal/mol are 1072 for $ScCl_3$, 998 for $YCl_3$, and 940 for $LaCl_3$, Krasnov [39].

### Thermodynamic Data of Formation

Two different values are given by Novikov [40] for the enthalpy of formation of gaseous $ScCl_3$ from the elements: $\Delta H° = -165$ and $-170$ kcal/mol.

For the formation of the gaseous trichlorides from the elements at 298 K, the Natl. Bur. Std. gives the following enthalpies of formation (in kcal/mol) for $YCl_3$ [41] and for some of the lanthanide trichlorides [42]:

| | | | | | | | | |
|---|---|---|---|---|---|---|---|---|
| $MCl_3$ . . . . | $YCl_3$ | $LaCl_3$ | $CeCl_3$ | $PrCl_3$ | $NdCl_3$ | $HoCl_3$ | $TmCl_3$ | $LuCl_3$ |
| $-\Delta H^\circ_{298}$ . . | 179.3 | 177 | 174 | 187 | 173 | 168 | 166 | 155 |

For $YCl_3$ at 0 K, $-\Delta H^\circ_0$ is 179 kcal/mol [41].

An empirical equation for transition metal trichlorides gives $\Delta H = -165$ kcal/mol for $YCl_3$ and $-179$ kcal/mol for $LaCl_3$, Filippenko et al. [22]. The value $\Delta H = -148 \pm 2.5$ kcal/mol for $YbCl_3$ (Dudchik [43]) was used to estimate $\Delta H \approx -145 \pm 2.5$ kcal/mol for $EuCl_3$ [22]. Feber [37] started with experimental data for the solid compounds to derive $\Delta H^\circ_{298}$ (in kcal/mol) $= -174.5$ for $CeCl_3$, $-173.8$ for $PrCl_3$, and $-172.2$ for $NdCl_3$. For the remaining trichlorides in the series, $\Delta H^\circ_{298}$ increases monotonically from $-172$ kcal/mol for $PmCl_3$ to $-152$ kcal/mol for $LuCl_3$ [37].

### Further Thermodynamic Data

For $ScCl_3$, $YCl_3$, and $LaCl_3$, experimental and estimated molecular parameters have been used to calculate the thermodynamic potential $\Phi^\circ_T$ (in $cal \cdot mol^{-1} \cdot K^{-1}$), the enthalpy $H^\circ_T - H^\circ_0$ (in cal/mol), and the entropy $S^\circ_T$ (in $cal \cdot mol^{-1} \cdot K^{-1}$), Krasnov, Danilova [44]:

| T in K | | 298.15 | 1000 | 2000 | 3000 |
|---|---|---|---|---|---|
| $ScCl_3$ | $\Phi^\circ_T$ | 63.164 | 81.937 | 94.385 | 101.983 |
| | $H^\circ_T - H^\circ_0$ | 3991 | 17296 | 37014 | 56835 |
| | $S^\circ_T$ | 76.549 | 99.233 | 112.892 | 120.928 |
| $YCl_3$ | $\Phi^\circ_T$ | 65.032 | 84.349 | 96.965 | 104.624 |
| | $H^\circ_T - H^\circ_0$ | 4150 | 17612 | 37372 | 57206 |
| | $S^\circ_T$ | 78.952 | 101.961 | 115.651 | 123.692 |
| $LaCl_3$ | $\Phi^\circ_T$ | 66.493 | 86.106 | 98.809 | 106.498 |
| | $H^\circ_T - H^\circ_0$ | 4243 | 17776 | 37554 | 57394 |
| | $S^\circ_T$ | 80.723 | 103.882 | 117.586 | 125.630 |

For $YCl_3$, the value $H^\circ_{298} - H^\circ_0 = 4400$ cal/mol is recommended, and the molar heat capacity of the gas is given as $C^\circ_p = 18$ cal·mol$^{-1}$·K$^{-1}$ [41].

Molecular data (partially estimated) of Filippenko et al. [22] were used by Chervonnyi [36] to calculate the thermodynamic potential and the entropy $S^\circ_T$ (T=1000 to 1600 K) of $EuCl_3$ and $YbCl_3$. As with the dichlorides (p. 9), interpolation and extrapolation were used to obtain values for the remaining trichlorides. Selected values for $S^\circ_T$ in cal·mol$^{-1}$·K$^{-1}$ [36]:

| $MCl_3$ | $LaCl_3$ | $CeCl_3$ | $PrCl_3$ | $NdCl_3$ | $PmCl_3$ | $SmCl_3$ | $EuCl_3$ | $GdCl_3$ |
|---|---|---|---|---|---|---|---|---|
| $S^\circ_{1000}$ | 111.3 | 112.1 | 112.6 | 113.0 | 113.3 | 113.6 | 113.9 | 114.0 |
| $S^\circ_{1600}$ | 120.6 | 121.4 | 121.9 | 122.3 | 122.6 | 122.9 | 123.2 | 123.3 |

| $MCl_3$ | $TbCl_3$ | $DyCl_3$ | $HoCl_3$ | $ErCl_3$ | $TmCl_3$ | $YbCl_3$ | $LuCl_3$ |
|---|---|---|---|---|---|---|---|
| $S^\circ_{1000}$ | 113.7 | 113.3 | 112.9 | 112.5 | 112.1 | 111.4 | 110.5 |
| $S^\circ_{1600}$ | 123.0 | 122.6 | 122.2 | 121.8 | 121.4 | 120.7 | 119.8 |

With the aid of experimental and estimated molecular parameters, Myers, Graves [45] calculated thermodynamic functions at temperatures of 298.15 to 2000 K. Selected values:

| | $H^\circ_T - H^\circ_0$ in kcal/mol | | | $S^\circ_T$ in cal·mol$^{-1}$·K$^{-1}$ | | | $-(G^\circ_T - H^\circ_{298.15})/T$ in cal·mol$^{-1}$·K$^{-1}$ | | |
|---|---|---|---|---|---|---|---|---|---|
| T in K | 298.15 | 1000 | 1500 | 298.15 | 1000 | 1500 | 500 | 1000 | 1500 |
| $LaCl_3$ | 4.61 | 18.30 | 28.20 | 85.21 | 108.72 | 116.74 | 87.37 | 95.02 | 101.01 |
| $CeCl_3$ | 4.60 | 18.62 | 29.06 | 88.57 | 112.51 | 120.98 | 90.74 | 98.49 | 104.67 |
| $PrCl_3$ | 4.60 | 18.66 | 29.33 | 89.35 | 113.34 | 121.98 | 91.51 | 99.28 | 105.49 |
| $NdCl_3$ | 4.59 | 18.73 | 29.41 | 89.45 | 113.56 | 122.22 | 91.62 | 99.42 | 105.67 |
| $PmCl_3$ | 4.59 | 18.95 | 29.75 | 89.19 | 113.64 | 122.40 | 91.38 | 99.28 | 105.62 |
| $SmCl_3$ | 4.61 | 19.32 | 30.24 | 88.43 | 113.50 | 122.35 | 90.66 | 98.78 | 105.26 |
| $EuCl_3$ | 4.99 | 20.34 | 31.28 | 86.90 | 113.24 | 122.12 | 89.31 | 97.89 | 104.59 |
| $GdCl_3$ | 4.57 | 18.25 | 28.15 | 88.76 | 112.24 | 120.27 | 90.91 | 98.56 | 104.54 |
| $TbCl_3$ | 4.57 | 18.58 | 29.08 | 89.68 | 113.59 | 122.15 | 91.84 | 99.58 | 105.76 |
| $DyCl_3$ | 4.56 | 18.31 | 28.51 | 90.02 | 113.57 | 121.84 | 92.17 | 99.82 | 105.87 |
| $HoCl_3$ | 4.56 | 18.25 | 28.24 | 90.10 | 113.59 | 121.69 | 92.25 | 99.90 | 105.90 |
| $ErCl_3$ | 4.56 | 18.24 | 28.16 | 89.89 | 113.37 | 121.41 | 92.04 | 99.69 | 105.68 |
| $TmCl_3$ | 4.55 | 18.23 | 28.19 | 89.39 | 112.86 | 120.93 | 91.54 | 99.18 | 105.17 |
| $YbCl_3$ | 4.55 | 18.22 | 28.12 | 88.40 | 111.87 | 119.89 | 90.55 | 98.19 | 104.18 |
| $LuCl_3$ | 4.54 | 18.22 | 28.12 | 84.22 | 107.68 | 115.70 | 86.37 | 94.01 | 99.99 |

**Chemical Reactions in the Gas Phase**

The formation of a number of rare earth halides and halide-$AlCl_3$ complexes in the vapor phase is described in a progress report by Schäfer [51] and by Øye, Gruen [52].

Such complexes with $AlCl_3$ enhance the vaporization of rare earth trichlorides as was first shown for $NdCl_3$. At 800 K in the presence of 1 atm of $Al_2Cl_6$, the apparent vapor pressure of $NdCl_3$ is increased by a factor of $3 \times 10^7$ and even $10^{13}$ at 600 K. The absorption spectrum and molar absorptivity of the Nd-Al-Cl complex (or complexes) were measured [46]. Other volatile lanthanide complexes (proposed suggested formula: $LnAlCl_6$) are formed in $Al_2Cl_6$ vapor ($p \approx 100$ Torr) with Ce, Pr, Pm, Gd, Tb, Dy, Tm, Yb, and Lu chlorides and could be transported in the investigated temperature range of 150 to 500 °C [53].

Transport of rare earth halides from a hot to a cool region of a sealed ampule via gaseous complex formation has been shown in the following systems; deposition zone temperatures are in parentheses: $YCl_3$ with 2.84 atm of $Al_2Cl_6$ (510 °C); $LaCl_3$ with 6.93 atm of $Al_2Cl_6$ (650 °C); $YCl_3$ with a mixture of 1.44 atm of $Fe_2Cl_6$ and 2.8 atm of $Cl_2$ (500 °C) [54]. According to Lascelles (private communication), there is a similar transport of $ScCl_3$ from a hot (500 °C) to a cool (400 °C) region with $Al_2Cl_6$ [54].

Vapor pressure measurements suggest that $NdCl_3$ in the presence of $Al_2Cl_6$ vapor forms two gaseous species: $Nd(AlCl_4)_3$ and $NdCl(Al_2Cl_7)_2$ [46]. With $HoCl_3$ the species $HoCl(AlCl_4)_2$ (g) is formed [55]. According to Caird et al. [56] four different temperature regions may be distinguished:

(a) The region below the triple point of $AlCl_3$ (465.8 K): The existence of volatile Tb-Al-Cl complexes has been demonstrated but little else is known.

(b) The region between the triple point of $AlCl_3$ and the point at which all of the $AlCl_3$ is in the vapor phase: Some or all of condensed $LnCl_3$ is dissolved in the molten $AlCl_3$.

(c) The region up to the point where all of the $LnCl_3$ has been complexed: The number of density of the complex in the gas phase continues to increase with temperature, and the vapor is saturated with respect to $LnCl_3$ at any particular temperature.

(d) The region in which the $LnCl_3$ has been completely volatilized: The vapor is unsaturated with respect to $LnCl_3$.

The possibility of using Ln-Al-Cl complexes in gas lasers to provide a high concentration of (bonded) rare earth ions in the vapor phase has been pointed out by Krupke [57] in a number of reports since 1974; cf. "Seltenerdelemente" C 6, 1978, p. 4. Investigations related to such an application have been carried out with complexes of Nd [48, 55, 57, 58, 59], Tb [49, 60, 56, 61, 62], and Er [47, 63].

The fluorescence of $NdCl_3(AlCl_3)_x(g)$ was studied by Jacobs et al. [48]. This is the first of the $LnCl_3(AlCl_3)_x(g)$ complexes for which an optical gain was observed (at 1.06 μm) [58].

The absorption spectrum of $SmCl_3(AlCl_3)_x(g)$ was studied between 600 and 800 K. For the reaction $SmCl_3(s) + 1.5\ Al_2Cl_6(g) \rightarrow SmAl_3Cl_{12}(g)$ in this temperature range, the enthalpy change $\Delta H = 6.7 \pm 0.2$ kcal/mol and the entropy change $\Delta S = -1.0 \pm 0.2\ cal \cdot K^{-1} \cdot mol^{-1}$ were derived. The Sm in the gaseous $SmAl_3Cl_{12}$ molecule seems to be ninefold coordinated [50].

Enhanced chemical transport of $EuCl_3$ was observed in the presence of $Al_2Cl_6$. But because of the low partial pressure of $EuCl_3(AlCl_3)_x$, its $AlCl_3 : EuCl_3$ ratio could not be evaluated [64].

The fluorescence of $TbCl_3(AlCl_3)_x(g)$ was first studied by Hessler et al. [49] and Jacobs, Krupke [60]. Later, Caird et al. made a more detailed study of the fluorescence between 700 and 350 nm [61] and measured the absorption spectra [56].

The intensities of $ErCl_3(AlCl_3)_x(g)$ absorption bands between 6000 and 44000 $cm^{-1}$ have been measured, Carnall et al. [47, 63].

In evaluating the various $LnCl_3(AlCl_3)_x$ complexes as potential optical gain media, it was planned to study the paths of excited state relaxation. This was done first for the Er complexes by an analysis of the absorption band intensities [47]. The results were also published in another paper [63] by the same authors in which comparisons were made with the $Er^{3+}$ spectra in other host systems. But since the excited free-ion states of $Er^{3+}$ are relatively close together, lifetimes in the vapor complexes are short even in the most favorable circumstances [63]. Therefore the Tb-Al-complexes were studied in more detail [56, 61, 62].

In the absorption spectrum of $TbCl_3(AlCl_3)_x$, the f→f transitions (observed at 401 °C) occur in the spectral range 390 to 270 nm, and the $4f^8 \rightarrow 4f^7 5d$ absorption band (observed at 25, 227, 244, and 262 °C) occurs between 260 and 200 nm [56]. The fluorescence spectrum consists of the following 15 transitions: seven $^5D_4-{}^7F_J$ transitions (three of which were observed earlier [49]), six transitions from the combined ($^5D_3$, $^5G_6$) manifold, and two transitions from the thermally excited $^5L_{10}$ level [61].

Stimulated emission of the strong $^5D_4 \rightarrow {}^7F_5$ transition has a cross section of $2.87 \times 10^{-21}$ $cm^2$, which is small compared with cross sections in most rare earth lasers. However, a population inversion density of $10^{18}$ $cm^{-3}$ should be attainable and would yield a small signal gain ($\approx$30% per meter) [61]. From these spectra, intensity parameters were derived which were then used to compute intensities of fluorescence transitions from the higher lying $^5D_3$, $^5G_6$, and $^5L_{10}$ levels [62]. See Section 11.2.6.12.3, p. 154, for reactions of $MCl_3$ with alkali and alkaline earth metals in the gas phase.

References:

[1] I.A. Rat'kovskii, L.N. Novikova, T.A. Pribytkova (Khim. Khim. Tekhnol. **9** [1975] 19/25; Zh. Fiz. Khim. **48** [1974] 2615; Russ. J. Phys. Chem. **48** [1974] 1551/2). – [2] H. Schäfer, M. Binnewies (Z. Anorg. Allgem. Chem. **410** [1974] 251/68). – [3] K. Wagner, H. Schäfer (Z. Anorg. Allgem. Chem. **430** [1977] 43/50). – [4] Yu.B. Patrikeev, V.A. Morozova, G.P. Dudchik, O.G. Polyachenok, G.I. Novikov (Zh. Fiz. Khim. **47** [1973] 266; Russ. J. Phys. Chem. **47** [1973] 152/3). – [5] J.D. McKinley (J. Chem. Phys. **42** [1965] 2245/6).

[6] J.W. Hastie, P. Ficalora, J.L. Margrave (J. Less-Common Metals **14** [1968] 83/91). – [7] A.D. Chervonnyi (Zh. Fiz. Khim. **51** [1977] 1144/9; Russ. J. Phys. Chem. **51** [1977] 672/5). – [8] C.E. Myers (Inorg. Chem. **14** [1975] 199/201, 2021/2). – [9] C.E. Myers (Inorg. Nucl. Chem. Letters **12** [1976] 575/9). – [10] C.E. Myers, L.J. Norman, L.M. Loew (Inorg. Chem. **17** [1978] 1581/4).

[11] P.A. Akishin, V.A. Naumov, V.M. Tatevskii (Kristallografiya **4** [1959] 194/200; Soviet Phys.-Cryst. **4** [1959] 174/80; Nauchn. Dokl. Vysshei Shkoly Khim. Khim. Tekhnol. **1959** 5/7, 229/32; C.A. **1959** 13702, 19493; Vestn. Mosk. Univ. Ser. Mat. Mekhn. Astron. Fiz. Khim. **14** No. 1 [1959] 229/36; C.A. **1960** 31). – [12] P.A. Akishin, V.A.

Naumov (Zh. Strukt. Khim. **2** [1961] 3/6; J. Struct. Chem. [USSR] **2** [1961] 1/3). – [13] K.S. Krasnov (Zh. Strukt. Khim. **4** [1963] 885/91; J. Struct. Chem. [USSR] **4** [1963] 814/20). – [14] V.N. Kupreev, E.Z. Zasorin, V.P. Spiridonov, G.I. Novikov, I.M. Zharskii (Khim. Khim. Tekhnol. **10** [1976] 174/9; C.A. **85** [1976] No. 170015). – [15] T.G. Danilova, G.V. Girichev, N.I. Giricheva, K.S. Krasnov, E.Z. Zasorin (Izv. Vysshikh Uchebn. Zavedenii Khim. Khim. Tekhnol. **22** [1979] 101/2).

[16] G.V. Girichev, T.G. Danilova, N.I. Giricheva, K.S. Krasnov, V.M. Petrov, A.N. Utkin, E.Z. Zasorin (Izv. Vysshikh Uchebn. Zavedenii Khim. Khim. Tekhnol. **21** [1978] 627/9). – [17] T.G. Danilova, G.V. Girichev, N.I. Giricheva, K.S. Krasnov, E.Z. Zasorin (Izv. Vysshikh Uchebn. Zavedenii Khim. Khim. Tekhnol. **20** [1977] 1069/71). – [18] G.V. Girichev, T.G. Danilova, N.I. Giricheva, K.S. Krasnov, E.Z. Zasorin (Izv. Vysshikh Uchebn. Zavedenii Khim. Khim. Tekhnol. **20** [1977] 1233/4). – [19] N.I. Giricheva, E.Z. Zasorin, G.V. Girichev, K.S. Krasnov, V.P. Spiridonov (Zh. Strukt. Khim. **17** [1976] 797/801; J. Struct. Chem. [USSR] **17** [1976] 686/9). – [20] K.S. Krasnov, N.I. Giricheva, G.V. Girichev (Zh. Strukt. Khim. **17** [1976] 667/70; J. Struct. Chem. [USSR] **17** [1976] 575/7).

[21] N.I. Giricheva, G.V. Girichev, K.S. Krasnov, E.Z. Zasorin, V.P. Spiridonov (Proc. 11th Rare Earth Res. Conf., Traverse City, Mich., 1974, Vol. 2, pp. 1087/94; C.A. **83** [1975] No. 69511). – [22] N.V. Filippenko, E.V. Morozov, N.I. Giricheva, K.S. Krasnov (Izv. Vysshikh Uchebn. Zavedenii Khim. Khim. Tekhnol. **15** [1972] 1416/8). – [23] K.S. Krasnov (Teplofiz. Vysokikh Temp. **5** [1967] 715/6; High Temp. [USSR] **5** [1967] 639/40). – [24] G.K. Selivanov, Yu.N. Sekachev, A.A. Mal'tsev (Zh. Fiz. Khim. **47** [1973] 2182; Russ. J. Phys. Chem. **47** [1973] 1239/40). – [25] P.A. Perov, S.V. Nedyak, A.A. Mal'tsev (Vestn. Mosk. Univ. **30** [1975] 281/3; Moscow Univ. Chem. Bull. **30** No. 3 [1975] 18/9).

[26] K.S. Krasnov, A.A. Zaitsev (Zh. Fiz. Khim. **39** [1965] 2477/81; Russ. J. Phys. Chem. **39** [1965] 1322/4). – [27] K.S. Krasnov (Izv. Vysshikh Uchebn. Zavedenii Khim. Khim. Tekhnol. **10** [1967] 997/1000). – [28] A. Phongsatha, S.J. Cyvin (Rev. Chim. Minerale **12** [1975] 218/22). – [29] J.C. Wells, J.B. Gruber, M. Lewis (Chem. Phys. **24** [1977] 391/7). – [30] S. Ciach, A.J.C. Nicholson, D.L. Swingler, P.J. Thistlethwaite (Inorg. Chem. **12** [1973] 2072/4).

[31] K.S. Krasnov, U.S. Timoshinin (Teplofiz. Vysokikh Temp. **7** [1969] 365/7; High Temp. [USSR] **7** [1969] 333/4). – [32] F.H. Spedding, J.P. Flynn (J. Am. Chem. Soc. **76** [1954] 1474/7). – [33] T.L. Allen (J. Chem. Phys. **26** [1957] 1644/7). – [34] G.P. Dudchik, O.G. Polyachenok, G.I. Novikov (Zh. Fiz. Khim. **45** [1971] 728/9; Russ. J. Phys. Chem. **45** [1971] 409/10). – [35] H. Wotke (9th Intern. Conf. Phenomena Ionized Gases Contrib. Papers, Bucharest 1969, p. 26; C.A. **73** [1970] No. 135022).

[36] A.D. Chervonnyi (Zh. Fiz. Khim. **51** [1977] 1308/12; Russ. J. Phys. Chem. **51** [1977] 772/5). – [37] R.C. Feber (LA-3164 [1965] 175, 177/9, 180; N.S.A. **19** [1965] No. 30148). – [38] G.I. Novikov, A.K. Baev (Izv. Vysshikh Uchebn. Zavedenii Khim. Khim. Tekhnol. **9** [1966] 180/4; C.A. **65** [1966] 9824). – [39] K.S. Krasnov (Teplofiz. Vysokikh Temp. **4** [1966] 139/41; High Temp. [USSR] **4** [1966] 128/30). – [40] G.I. Novikov (Probl. Sovrem. Khim. Koord. Soedin. **1970** No. 3, pp. 33/48, 37; C.A. **75** [1971] No. 10210).

[41] D.D. Wagman, W.H. Evans, V.P. Parker et al. (NBS-TN-270-5 [1971] 30). – [42] R.H. Schumm, D.D. Wagman, S. Bailey et al. (NBS-TN-270-7 [1973] 1, 9, 17, 43, 49, 54, 61). – [43] G.P. Dudchik (Diss. Minsk 1970). – [44] K.S. Krasnov, T.G.

Danilova (Teplofiz. Vysokikh Temp. **7** [1969] 1213/6; High Temp. [USSR] **7** [1969] 1131/3). – [45] C.E. Myers, D.T. Graves (J. Chem. Eng. Data **22** [1977] 436/9).

[46] H.A. Øye, D.M. Gruen (J. Am. Chem. Soc. **91** [1969] 2229/36), D.M. Gruen, H.A. Øye (Inorg. Nucl. Chem. Letters **3** [1967] 453/5). – [47] W.T. Carnall, J.P. Hessler, H.R. Hoekstra, C.W. Williams (J. Chem. Phys. **68** [1978] 4304/9). – [48] R.R. Jacobs, W.F. Krupke, J.P. Hessler, W.T. Carnall (Opt. Commun. **21** [1977] 395/8). – [49] J.P. Hessler, F. Wagner, C.W. Williams, W.T. Carnall (J. Appl. Phys. **48** [1977] 3260/2). – [50] G.N. Papatheodorou, G.H. Kucera (Inorg. Chem. **18** [1979] 385/9).

[51] H. Schäfer (Angew. Chem. **88** [1976] 775/89; Angew. Chem. Intern. Ed. Engl. **15** [1976] 713). – [52] H.A. Øye, D.M. Gruen (NBS-Spec. Publ.-561 [1979] 1517/37). – [53] T.S. Zvarova, I. Zvara (J. Chromatog. **44** [1969] 604/8). – [54] H. Schäfer, M. Binnewies, W. Domke, J. Karbinski (Z. Anorg. Allgem. Chem. **403** [1974] 116/26). – [55] E.J. Peterson, J.A. Caird, W.T. Carnall, J.P. Hessler, H.R. Hoekstra, C.W. Williams (CONF-790641-8 [1979]; C.A. **92** [1980] No. 117350; Rare Earths Modern Sci. Technol. **2** [1979/80] 111/6; C.A. **93** [1980] No. 81555).

[56] J.A. Caird, W.T. Carnall, J.P. Hessler, C.W. Williams (J. Chem. Phys. **74** [1981] 798/804). – [57] W.F. Krupke (UCID-16620 [1974]; C.A. **82** [1975] No. 177912; UCID-16933 [1976]; C.A. **85** [1976] No. 39146; UCRL-77033 [1975]; C.A. **84** [1976] No. 82338; UCRL-78419 [1976]; C.A. **87** [1977] No. 46419). – [58] R.R. Jacobs, W.F. Krupke (IEEE J. Quantum Electron. **13** [1977] 103; Appl. Phys. Letters **32** [1978] 31/3). – [59] E.J. Peterson, J.A. Caird, J.P. Hessler, H.R. Hoekstra, C.W. Williams (J. Phys. Chem. **83** [1979] 2458/62). – [60] R.R. Jacobs, W.F. Krupke (Appl. Phys. Letters **35** [1979] 126/8).

[61] J.A. Caird, J.P. Hessler, W.T. Carnall, C.W. Williams (J. Chem. Phys. **74** [1981] 805/12). – [62] J.A. Caird, W.T. Carnall, J.P. Hessler (J. Chem. Phys. **74** [1981] 3225/33). – [63] W.T. Carnall, J.P. Hessler, C.W. Williams, H.R. Hoekstra (J. Mol. Struct. **46** [1978] 269/84). – [64] M. Sørlie, H.A. Øye (J. Inorg. Nucl. Chem. **40** [1978] 493/8).

## 11.2 Solid Chlorides, Solutions, Melts

### 11.2.1 Comparative Data for M-$MCl_3$ Systems

#### 11.2.1.1 General Information and Phase Diagrams

The phase diagrams of the M-$MCl_3$ systems have been studied for M=Sc, Y, La, and all the lanthanides except Pm and Lu. Studies of M-$MCl_3$ systems at high temperatures require the rigorous exclusion of air and moisture and the use of Ta, Mo, or possibly W containers. Ceramic materials such as $Al_2O_3$, MgO, and $ThO_2$ are strongly attacked with the formation of oxide chlorides.

Most of the M-$MCl_3$ systems contain one or more phases in which M has an oxidation number less than 3. The chlorides of divalent Sm, Eu, and Yb have been known since the beginning of this century. Since then, the dichlorides of Nd, Dy, and Tm have also been characterized. The existence of one or more solid intermediate phases with average M oxidation numbers between 2 and 3 have been found for systems in which M=Pr, Nd, Sm, Dy, Ho, Tm, and Yb. The unusual "sesquichlorides" $MCl_x$ with $x \approx 1.6$ are observed in systems with M=Sc and Gd.

Apparently there is a long list of further intermediate phases, many of which form peritectically. But such phases are not yet understood because many of the systems have not been adequately characterized. This is especially true for those chlorides in which M has an oxidation number between ≈1.5 and 2. Attempts to prepare such chlorides by the reaction of liquid $MCl_3$ with metal are unsuccessful because a protective layer of higher intermediate chlorides forms on the metal. However, in recent years the reaction of low-pressure $MCl_3$ gas with metal foil or thin metal sheet at 800 to 900 °C has been used for the direct preparation of lower chlorides, including even the monochlorides of Tb, Ho, and Er. The existence of LaCl and CeCl is thought possible. The thermodynamic stability of certain of these phases is still an open question. All identified lower chlorides of the rare earth metals are listed in the following table:

| M | Cl/M | M | Cl/M |
|---|---|---|---|
| Sc | 1.0, 1.40 ?, 1.43, 1.45, 1.50, 1.58, 1.71 | Gd | 1.0, ≈1.6 |
| | | Tb | 1.0, 1.50 |
| Y | 1.0, 1.5 | Dy | 2.00, 2.11, 2.143, 2.167, 2.20 |
| La | 1.0 ? | Ho | 1.0, 2.20 |
| Ce | 1.0 ? | Er | 1.0 ?, 1.5 ? |
| Pr | 1.0 ?, 2.00 ?, 2.31 | Tm | 1.5 ?, 2.00, 2.040, 2.067, 2.074, 2.080, 2.090, 2.10, 2.11, 2.143, 2.167 |
| Nd | 2.00, 2.25, 2.27, 2.33 | | |
| Sm | 2.00, 2.20, 2.308, 2.333 | | |
| Eu | 2.00, 2.20, 2.25, 2.33 | Yb | 2.00, 2.167, 2.375 ? |
| | | Lu | 1.5 ? |

Corbett [1], Haschke [2], see also Poeppelmeier, Corbett [10], Simon et al. [11, 12], Mattausch et al. [13, 14] and the individual phase diagrams M-$MCl_3$ in "Rare Earth Elements" C 4 b.

In systems with M=Y, La, Ce, and Er, the molten $MCl_3$ dissolves significant amounts of M with the apparent formation of $M^{2+}$ ions in the melt. Nevertheless, when the melt is cooled the $M^{2+}$ ions do not separate as a solid lower chloride, but rather disproportionate back into $M^0$ and $M^{3+}$.

A general survey of the extent to which the metals react with their molten trichlorides is shown in **Fig. 1**. In this figure, the composition of the molten salt phase in equilibrium with the metal at ≈800 °C is expressed both as mol % "solubility" of M in $MCl_3$

Fig. 1

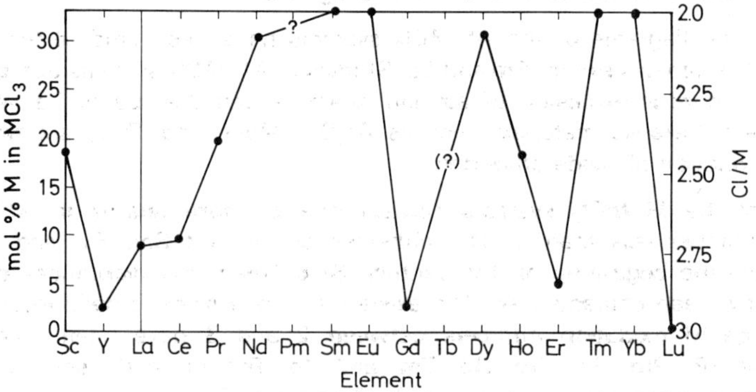


Composition of the molten chloride phase (expressed both as mol % metal dissolved in $MCl_3$ and as overall Cl/M ratio of the chloride phase) in equilibrium with the metal at ≈800 °C.

and, alternatively, as a Cl/M ratio; 33.3 mol % M (or a Cl/M ratio of 2) corresponds to complete reduction of the salt phase to the dichloride. In discussing this figure, Corbett [1] points out that the lanthanides (except for lanthanum itself, which shows some irregularity in behavior) (a) have equilibrium Cl/M ratios which roughly parallel the stability of the dipositive state (dihalide stability shows a regular increase from La to Eu, a rapid decrease at Gd, and then a much more irregular increase to Yb, with an intermediate maximum at Dy and an intermediate minimum at Er) and (b) show an approximate correspondence between the third ionization energy of the gaseous rare earth element and the negative heat for sublimation of the metal on the one hand (see the figure in [1]) and the stability of the dipositive states on the other.

The above survey is based largely on the detailed discussion of [1], with additional information from [2 to 14].

References:

[1] J.D. Corbett (Rev. Chim. Minerale **10** [1973] 239/57). – [2] J.M. Haschke (Inorg. Chem. **15** [1976] 298/303). – [3] J.D. Corbett, D.L. Pollard, J.E. Mee (Inorg. Chem. **5** [1966] 761/6). – [4] D. Brown (Halides of the Transition Elements, London-New York-Sydney 1968, pp. 149/70, 162/9). – [5] R.E. Thoma (in: L. Eyring, Progress in the Science and Technology of the Rare Earths, Vol. 2, Oxford - London - Edinburgh - New York - Toronto - Sydney - Paris - Braunschweig 1966, pp. 90/122, 104/6).

[6] G.I. Novikov, O.G. Polyachenok (Usp. Khim. **33** [1964] 732/6; Russ. Chem. Rev. **33** [1964] 342/50, 346, 348). – [7] J.D. Corbett (Preparative Inorganic Reactions, Vol. 3, New York 1966, pp. 1/33, 29; Proc. 12th Rare Earth Res. Conf., Vail, Colo., 1976, Vol. 1, pp. 396/403; C.A. **85** [1976] No. 151860). – [8] J.R. Peterson (Proc. 10th Rare Earth Res. Conf., Carefree, Ariz., 1973, Vol. 1, pp. 4/14, 10; C.A. **85** [1976] No. 27975). – [9] H. Bärnighausen (private communication to the Gmelin Institute, 1980). – [10] K.R. Poeppelmeier, J.D. Corbett (Inorg. Chem. **16** [1977] 294/7).

[11] A. Simon, H. Mattausch, N. Holzer (Angew. Chem. **88** [1976] 685/6; Angew. Chem. Intern. Ed. Engl. **15** [1976] 624). – [12] A. Simon (Angew. Chem. **93** [1981] 23/44, 32). – [13] H. Mattausch, J.B. Hendricks, R. Eger, J.D. Corbett, A. Simon (Inorg. Chem. **19** [1980] 2128/32). – [14] H. Mattausch, A. Simon, N. Holzer, R. Eger (Z. Anorg. Allgem. Chem. **466** [1980] 7/22).

### 11.2.1.2 Electrical Conductivity of M-$MCl_3$ Melts

Data on the electrical conductivity of M-$MCl_3$ melts have been published for M = La, Ce, Pr, Nd, Gd, Dy, Ho, and Er. The container material was Mo unless otherwise noted.

For all the systems studied, addition of metal to the trichloride melt causes an increase in electrical conductivity, Dworkin et al. [1], Bronstein et al. [2], Dworkin et al. [3, 4]; cf. Kuzyakin et al. [10] and data for the individual chlorides.

The specific conductivity $\varkappa$ versus metal concentration is shown in **Fig. 2**, p. 22, for M = La, Ce, Pr, and Nd; for comparison the curve for the K-KCl system (from [9]) is also shown, Bredig [5]. In the La-$LaCl_3$ and Ce-$CeCl_3$ systems, as in the K-KCl system, there is a sharp rise in $\varkappa$ with increasing metal concentration, which apparently indicates a considerable electronic conductivity in addition to the ionic conductivity. The electronic conduction is thought to be due to orbital overlapping and an increasing build up of an electron conduction band via dissociation of the metal: $M \rightarrow M^{3+} + 3e^-$. There may be a partial depletion of these conduction electrons with the formation of

Fig. 2

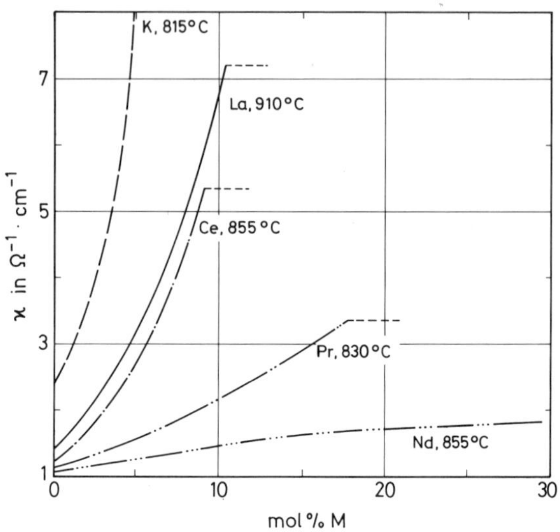


The specific conductivity ϰ versus mol % M for various M-$MCl_3$ systems and for the K-KCl system [5]. Original references: [1] for La and Nd, [2] for Ce, [3] for Pr, and [9] for K.

divalent metal ions: $e^- + M^{3+} \rightleftharpoons M^{2+}$ [1, 2]. An additional contribution to the electronic conductivity may result from an exchange of electrons between two ions having adjacent oxidation states (e.g., between $Ce^{2+}$ and $Ce^{3+}$) [2], see also Corbett [8]. For the La-$LaCl_3$ and Ce-$CeCl_3$ systems, the electronic part of the conductivity as a function of $x_M/z$ ($x_M$=mole fraction of the metal M; z=valence of the metal) resembles that of the K-KCl system in both shape and magnitude; see figure in the original publication, Bredig [5, 6].

In the Nd-$NdCl_3$ system there is only a small increase of ϰ with increasing metal concentration (Fig. 2). Such behavior is typical of a salt-like melt with very little or even no electronic conductance at all. Obviously, we are dealing with an $NdCl_2$-$NdCl_3$ melt in which the $Nd^{2+}$ ion is too stable to dissociate into $Nd^{3+} + e^-$. (As a solid, $NdCl_2$ is stable and nonconducting.) Thus the electrons necessary for electronic conduction are not present to a significant amount [1, 3]; see also Corbett [7]. The small positive deviation of the conductivity from the straight line suggests a limited electron exchange between $Nd^{2+}$ and $Nd^{3+}$, which is attributed to the 4f electrons [1, 7], cf. [8].

The conductivity behavior of the Pr-$PrCl_3$ system lies between those of the La-$LaCl_3$ and Ce-$CeCl_3$ systems and that of the Nd-$NdCl_3$ system, more closely resembling the latter. In the Pr-$PrCl_3$ system, as in the system with Nd, the relative stability of the divalent ion allows only limited electronic conductivity [3, 7]. The upward bend in the ϰ curve in the Pr-$PrCl_3$ system (Fig. 2) is ascribed to the gradual orbital overlap of the conduction electrons [3].

Experimental values for the specific conductivity ϰ in $\Omega^{-1}\cdot cm^{-1}$ for the systems with M = Gd, Dy, Ho, and Er have been gathered in the following table; uncertainties range from $\approx\pm1\%$ for lower values up to $\pm4\%$ for the highest values:

$Gd-GdCl_3$ melts at 650 °C:

| | | | | |
|---|---|---|---|---|
| mol % Gd | 0 | 0.47 | 0.98 | 1.64 |
| $\varkappa$ | 0.41 | 0.427 | 0.454 | 0.487 |

$Dy-DyCl_3$ melts at 700 °C:

| | | | | | | | |
|---|---|---|---|---|---|---|---|
| mol % Dy | 0 | 1.4 | 4.4 | 7.8 | 12.3 | 18.4 | 25.0 |
| $\varkappa$ | 0.415 | 0.46 | 0.54 | 0.65 | 0.76 | 0.90 | 0.98 |

$Ho-HoCl_3$ melts at 800 °C:

| | | | | | |
|---|---|---|---|---|---|
| mol % Ho | 0 | 1.1 | 5.2 | 9.9 | 16.4 |
| $\varkappa$ | 0.526 | 0.555 | 0.69 | 0.84 | 1.05 |

$Er-ErCl_3$ melts at 820 °C:

| | | | | |
|---|---|---|---|---|
| mol % Er | 0 | 1.46 | 4.78 | 6 |
| $\varkappa$ | 0.50 | 0.55 | 0.67 | 0.69 |

The conductivity behavior of the $Gd-GdCl_3$ system, with an initial increase of $\varkappa$ with increasing metal concentration, lies between that of the $La-LaCl_3$ system and that of the $Pr-PrCl_3$ system. In the $Dy-DyCl_3$ system, as in the $Nd-NdCl_3$ system, $\varkappa$ increases with increasing metal concentration but at a decreasing rate, and as in the $Nd-NdCl_3$ system (see above), this behavior is ascribed to that small fraction of the electronic conductivity which is presumably caused by an $M^{2+} \rightleftharpoons M^{3+}$ electron exchange. Plots of $\varkappa$ versus metal concentration for the $Ho-HoCl_3$ and $Er-ErCl_3$ systems are linear and roughly halfway between the plots for the $Pr-PrCl_3$ and $Nd-NdCl_3$ systems [4].

References:

[1] A.S. Dworkin, H.R. Bronstein, M.A. Bredig (Discussions Faraday Soc. **32** [1961] 188/96, 193, 195). – [2] H.R. Bronstein, A.S. Dworkin, M.A. Bredig (J. Phys. Chem. **66** [1962] 44/8). – [3] A.S. Dworkin, H.R. Bronstein, M.A. Bredig (J. Phys. Chem. **66** [1962] 1201/2). – [4] A.S. Dworkin, H.R. Bronstein, M.A. Bredig (J. Phys. Chem. **67** [1963] 2715/7). – [5] M.A. Bredig (in: M. Blander, Molten Salt Chemistry, New York - London - Sydney 1964, pp. 399/410, 405).

[6] M.A. Bredig (J. Chem. Phys. **37** [1962] 914/5). – [7] J.D. Corbett (in: B.R. Sundheim, Fused Salts, New York - San Francisco - Toronto - London 1964, pp. 341/407, 372). – [8] J.D. Corbett (Rev. Chim. Minerale **10** [1973] 239/57, 247). – [9] H.R. Bronstein, M.A. Bredig (J. Am. Chem. Soc. **80** [1958] 2077/81). – [10] E.B. Kuzyakin, E.V. Kuz'minskii, A.V. Gorodyskii (Ukr. Khim. Zh. **44** [1978] 249/52; C.A. **89** [1978] No. 15522).

### 11.2.2 Comparative Data for MCl

General Information

Based on thermodynamic calculations, the rare earth monochlorides were thought to be unstable in both the solid and molten state, Polyachenok, Novikov [1]. However, recent preparative and crystallographic studies of $M-MCl_3$ systems (M = rare earth) by Poeppelmeier, Corbett [2], Simon et al. [3], and Mattausch et al. [5, 9] establish the existence of solid MCl for M = Sc [2], Y [9], Gd, Tb [3], and Ho [5]. In addition, the existence of LaCl is regarded as possible by [2], and reported as fact together with CeCl, PrCl, and ErCl by Simon [8]; for "ErCl", however, see below. According to Lawson [4], the following preconditions are required for occurrence of MCl (M=Ce, Pr, Nd) in the M-Cl system: (1) the absence of a stable divalent state, (2) at least two outer electrons (or a very small excitation energy into such a configuration), and (3) a low value for the sublimation energy of the rare earth metal. The report of the existence of ErCl and LuCl (see [3]) is said to be obsolete by [5], but ErCl is reported as a fact by [8] (see above) in accordance with newest investigations by [14], which do indicate the existence of ErCl.

### Formation and Preparation

The monochlorides are obtained by prolonged reaction of higher chlorides (e.g., $MCl_3$) with the corresponding rare earth metals in closed containers (e.g., of tantalum) at temperatures of about 800 to 900 °C. Details are given under the individual compounds, see "Rare Earth Elements" C4b.

### Crystallographic Properties

The compounds crystallize as graphite-colored, readily cleaved hexagonal platelets [2, 3, 5]. The only known forms of ScCl [2], YCl [9], and the high-temperature modification of GdCl and TbCl crystallize in the so-called h-form which is isotypic with ZrBr, Mattausch et al. [5]. Also LaCl, CeCl, and PrCl are said to have this form, Simon [8]. The low-temperature modifications of GdCl and TbCl and the only known forms of HoCl [5] and ErCl [8] crystallize in the so-called t-form which is isotypic with ZrCl [5]. In GdCl and TbCl, the polymorphic transition is kinetically strongly hindered. Yet, there are peaks in the DTA-curve of GdCl at 993 and 855 K, when the compound is heated and cooled, respectively, at 1 K/min. With TbCl no peak due to this transition is observed. However, both forms of TbCl crystallize when the compound is slowly cooled from 900 K. Larger crystals of h-TbCl form on annealing (2d) at 1070 K [5].

Both the t- and the h-form have the space group $R\bar{3}m-D^5_{3d}$ (No. 166); Z=6. Lattice constants for a hexagonal setting (with the standard deviation of the last digit in parantheses) are as follows:

| MCl | a in Å | c in Å | Ref. | MCl | a in Å | c in Å | Ref. |
|---|---|---|---|---|---|---|---|
| ScCl | 3.473(2) | 26.71(4) | [2] | t-TbCl | 3.781(1) | 27.461(5) | [5] |
| YCl | 3.748(1) | 27.318(9) | [9] | h-TbCl | 3.786(1) | 27.461(4) | [5] |
| t-GdCl | 3.8208(8) | 27.477(6) | [5] | HoCl | 3.727(1) | 27.349(5) | [5] |
| h-GdCl | 3.819(1) | 27.482(6) | [5] | ErCl | 3.698(1) | 27.204(5) | [14] |

For rhombohedral setting a=9.421(2) Å, $\alpha$=23.40(1)° for t-GdCl, a=9.425(1) Å, $\alpha$=23.39(1)° for h-GdCl, a=9.415(2) Å, $\alpha$=23.18(1)° for t-TbCl, a=9.421(2) Å, $\alpha$=23.18(1)° for h-TbCl [5], and a=9.316(2) Å, $\alpha$=22.89(2)° for ErCl [14].

All atoms occupy the equivalent position 6c (0,0,z); for ScCl, z(Sc)=0.2137 and z(Cl)=0.3914, with R=8.8%. Interatomic distances are given in the original article, Poeppelmeier, Corbett [2]. For t-TbCl, z(Tb)=0.11629 and z(Cl)=0.38777, with R=11%, and for h-TbCl, z(Tb)=0.21716 and z(Cl)=0.38765, with R=5.2% [5]. In the four-layer sheet structure of all these compounds close-packed double layers of M atoms alternate with close-packed double layers of Cl atoms (ClMMCl... ClMMCl...) along [001]. In the h-form (ZrBr type), the layer geometry is |abca|cabc|bcab|... and the packing sequence of the four-layer sheets is ACB. Whereas the t-form (ZrCl type) has the layer geometry |abca|bcab|cabc|... and the packing sequence ABC [2]. The structures

Fig. 3

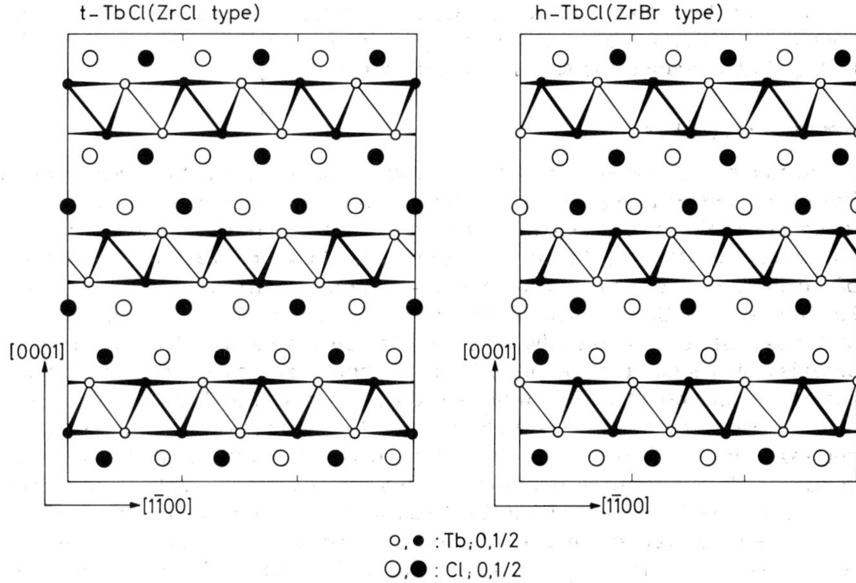


Crystal structures of t-TbCl and h-TbCl projected along [11$\bar{2}$0]; along [1$\bar{1}$00] three translation periods are shown.

of t-TbCl and h-TbCl projected along [11$\bar{2}$0] are shown in **Fig. 3** [5]. In both structures, the metal atoms are coordinated antiprismatically; the Cl atoms are coordinated prismatically in the h-form, and antiprismatically in the t-form. The M-M bonds within a layer are weaker (Sc↔Sc=3.47 Å [2]; Tb↔Tb=3.787 Å in t-TbCl, and 3.786 Å in h-TbCl [5]) than those between neighboring layers (Sc↔Sc=3.22 Å [2]; Tb↔Tb=3.515 Å in t-TbCl and 3.531 Å in h-TbCl [5]). Nevertheless, the double metal layer structure may be described approximately in terms of metal octahedra (trigonal antiprisms) each of which shares six edges with other octahedra. There are only weak van der Waals forces between Cl atoms of neighboring layers (Cl↔Cl=3.69 Å in ScCl [2], 3.700 Å in t-TbCl, and 3.689 Å in h-TbCl [5]), which accounts for the graphite-like properties of these compounds [2]. Whereas the distances between the metal atom and the next nearest Cl atoms are nearly alike in both structure types (Tb↔Cl=2.774 Å in t-TbCl and 2.772 Å in h-TbCl), there are striking differences in the distances to the second nearest Cl atoms (Tb↔Cl=5.174 Å in t-TbCl and 4.676 Å in h-TbCl), indicating stronger bonding interactions between the layer packets in the h-form. These may account for the fact that on grinding compounds of the t-form become severely disordered. This is not observed with the compounds of the h-form [5], which are less graphitic, more brittle, and easier to cleave than the t-form compounds [2].

In MCl each M atom has on the average two bonding electrons in the double layer [2, 10]. Values of the average Pauling metal-metal bond order n per metal valence electron (PBO/e) were derived from Pauling's equation $d_n = d_1 - 0.60 \log n$ [6], with $d_1 = 2.924$ Å for ScCl, 3.240 Å for YCl, and 3.203 Å for TbCl, and $d_n$ from the observed M-M distances (see [2] for M=Sc, [11] for Y, and [5] for Tb). PBO/e=0.84 for ScCl [10] (0.85 [7]), 0.95 for YCl, 0.77 for TbCl (ZrCl type), and 0.75 for TbCl (ZrBr type) [10]. In this connection compare the M-M bond relationships in other scandium-chlorides and in $M_2Cl_3$.

The sheet structure of MCl ($\triangleq M_{6/3}Cl_2$), which fits into the concept of so-called condensed clusters, is derived in a figure in an article by Simon et al. [3] from the $M_6Cl_8$ cluster through the $M_2X_3$ structure (see p. 27).

### Electrical Properties

According to non-self-consistent band calculations by the KKR (Korringa, Kohn, Rostoker) method using the Hartree-Slater ($\alpha=1$) approximation, which were performed to construct the crystal potential from the overlapping neutral atom charge densities, the energy bands of ScCl are very similar to those in ZrBr and ZrCl. The 3d bands of Sc in ScCl are narrower than the 4d bands of Zr in the monohalides, and the Fermi level of ScCl lies lower in the d band complex. Thus ScCl like the monohalides of Zr possesses d-band metallic properties, Marchiando et al. [12]. The (single sandwich) densities of electronic states in ZrCl, ScCl, and GdCl have been calculated using the linear-combination-of-atomic-orbitals (LCAO) method. The Cl 3s and 3p states are prominent in bands centered at −22.5 and −10 eV, respectively. The width of the band decreases from ZrCl to ScCl to GdCl. Atomic Gd 4f orbitals should form localized levels at about −12.2 eV. States between −8 and −2.5 eV are essentially composed of metal d orbitals. The Fermi level $E_F$ in ZrCl lies in a deep minimum in the density of states at −4.5 eV, separating metal-metal bonding states from higher d states. In ScCl and GdCl it lies near the trough at about −5 eV. The trough for ScCl and GdCl is about 0.2 eV narrower than that for ZrCl (a figure is given), Bullett [13].

The MCl compounds are strongly anisotropic. In single crystals of TbCl normal to the plane of the platelets, the specific electric resistivity is about $2\times10^{2}\ \Omega\cdot cm$. Pressed polycrystalline materials containing GdCl both of ZrCl and ZrBr type have $7.5\times10^{-2}\ \Omega\cdot cm$ between 50 and 298 K; there is a sharp rise at lower temperatures and a slight increase to about $3\times10^{-1}\ \Omega\cdot cm$ at 1000 K (a figure is given in the paper) [5].

### Chemical Reactions

GdCl and TbCl decompose in moist air and react with water to form $H_2$ [3].


References:

[1] O.G. Polyachenok, G.I. Novikov (Zh. Neorgan. Khim. **8** [1963] 1567/73; Russ. J. Inorg. Chem. **8** [1963] 816/20). – [2] K.R. Poeppelmeier, J.D. Corbett (Inorg. Chem. **16** [1977] 294/7). – [3] A. Simon, H. Mattausch, N. Holzer (Angew. Chem. **88** [1976] 685/6; Angew. Chem. Intern. Ed. Engl. **15** [1976] 624). – [4] L.D. Lawson (IS-T-828 [1978] 1/37 from C.A. **90** [1979] No. 214391). – [5] H. Mattausch, A. Simon, N. Holzer, R. Eger (Z. Anorg. Allgem. Chem. **466** [1980] 7/22).

[6] L. Pauling (The Nature of the Chemical Bond, 3rd Ed., Ithaca, N.Y., 1960, pp. 400/4). – [7] K.R. Poeppelmeier, J.D. Corbett (J. Am. Chem. Soc. **100** [1978] 5039/44). – [8] A. Simon (Angew. Chem. **93** [1981] 23/44; Angew. Chem. Intern. Ed. Engl. **20** [1981] 1). – [9] H. Mattausch, J.B. Hendricks, R. Eger, J.D. Corbett, A. Simon (Inorg. Chem. **19** [1980] 2128/32). – [10] J.D. Corbett (J. Solid State Chem. **37** [1981] 335/51).

[11] J. Ford, J.D. Corbett (unpublished research according to [10]). – [12] J.F. Marchiando, B.N. Harmon, S.H. Liu (Physica B+C **99** [1980] 259/63). – [13] D.W. Bullett (Inorg. Chem. **19** [1980] 1780/5). – [14] H. Mattausch (private communication to the Gmelin Institute 1981).

### 11.2.3 Comparative Data for $M_2Cl_3$

These $M_2Cl_3$ compounds are obtained by prolonged reaction of gaseous or molten $MCl_3$ with the corresponding rare earth metal. Details are given under the individual compounds in "Rare Earth Elements" C 4b.

#### Crystallographic Properties

$Sc_2Cl_3$ [1], $Y_2Cl_3$ [9], and $Gd_2Cl_3$ [2] form as bundles of needles, which on being cut split into progressively thinner, flexible fibers, McCollum et al. [1], Mattausch et al. [9], Lokken, Corbett [2]. The $Gd_2Cl_3$ needles have an approximately pentagonal cross section; the fiber axis coincides with the crystallographic b-axis [2].

Although the crystal habits of $Sc_2Cl_3$ and $Gd_2Cl_3$ are similar [1], the two compounds are not isotypic [2, 3]. X-ray powder patterns show $Sc_2Cl_3$ to be isostructural with $Sc_2Br_3$; d-values are listed in the original article [1]. $Gd_2Cl_3$ has a monoclinic $Gd_2Cl_3$-type structure, space group $C2/m-C^3_{2h}$ (No. 12) and Z=4 according to single crystal studies. This space group is in contrast to an earlier acentric description (see [2]) which is obsolete, Simon et al. [10]. $Y_2Cl_3$ [9] and $Tb_2Cl_3$ (and presumably $Er_2Cl_3$, $Tm_2Cl_3$, $Lu_2Cl_3$ [12]) also have a $Gd_2Cl_3$-type structure, Simon et al. [4]. Lattice constants are:

| $M_2Cl_3$ | a in Å | b in Å | c in Å | $\beta$ in ° | Ref. |
|---|---|---|---|---|---|
| $Y_2Cl_3$ | 15.144±0.003 | 3.825±0.001 | 10.077±0.002 | 118.24±0.02 | [9] |
| $Gd_2Cl_3$ | 15.231±0.003 | 3.895±0.001 | 10.175±0.002 | 117.64±0.01 | [10] |
| $Gd_2Cl_3$ | 15.237±0.004 | 3.896±0.001 | 10.179±0.003 | 117.66±0.03 | [5] |
| $Tb_2Cl_3$ | 15.184±0.009 | 3.869±0.003 | 10.135±0.006 | 118.07±0.05 | [4] |

The three dimensional structural refinement of $Gd_2Cl_3$ (R=5.2%) gives the following atom parameters [10]:

| | Gd(1) | Gd(2) | Cl(1) | Cl(2) | Cl(3) |
|---|---|---|---|---|---|
| x | 0.40983 | 0.13690 | 0.65905 | 0.76925 | 0.95035 |
| y | 0 | 0 | 0 | 0 | 0 |
| z | 0.05192 | 0.29308 | 0.51165 | 0.88245 | 0.26901 |

The $M_2Cl_3$ structure is characterized by infinite chains of M atom octahedra parallel to [010], with the octahedra attached to each other along trans edges as shown in **Fig. 4**, p. 28 [3] for $Gd_2Cl_3$. The octahedron edge dimensions are not equal. Using the atom designation given in **Fig. 5**, p. 28 [10], the bonding distances are: Gd(1)↔Gd(1′) (i.e., shared edge)=3.371 Å, Gd(1)↔Gd(2)=3.729 Å, Gd(1)↔Gd(2′)=3.783 Å, and Gd(1)↔Gd(1) (i.e., translation equivalent atoms along [010])=3.895 Å [10]. Thus the chains may also be described as a string of parallel M(1)-M(1′) dimers, which are separated from each other by the distance b and are capped by more distant M(2) and M(2′) atoms. Of the six designated Cl atoms in Fig. 5 [10], four are located over the free triangular faces of the octahedron, i.e., over the triangular faces which face away from the chain. The remaining two Cl atoms (designated Cl(1) and Cl(1′) in Fig. 5) are located between adjacent metal chains, alternately in front of and behind the triangular faces formed by the metal atoms at the apexes of octahedra of adjacent chains. The chains are displaced by b/2 with respect to each other. The bridged octahedron chains form sheets in the $(20\bar{1})$ plane of the structure; the sheets also have a relative

Fig. 4

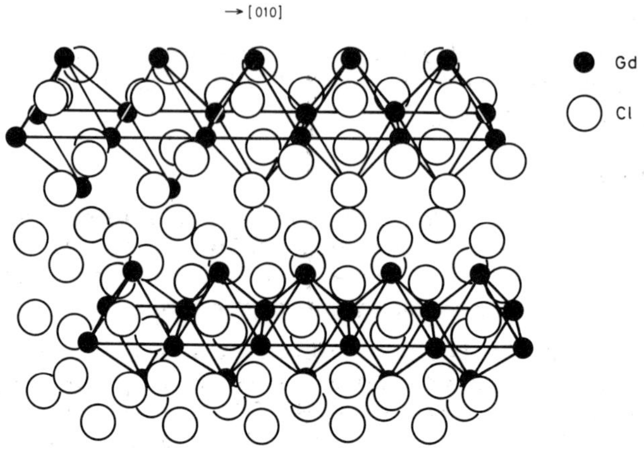


Crystal structure of $Gd_2Cl_3$. For a normal reading position, the reader's line of sight corresponds approximately to the [100] axis.

Fig. 5

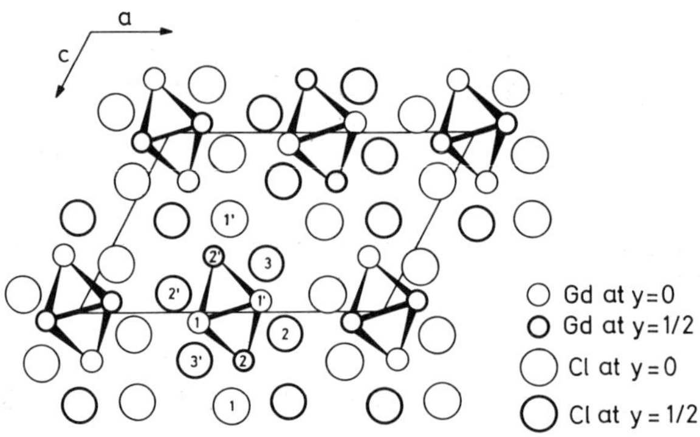


Crystal structure of $Gd_2Cl_3$ projected along [010].

displacement of b/2. The charge distribution of the chain polymer corresponds approximately to the structural formula $[M^{6+}(Cl^-)_6]_n$. The observed interatomic distances ($M \leftrightarrow Cl = 2.729$ to 2.857 Å in $Gd_2Cl_3$ [10]) are interpreted to mean that the M-M bonding occurs only within the chain and the metal atoms behave as $M^{3+}$ towards the surrounding chlorines. Nevertheless, interpretation of the interatomic distances is made somewhat uncertain by the unusual coordination number of the metal atoms in this structure (nine-or ten-coordinate, five of these neighbors are Cl's), Lokken, Corbett [2].

Nowadays, the $M_2Cl_3$ structure is preferably described in terms of shared $M_6$ octahedra, $M_2M_{4/2}Cl_4Cl_{4/2}$ [10], which emphasizes that the structure can be derived from the isolated $M_6Cl_8$ cluster by condensation [4, 10, 12]. In $M_2Cl_3$ there are 1.5 bonding electrons per metal site in the chain [7]. From Pauling's equation, $d_n = d_1 - 0.60 \log n$ [8], with $d_1 = 3.367$ Å as the average single bond distance, the metal-metal bond order n was calculated from the observed Gd-Gd distances $d_n$, n=0.98, 0.25, 0.20, and 0.13 [10]. The small Gd-Gd distances are interpreted as single bonds, whereas there are only weak but nevertheless important bonds between the other Gd atoms [12]. For $Y_2Cl_3$, with $d_1$=3.240 Å [11] and $d_n$ from [9], n=0.452, 0.431, 0.349, and 0.210, resulting in $\Sigma n/1.5$ (≙PBO/e, see p. 25)=0.961. Analogously for $Gd_2Cl_3$, with $d_1$=3.243 Å and $d_n$ from [10], PBO/e=0.63, Corbett [11].

The polyhedral volume cell of the Cl ions in $Gd_2Cl_3$ averages 32.69 $Å^3$. Thus it is somewhat larger than that of the Cl ions in $EuCl_2$ and $Dy_5Cl_{11}$, Carter [13]. Similarities of the $M_2Cl_3$ structures to those of the transition element compounds with $M_6X_8$ clusters are emphasized by Simon et al. [4, 12], while Haschke [6] emphasizes similarities to the anti NiAs structure.

**Electrical Properties**

For a single chain in the structure of $Gd_2Cl_3$, results from LCAO-calculations (linear-combination-of-atomic-orbitals) of the electron density of states in the Cl p bands and Gd d bands are given in a figure in the original article by Bullett [14]. The three lowest filled d bands situated between −5.8 and −4.6 eV hybridize to much lower energies than the remaining seventeen, which are separated from the former by a gap of 0.7 eV. The Fermi level is situated in this gap, i.e., a semiconducting gap is predicted, Bullett [14]. Very recent experimental evidence from photoemission and from temperature dependent conductivity studies provide evidence that $Gd_2Cl_3$ is a semiconductor, Bauhofer, Ebbinghaus [15].

References:

[1] B.C. McCollum, M.J. Camp, J.D. Corbett (Inorg. Chem. **12** [1973] 778/80). – [2] D.A. Lokken, J.D. Corbett (Inorg. Chem. **12** [1973] 556/9). – [3] J.D. Corbett (Rev. Chim. Minerale **10** [1973] 239/57, 252/3). – [4] A. Simon, H. Mattausch, N. Holzer (Angew. Chem. **88** [1976] 685/6; Angew. Chem. Intern. Ed. Engl. **15** [1976] 624). – [5] D.A. Lokken, J.D. Corbett (J. Am. Chem. Soc. **92** [1970] 1799/800).

[6] J.M. Haschke (J. Solid State Chem. **18** [1976] 205/16; Proc. 12th Rare Earth Res. Conf. 1976, Vol. 1, pp. 413/23). – [7] K.R. Poeppelmeier, J.D. Corbett (J. Am. Chem. Soc. **100** [1978] 5039/44). – [8] L. Pauling (The Nature of the Chemical Bond, 3rd Ed., Ithaca, N.Y., 1960, pp. 400/4). – [9] H. Mattausch, J.B. Hendricks, R. Eger, J.D. Corbett, A. Simon (Inorg. Chem. **19** [1980] 2128/32). – [10] A. Simon, N. Holzer, H. Mattausch (Z. Anorg. Allgem. Chem. **456** [1979] 207/16).

[11] J.D. Corbett (J. Solid State Chem. **37** [1981] 335/51, 340, 348). – [12] A. Simon (Angew. Chem. **93** [1981] 23/44; Angew. Chem. Intern. Ed. Engl. **20** [1981] 1). – [13] F.L. Carter (Proc. 13th Rare Earth Res. Conf., Wheeling, Va., 1977 [1978], pp. 225/31). – [14] D.W. Bullett (Inorg. Chem. **19** [1980] 1780/5). – [15] W. Bauhofer, G. Ebbinghaus (private communication to [14]).

## 11.2.4 Comparative Data for $MCl_2$

### 11.2.4.1 General Information

The dichlorides $MCl_2$ with M=Sm, Eu, Tm, and Yb are stable and, except for $TmCl_2$, have been known for a long time. More recently, the less stable Pr, Nd, and Dy dichlorides have been described, Brown [1], Polyachenok, Novikov [2]. The report of the existence of $ScCl_2$ is called in question by Brown [1]. However, some data are submitted concerning this compound, see e.g. Polyachenok, Novikov [7]. Studies of the Ce-$CeCl_3$ system by Mellors, Senderoff [4] and Cubicciotti [5] give no indication of the existence of $CeCl_2$, but equilibrium studies by Shchukarev, Novikov [6] on the reduction of $CeCl_3$ with $H_2$ show the formation of $CeCl_2$. An additional indication that this compound exists is given by Druding, Corbett [3]; but see also "Rare Earth Elements" C4b, Section 11.2.11.4.2. Other rare earth dichlorides mentioned in the following sections may be hypothetical compounds.

References:

[1] D. Brown (Halides of the Transition Elements, London - New York - Sydney 1968, pp. 149/70, 162). – [2] O.G. Polyachenok, G.I. Novikov (Zh. Neorgan. Khim. **8** [1963] 1567/73; Russ. J. Inorg. Chem. **8** [1963] 816/20, 818). – [3] L.F. Druding, J.D. Corbett (J. Am. Chem. Soc. **83** [1961] 2462/7). – [4] G.W. Mellors, S. Senderoff (J. Phys. Chem. **63** [1959] 1110/2). – [5] D. Cubicciotti (J. Am. Chem. Soc. **71** [1949] 4119/21).

[6] S.A. Shchukarev, G.I. Novikov (Zh. Neorgan. Khim. **1** [1956] 362/5; J. Inorg. Chem. [USSR] **1** No. 3 [1956] 12/5). – [7] O.G. Polyachenok, G.I. Novikov (Zh. Neorgan. Khim. **8** [1963] 2819/21; Russ. J. Inorg. Chem. **8** [1963] 1479/80).

### 11.2.4.2 Formation and Preparation

Surveys of the various methods of preparing lanthanide dichlorides are given by Corbett [1] and Novikov, Polyachenok [2].

#### Preparation from $MCl_3$

Most preparations of the dichlorides start with the trichlorides, which are obtained in excellent purity by vacuum distillation or vacuum sublimation. The long known dichlorides, $SmCl_2$, $EuCl_2$, and $YbCl_2$, were originally obtained by high-temperature reduction of the corresponding trichlorides with $H_2$ and/or $NH_3$, and this method continues to be the standard way of preparing these three dichlorides. But the method does have problems. It is time consuming, and numerous precautionary measures are required to prevent oxide chloride contamination and to insure complete conversion to the dichloride. Nevertheless, with refinements and modifications the procedure gives products which are up to 99.99% pure; see e.g. Il'in et al. [3]. On thermodynamic grounds, the procedure is limited to the easily reduced Sm, Eu, and Yb trichlorides, and in the case of $SmCl_3$ it is still doubtful whether complete reduction to the dichloride can be achieved. Reduction of the $SmCl_3$, $EuCl_3$, and $YbCl_3$ melts with $H_2$ is certainly theoretically possible, but the reaction is very slow. Wide differences are reported on the extent to which various rare earth trichlorides can be reduced with $H_2$. Exact thermodynamic data which could settle the question are not available; see for example Polyachenok, Novikov [4].

Early experimental work is considered invalid because of reaction with unsuitable crucible materials. Platinum and gold crucibles, for example, react with rare earth metals (formed by disproportionation of the dichlorides), with the result that $SmCl_2$ and $EuCl_2$ were falsely thought to have relatively narrow stability ranges; cf. surveys of Corbett [1, 5, 6].

In general, the key factor in dichloride stability is the amount of disproportionation: $3MCl_2$ (solid) $= 2MCl_3$ (solid, liquid) $+$ M (solid). This equilibrium is unfavorably shifted to the right by reaction of $MCl_3$ or M with the crucible. Also, some conventional reducing agents may lose their effectiveness by bonding to the rare earth metal or may reduce the trichloride to a lower valence level than two. Thus, $H_2$ forms very stable hydrides with some rare earth metals, and many rare earth trihalides are reduced to the metal by Na, Ca, K, and Ba. Exceptions are the reduction of $SmCl_3$, $EuCl_3$, and $YbCl_3$ to dichloride by Na, Ca, and, under certain conditions, K. Zinc seems to be an excellent reductant for preparing $EuCl_2$, $YbCl_2$, and, possibly, $SmCl_2$ since $ZnCl_2$ and excess Zn are readily removed by volatilization. In another method, $MCl_3$ (M = Sm, Eu, or Yb) is reduced by amalgams of Mg, Ca, or Zn; HCl – ethyl ether serves as both reaction medium and as a solvent for removing $MgCl_2$, $CaCl_2$, or $ZnCl_2$ and for removing unreacted $MCl_3$ from the relatively insoluble $MCl_2$.

The three "classical" dichlorides (of Sm, Eu, and Yb) are also prepared in tetrahydrofuran (THF) by reducing the trichloride with Li in the presence of naphthalene. $SmCl_2$ and $EuCl_2$ promptly precipitate as fine, solvent-free crystals which are relatively insoluble in THF. With $YbCl_3$, the procedure gives a solvated dichloride, $YbCl_2 \cdot THF$, which is moderately soluble in THF, Rossmanith [7]. With $NdCl_3$, $NdCl_2 \cdot 2THF$ is formed; this solvate is quite soluble in THF and readily gives up its THF molecules under high vacuum, Rossmanith [8].

Reaction of the rare earth metals with the corresponding trichlorides is the most universal procedure for preparing the dichlorides. All the known dichlorides may be obtained in very pure form by this method. The purest reactants must be used, and special care is needed to exclude air and moisture. Purified Ar or He serves as a protective atmosphere. The crucible may be of Ta, Mo, or W, all of which are exceptionally resistant to molten rare earth metals and their chlorides. Tantalum does dissolve slightly in melts of the heavier lanthanide metals and in molten Sc, but has the advantage of being available as ductile tubes or sheets, and is readily welded by a helium arc or by an electron beam in vacuum. Molybdenum and tungsten are much more difficult to work, although Mo sheet is readily folded into boats which may be used under a protective $H_2$ atmosphere. Molybdenum is said to be more suitable than tantalum because contaminant oxides can be removed more readily (by heating in $H_2$ at 1000 to 1100 °C).

Other than $H_2$ and $NH_3$, non-metallic reductants have been investigated only in isolated cases. The reaction of $PH_3$ with $SmCl_3$ gives $SmCl_2$, Matignon [9]; the reaction of $N_2H_4$ with $EuCl_3$ gives $EuCl_2$, Stubblefield, Taylor [10].

**Preparation from the Metal**

Attempts to prepare stoichiometric dichlorides by chlorination of the metal with $NH_4Cl$ (mole ratio 1:2) in liquid ammonia were successful in the case of Eu and Yb; ammoniates are formed, which on gentle heating under vacuum give the stoichiometric dichlorides in quantitative yield, Howell, Pytlewski [11], Howell [12]. Attempts were made to prepare $SmCl_2$ and $YbCl_2$ by reacting the metal at 850 to 900 °C with an $HCl$-$H_2$ mixture. With Sm, the product had a 0.9% Cl deficiency. With Yb, the product had the formula $3YbCl_3 \cdot 5YbCl_2$ corresponding to a 3.6% Cl excess, Johnson, Mackenzie [13].

References:

[1] J.D. Corbett (Rev. Chim. Minerale **10** [1973] 239/57, 241). – [2] G.I. Novikov, O.G. Polyachenok (Usp. Khim. **33** [1964] 732/6; Russ. Chem. Rev. **33** [1964] 342/50, 343). – [3] V.K. Il'in, A.D. Chervonnyi, V.A. Krenev (Zh. Neorgan. Khim. **21** [1976] 614/8; Russ. J. Inorg. Chem. **21** [1976] 332/4). – [4] O.G. Polyachenok, G.I. Novikov (Zh. Neorgan. Khim. **8** [1963] 1567/73; Russ. J. Inorg. Chem. **8** [1963] 816/20). – [5] J.D. Corbett (Preparative Inorganic Reactions, Vol. 3, New York 1966, pp. 1/33, 29).

[6] J.D. Corbett (Proc. 12th Rare Earth Res. Conf., Vail, Colo., 1976, Vol. 1, pp. 396/403). – [7] K. Rossmanith (Monatsh. Chem. **110** [1979] 109/14). – [8] K. Rossmanith

(Monatsh. Chem. **110** [1979] 1019/23). – [9] C. Matignon (Ann. Chim. Phys. [8] **8** [1906] 402/16, 415). – [10] C.T. Stubblefield, M.D. Taylor (in: J.F. Nachman, C.E. Lundin, Rare Earth Research, New York - London 1962, pp. 29/33).

[11] J.K. Howell, L.L. Pytlewski (J. Less-Common Metals **18** [1969] 437/9). – [12] J.K. Howell (Diss. Drexel Inst. Technol. 1970, pp. 1/121 from Diss. Abstr. Intern. B **31** [1970/71] 2539). – [13] K.E. Johnson, J.R. Mackenzie (J. Inorg. Nucl. Chem. **32** [1970] 43/8).

### 11.2.4.3 Crystallographic Properties

#### 11.2.4.3.1 Polymorphism

$DyCl_2$ has a tetragonal $SrBr_2$ structure with space group $P4/n-C_{4h}^3$ (No. 85). But other $MCl_2$ compounds (M=Nd, Sm, Eu, Tm, Yb) crystallize orthorhombic at room temperature, some with the $PbCl_2$-type structure, the rest with the $SrI_2$-type structure. At high temperatures there is a polymorphic transformation of $SmCl_2$, of $EuCl_2$, and quite probably of $DyCl_2$, see "Rare Earth Elements" C4b, Sections 11.2.11.8.2.2, 11.2.11.9.2.2, and 11.2.11.12.2, respectively.

#### 11.2.4.3.2 Crystal Structure

Data for $DyCl_2$ are in "Rare Earth Elements" C4b, Section 11.2.11.12.2.

##### 11.2.4.3.2.1 $MCl_2$ of the Orthorhombic $PbCl_2$ Type (M=Nd, Sm, Eu)

These compounds crystallize in the space group Pbnm (Pnma)-$D_{2h}^{16}$ (No. 62); Z=4. This was established by Druding, Corbett [1] for $NdCl_2$ and by Döll, Klemm [2] for $SmCl_2$ and $EuCl_2$ from X-ray powder patterns. The latter authors [2] use the setting Pmnb.

The following description of the structure from Bärnighausen [3] uses the well-studied $EuCl_2$ as the example. A refinement of the structure to R=0.037 using 815 single crystal reflections (MoKα radiation) and anisotropic temperature factors gives the following parameters for the atoms, which all occupy the special position 4(c), ±(x, y, 1/4), ±(1/2−x, 1/2+y, 1/4): x=0.11510 and y=0.24952 for Eu, x=0.42788 and y=0.14342 for Cl(1), and x=−0.16675 and y=0.02379 for Cl(2). The $Eu^{2+}$ ion is surrounded by nine $Cl^-$ ions which form a tetradecahedron (a tri-capped trigonal prism with 14 triangular faces). The Eu↔Cl(1) distances range from 2.916 to 2.994 Å; the Eu↔Cl(2) distances range from 3.046 to 3.440 Å. In comparison, the sum of the ionic radii is 3.10 Å. Since seven of the Eu↔Cl distances are relatively short (between 2.916 and 3.090 Å) it is possible, with qualification, to consider the coordination as (7+2)-fold, rather than nine-fold. The original article [3] tabulates the interatomic distances to near-neighbor $Cl^-$ ions and the angles of the coordination polyhedra. It also depicts the polyhedron around Eu (described above), around Cl(1) (the Cl(1) forms the peak of a flat four-sided pyramid whose base consists of four Eu ions), and around Cl(2) (which is located in the center of a four-sided pyramid formed by five Eu ions), taking into consideration the anisotropy of the thermal vibration. In $EuCl_2$, the volume of the polyhedral atomic volume cell for chlorine is, on the average, 31.43 $Å^3$. Thus it is less than that for chlorine in $Gd_2Cl_3$, Carter [8].

The following table gives lattice constants from X-ray powder patterns at room temperature. Assignment of axes corresponds to the space group symbol Pbnm:

| $MCl_2$ | a in Å | b in Å | c in Å | Ref. |
|---|---|---|---|---|
| $NdCl_2$ | 9.080±0.003 | 7.609±0.003 | 4.554±0.002 | [6] |
| | 9.06 | 7.59 | 4.50 | [4] [1)] |
| $SmCl_2$ | 8.993±0.002 | 7.556±0.001 | 4.517±0.001 | [3] |
| | 8.970±0.046 | 7.489±0.032 | 4.465±0.013 | [5] [1)] |
| $EuCl_2$ | 8.965±0.002 | 7.538±0.001 | 4.511±0.001 | [3] |
| | 8.961±0.007 | 7.528±0.005 | 4.505±0.004 | [7] |

[1)] a and c are interchanged in relation to the original paper.

The samples of Bärnighausen [3] have been purified by high-vacuum distillation and were measured in a Guinier camera, which had been modified to handle extremely air-sensitive substances. For earlier $SmCl_2$ and $EuCl_2$ lattice constants see [2].

References:

[1] L.F. Druding, J.D. Corbett (J. Am. Chem. Soc. **81** [1959] 5512). – [2] W. Döll, W. Klemm (Z. Anorg. Allgem. Chem. **241** [1939] 239/58, 246/50). – [3] H. Bärnighausen (Rev. Chim. Minerale **10** [1973] 77/92). – [4] L.F. Druding, J.D. Corbett (J. Am. Chem. Soc. **83** [1961] 2462/7). – [5] I.S. Astakhova, D.M. Laptev, V.F. Goryushkin (Zh. Neorgan. Khim. **22** [1977] 2590/1; Russ. J. Inorg. Chem. **22** [1977] 1405/6).

[6] H. Bärnighausen, W. Kuhn (private communication from H. Bärnighausen to the Gmelin Institute 1977). – [7] H.A. Eick (private communication from H. Bärnighausen to the Gmelin Institute 1977). – [8] F.L. Carter (Proc. 13th Rare Earth Res. Conf., Wheeling, Va., 1977 [1978], pp. 225/31).

### 11.2.4.3.2.2 $MCl_2$ of the Orthorhombic $SrI_2$ Type (M = Tm, Yb)

These compounds, like the orthorhombic form of $EuI_2$, crystallize in the space group Pbca – $D_{2h}^{15}$ (No. 61) with Z = 8, Bärnighausen et al. [1]. The isotypism of $TmCl_2$ was established by the powder data of Caro, Corbett [3] in conjunction with the data of Beck, Bärnighausen [2, pp. 224, 228] and by the single crystal work of Naterstad, Corbett [4]. For $YbCl_2$, the isotypism was established by the single crystal studies of Beck, Bärnighausen [2] and of Bärnighausen et al. [5].

In the structure, all atoms occupy the general position 8(c). For $YbCl_2$, refinement with anisotropic temperature factors (R = 0.025) gives the following atomic parameters: for Yb, x = 0.11251, y = 0.45705, and z = 0.26930; for Cl(1), x = 0.2051, y = 0.1151, and z = 0.1492; for Cl(2), x = −0.0289, y = 0.2638, and z = 0.0013, Bärnighausen et al. [5]. For $TmCl_2$ the parameters are less exact (R = 0.081), but, except for the z parameter of Cl(2) (z = 0.000), they differ only slightly from the parameters of $YbCl_2$; see Naterstad, Corbett [4]. These authors use the space group symbol Pcab, so x and y are correspondingly reversed in the original paper [4].

The M ion has seven neighboring Cl ions, which are so distributed into two approximately parallel planes that they form an approximate square in one of the planes and a triangle in the other (see **Fig. 6**, p. 34); in the case of $YbCl_2$, for example, normals to the two planes inclose an angle of 3.6°. The Cl(1) ion is coordinated by three M ions in such a way that it forms the peak of a flat, three-sided pyramid, whose M↔Cl(1) distances range from 2.768 to 2.863 Å in $TmCl_2$ [4] and from 2.757 to

Fig. 6

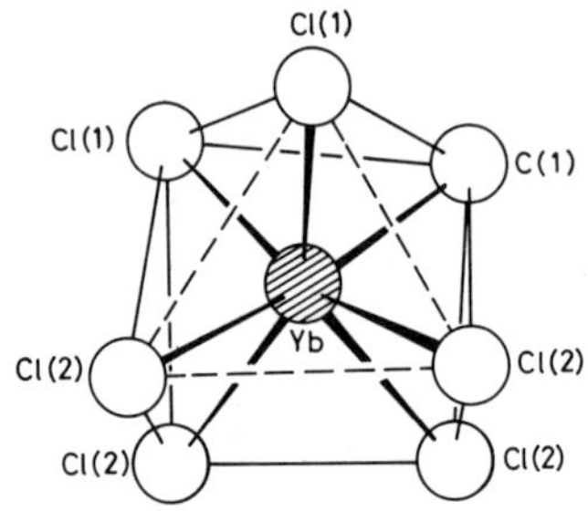


Crystal structure of $YbCl_2$: coordination polyhedron formed by $Cl^-$ ions around the $Yb^{2+}$ ion.

2.863 Å in $YbCl_2$ [5]. The Cl(2) is located within a three-sided pyramid formed by four M ions; M↔Cl(2) ranges from 2.857 to 2.917 Å in $TmCl_2$ [4] and from 2.846 to 2.911 Å in $YbCl_2$ [5]. Further internuclear distances are given in [4, 5]. Bond angles for $YbCl_2$ are given in [5], which also discusses the analogies of the $YbCl_2$ structure to that of $SrI_2$ with reference to thermal vibration anisotropies.

Lattice Constants

In 1970 and 1971, Fishel and Eick [6, 7] showed that earlier determinations of the $YbCl_2$ lattice constants gave only half the elementary cell. The need to double the a-dimension (axis assignment corresponds to the space group symbol Pbca) had been overlooked in earlier work [3, 8] because reflections with an odd number h index, almost without exception, are either very weak or are entirely missing because of special extinction of the hk0 reflection group [2]. An examination by Beck, Bärnighausen [2] of the Debye-Scherrer patterns of the $TmCl_2$ [3] and $YbCl_2$ [8] showed that these patterns can be completely indexed when a-doubling is assumed. Accordingly, the lattice dimensions in Å of powder samples (P) and of single crystals (C) are as follows (axis assignment corresponds to Pbca):

| $MCl_2$ | | a | b | c | Ref. |
|---|---|---|---|---|---|
| $TmCl_2$ | (P) | 13.10 | 6.93 | 6.68 | [3] a) |
| | | 13.176±0.002 | 6.975±0.002 | 6.710±0.001 | [10] c) |
| | (C) | 13.161±0.017 | 6.959±0.004 | 6.702±0.005 | [4] b) |
| $YbCl_2$ | (P) | 13.15±0.03 | 6.942±0.015 | 6.693±0.015 | [6, 7] c) |
| | (P) | 13.08 | 6.92 | 6.69 | [8] a) |
| | (C) | 13.14±0.02 | 6.948±0.009 | 6.698±0.009 | [9] c) |
| | (C) | 13.18±0.02 | 6.96±0.01 | 6.70±0.01 | [2] |

a) The a-value of [3, 8] has been doubled, and b and c have been interchanged. – b) The a and b of [4] have been interchanged. – c) The a, b, and c of [6, 7, 9, 10] have been cyclically interchanged.

References:

[1] H. Bärnighausen, H.P. Beck, H.-W. Grueninger (in: P.E. Field, Proc. 9th Rare Earth Res. Conf., Blacksburg, Va., 1971, Vol. 1, pp. 74/83, 76, 78). – [2] H.P. Beck, H. Bärnighausen (Z. Anorg. Allgem. Chem. **386** [1971] 221/8). – [3] P.E. Caro, J.D.

Corbett (J. Less-Common Metals **18** [1969] 1/10, 3). – [4] T. Naterstad, J.D. Corbett (J. Less-Common Metals **46** [1976] 291/5). – [5] H. Bärnighausen, H. Pätow, H.P. Beck (Z. Anorg. Allgem. Chem. **403** [1973] 45/55).

[6] N.A. Fishel, H.A. Eick (Abstr. Papers 160th Natl. Meeting Am. Chem. Soc., Chicago, Ill., 1970, INOR No. 170). – [7] N.A. Fishel, H.A. Eick (J. Inorg. Nucl. Chem. **33** [1971] 1198/201). – [8] W. Döll, W. Klemm (Z. Anorg. Allgem. Chem. **241** [1939] 239/58, 246/50). – [9] A.V. Hariharan, N.A. Fishel, H.A. Eick (High Temp. Sci. **4** [1972] 405/10). – [10] A.V. Hariharan (private communication from H. Bärnighausen to the Gmelin Institute 1977).

### 11.2.4.3.3 Lattice Energy

The lattice energies of the dichlorides of Nd, Sm, Eu, and Yb have been calculated using Born-Haber cycle and corrected to 0 K. These values and estimated $M^{2+}$ ionic radii ($r(Cl^-)=1.82$ Å) were used in an equation of Kapustinskii [2] to obtain $U_0$ values for the other lanthanide dihalides, Johnson [1]:

| M . . | (La) | (Ce) | (Pr) | Nd | Pm | Sm | Eu | (Gd) | (Tb) | Dy | (Ho) | (Er) | Tm | Yb |
|---|---|---|---|---|---|---|---|---|---|---|---|---|---|---|
| $U_0$ . . | 495 | 500 | 504 | 509 | 511 | 519 | 516 | 520 | 524 | 527 | 530 | 534 | 536 | 543 |

The parentheses in the above table indicate those dichlorides which are thermodynamically unstable with respect to disproportionation ($3\,MCl_2(solid) \rightarrow M\,(solid) + 2\,MCl_3(solid)$) at 25 °C [1].

Kapustinskii [2] reviews the calculation of U in ionic crystals.

References:

[1] D.A. Johnson (J. Chem. Soc. A **1969** 2578/80). – [2] A.F. Kapustinskii (Quart. Rev. [London] **10** [1956] 283/94).

### 11.2.4.4 Density and Molar Volume

X-ray densities $D_x$ at room temperature for $NdCl_2$, $SmCl_2$, and $EuCl_2$ (orthorhombic, $PbCl_2$-type structure), for $DyCl_2$ (tetragonal, $SrBr_2$-type structure), and for $TmCl_2$ and $YbCl_2$ (orthorhombic, $SrI_2$-type structure):

| $MCl_2$ . . . . | $NdCl_2$ | | $SmCl_2$ | | | $EuCl_2$ | | |
|---|---|---|---|---|---|---|---|---|
| $D_x$ in g/cm³ . | 4.542 | 4.62 a) | 4.788 | 4.81 | 4.93 | 4.856 | 4.871 | 4.90 |
| Ref. . . . . | [2] | [3] | [4] | [5] | [6] | [4] | [7] | [5] |

| $MCl_2$ . . . . | $DyCl_2$ | $TmCl_2$ | | $YbCl_2$ | | |
|---|---|---|---|---|---|---|
| $D_x$ in g/cm³ . | 5.025 | 5.166 | 5.190 a) | 5.272 | 5.299 | 5.303 a) |
| Ref. . . . . | [8] | [9] | [10] | [11] | [7] | [12] |

Pycnometric densities $D_p$ at room temperature:

| $MCl_2$ . . . . | $SmCl_2$ | | $EuCl_2$ | $YbCl_2$ |
|---|---|---|---|---|
| $D_p$ in g/cm³ . | 4.56±0.07 | 4.76±0.16 | 4.87 | 5.08 |
| Ref. . . . . | [13] | [6] | [14] | [15] |

a) Calculated by [1] from lattice constants of the reference given in the table.

The molar volumes calculated from $D_p$ and $D_x$ are 45.8 and 45.5 $cm^3$/mol for $SmCl_2$ and 48.0 and 46.0 $cm^3$/mol for $EuCl_2$, Döll, Klemm [5].

References:

[1] H. Bärnighausen (private communication to the Gmelin Institute 1977). – [2] H. Bärnighausen, W. Kuhn (from [1]). – [3] L.F. Druding, J.D. Corbett (J. Am. Chem. Soc. **83** [1961] 2462/7). – [4] H. Bärnighausen (Rev. Chim. Minerale **10** [1973] 77/92, 80). – [5] W. Döll, K. Klemm (Z. Anorg. Allgem. Chem. **241** [1939] 239/58, 248).

[6] I.S. Astakhova, D.M. Laptev, V.F. Goryushkin (Zh. Neorgan. Khim. **22** [1977] 2590/1; Russ. J. Inorg. Chem. **22** [1977] 1405/6). – [7] H.A. Eick (from [1]). – [8] H. Bärnighausen (Proc. 12th Rare Earth Res. Conf., Vail, Colo., 1976, Vol. 1, pp. 404/13, 405). – [9] A.V. Hariharan (from [1]). – [10] T. Naterstad, J.D. Corbett (J. Less-Common Metals **46** [1976] 291/5).

[11] H.P. Beck (Diss. Karlsruhe 1972 from [1]). – [12] N.A. Fishel, H.A. Eick (J. Inorg. Nucl. Chem. **33** [1971] 1198/201). – [13] W. Klemm, J. Rockstroh (Z. Anorg. Allgem. Chem. **176** [1928] 181/99, 195). – [14] G.P. Baxter, F.D. Tuemmler (J. Am. Chem. Soc. **60** [1938] 602/4). – [15] W. Klemm, W. Schüth (Z. Anorg. Allgem. Chem. **184** [1929] 352/8).

## 11.2.4.5 Thermal Properties

Data on gaseous $MCl_2$ molecules are given in 11.1.2, p. 6.

### 11.2.4.5.1 Melting Points

For judging the validity of the following values, compare them with the appropriate phase diagrams in "Rare Earth Elements" C4b.

Melting point $t_m$:

| $MCl_2$ . . | $ScCl_2$ | $NdCl_2$ | | $SmCl_2$ | | | |
|---|---|---|---|---|---|---|---|
| $t_m$ in °C . | 805 [a)] | 835 [a)] | 841 [a)] | 740 [b)] | 848 | 858±2 | 859 |
| Ref. . . . | [1] | [2] | [3] | [4] | [5] | [6] | [7, 8] |

| $MCl_2$ . . | $EuCl_2$ | | | | | $DyCl_2$ | $TmCl_2$ | | $YbCl_2$ | |
|---|---|---|---|---|---|---|---|---|---|---|
| $t_m$ in °C . | 731 | 738 | 757 | ≈850 | 854±2 [a)] | 721 [a)] | 697 | 718±3 | 702 | 708 |
| Ref. . . . | [5] | [7] | [9] | [12] | [10, 11] | [13] | [5] | [14] | [5] | [7] |

a) Melt with decomposition. – b) The low value of [4] is attributed by [8] to impurities in the sample.

References:

[1] O.G. Polyachenok, G.I. Novikov (Zh. Neorgan. Khim. **8** [1963] 2819/21; Russ. J. Inorg. Chem. **8** [1963] 1479/80; Zh. Obshch. Khim. **33** [1963] 2797; J. Gen. Chem. [USSR] **33** [1963] 2725). – [2] O.G. Polyachenok, G.I. Novikov (Zh. Neorgan. Khim. **8** [1963] 1567/73; Russ. J. Inorg. Chem. **8** [1963] 816/20, 817). – [3] L.F. Druding, J.D. Corbett (J. Am. Chem. Soc. **83** [1961] 2462/7). – [4] G. Jantsch, H. Rüping, W. Kunze (Z. Anorg. Allgem. Chem. **161** [1927] 210/6, 213). – [5] F.H. Spedding, A.H. Daane (Met. Rev. **5** [1960] 297/348, 340).

[6] D.M. Laptev, V.F. Goryushkin, I.S. Astakhova, G.G. Polyakova (Zh. Neorgan. Khim. **24** [1979] 1311/6; Russ. J. Inorg. Chem. **24** [1979] 730/3), D.M. Laptev, V.F. Goryushkin, N.M. Kulagin, E.S. Vorontsov (Zh. Neorgan. Khim. **21** [1976] 2616/20; Russ. J. Inorg. Chem. **21** [1976] 1438/41). – [7] O.G. Polyachenok, G.I. Novikov (Zh. Neorgan. Khim. **8** [1963] 2631/4; Russ. J. Inorg. Chem. **8** [1963] 1378/81). – [8] O.G. Polyachenok, G.I. Novikov (Zh. Neorgan. Khim. **8** [1963] 2818/9; Russ. J. Inorg. Chem. **8** [1963] 1478/9). – [9] L.H. Brixner, J.D. Bierlein (Mater. Res. Bull. **9** [1974] 99/104). – [10] D.M. Laptev, N.M. Kulagin, I.S. Astakhova (Zh. Neorgan. Khim. **20** [1975] 1987/9; Russ. J. Inorg. Chem. **20** [1975] 1108/9).

[11] N.M. Kulagin, D.M. Laptev (Zh. Fiz. Khim. **50** [1976] 810; Russ. J. Phys. Chem. **50** [1976] 483). – [12] W. Kapfenberger (Z. Anorg. Allgem. Chem. **238** [1938] 273/86, 282). – [13] J.D. Corbett, B.C. McCollum (Inorg. Chem. **5** [1966] 938/40). – [14] P.E. Caro, J.D. Corbett (J. Less-Common Metals **18** [1969] 1/10, 4).

### 11.2.4.5.2 Enthalpies and Entropies of Melting

For $SmCl_2$, $EuCl_2$, and $YbCl_2$, a critical evaluation of metal halide literature data and of theoretical relationships gives $6\pm1$ kcal/mol as the enthalpy of melting at 1000 K and $6\pm1$ $cal \cdot mol^{-1} \cdot K^{-1}$ as the entropy of melting, Brewer [1], cf. Gibson et al. [2], Hariharan et al. [3, 4]. Experimental enthalpies of melting (at the melting point): $3.4\pm0.4$ kcal/mol for $SmCl_2$, Laptev et al. [5]; $5.5\pm0.3$ kcal/mol for $EuCl_2$, Kulagin, Laptev [6].

References:

[1] L. Brewer (Natl. Nucl. Energy Ser. Div. IV **19** B [1950] 193/275, 196/9). – [2] J.A. Gibson, J.F. Miller, P.S. Kennedy, G.W.P. Rengstorff (The Properties of the Rare Earth Metals and Compounds, Battelle Memorial Institute, Columbus, Ohio, 1959, p. 10, Section 1). – [3] A.V. Hariharan, H.A. Eick (High Temp. Sci. **4** [1972] 91/8; Proc. 9th Rare Earth Res. Conf., Blacksburg, Va., 1971, Vol. 2, pp. 584/96). – [4] A.V. Hariharan, N.A. Fishel, H.A. Eick (High Temp. Sci. **4** [1972] 405/10). – [5] D.M. Laptev, V.F. Goryushkin, N.M. Kulagin, E.S. Vorontsov (Zh. Neorgan. Khim. **21** [1976] 2616/20; Russ. J. Inorg. Chem. **21** [1976] 1438/41).

[6] N.M. Kulagin, D.M. Laptev (Zh. Fiz. Khim. **50** [1976] 810; Russ. J. Phys. Chem. **50** [1976] 483).

### 11.2.4.5.3 Vaporization Data

Composition of the Vapor

The occurrence of the gaseous chlorides is also discussed in Section 11.1, p. 1.

Almost all published works report that $SmCl_2$, $EuCl_2$, and $YbCl_2$ vaporize congruently with the formation of monomeric molecules in gas phase. Only Hastie et al. [1], on the basis of mass spectrometric analyses of the vapor over molten $EuCl_2$ at pressures less than 0.1 Torr, report a small percentage (1%) of dimer molecules which is said to increase in the vicinity of the boiling point. But similar investigations done more recently on $SmCl_2$, $EuCl_2$, and $YbCl_2$ do not support these findings; see Il'in, Chervonnyi et al. [2 to 4], Hariharan et al. [5, 6].

Vapor Pressure. Boiling Point

Published vapor pressure values were used to derive the following constants for the general equation log p (p in Torr) = (A/T) + B − C log T:

| $MCl_2$ | temperature range in K | phase transition | −A | B | C | Ref. |
|---|---|---|---|---|---|---|
| $SmCl_2$ | 1473 to 1673 | liquid→vapor | 14770 | 17.89 | 2.5 | [7, 8] |
| | 923 to 1121 | solid →vapor | 13640±920 | 9.31±0.90 | – | [2 to 4] |
| $EuCl_2$ | 1473 to 1673 | liquid→vapor | 14770 | 17.36 | 2.5 | [7, 8] |
| | 998 to 1128 | liquid→vapor | 13040±280 | 8.30±0.26 | – | [2 to 4] |
| | 1135 to 1569 | liquid→vapor | 15400±40 | 9.79±0.11 | – | [5] |
| $YbCl_2$ | 1473 to 1673 | liquid→vapor | 14770 | 17.53 | 2.5 | [7, 8] |
| | 998 to 1278 | liquid→vapor | 13040±200 | 8.47±0.20 | – | [2 to 4] |
| | 1178 to 1506 | liquid→vapor | 15530±90 | 9.88±0.15 | – | [6] |

According to older estimates of Brewer [9], vapor pressures of $10^{-4}$ to $10^{-2}$ atm are reached at the following temperatures: approximately 1350±200 K for both $SmCl_2$ and $EuCl_2$; approximately 1300±200 K for $YbCl_2$. For the radiometric determination of the saturated vapor pressures (<$10^{-4}$ Torr) over the dichlorides by the use of the $^{153}Sm$, $^{154}Eu$, and $^{175}Yb$ isotopes see Il'in et al. [3].

Extrapolation of the above vapor pressures gives the following boiling points in K:

| | | | |
|---|---|---|---|
| $SmCl_2$ . . . . . . . . | – | 2223 [7, 8] | 2300 [9] |
| $EuCl_2$ . . . . . . . . | 2335±35 [5] | 2463 [7, 8] | 2300 [9] |
| $YbCl_2$ . . . . . . . . | 2306±25 [6] | 2378 [7, 8] | 2200 [9] |

References:

[1] J.W. Hastie, P. Ficalora, J.L. Margrave (J. Less-Common Metals **14** [1968] 83/91, 83, 91). – [2] V.K. Il'in, A.D. Chervonnyi, V.A. Krenev (Zh. Neorgan. Khim. **21** [1976] 614/8; Russ. J. Inorg. Chem. **21** [1976] 332/4). – [3] V.K. Il'in, A.D. Chervonnyi, A.V. Baluev, V.A. Krenev, V.I. Evdokimov (Deposited Publ. VINITI-5688-73 [1973] 1/25 from C.A. **85** [1976] No. 68410). – [4] A.D. Chervonnyi, V.K. Il'in, V.A. Krenev (V Splavy Redk. Met. Osobymi Fiz. Khim. Svoistv. **1975** 133/6; C.A. **85** [1976] No. 10960). – [5] A.V. Hariharan, H.A. Eick (High Temp. Sci. **4** [1972] 91/8, 93; Proc. 9th Rare Earth Res. Conf., Blacksburg, Va., 1971, Vol. 2, pp. 584/96, 588/9; C.A. **76** [1972] No. 145680).

[6] A.V. Hariharan, N.A. Fishel, H.A. Eick (High Temp. Sci. **4** [1972] 405/10, 407). – [7] O.G. Polyachenok, G.I. Novikov (Zh. Neorgan. Khim. **8** [1963] 2631/4; Russ. J. Inorg. Chem. **8** [1963] 1378/81). – [8] O.G. Polyachenok, G.I. Novikov (Vestn. Leningr. Univ. Fiz. Khim. **1963** No. 16, pp. 133/4; C.A. **60** [1964] 4875). – [9] L. Brewer (Natl. Nucl. Energy Ser. Div. IV B**19** [1950] 193/275, 198/9).

### 11.2.4.5.4 Enthalpies of Sublimation $\Delta H_s$

Second and third law treatments of the vapor pressures of [1 to 5] give the following $\Delta H_s$ values in kcal/mol at 298 K, Chervonnyi et al. [1]:

| $SmCl_2$ | $EuCl_2$ | $YbCl_2$ | method | Ref. |
|---|---|---|---|---|
| 79.1±2.7 | 81.4±2.7 | 78.5±2.7 | second law | [1] |
| | 84.4±2.5 | 84.8±2.5 | | [2, 3] |
| | 73.8±2.7 | – | | [4] |

| $SmCl_2$ | $EuCl_2$ | $YbCl_2$ | method | Ref. |
|---|---|---|---|---|
| 79.3±1.6 | 79.2±1.6 | 79.6±1.6 | third law | [1] |
| 79.5±1.6 | 80.6±1.6 | 79.6±1.6 | | [5] |
| | 83.3±2.5 | 83.7±2.3 | | [2, 3] |

The recommended $\Delta H_s$ values are 79.4±1.6 ($SmCl_2$), 79.9±1.6 ($EuCl_2$), and 79.6±1.6 kcal/mol ($YbCl_2$) [1].

Original values and additional sublimation and vaporization data are given under the individual dichlorides in "Rare Earth Elements" C4b.

References:

[1] A.D. Chervonnyi, V.K. Il'in, V.A. Krenev (V Splavy Redk. Met. Osobymi Fiz. Khim. Svoistv. **1975** 133/6; C.A. **85** [1976] No. 10960). – [2] A.V. Hariharan, H.A. Eick (High Temp. Sci. **4** [1972] 91/8; Proc. 9th Rare Earth Res. Conf., Blacksburg, Va., 1971, Vol. 2, pp. 584/96). – [3] A.V. Hariharan, N.A. Fishel, H.A. Eick (High Temp. Sci. **4** [1972] 405/10). – [4] J.W. Hastie, P. Ficarola, J.L. Margrave (J. Less-Common Metals **14** [1968] 83/91). – [5] O.G. Polyachenok, G.I. Novikov (Zh. Neorgan. Khim. **8** [1963] 2631/4; Russ. J. Inorg. Chem. **8** [1963] 1378/81).

### 11.2.4.5.5 Thermodynamic Data of Formation

#### 11.2.4.5.5.1 Standard Enthalpy of Formation $\Delta H^\circ_{298}$ in kcal/mol

| $MCl_2$ | $-\Delta H^\circ_{298}$ | method | Ref. | $MCl_2$ | $-\Delta H^\circ_{298}$ | method | Ref. |
|---|---|---|---|---|---|---|---|
| $ScCl_2$ | 145 | (a) | [1] | $SmCl_2$ | 191.8 | (e) | [16] |
| | 145 | (b) | [2, 15] | | 194.9 | (f) | [8] |
| | 149 | (c) | [6, p. 49] | | 195.6±1.0 | (e) | [5] |
| | 180 | (d) | [19] | | 203 | (g) | [1, 7] |
| $YCl_2$ | 160 | (c) | [6, p. 89] | $EuCl_2$ | 194 | (f) | [8] |
| | | | | | 195.8 | (e) | [21] |
| $PrCl_2$ | 163 | (b) | [1, 2] | | 197.0±1.3 | (e) | [22] |
| | 163±3 | (b) | [3] corrected by [20] | | 217 | (g) | [1, 7] |
| | | | | $DyCl_2$ | 236.4 | (e) | [16] |
| $NdCl_2$ | 163.2±1.9 | (e) | [20] | $ErCl_2$ | 150 | (a) | [1] |
| | 164±3 | (b) | [3] corrected by [20] | $TmCl_2$ | 169.5±1 | (e) | [4] |
| | | | | $YbCl_2$ | 181.8±3.6 | (h) | [23, 24] |
| | 169±1 | (e) | [4] | | 184.5±1.7 | (e) | [5] |
| | | | | | 186 | (g) | [1, 7] |
| | | | | | 191.1 | (f) | [8] |
| $PmCl_2$ | 195±10 | (c) | [9, p. 124] | | | | |

Methods: (a) From the solubility of the metal in the molten trichloride. (b) Thermoanalytical and tensimetric measurements of $M-MCl_3$ systems. (c) Estimated values. (d) Graphical application of an empirical relationship between the enthalpies of formation of compounds and ionization potentials of the constituent elements. (e) Solution calorimetry. (f) Critical evaluation of literature data. (g) Measurement of the decomposition pressure of the trichloride. (h) Second law calculations with the authors' own vapor pressure measurements.

Pre-1950 data for a few of the dichlorides are given by Brewer et al. [9].

Johnson [10] used a simple ionic-crystal model and a Born-Haber cycle to estimate $\Delta H^\circ_{298}$ for the dichlorides of La, Ce, Pr, Pm, Gd, Tb, Dy, Ho, Er, and Tm from experimental $\Delta H^\circ_{298}$ literature values (partly corrected) for the dichlorides of Nd, Sm, Eu, and Yb and from other thermodynamic data. These $\Delta H^\circ_{298}$ values, together with the values of Novikov, Polyachenok [11] (which were also calculated using the Born-Haber cycle) are shown in **Fig. 7**, Bayanov [12]. Not shown in the figure is the analogously calculated value of [11] for $\Delta H^\circ_{298}$ of $YCl_2$ which is $-119$ kcal/mol.

Fig. 7

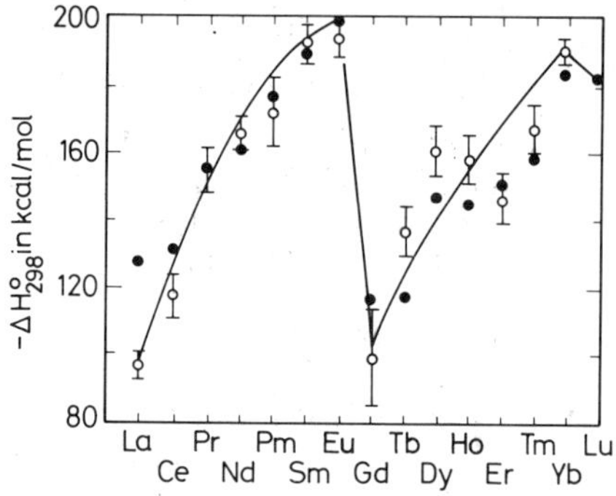


Standard enthalpy of formation $\Delta H^\circ_{298}$ of solid rare earth dichlorides. The empty circles represent data from [10], the full circles data from [11].

A compilation of estimated $\Delta H^\circ_T$ values (T = 298, 500, 1000, 1500 K) for the formation of $MCl_2$ (M = Sm, Eu, Yb) is given by Wicks, Block [14]. Glassner [25] (according to Gibson et al. [26]) gives an equation, valid from 298 to 1013 K, for the enthalpy of formation of solid $SmCl_2$ as function of temperature; the equation is based on a $\Delta H^\circ_{298}$ value of $-195.6$ kcal/mol.

Smirnov et al. [17, 18] derived the enthalpy of formation of liquid $LaCl_2$ ($-160.0$ kcal/mol [17]) and liquid $CeCl_2$ ($-163.7$ kcal/mol [18]) according to the reaction M(liquid) + $Cl_2$(gas) $\rightarrow$ $MCl_2$(liquid) at constant pressure from emf measurements over the temperature range 1150 to 1250 K. The following enthalpy of formation of liquid $CeCl_2$ was calculated from thermodynamic data in the literature: $\Delta H^\circ$ ($CeCl_2$, liquid) = $-161.6$ kcal/mol [18]. Glassner [25] (according to Gibson et al. [26]) gives equations for $\Delta H^\circ_T$ of formation of liquid $SmCl_2$ from solid Sm (1013 to 1623 K) and from liquid Sm (1623 to 1700 K).

References:

[1] O.G. Polyachenok, G.I. Novikov (Vestn. Leningr. Univ. Fiz. Khim. **1963** No. 16, pp. 133/4; C.A. **60** [1964] 4875). – [2] O.G. Polyachenok, G.I. Novikov (Zh. Obshch. Khim. **33** [1963] 2797; J. Gen. Chem. [USSR] **33** [1963] 2725). – [3] G.I. Novikov, O.G. Polyachenok (Zh. Neorgan. Khim. **8** [1963] 1053/9; Russ. J. Inorg. Chem. **8** [1963] 545/50, 548). – [4] L.R. Morss, M.C. McCue (Inorg. Chem. **14** [1975] 1624/7). – [5] G.R. Machlan, C.T. Stubblefield, L. Eyring (J. Am. Chem. Soc. **77** [1955] 2975/8).

[6] R.C. Feber (LA-3164 [1965] 1/185; N.S.A. **19** [1965] No. 30148). – [7] O.G. Polyachenok, G.I. Novikov (Zh. Neorgan. Khim. **9** [1964] 773/7; Russ. J. Inorg. Chem. **9** [1964] 429/31). – [8] R.H. Schumm, D.D. Wagman, S. Bailey, W.H. Evans, V.B. Parker (NBS-TN-270-7 [1973] 4, 34, 38, 43). – [9] L. Brewer, L.A. Bromley, P.W. Gilles, N.L. Lofgren (Natl. Nucl. Energy Serv. Div. IV B **19** [1950] 76/192, 105/6). – [10] D.A. Johnson (J. Chem. Soc. A **1969** 2578/80).

[11] G.I. Novikov, O.G. Polyachenok (Usp. Khim. **33** [1964] 732/6; Russ. Chem. Rev. **33** [1964] 342/50, 349). – [12] A.P. Bayanov (Usp. Khim. **44** [1975] 236/59; Russ. Chem. Rev. **44** [1975] 122/37, 123). – [13] M. Barber, J.W. Linnett, N.H. Taylor (J. Chem. Soc. **1961** 3323/32, 3325). – [14] C.E. Wicks, F.E. Block (U.S. Bur. Mines Bull. No. 605 [1963] 1/146, 46, 100/1, 136; C.A. **59** [1963] 14656). – [15] O.G. Polyachenok, G.I. Novikov (Zh. Neorgan. Khim. **8** [1963] 2819/21; Russ. J. Inorg. Chem. **8** [1963] 1479/80).

[16] L.R. Morss, J.A. Fahey (Proc. 12th Rare Earth Res. Conf., Vail, Colo., 1976, Vol. 1, pp. 443/50 from C.A. **85** [1976] No. 131498). – [17] M.V. Smirnov, P.M. Usov (Dokl. Akad. Nauk SSSR **151** [1963] 862/5; Dokl. Chem. Proc. Acad. Sci. USSR **148/153** [1963] 606/9). – [18] M.V. Smirnov, V.S. Lbov (Tr. Inst. Elektrokhim. Akad. Nauk SSSR Ural'sk. Filial No. 8 [1966] 3/6; Electrochemistry of Molten and Solid Electrolytes, Vol. 5, New York 1967, pp. 1/4, 3). – [19] L.N. Komissarova, L.K. Tarasov (Zh. Neorgan. Khim. **14** [1969] 621/3; Russ. J. Inorg. Chem. **14** [1969] 324/6). – [20] O.G. Polyachenok, G.I. Novikov (Zh. Neorgan. Khim. **8** [1963] 1567/73; Russ. J. Inorg. Chem. **8** [1963] 816/20).

[21] C.T. Stubblefield, J.L. Rutledge, R. Phillips (J. Phys. Chem. **69** [1965] 991/5). – [22] L.R. Morss, H.O. Haug (J. Chem. Thermodyn. **5** [1973] 513/24, 519, 521). – [23] N.A. Fishel (Diss. Michigan State Univ. 1970, pp. 1/118 from Diss. Abstr. Intern. B **31** [1970/71] 6477). – [24] N.A. Fishel, H.A. Eick (Abstr. Papers 160th Natl. Meeting Am. Chem. Soc., Chicago 1970, p. INOR 170). – [25] A. Glassner (ANL-5750 [1957] 1/70; N.S.A. **12** [1958] No. 2733).

[26] J.A. Gibson, J.F. Miller, P.S. Kennedy, G.W.P. Rengstorff (The Properties of the Rare Earth Metals and Compounds, Battelle Memorial Institute, Columbus, Ohio, 1959, Section 3, p. Sm-7).

### 11.2.4.5.5.2 Gibbs Free Energy and Entropy of Formation

$\Delta G$ and $\Delta S$ are in the units kcal/mol and $cal \cdot mol^{-1} \cdot K^{-1}$, respectively.

Values for the formation of solid $MCl_2$ from the elements under standard conditions at 298 K:

| $MCl_2$ . . | $ScCl_2$ | $PrCl_2$ | $NdCl_2$ | $SmCl_2$ | | | | $EuCl_2$ | | $YbCl_2$ | |
|---|---|---|---|---|---|---|---|---|---|---|---|
| $-\Delta G^\circ_{298}$ . | 170 [g)] | 156 ±4 [a)] | 158 ±4 [a)] | 183 [b)] | 184 [c)] | 187.1 | – | – | 174 [c)] | – | 175.3 ±3.6 [f)] |
| $-\Delta S^\circ_{298}$ . | – | 36 | 36 | 39 [b)] | 39 [d)] | – | 35 [e)] | 35 [e)] | 39 [d)] | 35 [e)] | 21.8 [f)] |
| Ref. . . | [12] | [1] | [1] | [2] | [3] | [4] | [6] | [6] | [3] | [6] | [13] |

a) Calculated from the authors' own thermographic and tensimetric measurements of the $M-MCl_3$ system. – b) Estimated value based on the $\Delta H^\circ_{298}$ value of [7]. – c) Recalculated from literature data. – d) Estimated value. – e) Calculated using $S^\circ_{298}$ = 14 $cal \cdot mol^{-1} \cdot K^1$ for the rare earth metal. – f) Second law calculations with the authors' own vapor pressure measurements. – g) Calculated according to various models.

A compilation of estimated $\Delta G^\circ_T$ values (T=298, 500, 1000, 1500 K) is given for M=Sm, Eu, and Yb in Wicks, Block [8] and for M=Sc in [10]. Pre-1950 values for $(\Delta G^\circ - \Delta H^\circ_{298})/T$ at 298, 500, 1000, and 1500 K for $SmCl_2$, $EuCl_2$, and $YbCl_2$ are compiled by Brewer et al. [9]. For $SmCl_2$, expressions valid over the range 298 to 1013 K are given in [4] for $\Delta G^\circ_T$ and $\Delta S^\circ_T$ as a function of temperature.

The emf measurements of Smirnov et al. [10, 11] in the temperature range 1150 to 1250 K were used to obtain the free energy of formation ΔG (in cal/mol) of liquid $MCl_2$ (M=La, Ce) according to the reaction M (liquid) + $Cl_2$(gas) → $MCl_2$(liquid) under constant pressure: $\Delta G(LaCl_2) = -160000 + 38.3\,T$ [10], and $\Delta G(CeCl_2) = -163700 + 35.5\,T$ [11]. Equations for calculating $\Delta G^\circ_T$ and $\Delta S^\circ_T$ for the formation of liquid $SmCl_2$ from solid Sm in the range 1013 to 1623 K and from liquid Sm in the range 1623 to 1700 K are given in Glassner [4].

References:

[1] G.I. Novikov, O.G. Polyachenok (Zh. Neorgan. Khim. **8** [1963] 1053/9; Russ. J. Inorg. Chem. **8** [1963] 545/50, 548). – [2] R.L. Montgomery, T.D. Hubert (U.S. Bur. Mines Rept. Invest. No. 5525 [1959] 1/8, 6; C.A. **1960** 4134). – [3] R.L. Montgomery (U.S. Bur. Mines Rept. Invest. No. 5468 [1959] 1/23, 11; C.A. **1959** 12815). – [4] A. Glassner (ANL-5750 [1957] pp. 1/70 from [5]). – [5] J.A. Gibson, J.F. Miller, P.S. Kennedy, G.W.P. Rengstorff (The Properties of the Rare Earth Metals and Compounds, Battelle Memorial Institute, Columbus, Ohio, 1959, Section 3, p. Sm-7).

[6] O.G. Polyachenok, G.I. Novikov (Zh. Neorgan. Khim. **9** [1964] 773/7; Russ. J. Inorg. Chem. **9** [1964] 429/31). – [7] G.R. Machlan, C.T. Stubblefield, L. Eyring (J. Am. Chem. Soc. **77** [1955] 2975/78). – [8] C.E. Wicks, F.E. Block (U.S. Bur. Mines Bull. No. 605 [1963] 1/146, 46, 100/1, 136; C.A. **59** [1963] 14656). – [9] L. Brewer, L.A. Bromley, P.W. Gilles, N.L. Lofgren (Natl. Nucl. Energy Ser. Div. IV B**19** [1950] 76/192, 105/6). – [10] M.V. Smirnov, P.M. Usov (Dokl. Akad. Nauk SSSR **151** [1963] 862/5; Dokl. Chem. Proc. Acad. Sci. USSR **148/153** [1963] 606/9).

[11] M.V. Smirnov, V.S. Lbov (Tr. Inst. Elektrokhim. Akad. Nauk SSSR Ural'sk. Filial No. 8 [1966] 3/6; Electrochemistry of Molten and Solid Electrolytes, Vol. 5, New York 1967, pp. 1/4). – [12] L.N. Komissarova, L.K. Tarasov (Zh. Neorgan. Khim. **14** [1969] 621/3; Russ. J. Inorg. Chem. **14** [1969] 324/6). – [13] N.A. Fishel (Diss. Univ. of Michigan 1970, pp. 1/118 from Diss. Abstr. Intern. B **31** [1970/71] 6477), see also N.A. Fishel, H.A. Eick (Abstr. Papers 160th Natl. Meeting Am. Chem. Soc., Chicago 1970, p. INOR 170).

### 11.2.4.5.6 Other Thermodynamic Functions

Solid Compounds

A compilation of values estimated from literature data is given in Wicks, Block [3] for the enthalpy increments $H^\circ_T - H^\circ_{298}$ and free energy functions $(G^\circ_T - H^\circ_{298})/T$ of $SmCl_2$, $EuCl_2$, and $YbCl_2$ at T=500, 1000, and 1500 K.

Standard entropy S° at 298 K in $cal \cdot mol^{-1} \cdot K^{-1}$:

| $MCl_2$ | S° | Ref. | $MCl_2$ | S° | Ref. |
|---|---|---|---|---|---|
| $ScCl_2$ | 25.9 [a)] | [4] | $EuCl_2$ | $32.9\pm2.4$ [e)] | [6] |
| | 29.5 [b)] | [4] | | 34.5 [f)] | [6] |
| $NdCl_2$ | 31 [c)] | [5] | $YbCl_2$ | $28.9\pm1.9$ [e)] | [7] |
| $SmCl_2$ | 30 [d)] | [2, 3] | | $29.3\pm0.2$ [f)] | [7] |
| | 32 [c)] | [1] | | 30 [d)] | [2, 3] |
| $EuCl_2$ | 30 [d)] | [2, 3] | | $30.7\pm2.5$ [e)] | [8] |
| | 32 [c)] | [1] | | 32 [c)] | [1] |

Methods: a) Latimer calculation. — b) Kireev calculation. — c) Calculated by additivity method. — d) Estimated values. — e) Calculated from the entropy of the gaseous dichloride and from the (second-law) sublimation entropy. — f) Latimer calculation which takes into account the magnetic contribution of the divalent metal ion.

For thermodynamic calculations of the Nd-$NdCl_3$ system, Novikov, Polyachenok [5] assumed a heat capacity of 19 $cal\cdot mol^{-1}\cdot K^{-1}$ for $NdCl_2$.

$EuCl_2$ is isostructural with $BaCl_2$, and $YbCl_2$ is isostructural with $CaCl_2$. On this basis, literature data for the molar heat capacities of $BaCl_2$ and $CaCl_2$ have been used to estimate the molar heat capacities of $EuCl_2$ [6] and $YbCl_2$ [7], respectively. At 298 K, $C_p^\circ$ $(EuCl_2)=18.13\pm0.15$ $cal\cdot mol^{-1}\cdot K^{-1}$, Hariharan, Eick [6], and $C_p^\circ$ $(YbCl_2)=17.75\pm0.15$ $cal\cdot mol^{-1}\cdot K^{-1}$, Hariharan et al. [7]. At the melting point (1000 and 975 K, respectively) $C_p^\circ$ $(EuCl_2)=C_p^\circ$ $(YbCl_2)=21.75$ $cal\cdot mol^{-1}\cdot K^{-1}$. Assuming a linear change of $C_p^\circ$ with temperature gives the following functions between 298 K and the melting point: $C_p^\circ$ $(EuCl_2)=(16.59\pm0.15)+5.16\times10^{-3}T$ [6] and $C_p^\circ$ $(YbCl_2)=(15.99\pm0.15)+5.91\times10^{-3}T$ [7].

### Molten Compounds

For liquid $CeCl_2$, $S_{298}^\circ=33.6$ $cal\cdot mol\cdot K^{-1}$ was estimated by Smirnov, Lbov [9] from literature data and for liquid $NdCl_2$, $C_p^\circ=24$ $cal\cdot mol^{-1}\cdot K^{-1}$, Novikov, Polyachenok [5]. Thermodynamic values (in $cal\cdot mol^{-1}\cdot K^{-1}$ and cal/mol) for liquid $EuCl_2$ are given in the following table:

| T in K | $C_p^\circ$ | $H_T^\circ-H_{298}^\circ$ | $S_T^\circ$ | $-(G_T^\circ-H_{298}^\circ)/T$ |
|---|---|---|---|---|
| 298 | — | 0.0 | 40.500 | 40.500 |
| 1000 | $25.60\pm0.7$ | 13994.4 | 64.198 | 50.204 |
| 1200 | $25.60\pm0.7$ | 19114.4 | 68.865 | 52.936 |
| 1400 | $25.60\pm0.7$ | 24234.4 | 72.812 | 55.502 |
| 1600 | $25.60\pm0.7$ | 29354.4 | 76.230 | 57.884 |

The free energy function has an uncertainty of 1.04 $cal\cdot mol^{-1}\cdot K^{-1}$. The value given for $C_p$ was assumed to be constant over the entire liquid range; it was estimated from literature data for $BaCl_2$, Hariharan, Eick [6]; and for liquid $YbCl_2$, $C_p^\circ=23.6\pm0.7$ $cal\cdot mol^{-1}\cdot K^{-1}$, Hariharan et al. [7].

References:

[1] O.G. Polyachenok, G.I. Novikov (Zh. Neorgan. Khim. **9** [1964] 773/7; Russ. J. Inorg. Chem. **9** [1964] 429/31). — [2] L. Brewer, L.A. Bromley, P.W. Gilles, N.L. Lofgren (Natl. Nucl. Energy Ser. Div. IV B**19** [1950] 76/192, 91/2). — [3] C.E. Wicks,

F.E. Block (U.S. Bur. Mines Bull. No. 605 [1963] 1/146, 46, 100, 136; C.A. **59** [1963] 14656). – [4] L.N. Komissarova, L.K. Tarasov (Zh. Neorgan. Khim. **14** [1969] 621/3; Russ. J. Inorg. Chem. **14** [1969] 324/6). – [5] G.I. Novikov, O.G. Polyachenok (Zh. Neorgan. Khim. **8** [1963] 1053/9; Russ. J. Inorg. Chem. **8** [1963] 545/50).

[6] A.V. Hariharan, H.A. Eick (High Temp. Sci. **4** [1972] 91/8; Proc. 9th Rare Earth Res. Conf., Blacksburg, Va., 1971, Vol. 2, pp. 584/96). – [7] A.V. Hariharan, N.A. Fishel, H.A. Eick (High Temp. Sci. **4** [1972] 405/10). – [8] N.A. Fishel (Diss. Univ. of Michigan 1970, pp. 1/118 from Diss. Abstr. Intern. B **31** [1970/71] 6477). – [9] M.V. Smirnov, V.S. Lbov (Tr. Inst. Elektrokhim. Akad. Nauk SSSR Ural'sk. Filial No. 8 [1966] 3/6; Electrochemistry of Molten and Solid Electrolytes, Vol. 5, New York 1967, pp. 1/4).

### 11.2.4.6 Color

Solid Compounds

The following colors are reported for $MCl_2$ crystals and their powders:

$NdCl_2$: shiny black [1, 2], dark green [1]

$SmCl_2$: shiny black [3 to 5], red-brown [6, 7], dark red-brown [8, 9], dark cherry red [10]

$EuCl_2$: colorless [9], red [4], pale rose [3], lustrous red-brown [11], very light gray (the stoichiometric compound) [3], white [11 to 14], gray [15]

$DyCl_2$: black [16]

$YbCl_2$: green [17], jade green (0.5% Cl deficiency) [3], dull green (3.6% Cl excess [3]) [17], colorless [9], light gray [8, 15]

Melts

| $SmCl_2$ | $EuCl_2$ | $YbCl_2$ |
|---|---|---|
| deep reddish brown [8] | light yellow [8] | bright green [8, 17] |

The melts are all clear [8].

References:

[1] L.F. Druding, J.D. Corbett (J. Am. Chem. Soc. **83** [1961] 2462/7). – [2] O.G. Polyachenok, G.I. Novikov (Zh. Neorgan. Khim. **8** [1963] 1567/77; Russ. J. Inorg. Chem. **8** [1963] 816/20). – [3] K.E. Johnson, J.R. Mackenzie (J. Inorg. Nucl. Chem. **32** [1970] 43/8). – [4] O.G. Polyachenok, G.I. Novikov (Zh. Neorgan. Khim. **8** [1963] 2631/4; Russ. J. Inorg. Chem. **8** [1963] 1378/81). – [5] K. Przibram (Nature **135** [1935] 100).

[6] G.R. Machlan, C.T. Stubblefield, L. Eyring (J. Am. Chem. Soc. **77** [1955] 2975/8). – [7] F.H. Spedding, A.H. Daane (J. Am. Chem. Soc. **74** [1952] 2783/5). – [8] C.W. De Kock, D.D. Radtke (J. Inorg. Nucl. Chem. **32** [1970] 3687/8). – [9] W. Klemm (Z. Anorg. Allgem. Chem. **184** [1929] 345/51). – [10] E. Staritzky, D.I. Walker (LA-1439 [1955] 1/43, 10; N.S.A. **10** [1956] No. 2391).

[11] A.V. Hariharan, H.A. Eick (Proc. 9th Rare Earth Res. Conf., Blacksburg, Va., 1971, Vol. 2, pp. 584/96, 585). – [12] L.R. Morss, H.O. Haug (J. Chem. Thermodyn. **5** [1973] 513/24, 515). – [13] C.T. Stubblefield, L. Eyring (J. Am. Chem. Soc. **77** [1955] 3004/5). – [14] W. Kapfenberger (Z. Anorg. Allgem. Chem. **238** [1938] 273/86, 281/2). – [15] J.K Howell, L.L. Pytlewski (J. Less-Common Metals **18** [1969] 437/9).

[16] J.D. Corbett, B.C. McCollum (Inorg. Chem. **5** [1966] 938/40). – [17] A.V. Hariharan, N.A. Fishel, H.A. Eick (High Temp. Sci. **4** [1972] 405/10).

### 11.2.4.7 Chemical Reactions

In absence of water and oxidants the "classical" dichlorides $MCl_2$ with M=Sm, Eu, and Yb are very stable at room temperature. The less stable of the dichlorides (e.g., M=Pr, Nd, Dy) disproportionate when heated [1]. In air, the compounds absorb water with the formation of hydrates. Except for $EuCl_2 \cdot 2H_2O$, these hydrates are not stable, but decompose more or less rapidly, depending on the particular rare earth element, to basic rare earth(III) chlorides and hydrogen. With the deeply colored dichlorides (p. 44), the speed of decomposition is readily observed from the change in appearance of sample. Decomposition in water is much more rapid than in moist air, and with deeply colored samples the reaction is especially rapid, with vigorous evolution of hydrogen. But in most cases, the color of the dissolved $M^{2+}$ ion is seen for a short period. Thus, for example, a solution of $TmCl_2$ retains its wine red color for approximately one half minute. As expected, the most stable aqueous solutions are those of $EuCl_2$, which in absence of air remain unoxidized for a considerable time. There is practically no oxidation of the $Eu^{2+}$ ion by water itself, because even the normal potential of the $Eu^{3+}/Eu^{2+}$ redox pair ($E^\circ = -0.34$ V) lies somewhat above the potential of the $H_3O^+$ ion of water ($E = -0.41$ V), Bärnighausen [2]. Solutions of $MCl_2$ (M=Sm, Eu, Yb) in organic solvents are discussed in "Rare Earth Elements" C4b, Sections 11.2.11.8.3, 11.2.11.9.4, and 11.2.11.16.2.5 respectively.

On Heating

Recent work shows that $SmCl_2$, $EuCl_2$, and $YbCl_2$ vaporize without decomposition; see p. 37. The following table gives calculated and estimated values for the enthalpy change $\Delta H^\circ_{298}$ in kcal for the reaction $3MCl_2(solid) \rightarrow M(solid) + 2MCl_3(solid)$ according to Johnson [3], Novikov, Polyachenok [4], and Polyachenok, Novikov [5]:

| M | Sc | Y | La | Ce | Pr | Nd | Pm | Sm |
|---|---|---|---|---|---|---|---|---|
| $\Delta H^\circ_{298}$ [3] | – | – | −222 ±18 | −153 ±21 | −39 ±21 | 0 ±12 | +24 ±30 | +96 ±15 |
| $\Delta H^\circ_{298}$ [4, 5] | +11 | +109 | −128 | −113 | −36 | −9 | +42 | +102 |

| M | Eu | Gd | Tb | Dy | Ho | Er | Tm | Yb | Lu |
|---|---|---|---|---|---|---|---|---|---|
| $\Delta H^\circ_{298}$ [3] | +138 ±12 | −186 ±45 | −66 ±21 | +6 ±21 | −3 ±21 | −36 ±21 | +30 ±21 | +114 ±12 | – |
| $\Delta H^\circ_{298}$ [4, 5] | +118 | −129 | −118 | −22 | −27 | −8 | +25 | +109 | −43 |

For this disproportionation reaction, the entropy change may be $-6 \pm 15$ cal·K$^{-1}$. If so, $\Delta G^\circ_{298} = \Delta H^\circ_{298}$ within ±6 kcal. Accordingly (and in agreement with experiment), $LaCl_2$, $CeCl_2$, $PrCl_2$, $GdCl_2$, $TbCl_2$, and $ErCl_2$ should be thermodynamically unstable and disproportionate to the metal and the trichloride. On the other hand, $PmCl_2$ and $TmCl_2$ should be quite stable and $NdCl_2$ should lie on the border between stability and instability. In agreement with these concepts, $TmCl_2$ melts congruently and $NdCl_2$ melts incongruently, Johnson [3]. The $\Delta H^\circ_{298}$ data of Novikov, Polyachenok [4] (which according to Brown [10] are less reliable than the data of [3]) together with an assumed $\Delta S$ value of $\approx -15$ cal·mol$^{-1}$·K$^{-1}$ would permit the existence of solid $ScCl_2$, $PrCl_2$, $HoCl_2$ and $ErCl_2$ at least at higher temperatures, although these compounds have not yet been experimentally confirmed. According to Barber et al. [6], $ScCl_2$ should be unstable with respect to the above mentioned disproportionation, but, non the less, possible to prepare. Newer measurements and recalculations give $\Delta H^\circ_{298}$ values of 37.1 kJ (≙8.87 kcal) for $NdCl_2$ and 154.1 kJ (≙36.83 kcal) for $TmCl_2$, Morss, McCue [7].

With Nonmetals and Their Compounds

Atmospheric oxygen oxidizes all the dichlorides more or less rapidly. With $EuCl_2$, for example: $12\,EuCl_2+3\,O_2\rightarrow 8\,EuCl_3+2\,Eu_2O_3$. With $SmCl_2$ there is a similar reaction, and also the reaction $2\,SmCl_2+O_2\rightarrow 2\,SmOCl+Cl_2$, Yost et al. [11]. The reaction of $SmCl_2$, $EuCl_2$, and $YbCl_2$ with $Cl_2$ is discussed on p. 143.

Reactions with water and with atmospheric moisture are qualitatively described in the opening paragraph of this section. Enthalpies of solution for $MCl_2$ in water at 25 °C (with no oxidation and apparently at infinite dilution) are given by Morss, McCue [7]:

| $MCl_2$ . . . . . | $NdCl_2$ | $SmCl_2$ | $EuCl_2$ | $DyCl_2$ | $TmCl_2$ | $YbCl_2$ |
|---|---|---|---|---|---|---|
| $-\Delta H$ in kJ/mol . | 29 | 36 | 37.7 | 59 | 67 | 71 |

Heats of solution for $NdCl_2$, $SmCl_2$, $EuCl_2$, $DyCl_2$, $TmCl_2$, and $YbCl_2$ in aqueous hydrochloric acid are given in the sections dealing with the individual chlorides in "Rare Earth Elements" C4b.

The reaction between $MCl_2$ (M=Sm, Eu, Yb) and HCl gas begins with physical adsorption which follows the Langmuir equation. Raising the temperature from 20 to 80 °C, decreases HCl coverage of the surface from 96 to 40% with $SmCl_2$, from 84 to 30% with $EuCl_2$, and from 96 to 36% with $YbCl_2$; see **Fig. 8**. The heat of adsorption in this temperature range changes only slightly with percent coverage and is approximately 4.5 kcal/mol. The mean adsorption coefficients of HCl on $SmCl_2$, $EuCl_2$, and $YbCl_2$ are 42.5, 48.2, and 47.4 $atm^{-1}$, respectively. With an increase in temperature, desorption of the physically adsorbed HCl is observed, which ends at 180 °C for $SmCl_2$, at 120 °C for $EuCl_2$, and at 140 °C for $YbCl_2$. A further raise in temperature is accompanied by transition of physical adsorption to chemisorption. In the region of chemisorption, surface coverage increases with increasing temperature and reaches a maximum at 300 °C for $SmCl_2$ and $YbCl_2$ and at 180 °C for $EuCl_2$. With further increase in temperature, the HCl reacts with the dichlorides with the formation of the corresponding trichlorides, Il'in et al. [8]. The same authors in [9] also studied the kinetics of trichloride formation with $SmCl_2$ at 560 to 660 °C, $EuCl_2$ at 280 to 360 °C, and $YbCl_2$ at 450 to 585 °C, and at HCl pressures of $1.35\times10^{-3}$ to $1.31\times10^{-2}$ atm. For all three dichlorides the

Fig. 8

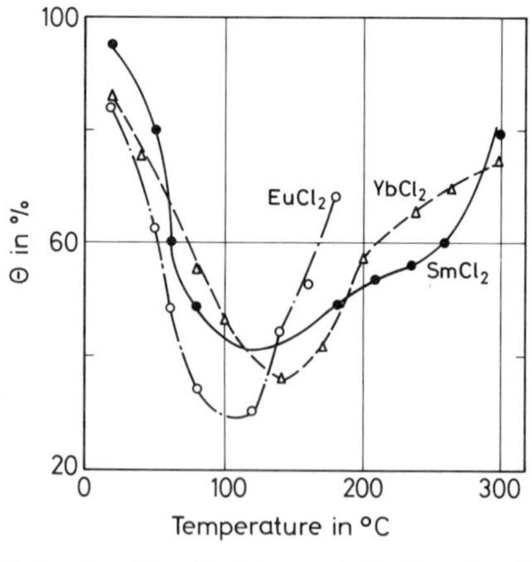


Surface coverage $\Theta$ by HCl for $SmCl_2$, $EuCl_2$, and $YbCl_2$. Studies were made at constant HCl pressures ranging from $1.35\times10^{-3}$ to $13.1\times10^{-3}$ atm, but the actual pressure used in obtaining data for these curves is not given.

weight change (trichloride formation) vs. time curves at various (constant) temperatures and HCl pressures may be described as more or less sigmoidal; see figures in the original article. The observed induction period decreased with increasing temperature and pressure. The activation energies for the reaction of HCl with $SmCl_2$, $EuCl_2$, and $YbCl_2$ were found to be 16.2, 11.3, and 16.0 kcal/mol, respectively, the reaction order with respect to HCl was 0.24, 0.41, and 0.20. The results indicate that the overall rate is determined by processes at the interface and that the entire surface of the sample takes part simultaneously in the reaction. Electron spectroscopic studies show that surface compounds of the type $MCl_2 \cdot HCl$ are not formed. Electron spin resonance studies of $EuCl_2$ samples, which had been treated with HCl for various periods of time at temperatures from 300 to 350 °C, showed that there was an irreversible transition from $Eu^{2+}$ to $Eu^{3+}$ as the interface moved into the bulk of the sample, Il'in et al. [9]. The reverse reaction of the trichlorides with $H_2$ is discussed on p. 144.

References:

[1] J.R. Peterson (Proc. 10th Rare Earth Res. Conf., Carefree, Ariz., 1973, Vol. 1, pp. 4/14, 5; C.A. **85** [1976] No. 27975). – [2] H. Bärnighausen (private communication to the Gmelin Institute 1977). – [3] D.A. Johnson (J. Chem. Soc. A **1969** 2578/80). – [4] G.I. Novikov, O.G. Polyachenok (Usp. Khim. **33** [1964] 732/6; Russ. Chem. Rev. **33** [1964] 342/50). – [5] O.G. Polyachenok, G.I. Novikov (Zh. Neorgan. Khim. **8** [1963] 1567/73; Russ. J. Inorg. Chem. **8** [1963] 816/20).

[6] M. Barber, J.W. Linnett, N.H. Taylor (J. Chem. Soc. **1961** 3323/32). – [7] L.R. Morss, M.C. McCue (Inorg. Chem. **14** [1975] 1624/7). – [8] V.K. Il'in, V.A. Krenev, V.I. Evdokimov (Izv. Akad. Nauk SSSR Neorgan. Materialy **11** [1975] 1127/8; Inorg. Materials [USSR] **11** [1975] 961/2). – [9] V.K. Il'in, V.A. Krenev, V.I. Evdokimov (Zh. Neorgan. Khim. **19** [1974] 2603/5; Russ. J. Inorg. Chem. **19** [1974] 1421/3). – [10] D. Brown (Halides of the Lanthanides and Actinides, London - New York - Sydney 1968, p. 170).

[11] D.M. Yost, H. Russel, C.S. Garner (The Rare Earth Elements and Their Compounds, New York 1947, pp. 51/69, 64, 66).

### 11.2.5 Phases of the Type $MCl_x$ and (M, M')$Cl_x$ with $2<x<3$

Review

No systematic investigations of these mixed valence compounds, which sometimes are also formulated as $(M^{2+}, M^{3+})Cl_{2+y}$, have been reported. Phases of this type have been obtained by reacting the trichloride with the corresponding elemental rare earth.

Line-rich Debye-Scherrer powder patterns have been reported for $NdCl_{2.27}$ and $NdCl_{2.37}$, but not indexed; d-values are given in Druding, Corbett [1]. Laptev et al. [2] refer to these authors [1] in indicating hexagonal symmetry for $NdCl_3 \cdot 2NdCl_2(NdCl_{2.33})$. Hexagonal symmetry has been proven for $SmCl_{2.33}$ and $EuCl_{2.33}$. $EuCl_{2.25}(\triangleq EuCl_3 \cdot 3EuCl_2)$ has orthorhombic symmetry. $EuCl_{2.20}$ $(\triangleq EuCl_3 \cdot 4EuCl_2)$ crystallizes tetragonal. Detailed data for $SmCl_{2.33}$ and the $EuCl_x$ phases are given in "Rare Earth Elements" C4b, Sections 11.2.11.8.4 and 11.2.11.9.5.

For most of the known Dy, Ho, Tm, and Yb chloride phases whose composition allows them to be considered members of the $M_nCl_{2n+1}$ series (see, e.g., Caro [3, 4] and Haschke [5]), there are indications that the structures have $MX_2$ superstructures. Thus at least some of the $TmCl_x$ phases in the range $x=2.04$ to 2.11 may be described as having a $TmCl_2$ superstructure, Naterstad, Corbett [6], namely those phases with x=2.04, 2.067, and 2.074 to 2.11, corresponding to n=25 (not certain), 15, and the integers 13 to 9, Haschke [5]. For $DyCl_{2.11}$, the Debye-Scherrer pattern was too weak to permit an analogous interpretation, Corbett, McCollum [7]. A phase which was chemically analyzed as $HoCl_{2.14}$ (Löchner, Corbett [8]) may be considered as $Ho_5Cl_{11}$ ($\triangleq HoCl_{2.20}$) according to structural studies of Löchner et al. [9]. Likewise, the light green material with the apparent composition $YbCl_{2.26}$ whose d-values are given by Fishel, Eick [10] is actually $YbCl_{2.167}$ ($\triangleq Yb_6Cl_{13}$) according to subsequent investigations of Lüke, Eick [11, p. 424].

Members of the series with n=5 ($MCl_{2.20}$, with M=Dy, Ho), with n=6 ($MCl_{2.167}$, with M=Dy, Tm, Yb), and with n=7 ($MCl_{2.143}$, with M=Dy, Tm) have a vernier type structure (see below) and may be regarded as having a one-dimensional superstructure of the fluorite ($CaF_2$) type, the members being built from n basic fluorite units, Bärnighausen [12, 16]. According to Lüke, Eick [11], the same is presumably true for higher members of the $M_nCl_{2n+1}$ series. Such phases have $MX_2$ superstructures in which the anion layers have variable packing densities so that the individual rows of anions lie one above the other as in a vernier scale. In this way excess anions (relative to formula $MX_2$) find a place in the lattice. A review article by Hyde et al. [13] discusses the vernier structure.

## $M_5Cl_{11}$

These compounds are monoclinic, space group $P2_1/m-C^2_{2h}$ (No. 11); Z=4, Bärnighausen [12]. This space group has been confirmed by investigations of single crystals of $Dy_5Cl_{11}$ which take twinning into account, Rinck [22]. Lattice constants (in Å) and X-ray densities $D_x$ (in g/cm³) calculated therefrom are as follows:

| $M_5Cl_{11}$ | a | b | c | β | $D_x$ | Ref. |
|---|---|---|---|---|---|---|
| $Dy_5Cl_{11}$ | 7.108(1) | 34.68(2) | 6.632(1) | 90.23(4)° | 4.884 | [16, 17] |
| $Ho_5Cl_{11}$ | 7.078(2) | 34.57(2) | 6.603(3) | 90.19(5)° | 4.994 | [9, 12] |

Slightly different lattice constants for $Dy_5Cl_{11}$ were measured earlier by Bärnighausen [12]. The d-values (Guinier camera) of $Ho_5Cl_{11}$ ($HoCl_{2.14}$ in the original article, see "Review" above) are given in Löchner, Corbett [8].

In the $M_5Cl_{11}$ structure, which has been determined for $Dy_5Cl_{11}$ (see **Fig. 9**), the Cl ions form networks of triangles and of quadrilaterals. The former lie above and below the latter between y=0 and 1/2 (at z≈3/4) and between y=1/2 and 1 (at z≈1/4), respectively. The M ions which lie between these networks (at z≈1/2 and z≈0) occupy six positions (4×4f, 2×2e). As with $TmCl_2$ and $YbCl_2$ (see p. 33), the M ions at two of the general positions are surrounded by seven Cl ions (polyhedron with one triangular and one quadrilateral base). For the M ions at the two other general positions, an eighth Cl ion caps the quadrilateral base of the Cl polyhedron (7+1 coordination). The M ions at the special positions are surrounded by eight Cl ions which form a bicapped trigonal prism [12, 22]. Positional parameters and interatomic distances are available from Bärnighausen [16], Rinck [22]. The interatomic distances indicate that one of the 2e positions is occupied mainly by $M^{3+}$ and that one of the 4f positions is occupied statistically by $M^{2+}$ and $M^{3+}$ in an approximately 1:1 ratio [12, 16, 22]. A similar $M^{3+}$ distribution is derived from the volumes of the Voronoi cells and the

Fig. 9

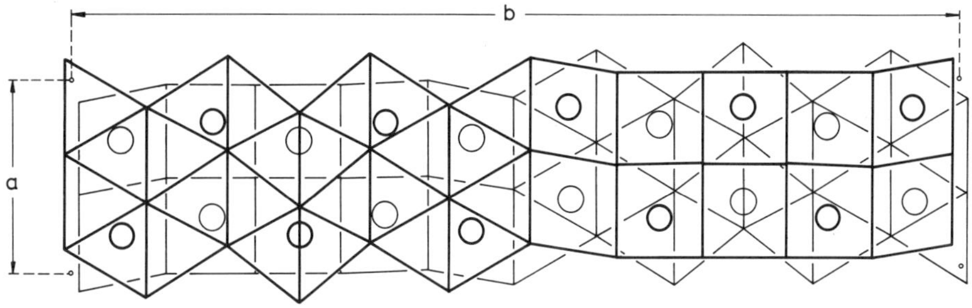


A schematic representation of the crystal structure of $Dy_5Cl_{11}$. The bold and light drawn circles are Dy atoms at $z \approx 1/2$ and $z \approx 0$, respectively. The Cl atoms are located at the intersection points of the somewhat puckered nets at $z \approx 3/4$ and $z \approx 1/4$.

polyhedral-atomic-volume (PAV) cells for the individual Dy sites. The average of the PAV chlorine atom volumes is 32.39 $Å^3$. It is thus somewhat less than that of the Cl ions in $Gd_2Cl_3$ and somewhat greater than that of the Cl ions in $Yb_5ErCl_{13}$, Carter [20].

### $M_6Cl_{13}$

These monoclinic compounds crystallize in space group $I2/a(C2/c)-C^6_{2h}$ (No. 15), which was confirmed by single-crystal studies of $Dy_6Cl_{13}$; Z=4. Lattice constants (in Å) and the calculated densities $D_x$ (in $g/cm^3$) are as follows:

| $M_6Cl_{13}$ | a | b | c | $\gamma$ | $D_x$ | Ref. |
|---|---|---|---|---|---|---|
| $Dy_6Cl_{13}$ | 7.099(1) | 41.41(2) | 6.667(1) | 91.35(3)° | 4.87 | [16, 17] |
| $Tm_6Cl_{13}$ | 7.009(3) | 41.037(6) | 6.557(3) | 91.30(3)° | 5.192 | [16, 18] |
| $Yb_6Cl_{13}$ | 6.958 | 40.98 | 6.539 | 91.19(20)° | 5.34 | [11] |

Slightly differing constants for $Dy_6Cl_{13}$ were determined earlier by Bärnighausen [12].

### $M_7Cl_{15}$

Crystals of these compounds are always twinned. The method of preparation causes the crystals to be topotactically intergrown with the dichloride crystals. Presumably, they have monoclinic symmetry, space group $P2_1/m-C^2_{2h}$ (No. 11); Z=4, Bärnighausen [16]. Lattice constants (in Å) and the calculated densities $D_x$ (in $g/cm^3$) are as follows:

| $M_7Cl_{15}$ | a | b | c | $\beta$ | $D_x$ | Ref. |
|---|---|---|---|---|---|---|
| $Dy_7Cl_{15}$ | 7.097(1) | 48.18(1) | 6.674(1) | ≈90° | 4.858 | [17, 19] |
| $Tm_7Cl_{15}$ | 7.001(1) | 47.677(5) | 6.571(3) | ≈90° | 5.190 | [18] |

### (Nd, M)$Cl_{2+y}$

There is no fcc phase for $NdCl_{2+y}$ in the range $0.0<y<0.25$ (Druding et al. [14]), but there are fcc phases (with a fluorite structure) in mixed phases of composition

$(Nd, M)Cl_{2.15-2.20}$, Corbett [15]. An example is $(Nd, Pr)Cl_{2.15}$ with Nd:Pr≈0.6 to 1, for which an a-value of 7.00 Å has been determined [14]. Other examples are $(Nd,Ce)Cl_{2.19}$, with Nd:Ce=6.05, and $(Nd,Ce)Cl_{2.20}$, with Nd:Ce=4.51, in which Ce presumably is present only as $Ce^{3+}$; for both compounds a=7.04 Å [1]. In the mixed phases, the more difficultly reducible hetero atoms (Ce or Pr) apparently prevent a rearrangement via electron exchange between Nd atoms, which occurs in the cooling down of pure neodymium chlorides [15].

**$Yb_5ErCl_{13}$**

Lattice constants a=41.44±0.02, b=6.537±0.005, c=7.004±0.005 Å, and $\beta$= 98.5°±0.1° and the calculated density $D_x$=5.128 g/cm³ have been reported for this monoclinic compound, Lüke, Eick [11]. Bärnighausen [12, 16] refers to these authors [11] in giving for a different setting a=7.004(5), b=41.00(2), c=6.537(5) Å, $\gamma$=91.25(10)°, and $D_x$=5.287 g/cm³. A preliminary structure refinement to R=5.8% was carried out for space group $C2/c-C_{2h}^6$ (No. 15), however, it appears likely that this vernier structure has a space group of lower symmetry than in $Yb_6Cl_{13}$, Lüke, Eick [11]. Yet the same space group has been reported by Eick [21], who calculated refined atom positions. The Er atom is presumed to be disordered on one or more of the three metal sites (so-called Yb(1) to Yb(3)) [21]. From the comparison of the volumes of the Voronoi cells and of the polyhedral-atomic-volume (PAV) cells it is suggested that the $Er^{3+}$ cation is primarily located on the Yb(3) site. The average of the PAV chlorine atom volumes is 32.17 Å³. It is thus less than that of the chlorine atoms in $Gd_2Cl_3$ and in $Dy_5Cl_{11}$, Carter [20].

References:

[1] L.F. Druding, J.D. Corbett (J. Am. Chem. Soc. **83** [1961] 2462/7). – [2] D.M. Laptev, V.F. Goryushkin, I.S. Astakhova, G.G. Polyakova (Zh. Neorgan. Khim. **24** [1979] 1311/6; Russ. J. Inorg. Chem. **24** [1979] 730/3). – [3] P.E. Caro (Natl. Bur. Std. [U.S.] Spec. Publ. No. 364 [1972] 367/83, 378/81). – [4] P.E. Caro (Compt. Rend. C **273** [1971] 1169/72). – [5] J.M. Haschke (Inorg. Chem. **15** [1976] 298/303).

[6] T. Naterstad, J.D. Corbett (J. Less-Common Metals **46** [1976] 291/5, 292). – [7] J.D. Corbett, B.C. McCollum (Inorg. Chem. **5** [1966] 938/40). – [8] U. Löchner, J.D. Corbett (Inorg. Chem. **14** [1975] 426/8). – [9] U. Löchner, H. Bärnighausen, J.D. Corbett (Inorg. Chem. **16** [1977] 2134/5). – [10] N.A. Fishel, H.A. Eick (J. Inorg. Nucl. Chem. **33** [1971] 1198/201).

[11] H. Lüke, H.A. Eick (Proc. 12th Rare Earth Res. Conf., Vail, Colo., 1976, Vol. 1, pp. 424/32). – [12] H. Bärnighausen (Proc. 12th Rare Earth Res. Conf., Vail, Colo., 1976, Vol. 1, pp. 404/13). – [13] B.G. Hyde, A.N. Bagshaw, S. Andersson, M. O'Keefe (Ann. Rev. Mater. Sci. **4** [1974] 43/92, 82/9). – [14] L.F. Druding, J.D. Corbett, B.N. Ramsey (Inorg. Chem. **2** [1963] 869/71). – [15] J.D. Corbett (Rev. Chim. Minerale **10** [1973] 239/57, 245/53).

[16] H. Bärnighausen (private communication to the Gmelin Institute 1980). – [17] C. Rinck (Thesis Univ. Karlsruhe 1978 from [16]). – [18] A. Beyer (Thesis Univ. Karlsruhe 1978 from [16]). – [19] H. Bärnighausen, W. Kuhn (unpublished results 1977 from [16]). – [20] F.L. Carter (Proc. 13th Rare Earth Res. Conf., Wheeling, Va., 1977, pp. 225/31).

[21] H.A. Eick (private communication to [20]). – [22] C. Rinck (from [16]).

## 11.2.6 Comparative Data on $MCl_3$

### 11.2.6.1 Formation. Preparation

General References:

M.D. Taylor, Preparation of Anhydrous Lanthanon Halides, Chem. Rev. **62** [1962] 503/11.
F.E. Block, T.T. Campbell, Rare Earth and Yttrium Halides for Metal Production-Chlorides, Bromides, Iodides, in: F.H. Spedding, A. Daane, The Rare Earths, New York - London 1961, pp. 89/112.
J.B. Reed, B.S. Hopkins, L.F. Audrieth, Anhydrous Rare Earth Chlorides, Inorg. Syn. **1** [1939] 28/33.

#### 11.2.6.1.1 General Review

The most important methods for the preparation of anhydrous rare earth chlorides do not differ substantially from those that are customary for the preparation of O- and $H_2O$-sensitive chlorides of other metals.

Starting materials are primarily the metals, the oxides, and the hydrated chlorides which are easily obtainable by evaporation of hydrochloric acid solutions of the oxides. Other compounds of the rare earths, for example carbides, sulfides, and salts with organic anions, have also been used as raw materials, but these are mostly not accessible with ease and sufficient purity. The corresponding preparation processes are troublesome and often suited only for preparation of the chlorides in laboratory quantities. In general no matter what the preparative method, obtaining completely pure preparations requires special precautions and much time. The highest purity reagents are likewise indispensable, and careful exclusion of air and moisture during the reaction. Also, the choice of suitable reaction containers can be decisive. At higher temperatures the trichlorides attack $SiO_2$ with formation of $SiCl_4$ and O-containing rare earth compounds, so that containers of quartz, glass, or porcelain are not usable for high purity syntheses. The rare earth metals themselves react with gold and platinum. They can be contained only in vessels of tungsten, molybdenum, or tantalum with special precautions to protect the vessels.

After procedures were developed to purify the rare earth chlorides very effectively even in kilogram quantities by vacuum distillation or sublimation, the choice of the preparation method lost much of its significance. Above all, the simplest and fastest methods are given priority in the technology.

#### 11.2.6.1.2 Preparation from Metals

Summary

The reaction of the metals with chlorine or hydrogen chloride at normal or elevated temperatures constitutes one of the first and most convenient methods for winning of the trichlorides when the metal is available in sufficient purity. Impurities contained in the metal like dissolved N and O carry over extensively into the end product. Attack on the commonly used Mo or W containers can be avoided by control of the $H_2$ partial pressure in the reaction zone; need for careful purification of the gases used is self-evident. In spite of the simplicity of this method it remains limited to laboratory purposes. While metals of good purity are available commercially, they in turn are commonly prepared from the chlorides.

### Chlorination of Metals with Gaseous HCl

Oxide-chloride-free $MCl_3$ compounds (M=Ce, Pr, Nd, Sm) for system studies are prepared in a quartz apparatus above the melting point of the trichlorides from 99.9% metal turnings with HCl gas generated from KCl and $H_2SO_4$. Ordinary commercial HCl gas is not suitable, since it is contaminated with hydrocarbons. The metal turnings are placed in an Mo boat enclosed in an Mo cylinder. The apparatus is initially flushed with helium. During the heating of the reaction tube up to the reaction temperature (800 to 900 °C) a mixture of He and a little $H_2$ is slowly passed through the tube, and then HCl gas is introduced with a moderately strong flow rate over the metal turnings. The complete reaction time is around 30 to 50 min. At the end the reaction product is further heated 15 to 30 min in an $HCl/H_2$ flow at the same temperature, and then is cooled down in an He flow, Druding, Corbett [1], Druding [2]; cf. also Corbett [3].

The preparation of pure $DyCl_3$ takes place in a similar manner, except commercial HCl gas is used and the product obtained is doubly distilled under high vacuum at 800 to 875 °C, Corbett, McCollum [4]. The method of [4] can also be used to produce 15 g $ScCl_3$ within 4 to 6 h in nearly quantitative yield, McCollum et al. [5], cf. also Poeppelmeier, Corbett [6], Corbett, Ramsey [7]. The preparation of pure $GdCl_3$ occurs in a corresponding manner. With pure HCl gas a single vacuum sublimation of the raw product in a Ta vessel at 850 °C is required, with commercial HCl gas a double sublimation is necessary, Mee, Corbett [8], Mee [9]. Pure $HoCl_3$ is obtained by reaction of pure metal with electronic grade HCl gas at 860 °C. The reaction product is sublimed twice under high vacuum in a Ta vessel, Löchner, Corbett [10].

$YCl_3$ can be prepared by reaction of Y with $Cl_2$ and a subsequent vacuum distillation at 950 °C, Pustil'nik et al. [11].

### Chlorination of Metals with $HgCl_2$

Laboratory quantities of anhydrous trichlorides can also be won by chlorination of the metals (M) with excess $HgCl_2$. The reactions are $2M+3HgCl_2 \rightarrow 2MCl_3+3Hg$ and $2M+6HgCl_2 \rightarrow 2MCl_3+3Hg_2Cl_2$. For preparation of $MCl_3$ of higher purity (M=La, Ce, Pr, Nd, Sm, Gd, Tb, Dy, Ho, Er) 2 to 5 mg of the metal M, after filing off the surface layers, is mixed with a fourfold excess of powdered $HgCl_2$ in a pyrex ampule and heated about 12 h in vacuum at 300 °C. Hg and Hg chlorides are distilled off. The $MCl_3$ is then purified by vacuum sublimation at 730 to 940 °C. A Ta crucible is used which has a protective Ta shield and a Ta cold finger. The sublimation temperatures t for the $MCl_3$ compounds are (pressure not stated):

| M | La | Ce | Pr | Nd | Sm | Gd | Dy | Ho | Er |
|---|---|---|---|---|---|---|---|---|---|
| t in °C | 940 | 930 | 880 | 850 | 825 | 800 | 925 | 730 | 785 |

Carter, Murray [12].

The first of the reactions cited above also proceeds readily in tetrahydrofuran (THF), giving a precipitate of Hg metal and a solution of the lanthanide trichloride ($ErCl_3$, $YbCl_3$), from which crystalline complexes $YbCl_3 \cdot 3THF$ and $ErCl_3 \cdot 3.5THF$ can be isolated by evaporation. The solvates lose THF on heating, Deacon, Koplick [13].

References:

[1] L.F. Druding, J.D. Corbett (J. Am. Chem. Soc. **83** [1961] 2462/7). – [2] L.F. Druding (Diss. Iowa State Univ. 1960, pp. 1/84, 11; Diss. Abstr. **21** [1961] 1751/2). –

[3] J.D. Corbett (Inorg. Nucl. Chem. Letters **8** [1972] 337/40). – [4] J.D. Corbett, B.C. McCollum (Inorg. Chem. **5** [1966] 938/40). – [5] B.C. McCollum, M.J. Camp, J.D. Corbett (Inorg. Chem. **12** [1973] 778/80).

[6] K.R. Poeppelmeier, J.D. Corbett (Inorg. Chem. **16** [1977] 294/7). – [7] J.D. Corbett, B.N. Ramsey (Inorg. Chem. **4** [1965] 260/2). – [8] J.E. Mee, J.D. Corbett (Inorg. Chem. **4** [1965] 88/93). – [9] J.E. Mee (Diss. Iowa State Univ. 1962, pp. 1/58, 16; N.S.A. **17** [1963] No. 22210). – [10] U. Löchner, J.D. Corbett (Inorg. Chem. **14** [1975] 426/8).

[11] A.I. Pustil'nik, T.S. Sedykh, R.V. Dubchak, S.I. Slatin (Zh. Fiz. Khim. **43** [1969] 1850/2; Russ. J. Phys. Chem. **43** [1969] 1036/8). – [12] F.L. Carter, J.F. Murray (Mater. Res. Bull. **7** [1972] 519/23; C.A. **77** [1972] No. 42529). – [13] G.P. Deacon, A.J. Koplick (Inorg. Nucl. Chem. Letters **15** [1979] 263/5).

### 11.2.6.1.3 Preparation from Oxides

Summary

The rare earth oxides are the cheapest and the most easily accessible starting materials for the winning of the anhydrous trichlorides, especially since many rare earth metals occur already in nature in the form of their oxides and since most of their salts with O-containing anions can be decomposed thermally into the oxides. Oxides of high purity are not difficult to obtain commercially.

Since the oxides are thermodynamically more stable than the chlorides, direct chlorination with $Cl_2$ or HCl is scarcely promising. Several reports on the preparation of anhydrous chlorides in this manner are, to be sure, present in the literature, however, in any case the results at best are milligram quantities. As shown by equilibrium studies of the reactions of the oxides with $Cl_2$ and HCl, see pp. 146 and 149, respectively, the formation of pure chlorides can be expected only under certain conditions obtainable only with difficulty. One will always obtain a more or less O-contaminated reaction product.

In order to improve the energy balance for the winning of the chloride from the oxide, one must add to the chlorinating agent a reducing agent. One can also use volatile compounds which combine in the same molecules chlorinating and reducing properties.

#### Chlorination of Oxides with Gaseous HCl

The preparation of anhydrous rare earth chlorides by chlorination of the heated oxides with gaseous hydrogen chloride has been reported by several authors: $YCl_3$ [55], $LaCl_3$ [66], $PmCl_3$ [56 to 58].

Other investigators were unable to bring about a complete conversion into the chlorides by this method in the cases of $Y_2O_3$ [59], $Pr_6O_{11}$, and $Sm_2O_3$ [51]. The behavior of some rare earth oxides against HCl at 475°C is shown in **Fig. 10**, p. 54. The start of the individual curves on the $M_2O_3$ plateau marks the beginning of HCl introduction into the furnace. The weight-gain curve of $La_2O_3$ exhibits a weakly indicated step at a composition corresponding to LaOCl, so does the curve for $Nd_2O_3$. More pronounced indication of oxide chloride formation is found with $Sm_2O_3$; dilution of the HCl with $N_2$ in the ratio 1:5 prolongs the arrest at the SmOCl step to 15 min. With $Er_2O_3$ the reaction does not proceed beyond the ErOCl stage. The oxide chlorides of Sm and heavier lanthanides, when formed by thermal decomposition of the hydrated chlorides

Fig. 10

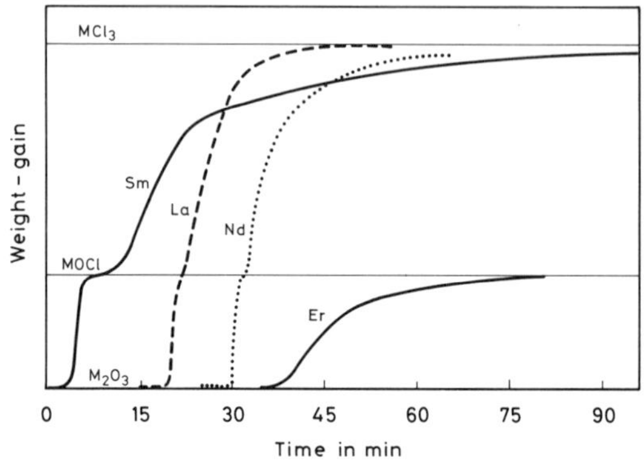


Formation of anhydrous chlorides from oxides in an HCl stream at 475 °C.

in an air/HCl stream (see p. 213/4), are not or only very reluctantly reconverted into the anhydrous chlorides after shutting off the air. Formation and chlorination of the oxide chlorides of the lighter lanthanides in air and HCl, respectively, proceed at nearly the same rate, Haeseler, Matthes [60, pp. 148/9]. Chlorination of the oxides by HCl starts at the following temperatures:

| | | | | | | |
|---|---|---|---|---|---|---|
| oxide . . . . | $La_2O_3$ | $CeO_2$ | $Pr_2O_3$ | $Nd_2O_3$ | $Sm_2O_3$ | $Eu_2O_3$ |
| t in °C . . . | 274 | 551 | 304 | 296 | 322 | 300 |

Most complete conversion of oxides to chlorides under HCl pressure was achieved for $La_2O_3$, $Pr_2O_3$, $Nd_2O_3$, $Sm_2O_3$ at 600 °C, $CeO_2$ at 650 °C, $Eu_2O_3$ at 550 °C. Data for the formal chlorination kinetics of $La_2O_3$, $Pr_2O_3$, and $Nd_2O_3$ are presented by Ivashentsev, Stoyanova [19]. The reducing power of a carbon tube significantly enhances the chlorinating action of HCl by permitting the reaction

$$M_2O_3 + 6HCl + 3C \rightarrow 2MCl_3 + 3CO + 3H_2,$$ Pettersson [2].

## Chlorination of Oxides with $Cl_2$ or HCl in the Presence of C

Although it has been known since the previous century [1, 2], this process has remained due to its simplicity one of the most interesting, especially on a technical scale. The reactions proceed smoothly following the equations:

$$M_2O_3 + 3Cl_2 + 3C \rightleftharpoons 2MCl_3 + 3CO \text{ and}$$
$$M_2O_3 + 6HCl + 3C \rightleftharpoons 2MCl_3 + 3CO + 3H_2$$

The yield of trichloride is mostly above 90 to 95%. However, since C is commonly used in 10 to 15% excess and since on unintentional (even if brief) exceeding of the melting point of the chloride a part of the oxide can be shielded from reaction, purification of the raw product is indispensable for most application purposes.

For preparation of $YCl_3$ and $CeCl_3$ as starting materials for winning of the metals the following procedure is proposed: A mixture of the appropriate oxide with excess C (see above) is formed into small pellets with the aid of some dextrin as a binding agent. These are placed into a vertical quartz reactor and heated initially to 600 °C in an Ar flow to drive off moisture and decompose the dextrin. Then $Cl_2$ or a mixture of $Cl_2$ and $CCl_4$ is passed through the charge, and the temperature is increased to the neighborhood of the melting point of the chloride. The optimal reaction temperature for $Y_2O_3$ is located at 650 °C, that for Ce oxide at 750 to 800 °C. The reaction product is purified by a double vacuum distillation, whereby the C content is lowered to below 0.092%. It is expedient to remove mechanically in advance any gross contamination by small solid particles, especially of C, which in the vacuum distillation are easily carried along by the evaporating gas. For this purpose the molten chloride is filtered through Mo wool under an Ar shield; Mo or Ni vessels are suitable as containers for the melts, U.S. Bureau of Mines [3, 4], Campbell et al. [5 to 7]. By reaction of an $Sc_2O_3$/charcoal mixture with $Cl_2$ at 1000 to 1100 °C Samodelov [8] prepared anhydrous $ScCl_3$. In an analogous manner Schäfer [9] obtained anhydrous $LaCl_3$, he reports no precise temperature data. Strzyzewska et al. [10] describe the chlorination of the oxides or oxalates with $Cl_2$ in the presence of C black as the best method for the preparation of anhydrous Y and Gd chlorides with small contents of oxide chlorides.

**Chlorination of Oxides with $NH_4Cl$**

While heating an intimate mixture of rare earth oxides and $NH_4Cl$ is a quite efficient method for producing anhydrous chlorides in quantity, it is time-consuming and not suitable for making materials of the highest purity. The reaction proceeds as $M_2O_3 + 6\,NH_4Cl \rightleftharpoons 2\,MCl_3 + 3\,H_2O + 6\,NH_3$ (M = rare earth element). Up to a temperature of about 190 °C the reaction appears to be very slow, but above this temperature it proceeds vigorously. If twice the theoretical amount of $NH_4Cl$ is used the reaction mixture soon becomes completely soluble in water except in the case of $CeO_2$ for which complete conversion to the chloride was not obtained. The excess $NH_4Cl$ is best removed by heating the mixtures at 300 to 320 °C in vacuum in an all-glass apparatus designed specifically for this purpose. In air, any attempt to drive it all off by heating results in the formation of some basic material. The anhydrous chlorides are obtained as finely divided powders. They are extremely hygroscopic and hiss on addition of water. The solutions in water are usually clear or only slightly cloudy. No ammonia can be detected. Yields of 85 to 95% are obtained, the purity is around 98%, Reed et al. [11]. The method and suitable apparatus are described in some detail by [12 to 17]. In order to remove the last traces of $NH_4Cl$, the products in many cases have to be heated up to 400 or 500 °C. The total procedure usually requires 24 to 36 h for completion.

A procedure recommended by Kiess [18] differs from that used by the earlier workers [11, 12] in two respects. First, the excess of $NH_4Cl$ is greater (four times the stoichiometric amount). Second, the reaction mixture is heated at atmospheric pressure for several hours longer while a stream of He or $N_2$ is passed over it. At the end of this time, most of the $NH_4Cl$ has evaporated. The prolonged heating at atmospheric pressure is necessary to produce salts pure enough to crystallize readily. Immediate melting and subsequent crystallization of the salt follow the removal of the $NH_4Cl$. In view of the extreme reactivity of these salts it is advisable not to recover them as powders. The slowly cooled, solidified material contains single crystals large enough to permit spectroscopic studies of the compounds.

Pure $YCl_3$, $CeCl_3$, $SmCl_3$, and $GdCl_3$ in amounts of 200 to 270 g are produced by heating at 160 to 170 °C an intimate mixture of the oxides with 65 to 100% excess $NH_4Cl$ for 48 to 72 h in a flat-bottomed aluminium pan. The mixtures are stirred intermittently, and heating is continued until evolution of ammonia vapors ceases. An amount of $NH_4Cl$ equivalent to that originally used is then added to the products, and final dehydration is accomplished in an all-nickel vacuum retort. The temperature of the retort is raised in increments over a 24 h period to a maximum of 330 °C and held at this point for 24 h or until no further ammonia is evolved. The absolute pressure in the retort ranges from <1 to 4 Torr during this time. Nitrogen analyses on the products indicate N contents of only 30 to 60 ppm. The yields are between 92 and 96%, Campbell et al. [6].

Chlorination of the oxides by $NH_4Cl$ starts at the following temperatures t:

| | | | | | | |
|---|---|---|---|---|---|---|
| oxide . . . . | $La_2O_3$ | $CeO_2$ | $Pr_2O_3$ | $Nd_2O_3$ | $Sm_2O_3$ | $Eu_2O_3$ |
| t in °C . . . | 170 | 170 | 168 | 135 | 130 | 130 |

Most complete conversion to chlorides is achieved at 300 °C for all chlorides, Ivashentsev, Stoyanova [19]. $CeO_2$ reacts with $NH_4Cl$ much more slowly than the oxides of Y, Nd, and Sm, but by adding a three- or fourfold excess of $NH_4Cl$ and continuing the reaction for 10 to 12 h it is possible to prepare a product containing only a small amount of unconverted oxide, West, Hopkins [13]. A 1 h heating of $CeO_2$ with 12 to 16 mol $NH_4Cl$ at 400 °C produces $CeCl_3$ in a yield of only 59%. Use of freshly prepared $Ce(OH)_3$ improves the yield to 71.2%. Anhydrous $LaCl_3$, $NdCl_3$, and $SmCl_3$ are obtained in almost quantitative yields by the reaction of 1 mol of the respective oxides with 12 to 18 mol $NH_4Cl$ at 300 to 350 °C for 30 to 60 min, followed by removal of unchanged $NH_4Cl$ by heating in vacuo at 300 °C for 2 h. Similarly, $Pr_6O_{11}$ is converted into $PrCl_3$ by heating with 44 mol $NH_4Cl$, Shiokawa, Ishino [20].

A patent has been granted for a variant of this preparation method, in which the rare earth oxides are treated with two to three times the stoichiometric amount of $NH_4Cl$ at 250 to 325 °C while an inert gas is passed through the reaction medium. Unreacted $NH_4Cl$ is then sublimed off at 500 to 600 °C. Final purification is accomplished by vacuum distillation of the rare earth chlorides at 700 to 1100 °C, Domning, Schechter [21]. In another variant of the method the rare earth oxides are heated in vacuum with a twofold excess of $NH_4Cl$ with intensive mixing. Up to 50-g amounts of $YCl_3$, $LaCl_3$, $PrCl_3$, $NdCl_3$, $SmCl_3$, $TbCl_3$, and $ErCl_3$ are prepared in a specially constructed apparatus. The products are free from oxide chlorides, Krestov et al. [22].

Hodgkinson [23] early described a method for the preparation of anhydrous chlorides by mixing the oxides with a small excess of $NH_4Cl$ and gradually dropping the mixture into a red hot crucible. Repeated trials of this method by Reed et al. [11] showed that complete reaction is never obtained in this way and that considerable loss of material is involved.

### Chlorination of Oxides with $S_2Cl_2$ and $Cl_2/S_2Cl_2$ Mixtures

The application of $Cl_2/S_2Cl_2$ mixtures for the chlorination of lanthanide oxides appears to have been first studied by Matignon, Bourion [24]. Comprehensive data concerning the utility of the method can be found in Taylor [25]. The procedure is not limited only to the oxide; the sulfate, carbonate, and all other O-containing salts are also convertible by it into the chlorides. It is also well suited to open up ores. Only $EuCl_3$ appears to be unobtainable by this method in pure form, as shown by the first studies of Bourion [26, 27]. Therefore, in this case the chloride hydrate is used as the starting material, Urbain, Bourion [28].

From the dissociation equilibria of $MCl_3$ into $MCl_2$ and $Cl_2$ (M=Sm, Eu, Yb) (see p. 143) Polyachenok, Novikov [29, 30] conclude that even in a $Cl_2$ stream saturated with $S_2Cl_2$ (or $CCl_4$) the molten trichlorides contain in the vicinity of their melting points the following quantities of dichloride in mol %: $SmCl_3$−0.005, $EuCl_3$−8, $YbCl_3$−0.1. While $SmCl_3$ and $YbCl_3$ thus under these conditions still form in satisfactory purity,

$EuCl_3$, which due to its low melting point is usually obtained in the molten state, must be post-chlorinated at lower temperatures. For this purpose it is cooled down, pulverized, and heated in a $Cl_2$ stream, without letting its melting point be exceeded. Even after such a post-treatment the preparations of Johnson, Mackenzie [31] still contained significant quantities of $EuCl_2$.

The reaction of the oxide with $S_2Cl_2$ and $Cl_2$ proceeds according to $4M_2O_3+3S_2Cl_2+9Cl_2\rightarrow 8MCl_3+6SO_2$. Sulfur monochloride and chlorine are very energetic chlorinating agents, so that in many cases the liberated heat of reaction leads to the melting of the reaction product. Thus oxide particles can be surrounded by the melt and be shielded from further reaction. Therefore, Bourion [27] early emphasized that always only small quantities of oxide are convertible. According to Klemm, Biltz [65] it is essential for a smooth conduct of the reaction that the oxide be spread out in a very thin layer and be chlorinated initially at the lowest possible temperature. $La_2O_3$ is, for example, heated in 2 h from 500 to 800 °C; in the end it is post-chlorinated for $^1/_2$ h at 950 °C. According to an earlier publication the chlorination of $M_2O_3$ (M = La, Pr, Nd) begins below a distinct red heat and can be completed in $^1/_2$ h with 10 g of oxide; the final temperature is 1000 °C. Unused $S_2Cl_2$ must be flushed out by $CO_2$ with the apparatus still at the reaction temperature, Voigt, Biltz [32]. Other authors report the following reaction temperatures for chlorination with $Cl_2/S_2Cl_2$ mixtures: 700 to 720 °C for $Pr_6O_{11}$ and $Nd_2O_3$ [33], 650 to 700 °C for $Nd_2O_3$ [34], 600 to 700 °C for $Sm_2O_3$, $Eu_2O_3$, and $Yb_2O_3$ [35], and $Er_2O_3$ [36, 37].

Taylor [25] proposes the following general procedure: A completely dry $Cl_2$ stream is passed through a wash bottle containing $S_2Cl_2$ and then over the heated oxide. The temperature in the reaction space is raised over 1.5 h to ≈400 °C, then increased in 4 to 5 h to 700 to 800 °C. Any melting of the reaction products must be avoided before complete conversion of the oxide. The preparation specifications of Brauer [38, p. 855] and Wetzel [61, pp. 1002/3] are very similar: 1 to 2 g of the oxide are heated in a porcelain boat in a $Cl_2$ stream saturated with $S_2Cl_2$ at 30 to 40 °C within about 5 h from 400 °C up to 20 °C below the melting point of the chloride. For the trichlorides of Sm, Eu, Gd, Tb, Dy, and Y which melt below 700 °C this process is regarded as unsuitable.

The chlorination of the oxide with $S_2Cl_2$ alone is associated with a smaller evolution of heat, so that the danger of exceeding the melting point – even in a part of the reaction product – is diminished. Incandescence and baking are not observed, thus the formation of oxide chloride is effectively avoided. A large quantity of the chloride can be prepared in one charge. The reaction proceeds as $4M_2O_3+12S_2Cl_2\rightarrow 8MCl_3+6SO_2+18S$. Attack by the $S_2Cl_2$ stream begins at 250 °C, the temperature is increased within about 2.5 h to 500 °C. Finally the product depending on the quantity is held 1 to 2 h at 50 to 100 °C below the melting point, Bourion [39].

### Chlorination of Oxides with $SOCl_2$

Darzens, Bourion [40] were the first who used $SOCl_2$ to convert lanthanide oxides into the anhydrous chlorides. By heating the respective oxides in a stream of $SOCl_2$ vapor at 400 °C, they obtained pure $LaCl_3$ and $SmCl_3$, while repeated attempts to prepare $GdCl_3$ only yielded products containing considerable amounts of GdOCl. The reaction is postulated to occur according to the equation $M_2O_3+3SOCl_2\rightarrow 2MCl_3+3SO_2$. Side reactions are likely to occur because $SOCl_2$ disproportionates above 440 °C completely into $SO_2$, $Cl_2$, and $S_2Cl_2$. The authors felt that $SOCl_2$ offered no essential chlorination advantage over a mixture of $S_2Cl_2$ and $Cl_2$.

Chlorination with a two- to threefold excess of $SOCl_2$ in a sealed bomb tube at 250 to 300 °C resulted after 2 days in complete transformation of $Y_2O_3$ and $La_2O_3$ into powdery chlorides. $NdCl_3$ produced in this manner after 20 h at 300 °C still contained some unchanged oxide. In chlorinating $PrO_2$, thionyl chloride at the same time acts as a reducing reagent, producing $SO_2Cl_2$. After 10 h at 300 °C some unreacted oxide was found in this case too, Hecht et al. [41].

Machlan et al. [42] designed a special apparatus to prepare small samples of very pure $SmCl_3$ and $YbCl_3$ by the interaction of $SOCl_2$ vapor and the sesquioxides at 450 °C for 24 h in a simple and effective cycling system. The yields were 100% as indicated by the complete solubility of the products in water. $EuCl_3$ produced by this method contained sulfide impurities, Stubblefield, Eyring [43].

## Chlorination of Oxides with $CCl_4$ and $Cl_2/CCl_4$ Mixtures

Chlorination of oxides with $CCl_4$ represents one of the oldest methods for the winning of chlorides. The reaction begins below a red heat at 400 to 500 °C and proceeds according to the equations

$$M_2O_3+3CCl_4 \rightarrow 2MCl_3+3Cl_2+3CO \text{ and}$$
$$M_2O_3+3CCl_4 \rightarrow 2MCl_3+3COCl_2$$

(M = rare earth element). At the reaction temperatures $CCl_4$ is in part catalytically decomposed into $C_2Cl_6$ and $Cl_2$. Use of $Cl_2$ as a carrier gas increases the chlorinating power of $CCl_4$.

In his review concerning the applicability of $CCl_4$ which takes into account also older literature results, Taylor [25] states that it represents a nearly universal chlorinating agent. In addition to oxides and oxalates, the carbonates, sulfates, phosphates, and silicates can also be converted into the chlorides with $CCl_4$. It is also suitable for opening up various ores. However, several oxides are said not to be attacked by it, others are converted only into the oxide chlorides. Nevertheless, apart from $PmCl_3$, all trichlorides of the rare earth metals have been obtained by chlorination with $CCl_4$. The formation of the highly poisonous phosgene as a by-product makes special precautions necessary in the use of $CCl_4$; however, it provides in several cases a suitable protective atmosphere for the sensitive chlorides.

The reaction is mostly carried out by a procedure thoroughly tested by Miller et al. [45], if necessary with modifications. In this process the oxides $M_2O_3$ (M = Y, Gd, Dy, and Ho, or mixtures of them) are heated in steps in a Vycor tube in a $CCl_4$ flow 6 h at 500 °C, 15 h at 550 °C, and finally 8 h at 600 to 650 °C. The best results are obtained when the end temperature lies closely below the melting temperature of the corresponding trichloride, however, never above 650 °C. At higher temperatures two deleterious effects are observed: (1) inhibition of the reaction by a film of molten chloride forming a barrier between the chlorinating agent and the unreacted oxide or oxide chloride, and (2) pyrolytic decomposition of organic compounds formed in the course of the reaction which results in contamination of the sample. For $M_2O_3$ with M = Sc, Y, Er, Tm, Yb, or Lu Revzin [46] states that because of the relatively high melting points of the chlorides chlorination temperatures around 700 °C are possible, at which the reaction can be completed in 2 to 6 h. For the chlorides of the Ce group with their lower melting points, this preparation method is not recommended. In the apparatus described it is said that the oxides can be chlorinated in kilogram quantities. The final products contain $\leqq$0.2 to 0.5% oxide chloride, the yields are 97 to 98%.

The reaction temperatures used by other authors for preparation of the chlorides in laboratory quantities mostly do not exceed the upper limit of 650 °C recommended by Miller et al. [45]. Higher temperatures are sometimes employed after conclusion of the reactions for compacting the often very spongy end products; cf. divergent data by Novikov et al. [29, 33, 35, 47, 48]. The chlorination temperature also depends on the nature of the oxide. Porous or very finely divided starting materials, as for example obtained by decomposition of the oxalate, favor the progress and the yield of the often kinetically inhibited reaction. Special difficulties are encountered in the preparation of $EuCl_3$, which forms at a satisfactory rate only in the vicinity of its relatively low melting point. Partial melting and the resulting isolation of the oxide particles from the chlorinating agent and also the partial dissociation into $EuCl_2$ and $Cl_2$ can scarcely be avoided even with stepwise heating. A satisfactory preparation is then obtained only by post-chlorination of the powdered primary product, see p. 65. Complete reaction was achieved by Stubblefield et al. [49] in a cycling system at 450 °C within 24 to 48 h.

### Chlorination of Oxides with Phosgene

Phosgene converts the oxides of Y, La, and Ce into the anhydrous chlorides at about 600 °C according to $M_2O_3 + 3\,COCl_2 \rightarrow 2\,MCl_3 + 3\,CO_2$, Chauvenet [50]. The reaction yields pure chlorides, but because of the poisonous nature of $COCl_2$ demands extra precautionary measures preventing its wider use, Taylor [25].

### Chlorination of Oxides with $PCl_5$

$PCl_5$ was early proposed as a chlorinating agent [51], but the method has not proved very effective. The main drawback of $PCl_5$ is its low thermal stability, which does not permit enhancement of the reaction rate by simply raising the temperature.

A method has been proposed to use instead of pure $PCl_5$ its eutectic and azeotropic mixtures with $AlCl_3$, $ZrCl_4$, or $FeCl_3$. These mixtures, unlike $PCl_5$, exhibit a large temperature range of the liquid state at normal pressure and, in addition, a significant solvent power for rare earth chlorides, which also contributes in speeding up the chlorination rate considerably. A special quartz apparatus was constructed which by chlorination of the oxides with a somewhat hypereutectic $PCl_5$-$AlCl_3$ mixture (46 to 58 wt % $PCl_5$) allows preparation of 300 to 400 g of chloride within 2 to 3 h. The purified chlorination mixture is held at 200 to 250 °C while the oxide is added in portions of 5 to 10 g. Reaction takes place at the surface of the melt. After all the oxide has been added, the liquid reaction products are distilled off through a side-tube under a vacuum of $\approx$10 Torr; the temperature is gradually raised to 450 to 500 °C. The friable chloride powder left in the chlorinator is subjected to a vacuum distillation at 0.1 Torr to bring down the P and Al content from an initial 0.1 to 0.5 wt % to below 0.001 wt % of each of these impurities. The distillation temperature for $YCl_3$ is 900 °C, Nisel'son et al. [52], and 900 to 950 °C for other ($LaCl_3$ to $GdCl_3$) rare earth chlorides, Nisel'son, Lyzlov [53]; see also Nisel'son et al. [54].

Temperature and thermal characteristics of the chlorination process were studied using a tenfold excess of a eutectic $PCl_5$-$AlCl_3$ mixture (43.5 wt % $PCl_5$). Thermograms taken at a heating rate of 10 K/min show that intense chlorination starts at 90 to 170 °C. They are characterized by two thermal effects, which can be explained by the rare earth chlorides being formed by a two-stage reaction via the oxide chlorides. Formation of oxide chlorides from $Y_2O_3$ and $Sc_2O_3$ does not take place, however. The chlorides from $TbCl_3$ to $LuCl_3$, and also $CeCl_3$, are formed from the corresponding oxide chlorides only at temperatures close to the normal boiling point of the chlorinating mixture

($\approx$380 °C). Investigations of the stoichiometry of the chlorination reaction by the example of $Y_2O_3$ furnished evidence that in certain circumstances the $POCl_3$ initially formed and the $AlCl_3$ may exert a chlorinating action in addition to that of $PCl_5$, Nisel'son et al. [52].

## Chlorination of Oxides with Other Reagents

Chlorinated hydrocarbons other than $CCl_4$ should also be good chlorinating reagents, but only few appear to have been tried. Freeman, Smith [62] failed to prepare rare earth chlorides with hexachloropropene.

Amine hydrochlorides $RNH_2 \cdot HCl$ and $R_2NH \cdot HCl$ ($R = C_6H_5$, $CH_3$, $C_2H_5$, etc.) with melting points <200 °C are used to prepare high-purity lanthanide chlorides in a single-step process. They simultaneously serve as solvent and chlorinating agent for the oxide. A double salt is formed from the reaction. The reaction mixture is heated to vaporize the solvent and decompose the double salt, leaving the anhydrous chloride product, Baldwin [63].

In a fused salt bath made up of chlorides of alkali metals, alkali earths, $NH_4^+$, $Zn^{2+}$, or $Fe^{3+}$, the rare earth oxides (or carbonates, sulfides, or sulfates) react with chlorine introduced as such or as supplied by a Cl donor ($FeCl_3$ and S chlorides) to give the corresponding chloride at low temperature and ambient pressure. It is necessary that the metal chloride formed be soluble in the fused salt mixture. Thus, $La_2O_3$ was reacted with Cl in a bath of $FeCl_3$ and NaCl without any reductant at 420 °C. The La all reacted to produce water-soluble $LaCl_3$ while a corresponding amount of Fe oxide was formed, Kruesi [64].

References:

[1] H.C. Oersted (Overs. Kgl. Danske Videnskab. Selskabs Forh. **1824** 251 from [25]). – [2] O. Pettersson (Z. Anorg. Allgem. Chem. **4** [1893] 1/9). – [3] Bureau of Mines (USBM-U-476 [1959] 1/9; N.S.A. **14** [1960] No. 1766). – [4] Bureau of Mines (USBM-U-494 [1959] 1/13; N.S.A. **14** [1960] No. 1766). – [5] T.T. Campbell, R. Mussler, G. Robidart, G. Engel (in: Bureau of Mines, USBM-U-550 [1958] 1/36; N.S.A. **14** [1960] No. 551).

[6] T.T. Campbell, F.E. Block, R.E. Mussler, G.B. Robidart (USBM-RI-5880 [1961] 1/29; C.A. **56** [1962] 5613). – [7] F.E. Block, T.T. Campbell (in: F.H. Spedding, A. Daane, The Rare Earths, New York - London 1961, pp. 89/101). – [8] A.P. Samodelov (Zh. Neorgan. Khim. **10** [1965] 1735/7; Russ. J. Inorg. Chem. **10** [1965] 946/8). – [9] H. Schäfer (Z. Anorg. Allgem. Chem. **403** [1974] 116/26). – [10] M. Strzyzewska, W. Szklarski, W. Szczepaniak, A. Bogacz (Rudy Metale Niezelazne **23** [1978] 444/8; C.A. **90** [1979] No. 139667).

[11] J.B. Reed, B.S. Hopkins, L.F. Audrieth (J. Am. Chem. Soc. **57** [1935] 1159/60). – [12] J.B. Reed, B.S. Hopkins, L.F. Audrieth (Inorg. Syn. **1** [1939] 28/33). – [13] D.H. West, B.S. Hopkins (J. Am. Chem. Soc. **57** [1935] 2185/6). – [14] A. Brukl (Angew. Chem. **52** [1939] 151/4). – [15] J.L. Moriarty (J. Chem. Eng. Data **8** [1963] 422/4; C. A. **59** [1963] 8145).

[16] A.S. Dworkin, M.A. Bredig (J. Phys. Chem. **67** [1963] 2499/500). – [17] A.S. Dworkin, H.R. Bronstein, M.A. Bredig (J. Phys. Chem. **67** [1963] 2715/7). – [18] N.H. Kiess (J. Res. Natl. Bur. Std. A **67** [1963] 343/5). – [19] Ya.I. Ivashentsev, M.I. Stoyanova (Obogashch. Met. Tsvetn. Metal. **1969** No. 3, pp. 103/12 from C.A. **75** [1971]

No. 91571). – [20] J. Shiokawa, T. Ishino (Denki Kagaku **32** [1964] 109/16 from C. A. **62** [1965] 6111/2).

[21] W.E. Domning, D.L. Schechter (U.S. 3153570 [1960/64] from C.A. **62** [1965] 257). – [22] G.A. Krestov, V.A. Kobenin, S.V. Semenovskii (Izv. Vysshikh Uchebn. Zavedenii Khim. Khim. Tekhnol. **14** [1971] 462/4 from C.A. **75** [1971] No. 29414). – [23] W.R. Hodgkinson (J. Soc. Chem. Ind. [London] **33** [1914] 445/6). – [24] C. Matignon, F. Bourion (Compt. Rend. **138** [1904] 631/3). – [25] M.D. Taylor (Chem. Rev. **62** [1962] 503/11).

[26] F. Bourion (Compt. Rend. **145** [1907] 62/4). – [27] F. Bourion (Ann. Chim. Phys. [8] **20** [1910] 547/73). – [28] G. Urbain, F. Bourion (Compt. Rend. **153** [1911] 1155/8). – [29] O.G. Polyachenok, G.I. Novikov (Zh. Neorgan. Khim. **9** [1964] 773/7; Russ. J. Inorg. Chem. **9** [1964] 429/31). – [30] G.I. Novikov, O.G. Polyachenok (Usp. Khim. **33** [1964] 732/47; Russ. Chem. Rev. **33** [1964] 342/50).

[31] K.E. Johnson, J.R. Mackenzie (J. Inorg. Nucl. Chem. **32** [1970] 43/8). – [32] A. Voigt, W. Biltz (Z. Anorg. Allgem. Chem. **133** [1924] 277/305). – [33] G.I. Novikov, O.G. Polyachenok (Zh. Neorgan. Khim. **8** [1963] 1053/9; Russ. J. Inorg. Chem. **8** [1963] 545/50). – [34] O.G. Polyachenok, G.I. Novikov (Zh. Neorgan. Khim. **8** [1963] 1567/73; Russ. J. Inorg. Chem. **8** [1963] 816/20). – [35] O.G. Polyachenok, G.I. Novikov (Zh. Neorgan. Khim. **8** [1963] 2631/4; Russ. J. Inorg. Chem. **8** [1963] 1378/81).

[36] F.G. Gavryuchenko, G.I. Novikov (Zh. Neorgan. Khim. **11** [1966] 2172/4; Russ. J. Inorg. Chem. **11** [1966] 1165/6). – [37] G.I. Novikov, F.G. Gavryuchenko (Zh. Neorgan. Khim. **10** [1965] 1668/74; Russ. J. Inorg. Chem. **10** [1965] 909/13). – [38] G. Brauer (Handbuch der Präparativen Anorganischen Chemie, Stuttgart 1954, pp. 841/64). – [39] F. Bourion (Ann. Chim. Phys. [8] **21** [1910] 49/131; Compt. Rend. **148** [1909] 170/1). – [40] Z. Darzens, F. Bourion (Compt. Rend. **153** [1911] 270/2).

[41] H. Hecht, G. Jander, H. Schlapmann (Z. Anorg. Allgem. Chem. **254** [1947] 255/64). – [42] G.R. Machlan, C.T. Stubblefield, L. Eyring (J. Am. Chem. Soc. **77** [1955] 2975/8). – [43] C.T. Stubblefield, L. Eyring (J. Am. Chem. Soc. **77** [1955] 3004/5). – [44] C.W. Koch, A. Broido, B.B. Cunningham (J. Am. Chem. Soc. **74** [1952] 2349/51). – [45] J.F. Miller, S.E. Miller, R.C. Himes (J. Am. Chem. Soc. **81** [1959] 4449/51).

[46] G.E. Revzin (Metody Poluch. Khim. Reaktivov Prep. **1967** No. 16, pp. 24/9 from C.A. **69** [1968] No. 48783). – [47] G.I. Novikov, A.K. Baev (Zh. Neorgan. Khim. **7** [1962] 1349/52; Russ. J. Inorg. Chem. **7** [1962] 694/6). – [48] O.G. Polyachenok, G.I. Novikov (Zh. Neorgan. Khim. **8** [1963] 2819/21; Russ. J. Inorg. Chem. **8** [1963] 1479/80). – [49] C.T. Stubblefield, J.L. Rutledge, R. Phillips (J. Phys. Chem. **69** [1965] 991/5). – [50] M.E. Chauvenet (Compt. Rend. **152** [1911] 87/9).

[51] C. Matignon (Ann. Chim. Phys. [8] **8** [1906] 364/86). – [52] L.A. Nisel'son, Yu.N. Lyzlov, K.V. Tret'yakova (Zh. Neorgan. Khim. **20** [1975] 2362/7; Russ. J. Inorg. Chem. **20** [1975] 1309/12). – [53] L.A. Nisel'son, Yu.N. Lyzlov (Dokl. Akad. Nauk SSSR **220** [1975] 608/9; Dokl. Chem. Proc. Acad. Sci. USSR **220/226** [1975] 103/4). – [54] L.A. Nisel'son, K.V. Tret'yakova, Yu.N. Lyzlov, F.N. Kozlov, Yu.A. Raikov, S.I. Timofeev (U.S.S.R. 381606 [1971/73] from C.A. **79** [1973] No. 147817). – [55] D.H. Templeton, G.F. Carter (J. Phys. Chem. **58** [1954] 940/4).

[56] F. Weigel, V. Scherer (Radiochim. Acta **7** [1967] 40/6). – [57] V. Scherer (BMWF-FBK-68-03 [1968] 1/91; C.A. **69** [1968] No. 102592). – [58] V.F. Wishnevsky (BMWF-FBK-68-42 [1968] 1/121; N.S.A. **23** [1969] No. 21678). – [59] O. Popp (Lie-

bigs Ann. Chem. **131** [1864] 179/201). – [60] G. Haeseler, F. Matthes (J. Less-Common Metals **9** [1965] 133/51).

[61] K. Wetzel (in: G. Brauer, Handbuch der Präparativen Anorganischen Chemie, 2nd Ed., Vol. 2, Stuttgart 1962, pp. 982/1013). – [62] J.H. Freeman, M.L. Smith (J. Inorg. Nucl. Chem. **7** [1958] 224/7). – [63] W.H. Baldwin (U.S. 3821356 [1973/74] from C.A. **81** [1974] No. 123815). – [64] P.R. Kruesi (U.S. 4179492 [1977/79] 1/6 from C. A. **92** [1980] No. 79016). – [65] W. Klemm, W. Biltz (Z. Anorg. Allgem. Chem. **152** [1926] 225/34).

### 11.2.6.1.4 Preparation from the Hydrated Chlorides

Summary

The hydrated chlorides are easily obtained by dissolution of the oxides in aqueous hydrochloric acid and concentration of the resulting solutions by evaporation or desiccation. Thermal dehydration of the hydrates seems to be a very practicable and convenient method to produce anhydrous chlorides, but cannot be effected without special precautions. Particular difficulties are experienced with the removal of the last mole of the water of crystallization remaining after careful dehydration at about 100 °C. Oxide chlorides are readily formed, even when the heat treatment is performed under $N_2$ or in a vacuum. In general, dehydration becomes increasingly difficult with increasing atomic weight of the lanthanide.

The dehydration process is greatly promoted by the presence of a high chloride activity. A variety of dehydration reagents have been used to prevent hydrolysis and in many cases are claimed to produce pure anhydrous chlorides. The most important of these are discussed below.

#### Dehydration in Presence of Hydrogen Chloride

The application of this method for the preparation of anhydrous rare earth chlorides $MCl_3$ (M=Y, Pr, Nd, Sm, Yb) seems to have been first described by Matignon [1 to 4]. The hydrated salts were heated at 100 to 110 °C until sufficient water had been removed to form the monohydrate. Then the temperature was slowly raised to 160 to 180 °C where the rest of the water was released. The procedure of later investigators (see for example [5 to 8]) follows the same route. The bulk of the crystal water is removed at temperatures as low as possible. Any kind of melting, partial or complete, has to be carefully avoided, since it inevitably causes the formation of basic salts which will be occluded in the melts and protected from further attack by HCl. In the end the anhydrous material is usually heated at higher temperatures (300 to 400 °C), retaining the HCl atmosphere, and in some cases fused to minimize the surface area. The final product is allowed to cool in carefully purified $N_2$ and thus freed from adsorbed HCl. Detailed descriptions of procedure and apparatus have been given by Kleinheksel, Kremers [9], Jantsch et al. [7], Brauer [10, pp. 841/2], Wetzel [11, p. 1002]. The method has been widely used to prepare anhydrous chlorides pure enough for chemical determinations of the atomic weights of the rare earth elements, see for example [12 to 16].

Harrison [17] has claimed that rare earth chlorides free of oxide chlorides can be obtained only if the final stages of dehydration are conducted at temperatures in excess of their melting points. When prepared in this manner (details of the technique are not reported), the chlorides are 100% volatile in a vacuum; preparations according to the method of [9] are said to be less than 50% volatile.

Kleinheksel, Kremers [9] find the following temperature ranges to be adequate for the removal of the first 5 (or 6) molecules of crystal water ($t_1$) and for the last molecule ($t_2$) of the hydrated chlorides $MCl_3 \cdot 6$ (or 7) $H_2O$:

| | | | | | |
|---|---|---|---|---|---|
| M . . . . | Y | La | Ce | Pr | Nd |
| $t_1$ in °C . | 110 to 120 | 85 to 90 | 90 to 110 | 110 to 115 | 120 to 125 |
| $t_2$ in °C . | 190 to 210 | 125 to 135 | 165 to 190 | 180 to 195 | 155 to 165 |
| M . . . . | Sm | Dy | Ho | Tm | |
| $t_1$ in °C . | 100 to 110 | 110 to 130 | 95 to 105 | 100 to 115 | |
| $t_2$ in °C . | 180 to 200 | 200 to 230 | 195 to 225 | 190 to 205 | |

With the necessity of keeping the temperature well below 100 °C during the removal of the first five molecules of water of hydration, dehydration is very slow and usually takes several days for completion. Operating at reduced pressure shortens the dehydration time to about 12 h. A gas flow is maintained through the reaction tube at the rate of about 1 mL/s and about 250 Torr. For the first 3 or 4 h the temperature is maintained at 70 °C, then gradually raised to 160 °C, held at this level for 1 to 2 h, and then raised slowly to 250 °C. At this point all of the water of hydration is removed [9]. According to Spedding, Daane et al. [18 to 20] the rare earth oxides, or for Ce the cerium ammonium nitrate, are dissolved in an excess of concentrated hydrochloric acid, and the resulting solution is evaporated to a boiling point of 125 to 130 °C. On pouring of the solution into a large porcelain dish and stirring while cooling, the hydrated chlorides crystallize as powders. These are dehydrated by slow heating in a stream of dry HCl to 400 °C at a pressure of 40 to 60 Torr. The dehydration of small amounts of the hydrates is complete in as short a time as 1.5 h at 450 °C and 30 Torr, Swanson et al. [21]. For the preparation of 1.5- to 2 g-amounts of the anhydrous chlorides, working under reduced pressure is found to be unnecessary, even with the chlorides of the less basic Eu and Sm, for which average dehydration times of 5 to 8 h are required, Jantsch et al. [22]. A modified procedure avoids the somewhat cumbersome adjustment of a constant reduced pressure: A steady stream of HCl at ordinary pressure is passed over 1 to 2 g of the hydrated chloride while the temperature is raised to 150 °C over a period of 1 to 1.5 h. The water evolved is removed by reducing the pressure inside the reaction tube for a few minutes before resuming passage of HCl. Heating is continued at the same rate, and water is again removed at 350 °C. This temperature is then maintained while the flow of HCl is continued for a period of 12 h. Dehydration is completed by raising the temperature to 400 to 500 °C for 2 h. Samples of $NdCl_3$, $SmCl_3$, and $TmCl_3$ gave clear solutions in water and showed Cl contents that corresponded within experimental error (0.2% Cl) to theoretical values. With $EuCl_3$ and $YbCl_3$, however, marked deficiencies in Cl content (3 to 6%) were apparent and turbid solutions were obtained in water, Johnson, Mackenzie [23].

The deleterious effects of $O_2$, $H_2O$, and melting during the process may be avoided by either adding an auxiliary agent, i.e., $NH_4Cl$, to the chloride to be dehydrated [14, 24 to 27] or mixing an auxiliary agent, i.e., $S_2Cl_2$ or $Cl_2$, with the HCl [4, 28].

#### Dehydration in Presence of Ammonium Chloride

$NH_4Cl$ is often used as an auxiliary reagent in the dehydration of hydrated chlorides with HCl, see above, and it has early been applied as the sole dehydrating reagent too, see [29 to 32]. The presence of $NH_4Cl$ prevents hydrolysis and oxidation of the lanthanide chlorides, oxide impurities in the starting materials are chlorinated. Because

$NH_4Cl$ is a more effective chlorinating agent than HCl, considerably less oxide chloride is present in the final chloride. Excess $NH_4Cl$ is easily removed by sublimation. Final dehydration and removal of $NH_4Cl$ are conducted under a vacuum or in a stream of $Cl_2$, HCl, or inert gas.

A general method suitable for the preparation of all anhydrous rare earth chlorides was developed by Taylor, Carter [33]. It consists of evaporating to dryness on a hot plate at ≦200 °C an HCl solution containing $NH_4Cl$ and the desired lanthanide chloride (5 to 10 g) in a mole ratio of 6:1, heating the remaining salt mixture slowly under vacuum to 200 °C to drive off the water of crystallization and finally raising the temperature to 300 °C or higher to expel the $NH_4Cl$. With $EuCl_3$ the final temperature must not be too high, or else a mixture of $EuCl_3$ and $EuCl_2$ will be obtained. The end products are completely free of $H_2O$ and $NH_4Cl$ and dissolve in water to yield a perfectly clear or at most slightly cloudy solution. The yields are generally greater than 60% and may exceed 95%. Most material is lost by passing over along with the subliming $NH_4Cl$. The purity is between 99.2 and 100%. The method is claimed to be the only one which is capable of producing absolutely pure anhydrous chlorides in quantity [33]. The entrainment of the rare earth chlorides by escaping gases occurs only if the temperature is increased too rapidly, Haschke [34]. According to Shimazaki, Niwa [35] the oxides $M_2O_3$ (M = La, Pr, Nd) are dissolved in the least possible quantity of hydrochloric acid, and $NH_4Cl$ is added in an amount corresponding to a molar ratio $NH_4Cl:M_2O_3 = 2:1$. The resulting solution is evaporated first on a water bath, then on a sand bath up to the point of fuming. The partially dehydrated, solid salt mixture is pulverized and in batches of ≈2 g slowly heated in a small silica container in high vacuum up to 350 °C, held at this temperature for 2 h, and then cooled. For the preparation of $CeCl_3$ the carbonate is used as starting material. The anhydrous chlorides are obtained in a finely grained form.

Material for growing optically pure single crystals is obtained by dissolving 10 g oxide (99.99%) plus 15 g $NH_4Cl$ (mole ratio $NH_4Cl:M^{3+} \approx 4:1$) in 100 mL reagent grade HCl on a hot plate and gently boiling the solution to dryness. The powdery residue is heated at 0.5 Torr in steps first for 6 h at 90 °C, then for another 6 h at 180 °C, and finally for 12 h at 300 °C. The relatively simple drying apparatus is described in detail. It is pointed out that the temperature and reaction times are not critical. At pressures significantly above 0.5 Torr, higher temperatures and longer reaction times are required for complete sublimation of the $NH_4Cl$; at pressures significantly below 0.5 Torr, $NH_4Cl$ may be carried throughout the entire vacuum line. The anhydrous chloride is melted down and purified by vacuum distillation (see p. 71). If a dopant ion is desired, the corresponding oxide is added to the starting mixture of oxide and $NH_4Cl$. The oxides of the heavier lanthanides Gd to Lu do not dissolve in HCl until the mixture is brought to boiling. If after a few minutes of boiling the solutions become cloudy, more acid should be added. $CeO_2$ and $Tb_4O_7$ dissolve in HCl only upon repeated addition of equal amounts of HCl and $H_2O_2$ under boiling conditions, Cox, Fong [36].

The following variant of the method has been used by experimenters of the U.S. Bureau of Mines: The oxides $M_2O_3$ (M = Y, La, Gd) are slowly dissolved in warm 6 N HCl, and 3 to 4 mol $NH_4Cl$ per mol $M_2O_3$ is added. Excess water is removed by heating the solution until the normal boiling temperature reaches 130 to 135 °C. This temperature range is quite critical because oxide chlorides are formed if the temperature exceeds 135 °C. The remaining thick, syrupy liquid upon cooling to room temperature solidifies to a crystalline mixture of $NH_4Cl$ and hydrated rare earth chloride. The bulk of the water is removed by heating from 100 to 120 °C, while the apparatus is evacuated with a water aspirator. The pressure varies from 20 to 100 Torr during this period. After the evolution of water has ceased, the water aspirator is replaced by a mechanical pump placed in series with a dry ice-acetone cold trap. The temperature is gradually increased from 120 to 350 °C, so that the absolute pressure in the apparatus does not exceed 0.5 Torr. Water is condensed in the cold trap while the $NH_4Cl$ is sublimed and condensed on a water condenser. A further purification of the crude chlorides can be effected by vacuum distillation in the manner described for $YCl_3$ (see "Rare Earth Elements" C4 b, Section 11.2.11.2.4.1), Campbell et al. [37, 38]. Anhydrous chlorides were prepared in this manner in batch sizes of up to 2 kg or up to 20 kg in the case of $YCl_3$. Complete dehydration at the low pressures employed required several days, but resulted in chloride suitable for the production of high-purity metals (with <0.1% O), Block, Campbell [39].

In order to hasten the evaporation of the solution of $NH_4Cl$ (0.75 mol) and rare earth oxide (0.05 mol) in concentrated hydrochloric acid (130 mL) a stream of $N_2$ may be directed over the mixture, while it is stirred on a hot plate. In view of the extensive loss of water occurring at the fairly low temperatures, extreme caution has to be taken in the latter stages of the evaporation to keep the temperature of the ever-thickening slurry below 90 °C. The moist solid is gradually heated to 200 °C over 4.5 h under vacuum, and this temperature is maintained for 1 h, although most of the water is removed at lower temperatures. The temperature is then raised to 427 °C over about 7 h, during which time the $NH_4Cl$ sublimes off leaving the trichloride as a polycrystalline material, Sommers, Westrum [40, 41]. The method is inadequate for $EuCl_3$. A chlorine deficiency corresponding to the composition $EuCl_x$ with $x=2.989 \pm 0.003$ was found in the initial preparation despite the fact that heating was stopped at 350 °C. After the product was heated at 360 °C under 1 atm of $Cl_2$, the composition was found to be $EuCl_x$ with $x=2.999 \pm 0.002$, Sommers, Westrum [41, 42].

A simplified, quick procedure has been applied for the preparation of $MCl_3$ (M = Gd, Dy, Er, Yb, and Lu) in amounts ranging from 50 to 300 mg for use in charging an isotope separator. The dissolution of the oxides in strong HCl is speeded up by the use of an infrared lamp and by pipetting a small amount of $H_2O_2$ into the mixture. Evaporation of the solution is also effected under the infrared lamp. For final dehydration and removal of the $NH_4Cl$ a Bunsen burner and a simple Pyrex test tube are used. The yield for conversion of the oxides into the chlorides is estimated to be 70 to 90% depending on circumstances, Jensen [43]. Increasing the molar ratio $NH_4Cl:MCl_3$ to 11.5:1 and lowering the final pressure to 0.1 to 0.05 Torr reduces the dehydration time to 3 to 3.5 h for batches of 3 to 4 g chloride. The quality of the final vacuum influences the contamination by oxide chloride, which under the just mentioned conditions was below 0.5% for the chlorides of La to Sm, about 0.5% for $EuCl_3$ and $GdCl_3$, and about 1 to 1.5% for the chlorides of Tb to Lu, Rossmanith [44].

The removal of $H_2O$ and $NH_4Cl$ from the salt mixtures may also be effected by heating in a stream of dry nitrogen for several hours. Good results were obtained with $NdCl_3$, preparations of $SmCl_3$ and $GdCl_3$ contained varying amounts of oxide and oxide chloride, Richards [45]. Heating mixtures of 1 mol $MCl_3 \cdot nH_2O$ (M = La, Ce, Pr, Nd) and 6 mol $NH_4Cl$ in a stream of inert gas at 200 to 400 °C results in anhydrous chlorides containing $\leq$3% oxide chloride, Mikheev et al. [46]. The dehydration of hydrated chlorides in the presence of $NH_4Cl$ in a $Cl_2/N_2$ stream has been studied by Quang et al. [47].

**Dehydration in Presence of Thionyl Chloride**

Several lanthanide chlorides have been prepared by dehydrating the hydrates under an atmosphere of $SOCl_2$. The procedure involves adding an excess of $SOCl_2$ to the finely ground hydrated chloride and refluxing the mixture. Care has to be taken not to overheat the flask as this causes some decomposition of the $SOCl_2$ and formation of oxide chloride. Deliquescent chlorides (e.g., $CeCl_3 \cdot nH_2O$) which cannot readily be ground are dried by treating with a small quantity of $SOCl_2$ before grinding. Concentrated solutions of the chlorides ($LaCl_3$, $GdCl_3$, $DyCl_3$, $ErCl_3$) in HCl can also be treated directly with an excess of $SOCl_2$ and refluxed as above. After the dehydration is complete, the excess $SOCl_2$ is removed by distillation or decantation. The difficulty in dehydrating rare earth chloride hydrates by this technique increases with increasing atomic weight of the lanthanides; only 1 h is required to dehydrate $LaCl_3$, whereas $ErCl_3$ requires 110 h. Nevertheless, the method is reported to be simple, effective, and suitable for large scale preparations, Freeman, Smith [48]. According to Johnson, Mackenzie [23], the elimination of water from hydrated samarium chloride by prolonged reflux distillation with thionyl chloride was unsatisfactory in that the product showed evidence of hydrolysis and was difficult to free from all traces of the reagent. Taylor [49] also reports, that the method did not prove especially satisfactory in his hands.

**Dehydration in Presence of Other Reagents**

Anhydrous chlorides $MCl_3$ (M=Y, La, Pr, Nd, Sm) are easily prepared by heating the hydrated chlorides in a $Cl_2/S_2Cl_2$ or $HCl/S_2Cl_2$ stream. The solid material remaining on evaporation of aqueous solutions on a sand bath at 130 to 140 °C can be used as starting material instead of isolated hydrates, Matignon [4]. $EuCl_3$ is obtained by predrying the hydrated salt at 100 °C and slowly heating it further in a $Cl_2/S_2Cl_2$ stream, avoiding fusion, Urbain, Bourion [50].

The crystal hydrates of the chlorides of La to Tb and Ho may also be dehydrated by heating in $CCl_4$ vapors, possibly mixed with $N_2$. The following process conditions are proposed: heating at 100 and 200 (to 250) °C for 1 h at each temperature and at 450 to 500 °C for 1 to 1.5 h. The yield is 95 to 96%. The oxide chloride content amounts to 0.5 to 1.0%, Revzin [51]. Alternatively, the evaporation residue of the aqueous chloride solution is heated in air to 500 °C, whereby the oxide chloride is formed. In a final stage at 550 to 650 °C in a $CCl_4$ atmosphere, the oxide chloride is converted into the anhydrous chloride, Repin [52]. A stepwise dehydration alternately in a vacuum and in a $Cl_2$ atmosphere at reduced pressure is recommended by Strzyzewska et al. [53].

Good results were also obtained by treating the hydrated chlorides with either $COCl_2$ or CO and $Cl_2$, Heap, Newberry [54]. Acetic anhydride and acetyl chloride did not stand a test as dehydration reagents, Taylor [49].

References:

[1] C. Matignon (Compt. Rend. **133** [1901] 289/91). – [2] C. Matignon (Compt. Rend. **134** [1902] 427/9). – [3] C. Matignon (Compt. Rend. **134** [1902] 1308/10). – [4] C. Matignon (Ann. Chim. Phys. [8] **8** [1906] 364/86). – [5] G. Jantsch, N. Skalla, H. Jawurek (Z. Anorg. Allgem. Chem. **201** [1931] 207/20).

[6] G. Jantsch, H. Jawurek, N. Skalla, H. Gawalowski (Z. Anorg. Allgem. Chem. **207** [1932] 353/67). – [7] G. Jantsch, N. Skalla, H. Grubitsch (Z. Anorg. Allgem. Chem. **212** [1933] 65/83). – [8] H. Bommer, E. Hohmann (Z. Anorg. Allgem. Chem. **248** [1941] 373/82). – [9] J.H. Kleinheksel, H.C. Kremers (J. Am. Chem. Soc. **50** [1928] 959/67). – [10] G. Brauer (Handbuch der Präparativen Anorganischen Chemie, Stuttgart 1954, pp. 841/64).

[11] K. Wetzel (in: G. Brauer, Handbuch der Präparativen Anorganischen Chemie, 2nd Ed., Vol. 2, Stuttgart 1962, pp. 982/1013). – [12] G.P. Baxter, O.J. Stewart (J. Am. Chem. Soc. **37** [1915] 516/36). – [13] H.C. Fogg, C. James (J. Am. Chem. Soc. **44** [1922] 307/16). – [14] O. Hönigschmid, H. Holch (Z. Anorg. Allgem. Chem. **177** [1928] 91/101). – [15] O. Hönigschmid, H. Auer von Welsbach (Z. Anorg. Allgem. Chem. **165** [1927] 284/8).

[16] O. Hönigschmid, A. Meuwsen (Z. Anorg. Allgem. Chem. **140** [1924] 341/56, 348). – [17] E.R. Harrison (J. Appl. Chem. [London] **2** [1952] 601/2). – [18] F.H. Spedding, A.H. Daane (J. Am. Chem. Soc. **74** [1952] 2783/5). – [19] F.H. Spedding, H.A. Wilhelm, W.H. Keller, D.H. Ahmann, A.H. Daane, C.C. Hack, R.P. Ericson (Ind. Eng. Chem. **44** [1952] 553/6). – [20] A.H. Daane, F.H. Spedding (J. Electrochem. Soc. **100** [1953] 442/4).

[21] H.E. Swanson, M.C. Morris, R.P. Stinchfield, E.H. Evans (NBS-Mon-25-1 [1962] 1/56). – [22] G. Jantsch, H. Grubitsch, F. Hoffmann, H. Alber (Z. Anorg. Allgem. Chem. **185** [1930] 49/64). – [23] K.E. Johnson, J.R. Mackenzie (J. Inorg. Nucl. Chem. **32** [1970] 43/8). – [24] C.W. Balke (U.S. 1289079 [1919] from C.A. **1919** 773). – [25] I.S. Morozov, B.G. Korshunov (Zh. Neorgan. Khim. **1** [1956] 2606/12; Russ. J. Inorg. Chem. **1** No. 11 [1956] 180/6).

[26] W. Muthmann, L. Stützel (Ber. Deut. Chem. Ges. **32** [1899] 3413/9). – [27] I.Z. Babievskaya, I.S. Morozov (Zh. Neorgan. Khim. **17** [1972] 571/2; Russ. J. Inorg. Chem. **17** [1972] 300/1). – [28] C. Matignon (Compt. Rend. **140** [1905] 1181/3). – [29] M.C. Marignac (Ann. Chim. Phys. [3] **38** [1853] 148/77, 160/1). – [30] R. Hermann (J. Prakt. Chem. **82** [1861] 385/408).

[31] A.G. Duboin (Ann. Sci. Ecole Normale Superieure [3] **5** [1888] 65/from [49]). – [32] W.R. Hodgkinson (J. Soc. Chem. Ind. [London] **33** [1914] 445/6). – [33] M.D. Taylor, C.P. Carter (J. Inorg. Nucl. Chem. **24** [1962] 387/91). – [34] J.M. Haschke (in: K.A. Gschneidner, L.R. Eyring, Handbook on the Physics and Chemistry of Rare Earths, Vol. 4, Amsterdam - New York - Oxford 1979, pp. 89/151). – [35] E. Shimazaki, K. Niwa (Z. Anorg. Allgem. Chem. **314** [1962] 21/34).

[36] D.E. Cox, F.K. Fong (J. Crystal Growth **20** [1973] 233/8). – [37] T.T. Campbell, F.E. Block, R.E. Mussler, G.B. Robidart (USBM-RI-5880 [1961] 1/29; C.A. **56** [1962] 5613). – [38] T.T. Campbell, R. Mussler, G. Robidart, G. Engel (USBM-U-550 [1958] 1/36; N.S.A. **14** [1960] No. 551). – [39] F.E. Block, T.T. Campbell (in: F.H. Spedding, A.H. Daane, The Rare Earths, New York - London 1961, pp. 89/101). – [40] J.A. Sommers, E.F. Westrum (J. Chem. Thermodyn. **8** [1976] 1115/36, 1116).

[41] J.A. Sommers, E.F. Westrum (Proc. 12th Rare Earth Res. Conf., Vail, Colo., 1976, Vol. 1, pp. 433/42, 435). – [42] J.A. Sommers, E.F. Westrum (J. Chem. Thermodyn. **9** [1977] 1/26, 2). – [43] B.S. Jensen (Nucl. Instr. **1** [1957] 323). – [44] K. Rossmanith (Monatsh. Chem. **100** [1969] 1484/8). – [45] C.E.C. Richards (private communication from [48]).

[46] N.B. Mikheev, A.N. Kamenskaya, N.A. Konovalova, T.A. Zhilina (Zh. Neorgan. Khim. **22** [1977] 1761/6; Russ. J. Inorg. Chem. **22** [1977] 955/8). – [47] Nguyen Thanh Quang, Chu Van Chinh, Pham Ngoc Tien (Tap Chi Hoa Hoc **14** No. 1 [1976] 1/4 from C. A. **87** [1977] No. 160916). – [48] J.H. Freeman, M.L. Smith (J. Inorg. Nucl. Chem. **7** [1958] 224/7). – [49] M.D. Taylor (Chem. Rev. **62** [1962] 503/11). – [50] G. Urbain, F. Bourion (Compt. Rend. **153** [1911] 1155/8).

[51] G.E. Revzin (Metody Poluch. Khim. Reaktivov Prep. **1967** No. 16, pp. 124/9 from C.A. **69** [1968] No. 48783). – [52] S.A. Repin (Redkozemel. Elementy Akad. Nauk SSSR Inst. Geokhim. Analit. Khim. **1963** 71/4 from C.A. **61** [1964] 3892). – [53] M. Strzyzewska, W. Szklarski, W. Szczepaniak, A. Bogacz (Rudy Metale Niezelazne **23** [1978] 444/8 from C.A. **90** [1979] No. 139667). – [54] W. Heap, E. Newberry (Brit. 130626 [1918]; Brit. 131039 [1918]; U.S. 1331257 [1920]; Can. 200907 [1920] from C.A. **1920** 99, 210, 1193, 2399).

### 11.2.6.1.5 Preparation from Other Starting Materials

#### Preparation from Compounds with Inorganic Anions

See pp. 53 and 146 for the formation of the anhydrous chlorides from the oxide chlorides.

Sulfides $M_2S_3$ (M = La, Ce, Pr, Nd) are readily converted into the chlorides by HCl gas at dull red heat, Muthmann, Stützel [1]. As early as 1827, Mosander [2] prepared $CeCl_3$ by heating the sulfide in a $Cl_2$ stream. According to Matignon [3, p. 367] only very pure sulfides may be used, which are not easily obtained; trace impurities of oxide sulfides are said to be not transformed into chloride.

The sulfates of Pr and Nd are only partially chlorinated in an $S_2Cl_2/Cl_2$ stream, Sm sulfate is completely chlorinated, Matignon [3, p. 378], Bourion [4]. Incomplete reaction of the sulfates is also found with gaseous HCl, Muthmann, Stützel [1].

When available, the carbides $MC_2$ (M = Y, Ce, Pr, Nd, Sm) can be converted into chlorides by their reaction with $Cl_2$ or HCl at elevated temperatures:

$$2\,MC_2 + 3\,Cl_2 \rightarrow 2\,MCl_3 + 4\,C, \quad MC_2 + 3\,HCl \rightarrow MCl_3 + 2\,C + 3/2\ H_2.$$

The reactions occur readily at dull red heat, Pettersson [5], Moissan [6 to 8], Moissan, Etard [9].

The preparation of anhydrous rare earth chlorides by high-temperature chlorination of monazite (and synthetic phosphate mixtures) has been investigated extensively by Hartley, Wylie [10, 11].

#### Preparation from Compounds with Organic Anions

Compounds of the lanthanides with weakly acidic organic anions are readily prepared. According to Taylor [12, p. 507], attempts to convert such compounds with gaseous HCl directly into the anhydrous chlorides usually do not yield pure products. The most common difficulty is that the chloride produced forms a low melting eutectic with the original salt and the melt is quite resistant to further attack by HCl. Another frequent difficulty is that at a temperature where the exchange rate is appreciable the organic anion is already beginning to decompose to produce carbonaceous materials. In order to avoid these difficulties various chlorination experiments were conducted in inert media like ether or benzene; the results were inconsistent.

Acetates could not be transformed completely into chlorides by treatment with HCl at the temperature of boiling benzene, Taylor [12, p. 507]. Attempts by Paul et al. [13] to prepare pure lanthanide chlorides by passing HCl gas through a suspension of the acetates in benzene or toluene likewise failed. A nearly quantitative yield of chloride was attained, however, by intermediate preparation of ammonium chlorocomplex salts, which in a second step were decomposed under high vacuum, see p. 70. Another method of chlorinating the acetates consists in stirring the finely ground hydrated starting compounds with excess acetyl chloride in benzene for about 30 min, washing the products with dry benzene, and drying them for about 3 h at 200 °C in an $N_2$ atmosphere, Watt et al. [14].

Thermal decomposition of the trichloroacetates under vacuum or in $N_2$ is a very convenient method for the preparation of the anhydrous chlorides ($LaCl_3$, $NdCl_3$, $DyCl_3$). The chloride product tends to absorb the other decomposition products strongly, but by prolonged pumping may be freed from them. In carefully dried air the same products as under vacuum are formed. When traces of moisture are present, the oxide chlorides are formed, Tilley, Roberts [15]. The decomposition temperatures of trichloroacetates determined by Mathur, Bhat [16] with a thermobalance at a heating rate of 6 °C/min and a gas flow rate of 15 to 20 mL/min range between 215 and 240 °C.

Oxalates of the rare earth elements are easily obtained by precipitation from aqueous solution. On heating in air they yield highly reactive oxide preparations which may be chlorinated according to the described methods. The direct chlorination has been preferred in some cases: Sm oxalate in 1 to 2 g amounts contained in vitreosil boats is chlorinated in a $Cl_2/CCl_4$ stream in a combustion tube. The temperature is maintained 1 h at 150 to 200 °C, raised by 50 °C an hour to 600 °C, kept there for 6 to 8 h, and finally raised again above the melting point of $SmCl_3$ to 750 °C, where chlorination is continued for a further 18 to 24 h. The product is cooled in $Cl_2$ alone. With Eu and Yb, complete conversion of oxalate to the trichloride was difficult to obtain; the samples always showed chlorine deficiency. The use of $S_2Cl_2$ in place of $CCl_4$ gave no improvement in the case of Yb, but with Eu it was eventually possible to obtain material completely soluble both in water and in an LiCl/KCl eutectic melt though Cl deficiency due to the presence of $EuCl_2$ was still apparent. The chlorination was first conducted below the melting point at 550 to 600 °C and then for a more extended period above the melting point. The product was ground up and again chlorinated for about half a day at 550 °C, Johnson, Mackenzie [17]. The chlorination of La and Er oxalates in a $Cl_2/CCl_4$ stream takes 2.5 to 3 h at 400 to 450 °C, Novikov, Tolmaecha [18]. On a technical scale, high-purity, anhydrous rare earth chlorides are prepared by chlorination of the oxalates by $Cl_2$ in a fluidized bed with dilution of the chlorinating agent by inert gases in a 2:1 ratio. Chlorides of light as well as heavy rare earth metals can be obtained in high yield, Bel'kova et al. [19]. Y and Gd oxalates are chlorinated with $CCl_4$ vapors, $Cl_2$, or $Cl_2$-$N_2$ mixtures in the presence of C black as a reducing agent, Strzyzewska et al. [20]. In an HCl stream the oxalates of the lighter lanthanides are only partially converted into the chlorides, Muthmann, Stützel [1].

Hydrated oxalates $M_2(C_2O_4)_3 \cdot 10\,H_2O$, however, with M = La, Ce, and possibly Pr, Nd are converted into anhydrous chlorides on heating in an HCl stream, Möbius, Matthes [21]. In a similar manner, Harmon, Wichers [22] obtained $CeCl_3$ free from oxide chloride and in close to theoretical yield. The method, which by the latter authors is claimed to be a general one for converting oxalates of the lanthanides into anhydrous chlorides, was early described by Robinson [23].

On heating of benzoates in an HCl stream fusion and charring occur, Taylor [12, p. 507]. Nearly quantitative yields of very pure chlorides ($LaCl_3$, $NdCl_3$, $SmCl_3$) are obtained, when the dehydrated benzoates are extracted with ether which has been freshly saturated with HCl gas. Adhering ether is removed from the products by flushing at 60 °C with dry HCl, which is in the end replaced by dry air, Brauman, Takvorian [24]. In a modification of this method dry HCl is passed into an ether suspension of the benzoates. The difficulties of preparing and dehydrating the benzoates and of handling large volumes of anhydrous ether, limit the use of this method, however, to small scale preparations, Freeman, Smith [25]. The procedure did not work well in the hands of

Taylor [12, p. 507]. Although chlorides were obtained with assay values of at least 99%, they dissolved in water with little or no hissing and the resulting solutions always were a little cloudy. In repeating the experiments of [24] with Ce benzoate, Hartley, Wylie [26] found, that the product of this operation is not anhydrous $CeCl_3$, but a monoetherate. The etherate was converted into the solvate-free chloride at 130 °C in a slow current of $N_2$ at 100 Torr.

When a suspension of lanthanum acetylacetonate in ether is treated with HCl, the addition compound $La(C_5H_8O_2)_3 \cdot 3HCl$ is formed. On heating in a vacuum this compound dissociates mainly into acetylacetone and $LaCl_3$, but before a temperature is reached where conversion is complete, charring occurs and a considerable amount of insoluble matter is formed, Taylor [12, p. 507].

**Preparation from Ammonium Chlorocomplexes**

Dry $NH_3$ gas was passed for about 20 min at room temperature through a suspension of 2 g lanthanide acetate in 50 mL benzene. After standing for about 4 h the suspension was saturated with gaseous HCl for about 1 h and then again allowed to stand for some time. The solvent was then decanted off, and the product after washing twice with petroleum ether was subjected to vacuum to remove traces of the solvent. Finally free as well as combined $NH_4Cl$ was sublimed off at $\approx 10^{-3}$ Torr and 250 to 270 °C. When ammonia was replaced by stoichiometric amounts of amines, such as pyridine and $\alpha$-picoline, complexes of the type $RNH \cdot MCl_4$ formed, which at $\approx 10^{-3}$ Torr and 250 to 260 °C also gave pure anhydrous chlorides in quantitative yields, Paul et al. [13]. The preparation of $PrCl_3$, $NdCl_3$, and $SmCl_3$ by this method is described by Paul et al. [27].

A recently reported method for preparing anhydrous $CeCl_3$ also involves the dehydration and subsequent decomposition of a double salt of $CeCl_3$ and pyridinium chloride (see "Rare Earth Elements" C 4b, Section 11.2.11.4.3.1), Didchenko [28].

References:

[1] W. Muthmann, L. Stützel (Ber. Deut. Chem. Ges. **32** [1899] 3413/9). – [2] G.G. Mosander (Ann. Physik Chem. [2] **11** [1827] 406/16). – [3] C. Matignon (Ann. Chim. Phys. [8] **8** [1906] 364/86). – [4] F. Bourion (Ann. Chim. Phys. [8] **20** [1910] 547/73). – [5] O. Pettersson (Ber. Deut. Chem. Ges. **28** [1895] 2419/22).

[6] H. Moissan (Compt. Rend. **122** [1895] 357/62). – [7] H. Moissan (Compt. Rend. **131** [1900] 595/600). – [8] H. Moissan (Compt. Rend. **131** [1900] 924/6). – [9] H. Moissan, Etard (Compt. Rend. **122** [1896] 573/7). – [10] F.R. Hartley, A.W. Wylie (J. Soc. Chem. Ind. [London] **69** [1950] 1/7).

[11] F.R. Hartley (J. Appl. Chem. [London] **2** [1952] 24/31). – [12] M.D. Taylor (Chem. Rev. **62** [1962] 503/11). – [13] R.C. Paul, M.S. Bains, J.S. Ghotra (J. Indian Chem. Soc. **44** [1967] 192/4). – [14] G.W. Watt, P.S. Gentile, E.P. Helvenston (J. Am. Chem. Soc. **77** [1955] 2752/3). – [15] G.P. Tilley, J.E. Roberts (Inorg. Chem. **2** [1963] 745/6).

[16] B.S. Mathur, T.R. Bhat (Z. Anorg. Allgem. Chem. **343** [1966] 220/4). – [17] K.E. Johnson, J.R. Mackenzie (J. Inorg. Nucl. Chem. **32** [1970] 43/8). – [18] G.I. Novikov, V.D. Tolmaecha (Zh. Prikl. Khim. **38** [1965] 1160/1; J. Appl. Chem. [USSR] **38** [1965] 1142/3). – [19] N.L. Bel'kova, N.M. Slastenova, I.M. Batyaev, M.A. Solov'ev (Izv. Vysshikh Uchebn. Zavedenii Khim. Khim. Tekhnol. **22** [1979] 1136/7 from C.A. **91** [1979]

No. 203454). – [20] M. Strzyzewska, W. Szklarski, W. Szczepaniak, A. Bogacz (Rudy Metale Niezelazne **23** [1978] 444/8 from C.A. **90** [1979] No. 139667).

[21] R. Möbius, F. Matthes (J. Less-Common Metals **10** [1966] 262/72). – [22] K.M. Harmon, E. Wichers (U.S. 2982603 [1961] from C.A. **1961** 21504). – [23] H. Robinson (Proc. Roy. Soc. [London] **37** [1884] 150/6). – [24] P. Brauman, S. Takvorian (Compt. Rend. Acad. Sci. URSS **194** [1932] 1579/80). – [25] J.H. Freeman, M.L. Smith (J. Inorg. Nucl. Chem. **7** [1958] 224/7).

[26] F.R. Hartley, A.W. Wylie (J. Chem. Soc. **1962** 679/81). – [27] R.C. Paul, G. Singh, J.S. Ghotra (Res. Bull. Panjab Univ. **19** Pt. 3-4 [1968] 261/3; C.A. **71** [1969] No. 108536). – [28] R. Didchenko (Trans. Met. Soc. AIME **215** [1959] 401/3).

### 11.2.6.1.6 Purification

Most of the preparation methods described above result in high quality products provided that they are carefully controlled and the necessary precautions against contamination are observed. In many cases, however, it is more advisable to content oneself with a lesser degree of purity in favor of a faster and simpler procedure and to subject the crude products to a purification operation. This aspect is, of course, of greater importance when large quantities are required.

Vacuum sublimation or distillation is a widely used and most effective means to obtain chlorides for different purposes of almost every desired degree of purity. Oxygen contamination is almost completely eliminated because the oxide chlorides vaporize incongruently with formation of the gaseous trichlorides and the nonvolatile oxides. Carbon and other impurities are likewise removed. At the same time powdery preparations are converted into a dense crystalline mass, which is less sensitive to atmospheric moisture and, consequently, more easily handled. Procedures and suitable apparatus for vacuum distillation have been described by [1 to 4] and are shortly outlined on p. 52 and in preparation descriptions for several individual chlorides in "Rare Earth Elements" C4b. According to Nisel'son, Lyzlov [5], the chlorides $MCl_3$ (M = La, Ce, Pr, Nd, Gd) are distilled at ≈0.1 Torr and ≈950 °C; $SmCl_3$, which has low thermal stability, is distilled at ≈800 °C. At $10^{-3}$ Torr the chlorides of La to Gd are distilled at or slightly above their melting points. However, pressures as high as 2 Torr have been used successfully, thus eliminating the need for a diffusion pump. The distillation temperatures corresponding to this pressure are listed in a table on p. 114. Pressures lower than $10^{-3}$ Torr can also be used with corresponding lower temperatures or shorter operation times. This is actually a vacuum sublimation and is preferred since it completely eliminates embrittling reactions between the chlorides and the walls of the quartz vessels, resulting in purer products, Cox, Fong [3].

Excess carbon present in preparations by the Oersted-Wöhler method is more effectively separated by filtration of the molten chloride (see p. 55), because solid impurities tend to be entrained in the gas stream during evaporation and require repeated distillations for complete removal. However, distillation is found to be more effective in removing lower-level impurities, Block, Campbell [1].

A purification via the gas phase is also effected by chemical transport reactions. $Al_2Cl_6$ and $Fe_2Cl_6$ form complex compounds with $YCl_3$ and $LaCl_3$, which are only stable in the gase phase (cf. p. 16), and in an adequate temperature gradient transport the rare earth chlorides from the hotter to the colder part of the reaction tube. The deposit consists of well-formed crystals and is free of O-containing impurities. The yield, however, is relatively low:

| starting material in mg | $p(Al_2Cl_6)$ in atm | temperature gradient | transported material in mg |
|---|---|---|---|
| 1000 $YCl_3$ | 2.84 | 600→510 °C | 270 $YCl_3$ |
| 410 $LaCl_3$ | 2.62 | 550→450 °C | 122 $LaCl_3$ |
| 1000 $LaCl_3$ | 6.93 | 750→650 °C | 520 $LaCl_3$ |

The heating time is 24 h; $p(Al_2Cl_6)$ corresponds to the total Al content of the gas phase at the average temperature disregarding the presence of other Al-containing species. Under the experimental conditions the $Al_2Cl_6$ is completely vaporized. $Fe_2Cl_6$ may also be used as a transport agent. From 190 mg $YCl_3$ about 135 mg are transported in 24 h in a temperature gradient 600→500 °C, when $p(Fe_2Cl_6) = 1.44$ atm and $p(Cl_2) = 2.8$ atm. The deposited crystals in this case have a slightly yellow color and evidently contain small amounts of Fe, Schäfer et al. [6]. Transport of $ScCl_3$ and $LaCl_3$ in the presence of $Al_2Cl_6$ and a temperature gradient 500→400 °C has also been reported by Lascelles [7]. In gas chromatographic separation experiments the formation of gaseous complex compounds with $Al_2Cl_6$ has earlier been used for the vaporization of chlorides $MCl_3$ (M = Ce, Pr, Pm, Gd, Tb, Dy, Tm, Yb, and Lu) in an $Al_2Cl_6$-containing carrier gas, Zvarova, Zvara [8].

Since rare earth trichlorides are soluble in various organic solvents, e.g., alcohols, acetone, and dimethylformamide, one might expect to use such solvents to extract the anhydrous chlorides from their mixtures with other materials. However, these solvents can never be removed completely from the products, Taylor [9].

A criterion for the purity of rare earth chlorides is their complete solubility in water and alcohol.

References:

[1] F.E. Block, T.T. Campbell (in: F.H. Spedding, A.H. Daane, The Rare Earths, New York - London 1961, pp. 89/101, 95). – [2] F.L. Carter, J.F. Murray (Mater. Res. Bull. **7** [1972] 519/23; C.A. **77** [1972] No. 42529). – [3] D.E. Cox, F.K. Fong (J. Cryst. Growth **20** [1973] 233/8). – [4] J. Kutscher, A. Schneider (Inorg. Nucl. Chem. Letters **7** [1971] 815/9). – [5] L.A. Nisel'son, Yu.N. Lyzlov (Zh. Neorgan. Khim. **21** [1976] 3344/7; Russ. J. Inorg. Chem. **21** [1976] 1842/4).

[6] H. Schäfer (Z. Anorg. Allgem. Chem. **403** [1974] 116/26). – [7] K. Lascelles (private communication from [6]). – [8] T.S. Zvarova, I. Zvara (J. Chromatog. **44** [1969] 604/8). – [9] M.D. Taylor (Chem. Rev. **62** [1962] 503/11, 507).

### 11.2.6.1.7 Preparation of Single Crystals

The major problem encountered in the growth of single crystals of rare earth chlorides is contamination of the melts. Entrapped foreign particles will distort the lattice field and interfere with normal regular growth. Therefore rigorous chemical purification procedures must precede the actual crystal growing operation.

If its purity is high enough, the slowly cooled, solidified material contains single crystals large enough to permit spectroscopic studies of the compounds, Kiess [4]. A general, consistently successful procedure for growing optically pure single crystals of the anhydrous chlorides has been developed by Cox, Fong [1]. It involves drying the hydrated chlorides with $NH_4Cl$, purification by vacuum distillation, and subsequent growth by a modified Bridgman-Stockbarger technique. The method is unable to produce single crystals of $EuCl_3$ due to partial transition of $Eu^{3+}$ into $Eu^{2+}$. However, $SmCl_3$, $YbCl_3$, and chloride hosts with $Sm^{3+}$, $Eu^{3+}$, and $Yb^{3+}$ dopants could be grown by substituting the normally used $N_2$ atmosphere by pure, dry $Cl_2$ gas [1]. When $Cl_2$ pressures of 2.5 to 3 atm were used, single crystals of essentially undecomposed $EuCl_3$ could be obtained in a standard Bridgman furnace, Birgeneau et al. [2], Mroczkowski [3]; see the preparation of $EuCl_3$, Section 11.2.19.9.2.1 in "Rare Earth Elements" C4b. In this volume similar methods for growing single crystals of $LaCl_3$, $CeCl_3$, $PrCl_3$, and $SmCl_3$ from the melts are also described in corresponding sections.

References:

[1] D.E. Cox, F.K. Fong (J. Cryst. Growth **20** [1973] 233/8). – [2] R.J. Birgeneau, M.T. Hutchings, W.P. Wolf (Phys. Rev. [2] **179** [1969] 275/88, 278). – [3] S. Mroczkowski (J. Cryst. Growth **6** [1969] 147/50). – [4] N.H. Kiess (J. Res. Natl. Bur. Std. A **67** [1963] 343/5).

### 11.2.6.2 Crystallographic Properties

#### 11.2.6.2.1 Polymorphism

$ScCl_3$ is rhombohedral ($BiI_3$ type), Klemm, Krose [1], and does not change its symmetry between room temperature and its melting point, according to the enthalpy studies of Dworkin, Bredig [2].

The usual form at room and higher temperatures of the trichlorides of La, Ce, Pr, Nd, Pm, Sm, Eu, and Gd is the hexagonal $UCl_3$ type, Zachariasen [3]. Among these $GdCl_3$ is dimorphic under ordinary conditions: in addition to the hexagonal modification an orthorhombic modification of the $PuBr_3$ type is occasionally observed. The exothermic peak of the DTA curve shows that the latter is stable below 368 to 373 K; however, it occurs only in a mixture with the hexagonal form, Harris, Veale [4]. The hexagonal symmetry of $LaCl_3$, $PrCl_3$, $NdCl_3$ [5], and $GdCl_3$ persists up to the melting points of the compounds, from enthalpy studies [2]. A phase transformation occurs in $PrCl_3$ at 0.42 K; at this temperature the NQR line of $^{35}Cl$ splits up into two components (see p. 91). The low temperature phase has a lower hexagonal ($P\bar{6}-C^1_{3h}$) or trigonal ($P3-C^1_3$) symmetry, Hessler, Carlson [6]. In $GdCl_3$ a phase transition of the second order takes place at the Curie point 2.20 K, from specific heat studies, Leask et al. [7].

$TbCl_3$ according to Forrester et al. [8] crystallizes orthorhombically in the $PuBr_3$ type and according to Spedding, Daane [9] monoclinically, see "Rare Earth Elements" C4b, Section 11.2.11.11.3.2. A polymorphic transition is deduced from the trend of the enthalpy between room temperature and the melting point, according to Dworkin, Bredig [2] at 783 K, to Spedding, Daane [10] as low as 770 K. This is connected with an enthalpy change of $\Delta H = 3.4$ kcal/mol and an entropy change of $\Delta S = 4.3$ cal·mol$^{-1}$·K$^{-1}$ [2]. The DTA curve yields 801 K as the temperature for the polymorphic transition, Lyzlov, Nisel'son [16]. On heating, the respective peak occurs at 790 K and on cooling at 765 K. The high temperature phase has the hexagonal $UCl_3$ structure, Mattausch et al. [17].

The trichlorides of Y, Dy, Ho, Er, Tm, Yb, and Lu ordinarily have the monoclinic $AlCl_3$ structure, Templeton, Carter [11], which from enthalpy studies on $YCl_3$, $DyCl_3$, $HoCl_3$ [2], and $ErCl_3$ [12] persists up to the melting point [2]. $YCl_3$ is according to Klemm et al. [13], $HoCl_3$ according to Bommer, Hohmann [14], probably dimorphic at room temperature; this would explain the different density values which are found for these compounds, see p. 102. DTA studies show a polymorphic transition in $DyCl_3$ at 632 K, Lyzlov, Nisel'son [16]. The existence of two polymorphic forms of $DyCl_3$ at room temperature was inferred by Bommer, Hohmann [14] from the different heats of solution and by Weigel, Wishnevsky [15] from the different temperature dependencies of the vapor phase hydrolysis data and X-ray studies conducted in parallel. While the monoclinic symmetry reported for the one phase is well established due to the sharp X-ray pattern, clarification of the possibly orthorhombic symmetry of the other phase presents difficulties. The available material is not sufficiently thoroughly crystallized, so that the comparison of the diffuse X-ray pattern with that of the orthorhombic $TbCl_3$ is inconclusive [15]. Nevertheless, Bommer, Hohmann [14] earlier determined by comparison of the powder patterns that two forms of $DyCl_3$ exist, of which the one crystallizes like the room temperature modification of $TbCl_3$ and the other like the trichlorides of Ho to Lu.

High-pressure modifications, which have the orthorhombic $PuBr_3$ structure, have been observed for $DyCl_3$ and $TmCl_3$, Beck, Gladrow [18].

In the molten state, the trichlorides have a quasi-crystalline structure. This was shown for $YCl_3$ by a Raman spectrum, Papatheodorou [19], and for $NdCl_3$ by an absorption spectrum, Barbanel' et al. [20].

References:

[1] W. Klemm, E. Krose (Z. Anorg. Allgem. Chem. **253** [1947] 218/25). – [2] A.S. Dworkin, M.A. Bredig (High Temp. Sci. **3** [1971] 81/90). – [3] W.H. Zachariasen (Acta Cryst. **1** [1948] 265/8). – [4] A.L. Harris, C.R. Veale (J. Inorg. Nucl. Chem. **27** [1965] 1437/9). – [5] A.S. Dworkin, M.A. Bredig (J. Phys. Chem. **67** [1963] 697/8).

[6] J.P. Hessler, E.H. Carlson (J. Appl. Phys. **42** [1971] 1316/7). – [7] J.M. Leask, W.P. Wolf, A.F.G. Wyatt (Proc. 8th Intern. Conf. Low Temp. Phys., London 1962 [1963], pp. 230/1). – [8] J.D. Forrester, A. Zalkin, D.H. Templeton, J.C. Wallmann (Inorg. Chem. **3** [1964] 185/8). – [9] F.H. Spedding, A.H. Daane (IS-350 [1961] 1/122, 31; N.S.A. **16** [1962] No. 14614). – [10] F.H. Spedding, A.H. Daane (Met. Rev. **5** [1960] 297/348, 339/40).

[11] D.H. Templeton, G.F. Carter (J. Phys. Chem. **58** [1954] 940/4). – [12] A.S. Dworkin, M.A. Bredig (J. Phys. Chem. **67** [1963] 2499/500). – [13] W. Klemm, K. Meisel, H.U. v. Vogel (Z. Anorg. Allgem. Chem. **190** [1930] 123/44, 130). – [14] H. Bommer, E. Hohmann (Z. Anorg. Allgem. Chem. **248** [1941] 373/82, 376; Naturwissenschaften **27** [1939] 583). – [15] F. Weigel, V. Wishnevsky (Chem. Ber. **102** [1969] 5/13, 8, 9).

[16] Yu.N. Lyzlov, L.A. Nisel'son (Zh. Neorgan. Khim. **22** [1977] 2245/7; Russ. J. Inorg. Chem. **22** [1977] 1215/6). – [17] H. Mattausch, A. Simon, N. Holzer, R. Eger (Z. Anorg. Allgem. Chem. **466** [1980] 7/22, 12/3). – [18] H.P. Beck, E. Gladrow (Z. Anorg. Allgem. Chem. **453** [1979] 79/92, 82/3, 91). – [19] G.N. Papatheodorou (J. Chem. Phys. **66** [1977] 2893/900). – [20] Yu.A. Barbanel', V.R. Klokman, V.P. Kotlin, V.V. Kolin (Radiokhimiya **21** [1979] 694/705; Soviet Radiochem. **21** [1979] 600/10).

### 11.2.6.2.2 Structure

For data regarding the rhombohedral $ScCl_3$ see "Rare Earth Elements" C4b, Section 11.2.11.1.11.2.

### 11.2.6.2.2.1 $MCl_3$ of the Hexagonal $UCl_3$ Type

The trichlorides of La, Ce, Pr, Nd, Pm, Sm, Eu, and (above 95 °C) Gd, like $Y(OH)_3$ (see "Seltenerdelemente" C 2, 1974, p. 8), have the hexagonal $UCl_3$ structure and the space group $P6_3/m-C_{6h}^2$ (No. 176) with $Z=2$, Zachariasen [1]. This is confirmed by single crystal X-ray studies on $GdCl_3$ by Au, Au [2] and in addition for $LaCl_3$, $NdCl_3$, and $EuCl_3$ by Morosin [3]. The lanthanide ions occupy the position 2(c): $\pm(^1/_3, ^2/_3, ^1/_4)$, the Cl ions the position 6(h): $\pm(x, y, ^1/_4)$, $\pm(\bar{y}, x-y, ^1/_4)$, $\pm(y-x, \bar{x}, ^1/_4)$ [1]. The structure refinement with anisotropic temperature factors for the Cl ion, with the use of three-dimensional intensity data (MoKα radiation) corrected for absorption, yields the following Cl parameters with the given R values [3]:

| | $LaCl_3$ | $NdCl_3$ | $EuCl_3$ | $GdCl_3$ |
|---|---|---|---|---|
| R | 0.033 | 0.034 | 0.025 | 0.036 |
| x | 0.38741 | 0.38777 | 0.38911 | 0.38929 |
| y | 0.30155 | 0.30167 | 0.30174 | 0.30153 |

Cl parameters interpolated from these data for $CeCl_3$, $PrCl_3$, and $SmCl_3$ can be found in [4]. Neutron diffraction data on $GdCl_3$ yield $x=0.3888$ and $y=0.3018$ for the Cl parameters at 4.2 K, Kötzler et al. [19].

In the structure trigonal prisms of Cl ions occur, in which the M (metal) ions are located. Each M ion is surrounded by these 6 Cl ions at a shorter distance and by 3 additional Cl ions at a greater distance (M↔Cl in Å for M=La 2.950, 2.953, for Nd 2.886, 2.923, for Eu 2.835, 2.919, and for Gd 2.822, 2.918). The latter Cl ions lie on the same mirror plane as the M ion and in $LaCl_3$ almost on the perpendicular to the prism faces; see Fig. 11, p. 77, in which the structure of these hexagonal phases is compared with that of the orthorhombic phases. The environment of the M ions is increasingly distorted in the trichlorides of Nd, Eu, and Gd. The Cl ions lying outside of the prism are no longer located almost on the perpendicular planes. In addition the prism faces running parallel to the c-axis in particular become smaller in this direction; the prism edges amount to 3.430 and 4.375 Å in $LaCl_3$ and 3.354 and 4.106 Å in $GdCl_3$. Additional bonding distances and angles are presented in the paper by Morosin [3]. The M↔Cl distances calculated by Zachariasen [5] for M=La, Ce, Pr, and Nd are, so far as comparable, larger.

Carter [18] with help from the data of Morosin [3] determined the shape and the volume of the Voronoi cells and the polyhedral-atomic-volume (PAV) cells for the $M^{3+}$ and $Cl^-$ ions in $MCl_3$ for M=La, Nd, Eu, and Gd, see table in [18]. From this work the PAV volume for the Cl ions in these compounds amounts on the average to 31.05 $Å^3$. It thus has the same order of magnitude as in $EuCl_2$, $Gd_2Cl_3$, $Dy_5Cl_{11}$, and $Yb_5ErCl_{13}$, but is intrinsically smaller than in the monoclinic $MCl_3$ compounds with M=Y, Dy, Ho, and Tm, see p. 78. The shape of this polyhedron is in the hexagonal compounds nevertheless unusually oblate, as a pictorial comparison of these cells for $GdCl_3$ and $DyCl_3$ in the paper shows. Such a flattening can be caused by the fact that the structure was not refined to the true minimum [18].

Lattice constants in Å at room temperature with uncertainties of the last positions in parentheses are compiled in Table 11.2.6-1, p. 76.

As the lattice constants interpolated by Current et al. [4] for $CeCl_3$, $PrCl_3$, and $SmCl_3$ from those of $LaCl_3$, $NdCl_3$, and $EuCl_3$ [3] distinctly show, c does not change linearly. In contrast, a follows two linear laws, one between $LaCl_3$ and $NdCl_3$, the other between $PmCl_3$ and $GdCl_3$, Templeton, Dauben [7]. The ratio c/a decreases from 0.583 in $LaCl_3$ [6] to 0.557 in $GdCl_3$ [7] approximately linearly. Interplanar spacings for the trichlorides of La, Ce, Pr, Nd, Sm, and Eu are given by Swanson et al. [6];

Table 11.2.6-1
Lattice Constants of Hexagonal Trichlorides

| | a in Å | c in Å | Ref. |
|---|---|---|---|
| $LaCl_3$ | 7.483 | 4.375 | [1] |
| | 7.483 | 4.364 | [6] |
| | 7.4779 (5) | 4.3745 (5) | [3] |
| $CeCl_3$ | 7.450 (4) | 4.315 (2) | [7] |
| | 7.451 | 4.313 | [1, 8] |
| | 7.445 (4) | 4.316 (4) | [9] |
| | 7.454 | 4.312 | [6] |
| $PrCl_3$ | 7.422 (5) | 4.275 (4) | [7] |
| | 7.42 | 4.26 | [1] |
| | 7.423 | 4.272 | [6] |
| $NdCl_3$ | 7.396 | 4.240 | [1] |
| | 7.400 | 4.240 | [6] |
| | 7.3988 (9) | 4.2423 (6) | [3] |
| $PmCl_3$ | 7.397 | 4.211 | [10, p. 206] |
| | 7.40 | 4.21 | [11] |
| $SmCl_3$ | 7.378 (7) | 4.171 (4) | [7] |
| | 7.380 | 4.169 | [6] |
| | 7.380 (7) | 4.175 (1) | [15] |
| | 7.348 (4) | 4.155 (1) | [16] |
| $EuCl_3$ | 7.370 (2) | 4.137 (4) | [15] |
| | 7.369 (4) | 4.133 (2) | [7] |
| | 7.375 | 4.134 | [6] |
| | 7.3746 (12) | 4.1323 (5) | [3] |
| | 7.324 (12) | 4.106 (11) | [17] |
| $GdCl_3$ | 7.363 (4) | 4.105 (2) | [7] |
| | 7.3663 (9) | 4.1059 (4) | [3] |
| | 7.367 (2) | 4.108 (1) | [15] |
| $TbCl_3$ [a] | 7.346 (7) | 4.034 (7) | [20] |

[a] High temperature modification.

cf. the results for $SmCl_3$ by Goryushkin et al. [16] and for $EuCl_3$ by Kulagin et al. [17].

The cleavage is perfect parallel to [001], Varsanyi [12]; {100} is reported by Hughes et al. [13] and (110) by Dawson, Schaack [14].

References:

[1] W.H. Zachariasen (Acta Cryst. **1** [1948] 265/8). – [2] C. Au, R. Au (Acta Cryst. **23** [1967] 1112). – [3] B. Morosin (J. Chem. Phys. **49** [1968] 3007/12). – [4] D.H. Current, C.L. Foiles, E.H. Carlson (Phys. Rev. [3] B **6** [1972] 737/47). – [5] W.H. Zachariasen (TID-5212 [1955] 157/64; C.A. **1957** 35; AECD-2091 (ANL-HDY-499) 1944/46 [1948] 1/14; N.S.A. **1** [1948] No. 588).

[6] H.E. Swanson, M.C. Morris, R.P. Stinchfield, E.H. Evans (Natl. Bur. Std. [U.S.] Monograph No. 25 [1962] 1/56). – [7] D.H. Templeton, C.H. Dauben (J. Am. Chem. Soc. **76** [1954] 5237/9; UCRL-2606 [1954] 1/13 from N.S.A. **8** [1954] No. 4919). – [8] T. Kojima, T. Inoue, T. Ishiyama (J. Electrochem. Soc. Japan **19** [1951] 383/6). – [9] J.P. Hessler, E.H. Carlson (J. Appl. Phys. **42** [1971] 1316/7). – [10] F. Weigel (Habilitationsschr. München 1965, pp. 1/296, 206; AD-687-223 [1965] 107/18, 113; C.A. **65** [1966] 8287; CONF-650804-7 [1963] 1/22 from N.S.A. **19** [1965] No. 43988).

[11] F. Weigel, V. Scherer (Radiochim. Acta **7** [1967] 40/6). – [12] F. Varsanyi (Appl. Phys. Letters **19** [1971] 169/71). – [13] W.E. Hughes, C.G. Montgomery, W.G. Moulton, E.H. Carlson (J. Chem. Phys. **41** [1964] 3470/3). – [14] P. Dawson, G. Schaack (J. Phys. Chem. Solids **34** [1973] 2193/9, 2194). – [15] J.A. Sommers, E.F. Westrum (J. Chem. Thermodyn. **9** [1977] 1/26, 3).

[16] V.F. Goryushkin, D.M. Laptev, I.S. Astakhova, E.S. Vorontsov (Zh. Neorgan. Khim. **21** [1976] 821/3; Russ. J. Inorg. Chem. **21** [1976] 445/7). – [17] N.M. Kulagin, D.M. Laptev, I.S. Astakhova (Zh. Neorgan. Khim. **21** [1976] 2549/51; Russ. J. Inorg. Chem. **21** [1976] 1401/3). – [18] F.L. Carter (Proc. 13th Rare Earth Res. Conf., Wheeling, Va., 1977 [1978], pp. 225/31). – [19] J. Kötzler, W. Scheithe, K. Knorr, W.B. Yelon

(J. Phys. C **9** [1976] 1291/302). – [20] H. Mattausch, A. Simon, N. Holzer, R. Eger (Z. Anorg. Allgem. Chem. **466** [1980] 7/22, 12/3).

### 11.2.6.2.2.2 $MCl_3$ of the Orthorhombic $PuBr_3$ Type

The orthorhombic phases of $GdCl_3$ and $TbCl_3$ crystallize in the $PuBr_3$ type with space group $Cmcm-D_{2h}^{17}$ (No. 63), Z=4, Forrester et al. [2]. The orthorhombic $GdCl_3$ phase is the stable modification below about 368 K, and always occurs in a mixture with the hexagonal high temperature form, Harris, Veale [1]. The $TbCl_3$ phase is observed as long thin platelets with a fibrous structure [2]. The tribromides of Nd, Sm, and Eu and the triiodides of La, Ce, Pr, and Nd are, for example, isostructural with these two phases, see Zachariasen [3] and Templeton, Carter [4]. The atom positions (for M=Tb or Gd) are:

4 M and 4 Cl(1) in (c): $\pm(0, y, {}^1/_4)$, $\pm({}^1/_2, {}^1/_2+y, {}^1/_4)$;
8 Cl(2) in (f): $\pm(0, y, z)$, $\pm(0, y, {}^1/_2-z)$, $\pm({}^1/_2, {}^1/_2+y, z)$, $\pm({}^1/_2, {}^1/_2+y, {}^1/_2-z)$.

In the case of $TbCl_3$ the refinement with isotropic temperature factors to R=0.18 yields y=0.244 for Tb, y=0.583 for Cl(1) and y=0.145, z=0.569 for Cl(2) [2]. The structure is built up from three sided prisms with M in the centers and Cl at the corners. Each M atom has in addition to these 6 Cl two further Cl neighbors on two of the three prism faces (distances in Å: Tb↔Cl, 2.70 (2×), 2.79 (4×), 2.95 (2×); shortest Cl↔Cl distance 3.07 Å). The prisms are linked together through the triangular faces into chains along the a-axis. While in the hexagonal phases with similar chain structures all three neighboring chains contribute to the coordination of each M atom, in these orthorhombic phases only Cl atoms from two of the four neighboring chains participate, see **Fig. 11.** The distance of the Cl atoms from the third chain, which could contribute to the coordination, is somewhat more than 1 Å greater than that to the nearest Cl neighbors, Forrester et al. [2].

Fig. 11

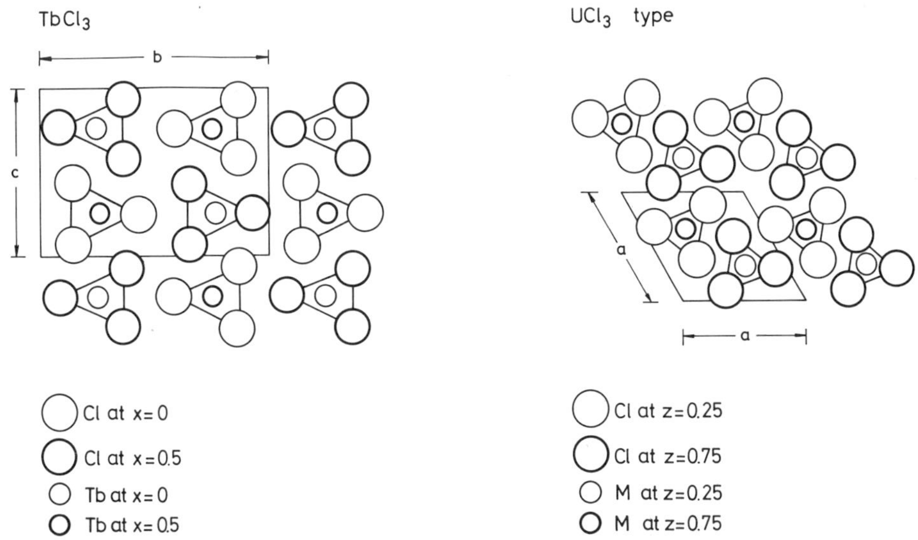


Crystal structure of the orthorhombic $TbCl_3$ ($MCl_3$ of the $PuBr_3$ type) projected along [100], compared to that of the hexagonal $UCl_3$ type of $MCl_3$, M=La through Gd, projected along [001].

Lattice constants in Å at room temperature:

| $MCl_3$ | a | b | c | Ref. |
|---|---|---|---|---|
| $GdCl_3$ | 3.88 | 11.73 | 8.52 | [1] |
| $TbCl_3$ | 3.86±0.02 | 11.71±0.03 | 8.48±0.03 | [2] |

References:

[1] A.L. Harris, C.R. Veale (J. Inorg. Nucl. Chem. **27** [1965] 1437/9). – [2] J.D. Forrester, A. Zalkin, D.H. Templeton, J.C. Wallmann (Inorg. Chem. **3** [1964] 185/8). – [3] W.H. Zachariasen (Acta Cryst. **1** [1948] 265/8). – [4] D.H. Templeton, G.F. Carter (J. Phys. Chem. **58** [1954] 940/4).

### 11.2.6.2.2.3 $MCl_3$ of the Monoclinic $AlCl_3$ Type

The trichlorides of Y, Dy, Ho, Er, Tm, Yb, and Lu are isomorphic with each other, as first established by Bommer, Hohmann [1]. They crystallize in the $AlCl_3$ type, space group $C2/m-C^3_{2h}$ (No. 12), Z=4, Templeton, Carter [2]. The structure of the compounds is described for the example of $YCl_3$. The atom positions are [2]:

4 Y in (g): $\pm(0, y, 0), \pm(^1/_2, ^1/_2+y, 0)$, with y=0.166;
4 Cl(1) in (i): $\pm(x, 0, z), \pm(^1/_2+x, ^1/_2, z)$ with x=0.211, z=0.247;
8 Cl(2) in (j): $\pm(x, y, z), \pm(^1/_2+x, ^1/_2+y, z), \pm(x, \bar{y}, z)$,
$\pm(^1/_2+y, ^1/_2-y, z)$, with x=0.229, y=0.179, z=0.760.

The structure determined by means of Fourier syntheses can be described as a slightly distorted NaCl type, in which $^2/_3$ of the metal atoms are missing. The Cl ions are arranged approximately in a cubic close packing. Y occupies octahedral holes between alternate pairs of close-packed layers of Cl ions. The coordination octahedra are linked by common edges into pseudohexagonal sheets parallel to (001). Interatomic distances in the octahedra are Y↔Cl=2.58 to 2.69 Å and Cl↔Cl 3.45 to 3.91 Å, in comparison to Cl↔Cl=3.69 to 4.16 Å between the layers and Cl↔Cl=4.27 to 4.42 Å between octahedra in the same layer [2].

From the atomic parameters for $YCl_3$, Carter [3] determined the shape and the volume of the Voronoi cells and of the polyhedral-atomic-volume (PAV) cells for the $M^{3+}$ and $Cl^-$ ions in $DyCl_3$, $HoCl_3$, and $TmCl_3$, see the table in [3]. From this work the PAV volume for the two Cl ions in these compounds averages 38.89 Å. Thus it unexpectedly falls in the same order of magnitude as that of the Br ions in the orthorhombic $NdBr_3$ of the $PuBr_3$ type and is around 30% larger than in the hexagonal $MCl_3$ compounds (M=La, Nd, Eu, and Gd) of the $UCl_3$ type, see p. 75. The shape of the polyhedra in $DyCl_3$ is pictured in the paper [3].

Lattice constants in Å at room temperature [2]:

| $MCl_3$ | a | b | c | β in ° |
|---|---|---|---|---|
| $YCl_3$ | 6.92±0.02 | 11.94±0.04 | 6.44±0.02 | 111.0±0.5 |
| $DyCl_3$ | 6.91 | 11.97 | 6.40 | 111.2 |
| $HoCl_3$ | 6.85 | 11.85 | 6.39 | 110.8 |
| $ErCl_3$ | 6.80 | 11.79 | 6.39 | 110.7 |
| $TmCl_3$ | 6.75 | 11.73 | 6.39 | 110.6 |
| $YbCl_3$ | 6.73 | 11.65 | 6.38 | 110.4 |
| $LuCl_3$ | 6.72 | 11.60 | 6.39 | 110.4 |

The constants for $DyCl_3$ through $LuCl_3$ are determined from lines at relatively small angles and therefore are not very precise, Templeton, Carter [2]. For $ErCl_3$ the constants a=6.813 (15), b=11.804 (24), c=6.398 (10) Å, and $\beta$=110.76 (13)° are given later by Fuger et al. [4].

The cleavage is good parallel to (001) [2].

References:

[1] H. Bommer, E. Hohmann (Z. Anorg. Allgem. Chem. **248** [1941] 373/82). – [2] D.H. Templeton, G.F. Carter (J. Phys. Chem. **58** [1954] 940/4). – [3] F.L. Carter (Proc. 13th Rare Earth Res. Conf., Wheeling, Va., 1977 [1978], pp. 225/31). – [4] J. Fuger, L. Morss, D. Brown (J. Chem. Soc. Dalton Trans. **1980** 1076/8).

### 11.2.6.2.3 Lattice Energy and Related Data

**Lattice Energy** U in kcal/mol

Values for the lattice energies of the rare earth trichlorides under standard conditions are determined by Karapet'yants [1] as the differences between the enthalpies of formation of the particular solid salts and the sums for their gaseous ions for $MCl_3$ with M=Sc, Y, and La and by Krestov, Krestova [2] for $MCl_3$ with M=La through Lu. The values of [1] are based on the thermochemical data assembled by Rossini et al. [3], those of [2] on the data of Wicks, Block [4]. They are, as the following table shows, consistently larger than the lattice energies U (theor.) calculated by Ladd, Lee [5] by the Born theory from the Madelung constants A′ given on p. 80 and the repulsion coefficient n=9:

| $MCl_3$ | −U | −U (theor.) | $MCl_3$ | −U | −U (theor.) |
|---|---|---|---|---|---|
| $ScCl_3$ | 1163 [1], 1160 [6] | – | $GdCl_3$ | 1072 [2] | 1053 [5] |
| $YCl_3$ | 1076 [1] | – | $TbCl_3$ | 1078 [2] | 1036 [7] |
| $LaCl_3$ | 1016 [1], 1018 [2] | 1019 [5] | $DyCl_3$ | 1084 [2] | 1071 [5] |
| $CeCl_3$ | 1048 [2] | 1027 [5] | $HoCl_3$ | 1095 [2] | 1078 [5] |
| $PrCl_3$ | 1046 [2] | 1033 [5] | $ErCl_3$ | 1098 [2] | 1082 [5] |
| $NdCl_3$ | 1046 [2] | 1038 [5] | $TmCl_3$ | 1104 [2] | 1087 [5] |
| $PmCl_3$ | 1056 [2] | – | $YbCl_3$ | 1110 [2] | 1091 [5] |
| $SmCl_3$ | 1062 [2] | 1046 [5] | $LuCl_3$ | 1115 [2] | 1092 [5] |
| $EuCl_3$ | 1068 [2] | 1050 [5] | | | |

The value U (theor.) reported for $TbCl_3$ is derived by Gopal, Husain [7] from the average value of the ratio U (theor.)/$T_m \approx 1.04$ or U (theor.)/$T_b$=0.54 for the other rare earth trichlorides ($T_m$=melting point, $T_b$=boiling point in K, U (theor.) from [5]). Lattice energies calculated by the Born-Landé equation (presumably, with n=10.5) are plotted against the reciprocals of the characteristic distance $r_0$ in a figure by Kim, Johnson [12]. Values for hexagonal $LaCl_3$ through $GdCl_3$ and orthorhombic $TbCl_3$ lie on a straight line ($-U_0 \approx 4246$ to 4465 kJ/mol, ≈1014 to 1066 kcal/mol). Values for monoclinic $DyCl_3$ through $YbCl_3$ lie on a second straight line ($-U_0 \approx 4470$ to 4593 kJ/mol, ≈1068 to 1097 kcal/mol). However, the value for $LuCl_3$ ($-U_0 \approx 4548$ kJ/mol, ≈1086 kcal/mol) is about as low as that for $TmCl_3$ [12]. Calculated lattice energies for the chlorides of La, Ce, Pr, Sm, and Gd based on X-ray determined Madelung constants are given by Koch [8]. Comparison of the values of U of $LaCl_3$ through $LuCl_3$ with the corresponding values for the other halides and oxides of the lanthanides as a function of the number of 4f electrons is found in a figure in [2]. Values obtained by an interpolation process (see the paper for details) for $ScCl_3$, $YCl_3$, and $LaCl_3$ differ from U in the foregoing table by about 2 kcal/mol [1]. Additional data for $LaCl_3$ can be found in Shewchuck [9].

### Madelung Constant

The Madelung constant A′ of $YCl_3$, representative of the trichlorides of the rare earth elements which crystallize in the $AlCl_3$ type (see "polymorphism", p. 73), is calculated to be 9.5 from the lattice energies determined from thermodynamic data. Correspondingly, A′=10.3 for $LaCl_3$, representative of the $UCl_3$-type compounds, when the exponent of the repulsion term for the interionic distances n=9, Ladd, Lee [5]. The $\alpha$ parameter (the Madelung constant per valence bond) of $LaCl_3$ is calculated to be about 1.55 by C.W. Koch, according to Templeton [10] (read from a figure in [10]). This value of $\alpha$ corresponds to about A′=9.6 and is regarded as imprecise [5].

### Crystal Field Stabilization Energy

The crystal field stabilization energies have been calculated for the trichlorides of Ce, Pr, Nd, Sm, Dy, Er, and Yb with and without consideration of the spin-orbit interaction in fields of cubic or of octahedral symmetry. Within the lanthanide series the crystal field stabilization energy in the presence of a spin-orbit interaction shows in the two fields a maximum at $CeCl_3$ in the octahedral field (E=111 $cm^{-1}$) and at $PrCl_3$ in the cubic field (E=68 $cm^{-1}$), smaller maxima at $YbCl_3$ and $DyCl_3$, and a minimum at $GdCl_3$. In the absence of this interaction only two maxima occur. These are well defined in a field of cubic symmetry, at $NdCl_3$ (E=156 $cm^{-1}$) and $ErCl_3$ (E=83 $cm^{-1}$). In a field of octahedral symmetry the second maximum also occurs at $ErCl_3$ (E=105 $cm^{-1}$), while the plateau of the first stretches out very flatly from $CeCl_3$ through $PrCl_3$ to $NdCl_3$ (E=156 $cm^{-1}$). Additional numerical values and a graphical representation of the changes in the crystal field stabilization energy in the lanthanide trichloride series in comparison with the corresponding trifluorides, tribromides, and triiodides are presented by Yatsimirskii, Kostromina [11].

References:

[1] M.Kh. Karapet'yants (Zh. Fiz. Khim. **28** [1954] 1136/52, 1149). – [2] G.A. Krestov, N.V. Krestova (Radiokhimiya **11** [1969] 62/74; Soviet Radiochem. **11** [1969] 55/66, 56). – [3] F.D. Rossini, D.D. Wagman, W.H. Evans, S. Levine, I. Jaffe (Natl. Bur. Std. [U.S.] Circ. No. 500 [1952]). – [4] C.E. Wicks, F.E. Block (U.S. Bur. Mines Bull. No. 605 [1963] 1/146; C.A. **59** [1963] 14656). – [5] M.F.C. Ladd, W.H. Lee (J. Inorg. Nucl. Chem. **23** [1961] 199/205).

[6] K.B. Yatsimirskii (Zh. Neorgan. Khim. **3** [1958] 2244/52, 2245; Russ. J. Inorg. Chem. **3** No. 10 [1958] 26/36, 28/30). – [7] R. Gopal, M.M. Husain (J. Indian Chem. Soc. **44** [1967] 726/7). – [8] C.W. Koch (UCRL-2286 [1953] 1/101 from C.A. **1954** 4940). – [9] S. Shewchuck (UCRL-1606 [1951] 1/9 from N.S.A. **6** [1952] No. 1753). – [10] D.H. Templeton (J. Chem. Phys. **21** [1953] 2097/8).

[11] K.B. Yatsimirskii, N.A. Kostromina (Teor. i Eksperim. Khim. **2** [1966] 583/8; Theor. Exptl. Chem. [USSR] **2** [1966] 436/9). – [12] K.-Y. Kim, C.E. Johnson (J. Chem. Thermodyn. **13** [1981] 13/25, 19/21).

### 11.2.6.3 Lattice Vibrations

For data regarding the rhombohedral $ScCl_3$ and the orthorhombic $TbCl_3$ see "Rare Earth Elements" C4b, Sections 11.2.11.1.11.2 and 11.2.11.11.3.2.

**Hexagonal $MCl_3$**

For the hexagonal space group $P6_3/m-C_{6h}^2$ (Z=2) of $MCl_3$ with M=La, Ce, Pr, Nd, Eu, and Gd 6 Raman-active ($2A_{1g}+E_{1g}+3E_{2g}$), 3 IR-active ($A_u+2E_{1u}$), and 5 inactive ($2B_g+2B_u+1E_{2u}$) vibrational modes are possible, Murphy et al. [1].

Raman Spectra

All six expected lines are observed by Schaack, Koningstein [2] only in the case of $LaCl_3$ and $GdCl_3$ in the Raman spectra at 80 K of single crystals of $LaCl_3$, $CeCl_3$, $NdCl_3$, $EuCl_3$, and $GdCl_3$, excited by an He-Ne laser at 632.8 nm or by an Ar laser at 514.5, 488.0, and 457.9 nm.

Wave numbers in $cm^{-1}$:

| mode | $LaCl_3$ | $CeCl_3$ | $PrCl_3$ [a] | $NdCl_3$ | $EuCl_3$ | $GdCl_3$ |
|---|---|---|---|---|---|---|
| $A_g$ | 181.6 | – | 183 | – | – | 179.4 |
| $A_g$ | 210.9 | 216.8 | 214 | 218.6 | 211.2 | 189.8 |
| $E_{1g}$ | 187.0 | 194.8 | 198 | 192.6 | 187.4 | 203.4 |
| $E_{2g}$ | 108.0 | 108.4 | 106 | 103.6 | 107.5 | 99.3 |
| $E_{2g}$ | 179.8 | 183.1 | 179 | 183.5 | 180.3 | 193.2 |
| $E_{2g}$ | 217.6 | 221.5 | 217 | 225.5 | 217.5 | 234.4 |

[a] Measured at 10 K by Damen et al. [3].

The occasionally not observed lower frequency $A_g$ vibration forms with one of the $E_{2g}$ vibrations a nearly degenerate line pair [3]. In earlier studies on $LaCl_3$ and $PrCl_3$ at 77 K and room temperature as well as on $NdCl_3$ at 77 K (excited by the 253.7 nm Hg line) Hougen, Singh [4] observed throughout only five lines, from which they erroneously interpreted the high frequency $A_g$ vibration as an $E_{2g}$ vibration and the single line (in $LaCl_3$ at 77 K a doublet) observed in place of the nearly degenerate line pair as an $A_g$ vibration. The peculiarities of the $GdCl_3$ spectrum with regard to the intensities (see figure in [2]) and assignment of the lines are due to its structurally special position as the last member in the series of the hexagonal compounds. The effect of the mass of the trivalent rare earth ion and the volume decrease in the different types are discussed in [2], and the wave numbers of the lattice vibrations are portrayed graphically as a function of the atomic number of the lanthanide element and the volume of the unit cell. The broad band observed in $LaCl_3$, $CeCl_3$, $NdCl_3$, $EuCl_3$, and $GdCl_3$ at 83, 74, 50 (?), 62, and 66 $cm^{-1}$, respectively, with $E_{1g}$ polarization with a comparatively very small and temperature independent intensity is assigned to a two phonon transition within the acoustic branch of the crystal [2].

In the Raman spectra at 10 K the lines in $CeCl_3$ and $GdCl_3$ are displaced slightly with respect to those at 77 K, in $EuCl_3$ more strongly (up to 4 $cm^{-1}$ to higher frequencies), see the table in Schaack, Koningstein [2]. The lower frequency vibration of the $A_g$ mode is not observed by these authors for $CeCl_3$, $NdCl_3$, and $GdCl_3$. In the spectrum of Damen et al. [3] the lower frequency line of the nearly degenerate line pair in $CeCl_3$ at 176 $cm^{-1}$ corresponds to the $A_g$ vibration and the high frequency line at 180 $cm^{-1}$ to the $E_{2g}$ vibration. Thus the relative location of these vibrations in $LaCl_3$ at 77 K [2] and in $PrCl_3$ at 10 K is interchanged [3]. The Raman line in $CeCl_3$ at 193 $cm^{-1}$ assigned to $E_{1g}$ has a distinctly asymmetric intensity distribution, which possibly is due to interactions between phonons and electrons in the 4f level, Kiel et al. [5]. This interaction is probably also responsible for the excessive broadening of this line with increasing temperature. This was not observed in the spectra of $PrCl_3$ and $NdCl_3$, for which in contrast to $LaCl_3$ (without 4f levels) and $EuCl_3$ (the 4f levels of $Eu^{3+}$ do not lie in the IR region), several electronic levels lie in the phonon region, Schaack, Koningstein [2].

### Raman Spectra in Magnetic Fields

At 1.7 to 2 K in an external magnetic field B ($B \leqq 6$ T) parallel to the optical axis of the crystal, the $E_{1g}$ mode and the lower frequency $E_{2g}$ mode at $\approx 108$ $cm^{-1}$ split in $CeCl_3$ [15], $PrCl_3$, and $NdCl_3$; no splitting is observed in $EuCl_3$ and $GdCl_3$ (excitation by an Ar laser at 514.5 nm for Ce, Pr, and Gd, at 488.0 nm for Nd and Eu trichlorides), Schaack [16]. The splitting initially increases linearly with the field strength and attains its temperature independent saturation value $\Delta W_\infty (\equiv W(B \to \infty) - W(B=0))$ for both modes at about 3 T in $CeCl_3$ [15]; $\Delta W_\infty (E_{1g}) = 9$ $cm^{-1}$, $\Delta W_\infty (E_{2g}) = 3.5$ $cm^{-1}$. $PrCl_3$ behaves similarly: $\Delta W_\infty (E_{1g}) = 3.25$ $cm^{-1}$, $\Delta W_\infty (E_{2g}) = 0.7$ $cm^{-1}$. In $NdCl_3$ only the $E_{1g}$ mode reaches its saturation value (3.1 $cm^{-1}$) at 4 T, while the splitting of the $E_{2g}$ mode is noticeable at 6 T only through a line broadening, from which $\Delta W_\infty (E_{2g}) \leqq 0.3$ $cm^{-1}$ is inferred [16]. The line widths of the split components, which have different intensities, decrease with increasing magnetic field [15]. The light scattered in the direction of the optical axis by these phonon states, with linearly polarized incident light, is not purely circularly polarized, but elliptically polarized because of the depolarization phenomena in these compounds, Ahrens, Schaack [17]. A magnetic field oriented perpendicular to the optical axis of the crystal effects a distinct splitting of the modes in $CeCl_3$; this increases proportionally to $B^2$ and is still increasing even at 6 T. The components are linearly polarized [15]; $\Delta W (\equiv W(B=6\ T) - W(B=0)) = 6.5$ $cm^{-1}$ for $E_{1g}$ and 2.0 $cm^{-1}$ for $E_{2g}$. In the Raman spectrum of $NdCl_3$ only for the $E_{1g}$ mode a weak splitting occurs ($\Delta W = 0.5$ $cm^{-1}$) in the form of a shoulder, while in the spectrum of $PrCl_3$, $EuCl_3$, and $GdCl_3$ no changes of any sort are observed [16]. The observations made were predicted from symmetry considerations for uniaxial crystals in a magnetic field by Anastassakis et al. [18, 19, 20]. They were explained by Thalmeier, Fulde [21] in $CeCl_3$ by a coupling of the 4f electrons of the rare earth ions to the optical phonons. They are discussed by Schaack [15, 16, 17] in part in connection with similar observations on rare earth fluorides.

### IR spectra

The transverse (T) and longitudinal (L) optical lattice vibrations of the $A_u$ type (for field directions parallel to the c axis) and of the $E_{1u}$ type (for field directions perpendicular to the c axis) at room temperature have been derived by dispersion analyses of the polarized reststrahlen reflectance bands on $LaCl_3$ and $PrCl_3$ [6] and through Kramers-Kronig transformation of the reflection curves of $GdCl_3$ (T=77 K, region 40 to 400 $cm^{-1}$) and classical oscillator fitting procedures [7]:

| $MCl_3$ . . . . . | $LaCl_3$ [6] | | | $PrCl_3$ [6] | | | $GdCl_3$ [7] | | |
|---|---|---|---|---|---|---|---|---|---|
| mode . . . . . | $A_u$ | $E_{1u}$ | $E_{1u}$ | $A_u$ | $E_{1u}$ | $E_{1u}$ | $A_u$ | $E_{1u}$ | $E_{1u}$ |
| $\nu_{TO}$ in $cm^{-1}$ . . . | 157.6 | 167.3 | 195.6 | 156.2 | 164.0 | 201.9 | 158 | 154 | 214 |
| $\nu_{LO}$ in $cm^{-1}$ . . . | 256.3 | 191.0 | 260.0 | 257.0 | 195.5 | 257.0 | – | – | – |

In all three compounds the vibrations of the $A_u$ type show at higher wave numbers a shoulder, whose occurrence is connected with anharmonic effects [7]. In $LaCl_3$ the analysis mentioned yields additional wave numbers (poles and zeros in the complex-frequency plane of the dielectric dispersion function) which, however, are not assigned to fundamental vibrations; wave numbers for the vibrations and the associated damping constants are given for 293 and 82 K in [6].

In the IR spectra of powdered $PrCl_3$, $NdCl_3$, $SmCl_3$, $EuCl_3$, and $GdCl_3$ as Nujol suspensions, at room temperature the three absorption bands corresponding to the expected lattice vibrations are observed by Basile et al. [8], wave numbers in $cm^{-1}$:

| $LaCl_3$ | $PrCl_3$ | $NdCl_3$ | $SmCl_3$ | $EuCl_3$ | $GdCl_3$ |
|---|---|---|---|---|---|
| 210 | 210 (s) | 210 (s) | 216 (s) | 218 (s) | 220 (m) |
| 165 | 180 (s) | 178 (m) | 176 (m) | 176 (m) | 174 (m, b) |
| (138) | 167 (s) | 167 (m) | 165 (m) | 163 (m) | 166 (m) |

(s = strong, m = medium, b = broad)

The results for $LaCl_3$ come from Murphy et al. [1] for a film evaporated onto a polyethylene substrate, for which the absorption band at 138 $cm^{-1}$ is, however, possibly due to oxide chloride contamination [8].

Stedman, Newman [9] calculate from the transverse and longitudinal lattice vibrations of Berreman, Unterwald [6] the single frequencies (predicted by the harmonic force model) in $LaCl_3$ and $PrCl_3$. In earlier work the absorption bands at 165 $cm^{-1}$ [1] and at 210 $cm^{-1}$ were from analysis of the vibronic spectrum [10] assigned to vibrations of the $A_u$ type. The IR spectrum of $PrCl_3$, representing these compounds, is graphically presented in [8] in comparison with those of the orthorhombic $TbCl_3$ and monoclinic $LuCl_3$.

At 77 K the following absorption maxima are observed for $PrCl_3$, $NdCl_3$, and $GdCl_3$ and assigned to phonon vibrations [8]:

$PrCl_3$: 240 (s), 210 (s, very broad), 187 (s, very broad)
$NdCl_3$: 239 (shoulder), 209 (s), 186 (s), 169 (s, broad), 125 (w)
$GdCl_3$: 229 (s), 221 (s), 188 (s, broad)

(s = strong, w = weak)

The resolution is slight because of measurement in a Dewar cell. Very weak absorptions are observed occasionally between 300 and 500 $cm^{-1}$ at 298 and 77 K and assigned to multiphonon vibrations [8].

The sharp vibronic lines with wave numbers of about 70 to 270 $cm^{-1}$, which accompany the electronic transitions within the trivalent rare earth ions in the visible region in the polarized absorption spectra of $LaCl_3$ doped with $Pr^{3+}$, $PrCl_3$, $GdCl_3$ [11], and $NdCl_3$ [12] at 1.5 K, are assigned to optical phonon branches with vectors k in various regions of the Brillouin zone. The broad absorption band between 0 and 75 $cm^{-1}$ belongs to acoustic phonon branches around k = 0 [11, 12]. The vibronic spectrum of $LaCl_3$ with $Pr^{3+}$ doping was earlier analyzed by Richman et al. [10], however, under the assumption that yield exclusively k = 0 contributions; see the tables and discussions in the papers [10, 11, 12]. See Hellwege et al. [22] for data for $LaCl_3$ doped with $Ce^{3+}$.

Cohen et al. [13] have calculated the distribution function of the phonon frequencies from the polarized vibronic absorption spectrum of $NdCl_3$ at 1.5 K in the region of the transition $^4I_{9/2}$ ($\mu = \pm 5/2$) $\rightarrow$ $^2P_{1/2}$ ($\mu = \pm 1/2$) of $Nd^{3+}$; this yields the relative density of states as a function of the phonon frequency, see figure in the paper. It shows that the area under the curve in the optical phonon region (80 to 270 $cm^{-1}$) is about 40 times larger than that for the acoustic phonon region; this is said to break off at about 80 $cm^{-1}$ [13].

If during the spectroscopic investigations water vapor is not sufficiently eliminated, then there occur in the absorption spectra of the trichlorides of La, Nd, Sm, and Eu as Nujol suspensions absorption bands at about 210, 280, and 360 $cm^{-1}$. In addition the maxima observed at higher frequencies are due to structurally bound and absorbed water, Taylor et al. [14].

A splitting of the IR-active $E_u$ modes, for example in $CeCl_3$, is said to be observable, if an external magnetic field is applied to the crystal parallel to the optical axis. This hypothesis is based on the theory that in an external magnetic field observed splittings of the Raman-active $E_g$ modes is based upon the magnetoelastic coupling of the optical phonons and the 4f states of the rare earth ions split in the crystal field (see p. 82). The splitting of the $E_u$ modes is said to be determinable by transmission and reflectance measurements, see the formulations of Thalmaier, Fulde [23].

Symmetry coordinates for the 24 k=0 modes are derived by Murphy et al. [1] and corrected by Stedman, Newman [9].

### Monoclinic $MCl_3$

Based on the symmetry of the lattice (space group $C2/m-C_{2h}^3$, Z=4), in the monoclinic trichlorides of Y and Dy through Lu there are 12 Raman-active ($6\ A_g+6\ B_g$) and 9 IR-active ($4\ A_u+5\ B_u$) vibrations expected, Basile et al. [8]. The following absorption maxima are found in the IR spectra of the powdered substances as Nujol suspensions at 298 and 78 K, wave numbers in $cm^{-1}$ [8]:

| | | |
|---|---|---|
| 298 K | $DyCl_3$ | 262 (sh), 220 (s), 207 (m), 165 (sh), 128 (m), 116 (sh), 107 (m), 65 (w) |
| | $HoCl_3$ | 254 (sh), 221 (m, b), 209 (m), 106 (m), 102 (sh), 68 (w) |
| | $ErCl_3$ | 262 (sh), 222 (s, b), 210 (s), 108 (w) |
| | $TmCl_3$ | 262 (sh), 226 (s, b), 212 (s), 136 (w, sh), 109 (m), 72 (w) |
| | $YbCl_3$ | 269 (m), 229 (s), 222 (sh), 138 (vw), 120 (w), 112 (m), 73 (w) |
| | $LuCl_3$ | 272 (m), 230 (s), 217 (s), 140 (vw), 113 (m), 75 (w) |
| 78 K | $TmCl_3$ | 271 (s), 237 (sh), 228 (s), 214 (s), 168 (vw), 110 (s), 78 (vw) |
| | $YbCl_3$ | 259 (w), 226 (s), 210 (s), 175 (s), 143 (s, b) |
| | $LuCl_3$ | 269 (s), 233 (s), 216 (s), 108 (s) |

(sh=shoulder, s=strong, m=medium, w=weak, vw=very weak, b=broad)

Very weak absorptions are occasionally observed between 300 and 500 $cm^{-1}$ and assigned to multiphonon vibrations. The spectrum of $LuCl_3$ as a representative of this group is compared in a figure in [8] with that of the hexagonal $PrCl_3$ and the orthorhombic $TbCl_3$. The spectra of $HoCl_3$ and $TmCl_3$ in the region from 30 to 400 $cm^{-1}$ are pictured without data for vibration frequencies in figures in Berringer et al. [24] together with spectra for the corresponding hexahydrates.

References:

[1] J. Murphy, H.H. Caspers, R.A. Buchanan (J. Chem. Phys. **40** [1964] 743/53). – [2] G. Schaack, J.A. Koningstein (J. Phys. Chem. Solids **31** [1970] 2417/20). – [3] T.C. Damen, A. Kiel, S.P.S. Porto, S. Singh (Solid State Commun. **6** [1968] 671/3). – [4] J.T. Hougen, S. Singh (Proc. Roy. Soc. [London] A **277** [1964] 193/205, 196/7). – [5] A.

Kiel, T. Damen, S.P.S. Porto, S. Singh, F. Varsanyi (Phys. Rev. [2] **178** [1969] 1518/24, 1523/4).

[6] D.W. Berreman, F.C. Unterwald (Phys. Rev. [2] **174** [1968] 791/9). – [7] P. Dawson, G. Schaack (J. Phys. Chem. Solids **34** [1973] 2193/9). – [8] L.J. Basile, J.R. Ferraro, D. Gronert, A. Quattrochi (J. Chem. Phys. **55** [1971] 3729/33). – [9] G.E. Stedman, D.J. Newman (J. Phys. Chem. Solids **31** [1970] 1587/93). – [10] I. Richman, R.A. Satten, E.Y. Wong (J. Chem. Phys. **39** [1963] 1833/46; Erratum: J. Chem. Phys. **40** [1964] 1451/2).

[11] E. Cohen, H.W. Moos (Phys. Rev. [2] **161** [1967] 268/71). – [12] E. Cohen, H.W. Moos (Phys. Rev. [2] **161** [1967] 258/68). – [13] E. Cohen, L.A. Riseberg, H.W. Moos (Phys. Rev. [2] **175** [1968] 521/5). – [14] M.D. Taylor, T.T. Cheung, M.A. Hussein (J. Inorg. Nucl. Chem. **34** [1972] 3073/9). – [15] G. Schaack (Z. Physik B **26** [1977] 49/59).

[16] G. Schaack (Physica B+C **89** [1977] 195/200). – [17] K. Ahrens, G. Schaack (Indian J. Pure Appl. Phys. **16** [1978] 311/8). – [18] E. Anastassakis, E. Burstein, A.A. Maradudin, R. Minnick (J. Phys. Chem. Solids **33** [1972] 519/31). – [19] E. Anastassakis, E. Burstein, A.A. Maradudin, R. Minnick (J. Phys. Chem. Solids **33** [1972] 1091/103). – [20] E. Anastassakis, E. Burstein (J. Phys. C **5** [1972] 2468/81).

[21] P. Thalmeier, P. Fulde (Z. Physik B **26** [1977] 323/8). – [22] K.H. Hellwege, E. Orlich, G. Schaack (Physik Kondensierten Materie **4** [1965] 196/206). – [23] P. Thalmeier, P. Fulde (Z. Physik B **32** [1978] 1/3). – [24] B.W. Berringer, J.B. Gruber, D.N. Olsen, J. Stöhr (J. Inorg. Nucl. Chem. **34** [1972] 373/5).

### 11.2.6.4 Force Constants

On the basis of a simple rigid ion model, which does not consider Coulombic forces, the observed Raman-active and transverse IR-active lattice vibrations yield the following force constants a (in $10^{-19}$ g/cm$^2$):

a($M^{3+}$-$Cl^-$): 8.3 ($LaCl_3$) 8.25 ($PrCl_3$) 9.1 ($GdCl_3$)

a($Cl^-$-$Cl^-$): 4.3 ($LaCl_3$) 4.6 ($PrCl_3$) 5.8 ($GdCl_3$)

A refinement with four force constant parameters (two for $M^{3+}$-$Cl^-$ and two for $Cl^-$-$Cl^-$) yields no significantly better agreement between the calculated and observed wave numbers, Dawson, Schaack [1]. Various harmonic force field models have been used by Stedman, Newman [2] for calculation of the force constants in $LaCl_3$, $CeCl_3$, and $PrCl_3$, from which again a model with only two parameters yields the best results; for numerical values see [2]. These authors [1, 2] find different relative sizes for the $M^{3+}$-$Cl^-$ and $Cl^-$-$Cl^-$ force constants; these also differ from the first results of Murphy et al. [3], who based their calculations on a central force field with four force constants.

References:

[1] P. Dawson, G. Schaack (J. Phys. Chem. Solids **34** [1973] 2193/9, 2196/7). – [2] G.E. Stedman, D.J. Newman (J. Phys. Chem. Solids **31** [1970] 1587/93). – [3] J. Murphy, H.H. Caspers, R.A. Buchanan (J. Chem. Phys. **40** [1964] 743/53, 749/53).

### 11.2.6.5 Bonding

The bonds in the anhydrous rare earth trichlorides are substantially ionic, but there are nevertheless various indications of a covalent bonding component, i.e., a small amplitude coefficient with which ligand wave functions are mixed with the 4f functions of the rare earth ions. This mixing reduces the orbital angular momentum of the rare earth ions, which becomes noticeable in the spin-orbit coupling constant [1] and in the electronic g-factor [2]. Also, the decrease determined in the electronic spectrum for the Slater-Condon parameters with respect to the values for the free ions ("nephelauxetic effect") is traced to a covalent bonding component [3 to 5]; more recent MO calculations for $PrCl_3$, however, show that the overlap and covalent contributions to the Slater-Condon parameters are small and their reduction is attributed rather to the screening of the Coulomb interaction by the dielectric medium of the crystal [6].

The covalent share of the bonding between the rare earth ions and the Cl ligands also shows itself in magnetic fields at the nuclei of the Cl atoms, which are due to the spins of the rare earth ions and cause at low temperatures a hyperfine structure in the $^{35}Cl$ quadrupole resonance spectra [7, 8].

Detailed information concerning the changes which the wave functions of the free rare earth ions undergo on introduction into a crystalline environment is obtained by comparison of calculated and measured crystal field parameters. These parameters characterize the Stark splitting of the electronic levels of the rare earth ions under the action of the ligand field. For 4f electrons in a field of $C_{3h}$ symmetry which is the site symmetry of the rare earth ions in the hexagonal trichlorides, the crystal field splitting to first order is determined by four parameters, $A_2^0\langle r^2\rangle$, $A_4^0\langle r^4\rangle$, $A_6^0\langle r^6\rangle$, and $A_6^6\langle r^6\rangle$. The coefficients $A_n^m$ are defined by an expansion of multipole potentials, and $\langle r^n\rangle$, n=2, 4, 6, are the mean values of $r^n$ averaged over the 4f functions, see for example [9]. The following table contains a selection of experimental and calculated crystal field parameters in $cm^{-1}$ for $PrCl_3$ and $Pr^{3+}$ in $LaCl_3$:

| | experimental | | theoretical | |
|---|---|---|---|---|
| | $PrCl_3$ | $LaCl_3:Pr^{3+}$ | $PrCl_3$ | $PrCl_3$ |
| $A_2^0\langle r^2\rangle$ | 62 | 47.26 | 168 | 139 |
| $A_4^0\langle r^4\rangle$ | −43 | −40.58 | −17.54 | −41 |
| $A_6^0\langle r^6\rangle$ | −44 | −39.62 [a)] | 2.80 | −36 |
| $A_6^6\langle r^6\rangle$ | 440 | 405.4 [b)] | −33.3 | 437 |
| Ref. | [2] | [10] | [16] | [20] |

a) Positive sign for $A_6^0\langle r^6\rangle$ in the paper [10] obviously in error.
b) Corrected value from [11].

The experimental values for $PrCl_3$ are obtained from the magnetic susceptibility [2], those for $Pr^{3+}$ in $LaCl_3$ from optical spectra [10]. The theoretical calculations are based on two different methods: Raychaudhuri et al. [16] have used a point charge model [12], but Ellis et al. [20] an MO model [17, 18] instead. The point charge model [12] is improved by consideration of the polarizability of the ligands [12] as well as by consideration of the screening effect which the remaining $Pr^{3+}$ orbitals exert on the 4f electrons [13, 14], and the modification of the Coulomb interaction by the overlap between the 4f orbitals and the ligand orbitals [14 to 16]. It is found that it is not possible to achieve agreement with the experimental crystal field parameters using this improved point charge model; especially serious discrepancies remain for the $A_6^0$ and $A_6^6$ parameters [16].

The MO calculations differ from the formulation of the point charge model initially only through the consideration of 4f overlap and of exchange effects [17, 18]. In further approximations then the additional following effects were investigated: The modification of the exchange interaction by the $Pr^{3+}$5s5p-ligand and ligand-ligand overlap [19], the modification of the Coulomb interaction by the 4f-ligand overlap [20], the contribution of the covalency, which is taken into account by inclusion of charge transfer states [20], and the influence of the ligand-ligand overlap on the superposition approximation, for which an additivity of the potentials of individual ligands is assumed [21]. The most important result of the calculations of Ellis, Newman et al. [17 to 21] is that overlap and covalency contributions alone give very nearly the correct relative magnitudes of all the crystal field parameters [20]. This was also demonstrated for $SmCl_3$ with use of a highly simplified MO model [22].

Evidence concerning the occupied valence bands and the core levels has been obtained from measurements of the absorption and X-ray photoelectron spectroscopy (XPS) of $LaCl_3$ and $CeCl_3$ thin films. The valence bands have similar profiles, a half-width of about 3 eV, and consist of two partial bands. Their formation obviously comes about only through the outermost p-electrons of the halogen atoms. In $CeCl_3$ the occupied 4f level partially overlaps the valence band, while it lies above the valence band in $CeF_3$ and in $CeBr_3$ completely overlaps the valence band. Interpretations of the other peaks in the photoelectron spectra up to binding energies of 22 eV are also given, Sato [23].

References:

[1] K.H. Hellwege, E. Orlich, G. Schaack (Physik Kondensierten Materie **4** [1965/66] 196/206). – [2] B. Schneider (Z. Physik **183** [1965] 504/20). – [3] E.Y. Wong, O.M. Stafsudd, D.R. Johnston (J. Chem. Phys. **39** [1963] 786/93). – [4] S.P. Tandon, P.C. Mehta (J. Chem. Phys. **52** [1970] 5417/20). – [5] S.P. Sinha (Spectrochim. Acta **22** [1966] 57/62).

[6] D.J. Newman (J. Phys. Chem. Solids **34** [1973] 541/5). – [7] E.H. Carlson, H.S. Adams (J. Chem. Phys. **51** [1969] 388/92). – [8] W.E. Hughes, C.G. Montgomery, W.G. Moulton, E.H. Carlson (J. Chem. Phys. **41** [1964] 3470/3). – [9] R.J. Elliott, K.W.H. Stevens (Proc. Roy. Soc. [London] A **215** [1952] 437/53). – [10] J.S. Margolis (J. Chem. Phys. **35** [1961] 1367/73).

[11] J.S. Margolis (from [12]). – [12] M.T. Hutchings, D.K. Ray (Proc. Phys. Soc. [London] **81** [1963] 663/76). – [13] D.K. Ray (Proc. Phys. Soc. [London] **82** [1963] 47/57). – [14] A.K. Raychaudhuri, D.K. Ray (Proc. 9th Nucl. Phys. Solid State Phys. Symp., Calcutta 1965, pp. 53/61). – [15] A.K. Raychaudhuri, D.K. Ray (Proc. Phys. Soc. [London] **86** [1965] 891/2).

[16] A.K. Raychaudhuri, D.K. Ray (Proc. Phys. Soc. [London] **90** [1967] 839/46). – [17] M.M. Ellis, D.J. Newman (Phys. Letters **21** [1966] 508/10). – [18] M.M. Ellis, D.J. Newman (J. Chem. Phys. **47** [1967] 1986/93). – [19] S.S. Bishton, M.M. Ellis, D.J. Newman, J. Smith (J. Chem. Phys. **47** [1967] 4133/6). – [20] M.M. Ellis, D.J. Newman (J. Chem. Phys. **49** [1968] 4037/43).

[21] M.M. Curtis, D.J. Newman (J. Chem. Phys. **52** [1970] 1340/4). – [22] E. Borchi, S. De Gennaro, M. Mancini (Lettere Nuovo Cimento **2** [1969] 810/6). – [23] S. Sato (J. Phys. Soc. Japan **41** [1976] 913/20).

### 11.2.6.6 Resonance Spectroscopy

#### 11.2.6.6.1 Mössbauer Spectroscopy

The results of Mössbauer studies of the lanthanide chlorides $MCl_3$ are collected in the following Table 11.2.6-2.

Table 11.2.6-2
Mössbauer Parameters for $MCl_3$, M = Nd, Sm, Eu, Gd, Tm, Yb.
$E_M$ denotes the resonance energy, $T_A$ and $T_S$ the temperatures of the absorber and the source (RT room temperature), $\delta$ the isomer shift, eqQ the quadrupole coupling constant of the excited nuclear level; however for $^{155}Gd$ that of the nuclear ground state.

| $MCl_3$ | isotope | $E_M$ in keV | source | $T_A$ in K | $T_S$ in K | $\delta$ in mm/s | referred to | eqQ in mm/s | notes | Ref. |
|---|---|---|---|---|---|---|---|---|---|---|
| $NdCl_3$ | $^{145}Nd$ | 72.50 | $^{145}Pm$ in $^{144}Nd_2O_3$ (A-form) | 4.2 | 4.2 | $0.10 \pm 0.03$ | source | – | – | [1] |
| $SmCl_3$ | $^{149}Sm$ | 22.5 | $^{149}Eu$ in $EuF_3$ | RT | RT | $-0.01 \pm 0.04$ | $SmF_3$ | – | – | [2] |
| $EuCl_3$ | $^{151}Eu$ | 21.6 | $^{151}Sm$ in $SmF_3$ | RT | RT | $0.049 \pm 0.005$ | $EuF_3$ | – | a) | [3] |
| $GdCl_3$ | $^{155}Gd$ | 86.5 | $^{155}Eu$ in $Sm_2Sn_2O_7$ | 4.2 | 4.2 | $0.59 \pm 0.02$ | $^{155}Eu$ in Pd | $3.86 \pm 0.06$ | b) | [4] |
| $GdCl_3$ | $^{155}Gd$ | 86.5 | $^{155}Eu$ in $Sm_2O_3$ | 78 | $\approx 85$ | 0.66 | $Gd_2O_3$ | 4.08 | b) | [5] |
| $GdCl_3$ | $^{158}Gd$ | 79.5 | $^{158}Gd$ in $Gd_{0.03}Y_{0.97}Al_2$ | $\approx 25$ | $\approx 25$ | $0.047 \pm 0.054$ | Gd metal | $5.92 \pm 0.38$ | c) | [6] |
| $TmCl_3$ | $^{169}Tm$ | 8.42 | $ErF_3$ | 77; RT | RT | 42; 37(±2) | source | 17; 25 (±2) | – | [7] |
| $YbCl_3$ | $^{170}Yb$ | 84.3 | $^{170}Tm$ in $TmAl_2$ | 4.2 | 4.2 | $0.11 \pm 0.07$ | $YbAl_2$ | $< 1.2$ | d) | [8, 9] |

a) Additional results for $^{151}EuCl_3$ [10, 11] probably are for hydrates of unidentified composition, cf. [3].

b) Reference to the common standard $Gd_2O_3$ produces no agreement between the $\delta$ values of Cashion et al. [4] and Balabanov et al. [5].

c) In the paper [6] $eqQ = (15.7 \pm 1.0) \times 10^{-7}$ eV.

d) The upper limit for eqQ (in paper [9] 80 MHz) is estimated from analysis of an unsplit line. From the absence of a magnetic hyperfine structure it is concluded that the electronic spin relaxation time is smaller than $10^{-12}$ s [9].

References:

[1] G. Kaindl (Z. Physik **240** [1970] 100/22). – [2] J.M.D. Coey (J. Inorg. Nucl. Chem. **38** [1976] 1139/40). – [3] G.W. Dulaney, A.F. Clifford (Moessbauer Eff. Methodol., Proc. Symp. **5** [1969/70] 65/76). – [4] J.D. Cashion, D.B. Prowse, A. Vas (J. Phys. C **6** [1973] 2611/24). – [5] A.E. Balabanov, N.N. Delyagin, E.N. Kornienko, L.N. Pankratova (Yadern. Fiz. **9** [1969] 1116/8; Soviet J. Nucl. Phys. **9** [1969] 653/4).

[6] J. Fink (Z. Physik **207** [1967] 225/34). – [7] C.I. Wynter, C.H. Cheek, M.D. Taylor, J.J. Spijkerman (Nature **218** [1968] 1047). – [8] U. Atzmony, E.R. Bauminger, J. Hess, A. Mustachi, S. Ofer (Phys. Rev. Letters **18** [1967] 1061/3). – [9] I. Nowik, S. Ofer (J. Phys. Chem. Solids **29** [1968] 2117/9). – [10] F.A. Deeney, J.A. Delaney, V.P. Ruddy (Phys. Letters A **25** [1967] 370/1).

[11] G. Gerth, P. Kienle, K. Luchner (Phys. Letters A **27** [1968] 557/8).

### 11.2.6.6.2 Nuclear Quadrupole Resonance (NQR)

#### $^{35}$Cl-$^{37}$Cl NQR

Introductory Remarks

The $^{35}$Cl and $^{37}$Cl NQR spectra reflect the differences in the crystallographic forms of the anhydrous chlorides: In the hexagonal chlorides $LaCl_3$ through $GdCl_3$ there are two physically nonequivalent positions, while in contrast in the monoclinic chlorides $YCl_3$, $DyCl_3$ (high temperature modification), and $HoCl_3$ through $YbCl_3$, as well as in the orthorhombic $TbCl_3$, there are two chemically nonequivalent positions for the Cl atoms. (At physically nonequivalent lattice sites the tensor of the electric field gradient has the same values for its principal components, but different orientations of the principal axes; at chemically nonequivalent sites the principal field gradient tensor components are different, see for example [1].) Correspondingly, the pure NQR spectra of the hexagonal chlorides consist of one line, whereas for the monoclinic and orthorhombic chlorides they consist of two lines. The resonance frequencies are given by the relation $\nu_Q = (^1/_2) \cdot eQV_{zz} \cdot (1+\eta^2/_3)^{1/2}/h$, where Q is the nuclear quadrupole moment, $\eta = (V_{xx}-V_{yy})/V_{zz}$ is the asymmetry parameter of the field gradient tensor in the principal axis system, and h is Planck's constant. In the hexagonal chlorides from La through Gd on symmetry grounds the threefold axis must coincide with one of the principal axes, namely with the x-axis if the axis setting is chosen as is customary, so that $|V_{xx}| \leqq |V_{yy}| \leqq |V_{zz}|$ [6, 7].

Resonance Frequencies and Asymmetry Parameters

Values of $\nu_Q$ and $\eta$ are listed in Table 11.2.6-3.

Table 11.2.6-3
$^{35}$Cl (in Parentheses $^{37}$Cl) NQR Frequencies $\nu_Q$ in kHz at Various Temperatures and Asymmetry Parameters $\eta$ for the Rare Earth Trichlorides $MCl_3$.
single = monocrystalline, poly = polycrystalline samples. RT = room temperature.

| $MCl_3$ | sample | $\nu_Q$(4.2 K) | $\nu_Q$(77 K) | $\nu_Q$(RT) | $\eta$ | Ref. |
|---|---|---|---|---|---|---|
| $YCl_3$ | poly | 4212.2 ± 0.5<br>4273.1 ± 0.5 | | | | [2] |
| $LaCl_3$ | single | 4167 ± 1 | | | 0.50 ± 0.02 | [2] |
| $CeCl_3$ | single | 4387 ± 1 | 4377.2 ± 0.3 | 4341.0 ± 1.0 | | [3 to 5] |

Table 11.2.6-3 [continued]

| $MCl_3$ | sample | $\nu_Q$(4.2 K) | $\nu_Q$(77 K) | $\nu_Q$(RT) | $\eta$ | Ref. |
|---|---|---|---|---|---|---|
| $PrCl_3$ | single | 4568.5 ± 0.5<br>(3600.1 ± 0.1) | 4561.9 ± 0.8<br>(3595.5 ± 0.2) | | 0.4937<br>± 0.0001 | [3, 6, 10] a) |
| $NdCl_3$ | single | 4729.0 ± 0.3 | 4722.3 ± 0.3 | 4676.4 ± 0.3 | | [3 to 5] |
| $SmCl_3$ | poly | 5033.0 ± 0.5 | 5026.8 ± 0.5 | 4975.6 ± 0.4 | | [3 to 5] |
| $GdCl_3$ | single | 5315 ± 1<br>(4190 ± 3) | 5307.6 ± 0.3 | 5248.4 ± 0.3 | 0.4265<br>± 0.0001 | [3, 7, 8, 10] |
| $TbCl_3$ | poly | | 4300 ± 20<br>7450 ± 20 | | | [9] b) |
| $DyCl_3$ | poly | | | 4203 ± 2 | | [2] c) |
| $HoCl_3$ | poly | 4307.1 ± 0.3<br>4390 ± 10 | 4302.7 ± 0.4<br>4364.7 ± 1.0 | 4276.6 ± 0.3<br>4334.0 ± 1.0 | | [2] |
| $ErCl_3$ | single | 4456.6 ± 0.2 d)<br>4515.7 ± 0.2 e) | 4449.5 ± 0.3<br>4506 ± 5 | 4424.3 ± 0.5<br>4445 ± 10 | 0.53 ± 0.02 d)<br>0.52 ± 0.03 e) | [2] |
| $ErCl_3$ | single | | 4449.2 ± 0.3<br>4508.5 ± 0.3 | | | [3] |
| $YbCl_3$ | poly | 4790.3 ± 0.2<br>4846.4 ± 0.2 | 4781.7 ± 0.3<br>4836.6 ± 0.5 | 4738.2 ± 0.5<br>4787.7 ± 0.5 | | [2] |
| $YbCl_3$ | single | | 4781.1 ± 0.3<br>4836.0 ± 0.4 | | | [3] |

a) Slightly different values were reported elsewhere [4, 5].

b) For orthorhombic $TbCl_3$ the smaller resonance frequency is from Cl at f sites, the larger one from Cl at c sites [9], see p. 77.

c) Attempts to obtain at 77 K a spectrum of the orthorhombic low temperature modification of $DyCl_3$ yielded nonreproducible results [9]. At room temperature with a special spectrometer a second weak line about 30 kHz above the principal line was found [2].

d) Values belong to the Cl(2) atoms of the monoclinic $ErCl_3$, space group C2/m.

e) Values belong to the Cl(1) atoms of the monoclinic $ErCl_3$ structure, space group C2/m. The 4 Cl(1) atoms are located at special positions (i) with point symmetry m, the 8 Cl(2) atoms at general positions (j) with point symmetry 1, see p. 78 [2].

For $ErCl_3$ the measured field gradient $|V_{zz}/e| = 3.103$ Å$^{-3}$ at the Cl(1) positions is explained by a bond-switching model, for which the field gradient is due only to the two nearest $Er^{3+}$ neighbors [2].

## Effects of Magnetic Ordering Phenomena

The magnetic field present in a magnetically ordered crystal leads to a Zeeman splitting of the pure NQR lines. Observation of this splitting as well as of line shapes and intensities can in combination with measurements of the susceptibility and heat capacity yield conclusions regarding the magnetic ordering of the hexagonal rare earth chlorides which occurs at low temperatures. The NQR spectra yield the following results:

In $CeCl_3$ at $T \leqq 4.2$ K an unsplit line is observed, whose intensity is proportional to $T^{-1}$. Below 0.20 K this intensity diminishes without line broadening and becomes unmeasurably small for $T < 0.110$ K. From this it is concluded that an internal magnetic field at the nucleus of $^{35}Cl$ averages out to zero within a time of the order of magnitude of the reciprocal line width ($\approx 10^{-4}$ s), Colwell et al. [4]; for preliminary results see [5, 11].

In $PrCl_3$ the Zeeman effect of the NQR line in an external magnetic field shows the absence of any magnetic ordering for $T \geqq 1.7$ K [6]. Below 0.4 K the NQR line splits into two components, which at 0.31 K have a separation of 80 kHz [12]. A precise measurement of the transition temperature yields $T = 0.428$ K [4]. From Zeeman studies in an external magnetic field Hessler et al. [13] conclude that the splitting is due to a transition into a new crystal modification with two chemically nonequivalent Cl positions. The surmise by Hessler et al. [12] that the splitting is attributable to a magnetic ordering cannot be confirmed [13]. Doubts about this conjecture are also expressed by Colwell et al. [4], who show that the earlier data from [12] would result in an orientation of the internal magnetic field relative to the threefold axis, from which in any case a spectrum with four lines instead of the measured two-line spectrum would have to occur.

In $NdCl_3$ on cooling below 4.2 K initially no change in the NQR frequency is observed. The resonance vanishes at 0.50 K and appears again at 0.17 K, where a splitting occurs into three lines at 4.67, 5.26, and 6.73 MHz with an intensity ratio of 1:1.5:10. In the $^{37}Cl$ NQR spectrum the two most intense lines are at 3.82 and 4.26 MHz. The internal magnetic field at the nucleus of $^{35}Cl$ is determined to be $B = 2520$ G; its orientation in the principal axis system of the electric field gradient is defined by the polar angle $\Theta = 46° \pm 1°$ and the azimuth $\Phi = 20° \pm 2°$. The NQR spectra together with measurements of the heat capacity show that both short and long range order are present and that a short range order goes through fluctuations in time intervals which are smaller than the time required to sweep through a resonance, i.e., a few seconds, but greater than $10^{-6}$ s [4].

In $SmCl_3$ the NQR line below 0.32 K becomes broader and weaker and vanishes at 0.21 K, which as for $CeCl_3$ and $NdCl_3$ is attributed to a fluctuating magnetic ordering, Colwell et al. [4]; for preliminary results see [5].

In $GdCl_3$, which becomes ferromagnetic below $T_C = 2.214$ K, see "Rare Earth Elements" C4b, Section 11.2.11.10.4.4, the NQR line splits at the Curie point into three components, whose frequency spacing increases with lowering temperature. Measurements of the resonance frequencies as functions of an external magnetic field at various temperatures show that the internal field at the Cl nucleus is exactly proportional to the bulk magnetization and that the internal field produced by spin transfer is oriented antiparallel to the magnetization along the $C_3$ axis. The component $A_{zz}$ of the coupling tensor summed over all $Gd^{3+}$ sites is calculated at 685.0 kHz for $^{35}Cl$, which corresponds to a hyperfine field at the $^{35}Cl$ nucleus of 1644 G, Hessler [7]; for preliminary results see [8, 14].

## Pressure Dependence

On application of pressures up to 5000 kg/cm² the $^{35}Cl$ NQR frequencies measured at 77 K decrease linearly for the hexagonal chlorides $LaCl_3$ through $GdCl_3$, those for the monoclinic chlorides $ErCl_3$ and $YbCl_3$ increase linearly. For the hexagonal chlorides the normalized pressure coefficients $\nu_0^{-1} \cdot (\delta\nu/\delta P)_T$ vary smoothly between the extremes $-3.855 \cdot 10^{-6}$ cm²/kg and $-5.586 \cdot 10^{-6}$ cm²/kg for $GdCl_3$ and $CeCl_3$, respectively. For both $ErCl_3$ and $YbCl_3$ the pressure coefficients for the two nonequivalent Cl sites differ. The decrease of the frequency on compression of the hexagonal chlorides shows that the electric field gradient at the nuclei of $^{35}Cl$ cannot be explained with a point charge model, since for that model in any case the frequency would be expected to increase with pressure. An explanation of the measured pressure dependence can be effected, with certain assumptions regarding the modulus of compression and its anisotropy, by a "soft-sphere model", in which interaction potentials between ion pairs are constructed in analogy to Born-Mayer potentials. The soft-sphere model is not applied to $ErCl_3$ and $YbCl_3$, since for them the atomic coordinates are less well known, Current et al. [3]. The insufficiency of a point charge model is also demonstrated by calculations of Carlson et al. [2], where for both the hexagonal and the monoclinic chlorides the calculated $^{35}Cl$ NQR frequencies in the series $LaCl_3$ through $GdCl_3$, or $DyCl_3$ through $YbCl_3$, increase much more slowly than the measured frequencies. The disappearance of the resonance lines in $ErCl_3$ at 3700 kg/cm² and 77 K is supposed to be due to a phase change [3].

**La and Gd NQR**

The $^{139}La$ NQR spectrum of $LaCl_3$, in view of the nuclear spin $I = {}^7/_2$ of $^{139}La$, on application of a magnetic field parallel to the hexagonal c axis consists of seven equidistant lines. The separation of neighboring lines amounts to 2P, where P is a quadrupole coupling constant defined by $P = 3 \cdot Q \cdot A_2^0 \cdot \gamma_N / I \cdot (2I-1)$, with Q = nuclear quadrupole moment, $A_2^0$ = crystal field parameter of the second order (see p. 86), and $\gamma_N$ = Sternheimer factor. P is measured to be $\pm(1.826 \pm 0.010) \times 10^{-5}$ cm$^{-1}$ [15]. P calculated by a point charge model including screening effects agrees approximately with this value [16]. Attempts to obtain Gd NQR spectra of $GdCl_3$ in the ferromagnetic state were unsuccessful [17].

References:

[1] T.P. Das, E.L. Hahn (Solid State Phys. Suppl. No. 1 [1958] 1/209, 22/3). – [2] E.H. Carlson, H.S. Adams (J. Chem. Phys. **51** [1969] 388/92). – [3] D.H. Current, C.L. Foiles, E.H. Carlson (Phys. Rev. [3] B **6** [1972] 737/47). – [4] J.H. Colwell, B.W. Mangum, D.B. Utton (Phys. Rev. [2] **181** [1969] 842/54). – [5] B.W. Mangum, D.B. Utton (Bull. Am. Phys. Soc. [2] **12** [1967] 1043).

[6] W.E. Hughes, C.G. Montgomery, W.G. Moulton, E.H. Carlson (J. Chem. Phys. **41** [1964] 3470/3). – [7] J.P. Hessler (Phys. Rev. [3] B **8** [1973] 3151/60). – [8] E.H. Carlson (Bull. Am. Phys. Soc. [2] **11** [1966] 377). – [9] E.H. Carlson (Phys. Letters A **29** [1969] 696/7). – [10] J.P. Hessler (Diss. Michigan State Univ. 1971, pp. 1/130, from Diss. Abstr. Intern. B **32** [1972] 5393).

[11] B.W. Mangum, R.J. Soulen, D.B. Utton (Proc. 11th Intern. Conf. Low Temp. Phys., St. Andrews, Scot., 1968, Vol. 1, pp. 1361/3, 1367). – [12] J.P. Hessler, E.H. Carlson (J. Appl. Phys. **39** [1968] 1014). – [13] J.P. Hessler, E.H. Carlson (J. Appl. Phys. **42** [1971] 1316/7). – [14] J.P. Hessler (Bull. Am. Phys. Soc. [2] **18** [1973] 113/4). – [15] D.T. Edmonds (Phys. Rev. Letters **10** [1963] 129/31).

[16] D.K. Ray (Proc. Phys. Soc. [London] **82** [1963] 47/57). – [17] L.W. Johnson (Diss. Univ. Nottingham, Engl., from Phys. Abstr. **74** [1971] No. 21956).

### 11.2.6.6.3 Electron Paramagnetic Resonance (EPR)

General References:

A. Abragam, B. Bleaney, Electron Paramagnetic Resonance of Transition Ions, Oxford 1970.
S.A. Al'tshuler, B.M. Kozyrev, Electron Paramagnetic Resonance of Compounds of Transition Elements, Jerusalem - London 1974.

Review

EPR spectra of rare earth chlorides have been obtained for samples in which the ions under study were added in small concentrations to the diamagnetic $MCl_3$ compounds of Y, La, Eu, and Lu. In the pure compounds EPR spectra are rather featureless and often unobservable, as a result of line broadening due to spin-spin interaction. On dilution of the resonant ions in diamagnetic hosts, however, this broadening is substantially reduced. Measurements have mostly been performed at temperatures in the liquid hydrogen or liquid helium range, since at higher temperatures considerable line broadening occurs due to enhanced spin-lattice relaxations. An exception is provided by $Gd^{3+}$, and also by $Eu^{2+}$, which are S-state ions and thus have a spin-lattice relaxation rate slow enough to enable EPR observation even at room temperature.

Results are compiled in Table 11.2.6-4. The table is designed to provide rapid access to the pertinent literature, not to give a full account of the great variety of EPR phenomena. Therefore, no explicit numerical results are presented, except for the basic spectroscopic splitting factors, which for briefness are given without error limits.

The following symbols and abbreviations are used in Table 11.2.6-4:

| | |
|---|---|
| CF | crystal field |
| e | (preceding symbol of rare earth element): even isotope |
| ENDOR | electron nuclear double resonance |
| $g_{\parallel}$, $g_{\perp}$ | spectroscopic splitting factors in hexagonal $LaCl_3$ and $EuCl_3$ for magnetic fields parallel and perpendicular to the hexagonal c axis, respectively |
| $g_x$, $g_y$, $g_z$ | spectroscopic splitting factors in monoclinic $YCl_3$ and $LuCl_3$, for magnetic fields along the principal magnetic axes x, y, and z, respectively. Symmetry requires one of these axes to coincide with the crystal b axis, which is taken here as the z axis following the convention adopted in reference [1] |
| ∢xa | the acute angle enclosed by the x axis and the positive a axis |
| H | magnetic field strength |
| Hfs | hyperfine structure |
| $\mu$ | nuclear magnetic moment |
| $\nu$ | microwave frequency |
| Q | nuclear quadrupole moment |
| SI | spin interaction |
| T | sample temperature; room temperature = 298 K |
| $\tau_1$ | spin-lattice (longitudinal) relaxation time |
| $\tau_2$ | spin-spin (transverse) relaxation time |

Table 11.2.6-4
EPR Results for Rare Earth Impurity Ions in Host Lattices of $MCl_3$, M=Y, La, Eu, Lu. Impurity ions are tripositive unless stated otherwise.
For symbols and abbreviations see p. 93. References see p. 100.

| impurity ion | conc. (%) [a)] | M of host | T in K | $\nu$ in GHz | observed properties and derived parameters | Ref. |
|---|---|---|---|---|---|---|
| Ce | a: 1 | Y | 4.2 | 9.6 | $g_x$=1.60, $g_y$=0.66, $g_z$=1.708, ∢xa=0° | [1] |
| Ce | a: 1 | La | 1.3 to 4.2 | 25.0; 36.2 | $g_\parallel$=4.037, $g_\perp$=0.23; SI constants | [2, 3] [b)] |
| Ce | a: 2.5 to 20 | La | 1.3 to 4.2 | 25.0; 36.2 | unanalyzed spectrum, partly originating from impurities | [2] |
| Ce | a: 1 | La | – | 9; 35 | SI constants | [6] |
| Ce | a: 2 | La | 4 | 23.6 | $g_\parallel$=4.0366, $g_\perp$=0.17 | [7] |
| Ce | a: 1.1; 1.6 [c)] | La | 8.0 to 11.4 | 15.7; 16.2 | $\tau_1$, $\tau_2$ | [8] |
| Ce | a: 2 | La | 1.5 to 3.3 | 9.22 to 9.24 | $\tau_1$ | [9] |
| Ce | 0.1 to 2 | La | 1.1 to 3.6 | 9.5 | $\tau_1$ | [10] |
| Ce | a: 1; 2 | La | 1.6 to 3.3 | 9.15 | $\tau_1$ | [11] |
| Ce | a: 1 | La | 1.6 to 4.2 | 8.7 | no SI observed | [12] |
| Pr | a: 1 | Y | 4.2 | 9.6 | no resonance observed for H≦12 kG | [1] |
| Pr | 1 | La | 1.5 to 4.2 | 36.63 | $g_\parallel$=1.029; Hfs and SI constants [d)] | [13] |
| Pr | a: 2 | La | 4 | 9.20 | $g_\parallel$=1.035, $g_\perp$=0.1 [d)]; Hfs constants; $\mu(^{141}Pr)$ | [7] |
| Pr | m: 11 [e)] | La | 4.2 | 23.2 | $g_\parallel$=1.79, $g_\perp$=3.98 [f)] | [14] |
| Pr | 0.3; 1 | La | 1 to 4.2 | 15.44 | $\tau_1$ | [15] |
| Pr | a: 0.3; 1 | La | 1.6 to 3.3 | 9.17 | $\tau_1$ | [11] |
| $^{143,145}Nd$ | m: 1 | Y | 1.5 | 9.38 | ENDOR ($^{143,145}Nd$); $g_x$=1.1729, $g_y$=2.6359, $g_z$=1.6517 for $^{143}Nd$; $g_x$=1.1633, $g_y$=2.6328, $g_z$=1.6540 for $^{145}Nd$; Hfs constants; $\mu$,Q($^{143,145}Nd$) | [16] |

Table 11.2.6-4 [continued]

| impurity ion | conc. (%) [a)] | M of host | T in K | $\nu$ in GHz | observed properties and derived parameters | Ref. |
|---|---|---|---|---|---|---|
| Nd | a: 1 | Y | 4.2 | 9.6 | $g_x=1.16$, $g_y=2.63$, $g_z=1.653$, $\sphericalangle xa=0°$ | [1] |
| e Nd | a: 0.2; 2 | La | 4 | 9.2 | $g_\parallel=3.9903$, $g_\perp=1.7635$ | [17] |
| e Nd, $^{143}Nd$, $^{145}Nd$ | a: 0.2; 2 | La | 4 | 23.9 | $g_\parallel=3.996$, $g_\perp=1.763$; Hfs constants; $\mu(^{143,145}Nd)$ | [7] |
| Nd | a: 2 | La | 4.2 | 9.03 to 9.04 | SI constants | [12] |
| Nd | a: 1 | La | – | 9.1816; 35.35 | g-values and SI constants for nearest neighbor and for second nearest neighbor Nd pairs | [6] |
| $^{143}Nd$, $^{145}Nd$ | a: 0.2; 2 | La | 4 | 9.2 | ENDOR ($^{143,145}Nd$); $g_\perp=1.754$; $\mu$, $Q(^{143,145}Nd)$ | [17] [g)] |
| $^{143}Nd$, $^{145}Nd$ | m: 0.1 | La | 4; 14 | 1.00 to 4.60 | frequency-modulated EPR at $H=0$; Hfs constants; $\mu(^{143,145}Nd)$ | [19] |
| $^{143}Nd$ [h)], $^{145}Nd$ [h)] | m: 0.3, m: 0.2 | La | 1.5 to 1.8 | 9.38; 23.9 to 24.4 | $g_\parallel=3.99458$, $g_\perp=1.76133$; Hfs constants; EPR also in excited states $^4I_{13/2}$ ($g_\parallel=10.1892$, $g_\perp=1.3447$) and $^4I_{15/2}$ ($g_\parallel=9.5475$, $g_\perp=5.1212$) [o)]; Hfs constants | [20, 43] |
| Nd [h)] | 1 | La | 4.2 | 9.15 | EPR in excited states $^4I_{13/2}$ and $^4I_{15/2}$ | [21] |
| $^{145}Nd$ [h)] | m: 0.2 | La | 1.5 | 23.8 | EPR in excited states $^4I_{13/2}$ ($g=10.180$, $g_\perp=1.347$) and $^4I_{15/2}$ ($g_\parallel=9.554$, $g_\perp=5.127$); Hfs constants | [22] |
| $^{143,145}Nd$ [h)] | – | La | – | – | ENDOR in excited state $^4I_{15/2}$; Hfs constants | [44] |
| $^{144}Nd$ [h)] | m: 1 | La | 1.5 | 24.638 | EPR of coupled $Nd^{3+}$ pairs in electronic ground state and in excited state $^4I_{15/2}$ | [45] |

Table 11.2.6-4 [continued]

| impurity ion | conc. (%) [a)] | M of host | T in K | $\nu$ in GHz | observed properties and derived parameters | Ref. |
|---|---|---|---|---|---|---|
| $^{143}Nd$, $^{144}Nd$ | m: 1, m: 0.3 | La | 2.0 to 2.1 | 24.35; 33.93 | EPR of ground state detected by change of fluorescence intensity from excited $Nd^{3+}$ level on sweeping H through resonance | [23, 46] |
| eNd | a: 0.073 [i)] | La | 2.7 to 9; 1.6 to 2.9 | 14.800; 15.784 to 16.274 | $\tau_1$; $\tau_2$ | [8] |
| Nd | 0.25; 2.0 | La | 1.1 to 10 | 9.52 | $\tau_1$ | [10] |
| Nd | a: 0.1; 2 | La | 1.75 to 4.2 | 9.00; 9.15 | $\tau_1$ | [11] |
| $^{143}Nd$ | m: 0.011; 0.72 | La | 1.3 to 4.2 | 1.99801 | $\tau_1$ at H=0 | [24] |
| $^{143}Nd$ | m: 0.0114 to 0.271 | La | 1.4 to 4.2 | 1.998 | population inversion, nutation, free induction decay, and spin echoes, observed at H=0 | [25] |
| Sm | a: 1 | Y | 4.2 | 9.6 | g value so small that spectrum was partly outside of available H≦12 kG | [1] |
| eSm | m: 0.05; 0.5 | La | 2.07 | 9.65 to 9.71 | $g_{\parallel}=0.584288$ | [26] |
| eSm, $^{147}Sm$, $^{149}Sm$ | a: 2 | La | 4 | 9.2 | $g_{\parallel}=0.5841$, $g_{\perp}=0.6127$; Hfs constants; $\mu(^{147,149}Sm)$ | [7] |
| $^{147}Sm$, $^{149}Sm$ | m: 0.05; 0.5 | La | 2.07 | 9.65 to 9.71 | ENDOR ($^{147,149}Sm$); Hfs constants; $\mu$, $Q(^{147,149}Sm)$ | [26] |
| Sm | 0.2; 0.5 | La | 1.0 to 4.2 | 9.52 | $\tau_1$ | [10] |
| Sm | a: 0.1 | La | 1.6 to 3.6 | 9.15 | $\tau_1$ | [11] |
| Sm | – | La | 2.0 | 9.381 | EPR of ground state detected by change of fluorescence intensity from excited $Sm^{3+}$ level on sweeping H through resonance | [47] |

Table 11.2.6-4 [continued]

| impurity ion | conc. (%) [a)] | M of host | T in K | $\nu$ in GHz | observed properties and derived parameters | Ref. |
|---|---|---|---|---|---|---|
| $Eu^{2+}$ | 1 | La | 4.2 to 298 | X-band | $g_{\parallel}=1.9924$, $g_{\perp}=1.9927$ (T=298 K); CF and Hfs constants (T=298 K) | [27] |
| Gd | a: 1 | Y | 4.2 | 9.6 | unanalyzed complex spectrum | [1] |
| Gd | – | La | 77 | – | $g_{\parallel}=1.9915$; CF constants; line shifts and splittings in external electric field | [28] |
| Gd | w: <0.001 [k)] | La | 150 to 294 | 9 | $g_{\parallel}=1.9917$ (T=293.5 K); CF constants | [29] |
| Gd | w: 0.5 | La | 2 to 293.5 | X-band | $g_{\parallel}=1.9919$ (77 K); CF constants | [30] |
| Gd | 1 | La | 1.6 to 360 | 25.6; 9.51 | $g_{\parallel}=1.9915$ (20 and 77 K); CF and SI constants | [31] [l)] |
| e Gd, $^{155,157}Gd$ | a: 0.2 | La | 4 | 9.2 | $g_{\parallel}=g_{\perp}=1.991$; Hfs constants | [7] |
| Gd | m: 0.01 | La | 20; 90; 298 | 1 | $g_{\parallel}=g_{\perp}=1.991$ (T=90 and 290 K); CF constants | [33] |
| $^{155}Gd$, $^{157}Gd$ | w: 0.01 | La | 1.6 | X-band | ENDOR ($^{155,157}Gd$); Hfs constants; $\mu(^{155,157}Gd)$ | [34] |
| Gd | w: 0.01 | La | 4.2 | 34.5 | ENDOR ($^{139}La$); Hfs of $^{139}La$ transferred from Gd; Hfs constants | [35] |
| Gd | 0.1 | Eu | 2; 4.2; 77 | 25.6 | $g_{\parallel}=g_{\perp}=1.9971$ (77 K); SI constants Gd-Gd and Gd-Eu | [36] [m)] |
| Tb | a: 0.5; 2 | La | 4 | 9.2; 24.2 | $g_{\parallel}=17.78$, $g_{\perp}<0.1$ [d)]; Hfs constants; $\mu(^{159}Tb)$ | [7] |
| Tb | 1 | La | 0.24 to 4.2 | 9.2; 15.4 | $\tau_1$ | [15] |
| Tb | a: 0.1; 1 | La | 1.6 to 4.2 | 8.85; 9.17 | $\tau_1$ | [11] |
| Dy | a: 0.1; 1 | Y | 4.2 | 9.6 | $g_x=6.86$, $g_y=3.02$, $g_z=9.34$, $\measuredangle xa=9°$ | [1] |

Table 11.2.6-4 [continued]

| impurity ion | conc. (%) [a)] | M of host | T in K | $\nu$ in GHz | observed properties and derived parameters | Ref. |
|---|---|---|---|---|---|---|
| eDy | a: 1 | La | 1.67 to 4.58 | 8.911; 8.916 | EPR from $Dy^{3+}$ in distorted environment, $g_{\parallel}=10.43$, $g_{\perp}<1.0$ [d)]; EPR of low lying excited CF doublets | [38] |
| Dy | a: 0.2; 0.5 | La | 4 | 9.2; 24 | no resonance observed [d)] | [7] |
| Dy [h)] | m: 0.2 | La | 2.1 | 24.14 to 24.18 | EPR in excited state $^6H_{9/2}$, $g_{\parallel}=6.15$, $g_{\perp}=3.3270$ | [48] |
| Dy | a: 1 | Lu | 4.2 | 9.6 | $g_x=7.21$, $g_y=3.16$, $g_z=8.84$, $\measuredangle xa=7°$ | [1] |
| Ho | a: 0.5 | La | 4 | 24.6 | $g_{\parallel}=16.01$, $g_{\perp}\approx 0$ [d)]; Hfs constants; $\mu(^{165}Ho)$ | [7] |
| Ho | – | La | – | X-band | Hfs constants | [39] |
| Ho | 0.1 to 2.0 | La | $\leqq 1.14$ | 9.52 | $\tau_1$ | [10] |
| Er | a: 1 | Y | 4.2 | 9.6 | $g_x=8.27$, $g_y=3.78$, $g_z=7.70$, $\measuredangle xa=0°$ | [1] |
| Er | 0.5 | Y | 1.5 to 2.1 | 23.943 | EPR and $\tau_1$ in optically excited state $^4S_{3/2}$ ($g_{\parallel}=3.350$, $g_{\perp}=2.857$), monitored by modulation of fluorescence intensity on sweeping H through resonance | [40] |
| eEr, $^{167}Er$ | a: 0.5 | La | 4 | 9.25; 24 | $g_{\parallel}=1.989$, $g_{\perp}=8.757$; Hfs constants; $\mu(^{167}Er)$ | [7] |
| eEr | a: 0.002; 0.01 [n)] | La | 2.6 to 4.2; 10 to 24 | 15.47 to 15.78 | $\tau_1$; $\tau_2$ | [8] |
| eEr, $^{167}Er$ | a: 1 | La | $\leqq 4.2$ | 9.1 | $\tau_1$ | [41] |
| Er, eEr | 0.1; 0.2; 1 | La | 0.24 to 4.2 | 15.44; 16.6 | $\tau_1$ | [15] |
| $^{170}Er$ | a: 0.1; 1 | La | 1.5 to 3.3 | 9.22 to 9.24 | $\tau_1$ | [9] |
| eEr | a: 1.0 | La | 1.6 to 3.3 | 8.76; 8.89 | $\tau_1$ | [11] |

Table 11.2.6-4 [continued]

| impurity ion | conc. (%) [a)] | M of host | T in K | $\nu$ in GHz | observed properties and derived parameters | Ref. |
|---|---|---|---|---|---|---|
| Er | 0.2 to 2 | La | 1.0 to 4.2 | 9.52 | $\tau_1$ | [10] |
| eEr | a: 1 | La | 1.67 | 9.140 | no SI observed | [12] |
| ($^{166}Er$, $^{144}Nd$) | m: (1, 1); (0.2, 1); (0.2, 0.2) | La | 1.5; 1.5; 2.1 | 24.684; 24.744; 24.531 | EPR of $Er^{3+}$ and $Nd^{3+}$ ground state detected by change of fluorescence intensity from excited $Nd^{3+}$ level on sweeping H through resonance | [49] |
| Er | a: 1 | Lu | 4.2 | 9.6 | $g_x=8.34$, $g_y=3.51$, $g_z=7.92$, $\sphericalangle xa=0°$ | [1] |
| Tm | a: 0.5 | La | 4 | 9.2; 24 | no resonance observed [d)] | [7] |
| Yb | a: 1 | Y | 4.2; 20 | 9.6 | $g_x=2.99$, $g_y=1.53$, $g_z=2.957$, $\sphericalangle xa=0°$ | [1] |
| Yb | a: 0.2 | La | 4 | 9.2 | no resonance observed [d)] | [7] |
| Yb | a: 1 | Lu | 4.2; 20 | 9.6 | $g_x=2.95$, $g_y=1.58$, $g_z=2.96$, $\sphericalangle xa=0°$ | [1] |

[a)] To the extent specified in the original papers, concentrations of impurity ions are given as a=at.%, m=mol%, or w=wt%.
Impurity concentrations always are nominal values obtained from the stoichiometry of the starting materials unless stated otherwise. Actual concentrations may be quite different due to ion selectivity during sublimation and crystallization [11]. In particular, $Er^{3+}$ is strongly rejected by the $LaCl_3$ lattice during crystal growth [8].

[b)] SI constants of [2, 3] have been reanalyzed and revised, see [4, 5].

[c)] Ce concentration determined by titration [8].

[d)] Magnetic dipole transitions are forbidden for the ground states of $Pr^{3+}$, $Tb^{3+}$, $Dy^{3+}$, $Ho^{3+}$, $Tm^{3+}$, and $Yb^{3+}$ in a regular crystalline environment of site symmetry $C_{3h}$ ($LaCl_3$ lattice), since $Tm^{3+}$ has a singlet ground state and for the other ions the theoretical value of $g_\perp=0$. A finite value of $g_\perp$ is brought about by lattice distortions, see, e.g., [7].

[e)] Pr concentration determined by spectrophotometry [14].

[f)] According to Hutchison, Wong [7], the splitting factors given in [14] for $Pr^{3+}$ in $LaCl_3$ originate from an unintended 0.07 mol% $Nd^{3+}$ impurity.

[g)] For preliminary ENDOR($^{143,145}Nd$) results in $LaCl_3$ see [18].

[h)] Samples for EPR studies in excited electronic states of $Nd^{3+}$ or $Dy^{3+}$ are cosubstituted with $U^{3+}$ ions which can be photoexcited and subsequently transfer excitation energy to a tripositive rare earth ion, see, e.g., [42, p. 67].

[i)] Nd concentration determined by titration and mass spectrometry [8].

[k)] Observed resonance originates from an unintended $Gd^{3+}$ impurity in a nominally 99.999% pure $LaCl_3$ sample [29].

[l] For preliminary results for CF and SI constants of $Gd^{3+}$ in $LaCl_3$ see [32].
[m] For preliminary results for SI constants of $Gd^{3+}$ in $EuCl_3$ see [37].
[n] Er concentration determined by titration and mass spectrometry [8].
[o] In [43] $g_{\parallel}=9.54949$, $g_{\perp}=5.1211$ for $^4I_{15/2}$.

References:

[1] G. Garton, M.T. Hutchings, R. Shore, W.P. Wolf (J. Chem. Phys. **41** [1964] 1970/4). – [2] R.J. Birgeneau, M.T. Hutchings, R.N. Rogers (Phys. Rev. [2] **175** [1968] 1116/33). – [3] R.J. Birgeneau, M.T. Hutchings, R.N. Rogers (Phys. Rev. Letters **16** [1966] 584/6). – [4] J.M. Baker, R.J. Birgeneau, M.T. Hutchings, J.D. Riley (Phys. Rev. Letters **21** [1968] 620/3). – [5] R.J. Birgeneau, M.T. Hutchings, J.M. Baker, J.D. Riley (J. Appl. Phys. **40** [1969] 1070/9).

[6] J.D. Riley, J.M. Baker, R.J. Birgeneau (Proc. Roy. Soc. [London] A**320** [1970] 369/95). – [7] C.A. Hutchison, E. Wong (J. Chem. Phys. **29** [1958] 754/60). – [8] W.T. Gray, H.J. Stapleton (Phys. Rev. [3] B**9** [1974] 2863/9). – [9] B.A. Young, H.J. Stapleton (Phys. Rev. [2] **176** [1968] 502/9). – [10] B.W. Mangum, R.P. Hudson (J. Chem. Phys. **44** [1966] 704/13).

[11] R.C. Mikkelson, H.J. Stapleton (Phys. Rev. [2] A**140** [1965] 1968/82). – [12] K.L. Brower, H.J. Stapleton, E.O. Brower (Phys. Rev. [2] **146** [1966] 233/43). – [13] J.W. Culvahouse, D.P. Schinke, L.G. Pfortmiller (Phys. Rev. [2] **177** [1969] 454/64). – [14] J.H. Anderson, C.A. Hutchison (Phys. Rev. [2] **97** [1955] 76/8). – [15] T.L. Bohan, H.J. Stapleton (Phys. Rev. [2] **182** [1969] 385/94).

[16] C.A. Hutchison, S.S. Kim (Phys. Rev. [3] B**19** [1979] 4454/9). – [17] D. Halford (Phys. Rev. [2] **127** [1962] 1940/8). – [18] D. Halford, C.A. Hutchison, P.M. Llewellyn (Phys. Rev. [2] **110** [1958] 284/6). – [19] L.E. Erickson (Phys. Rev. [2] **143** [1966] 295/303). – [20] J.P. Hessler, C.A. Hutchison (Phys. Rev. [3] B**8** [1973] 1822/6).

[21] R.H. Clarke, R.E. Connors (Phys. Letters A**47** [1974] 273/4). – [22] R.H. Clarke, C.A. Hutchison (Phys. Rev. Letters **27** [1971] 638/40). – [23] C.A. Hutchison, E.-D. Liu (J. Lumin. **12/13** [1976] 665/8). – [24] C.A. Hutchison, M.D. Kemple, Y.-T. Yen (Phys. Rev. Letters **33** [1974] 937/40). – [25] C.E. Davoust, C.A. Hutchison, M.D. Kemple, H.-J. Kim, Y.-T. Yen (Phys. Rev. [3] B**15** [1977] 5166/80).

[26] I.Y. Chan, C.A. Hutchison (Phys. Rev. [3] B**5** [1972] 3387/95). – [27] B.F. Kim, H.W. Moos (Phys. Rev. [2] **161** [1967] 869/76). – [28] K.E. Roelfsema, H.W. den Hartog (Phys. Rev. [3] B**19** [1979] 2440/6). – [29] R.W. Reynolds, L.A. Boatner (J. Chem. Phys. **54** [1971] 1827/9). – [30] L.A. Boatner, M.M. Abraham (Phys. Rev. [2] **163** [1967] 213/9).

[31] M.T. Hutchings, R.J. Birgeneau, W.P. Wolf (Phys. Rev. [2] **168** [1968] 1026/44). – [32] R.J. Birgeneau, M.T. Hutchings, W.P. Wolf (Phys. Rev. Letters **17** [1966] 308/12). – [33] C.A. Hutchison, B.R. Judd, D.F.D. Pope (Proc. Phys. Soc. B**70** [1957] 514/20). – [34] D. van Ormondt, R. de Beer, C.M. de Jong, M.G. van der Oord, M.H. Homs, H.W. den Hartog (Physica B+C **84** [1976] 110/4). – [35] R. de Beer, D. van Ormondt, M.H. Homs, H.W. den Hartog (Physica B+C **84** [1976] 367/70).

[36] R.J. Birgeneau, M.T. Hutchings, W.P. Wolf (Phys. Rev. [2] **179** [1969] 275/88). – [37] R.J. Birgeneau, M.T. Hutchings, W.P. Wolf (J. Appl. Phys. **38** [1967] 957/8). – [38] K.L. Brower, H.J. Stapleton (J. Chem. Phys. **46** [1967] 888/90). –

[39] J. Pelzl, S. Hüfner, S. Scheller (Z. Physik **231** [1970] 377/96). – [40] R. Furrer, J. Pelzl, S. Hüfner (Z. Physik **247** [1971] 267/74).

[41] H.J. Stapleton, R.L. Marchand, E.R. Lemar (Phys. Rev. [3] B**10** [1974] 2687/96). – [42] C.A. Hutchison (Semicond. Insul. **3** [1978] 61/103). – [43] J.P. Hessler, E.-D. D. Liu (from [42, pp. 67/8]). – [44] J.P. Hessler (from [42, pp. 67/73]). – [45] J.M. Clemens, E.-D. D. Liu (from [42, pp. 76/81]).

[46] E.-D. D. Liu, J.M. Clemens (from [42, pp. 81/9]). – [47] E.-D. D. Liu (from [42, pp. 88/90]). – [48] E.-D. D. Liu (from [42, pp. 74/5]). – [49] E.-D. D. Liu, J.M. Clemens, C.A. Hutchison (from [42, pp. 91/101]).

### 11.2.6.7 Mechanical Properties

**Density of Crystalline Compounds** in g/cm$^3$

Values are tabulated below for the X-ray densities, $D_{calc}$, and experimentally determined densities, $D_{exp}$, found by flotation [15], air displacement [8], or pycnometric methods, at room temperature:

Rhombohedral $ScCl_3$: $D_{calc}$=2.39 [1], 2.42, [2], $D_{exp}$=2.39 [3].

Hexagonal Trichlorides

| | $LaCl_3$ | $CeCl_3$ | $PrCl_3$ | $NdCl_3$ |
|---|---|---|---|---|
| $D_{calc}$ | 3.848 [4] | 3.946 [9] | 4.027 [4, 9] | 4.138 [4] |
| | 3.858 [5] | 3.944 [4] | 4.063 [6] | 4.139 [5] |
| | 3.863 [6] | | | 4.169 [6] |
| $D_{exp}$ | 3.842 [7] | 3.92 [7] | 4.02 [7] | 4.134 [7] |
| | 3.82 [3, 8] | 3.97 [8] | 4.12 [10] | 4.17 [8] |
| | | | 4.15 [8] | |

| | $PmCl_3$ | $SmCl_3$ | $EuCl_3$ | $GdCl_3$ |
|---|---|---|---|---|
| $D_{calc}$ | 4.19 [11] | 4.330 [13] | 4.405 [4] | 4.412 [5] |
| | 4.23 [12] | 4.336 [9] | 4.410 [5] | 4.542 [9, 13] |
| | | 4.334 [4] | 4.414 [9, 13] | |
| | | 4.413 [14] | 4.52 [18] | |
| $D_{exp}$ | | 4.30 [8] | 4.50 [18] | |
| | | 4.31 [15, 16] | | |
| | | 4.45 [17] | | |

Orthorhombic Trichlorides

$GdCl_3$: $D_{calc}$=4.515 [19], from the lattice constants of [20]. $TbCl_3$: $D_{calc}$=4.60 [21].

Monoclinic Trichlorides

| | $YCl_3$ | $DyCl_3$ | $HoCl_3$ | $ErCl_3$ | $TmCl_3$ | $YbCl_3$ | $LuCl_3$ |
|---|---|---|---|---|---|---|---|
| $D_{calc}$ | 2.61 [15] | 3.59 [15] | 3.69 [15] | 3.77 [15] | 3.83 [15] | 3.94 [15]<br>4.06 [24] | 3.98 [15] |
| $D_{exp}$ | 2.55 [15]<br>2.61 [22]<br>2.67 [3]<br>2.81 [8] | 3.60 [8]<br>– | 3.80 [23]<br>4.25 [8] | –<br>– | 4.34 [8]<br>– | –<br>– | –<br>– |

The $TmCl_3$ sample of [8] was contaminated by Er and probably also by some Y and Lu. The values reported for $YCl_3$, $HoCl_3$, and $TmCl_3$, which spread considerably, were determined on samples made from the oxides with gaseous HCl at 550 °C [15] or by dehydration of the hydrated trichlorides at reduced pressure [8]. Regarding the possibility that these samples were in a form other than monoclinic, see p. 73. For a review of additional density values of the trichlorides of La, Ce, Pr, Nd, Dy, and Y from pre-1928 works see [8]; compare also [25].

**Density of Melts** in g/cm³

The following densities $D_l$ have been determined dilatometrically by Cho et al. [26] for molten rare earth chlorides at various temperatures t in °C (selected values):

$PrCl_3$

| | | | | | | | | |
|---|---|---|---|---|---|---|---|---|
| t . . . . . | 839 | 855 | 882 | 911 | 934 | 954 | 969 | 984 |
| $D_l$ . . . . . | 3.192 | 3.179 | 3.151 | 3.128 | 3.116 | 3.103 | 3.097 | 3.082 |

$NdCl_3$

| | | | | | | | | |
|---|---|---|---|---|---|---|---|---|
| t . . . . . | 809 | 849 | 861 | 899 | 929 | 959 | 969 | 999 |
| $D_l$ . . . . . | 3.254 | 3.212 | 3.208 | 3.188 | 3.148 | 3.129 | 3.114 | 3.062 |

$GdCl_3$

| | | | | | | | | |
|---|---|---|---|---|---|---|---|---|
| t . . . . . | 659 | 708 | 754 | 805 | 859 | 919 | 969 | 1007 |
| $D_l$ . . . . . | 3.533 | 3.485 | 3.453 | 3.424 | 3.396 | 3.354 | 3.316 | 3.285 |

$DyCl_3$

| | | | | | | | | |
|---|---|---|---|---|---|---|---|---|
| t . . . . . | 689 | 739 | 764 | 804 | 879 | 909 | 959 | 996 |
| $D_l$ . . . . . | 3.631 | 3.602 | 3.586 | 3.553 | 3.478 | 3.455 | 3.422 | 3.396 |

At an identical temperature, the densities decrease in the sequence $DyCl_3 > GdCl_3 > NdCl_3 > PrCl_3$. The values in the table are very precisely described by linear equations $D_l = a - bt$. The constants a and b have been derived by the least squares method [26]. They are presented in the following table along with data for $YCl_3$ and $LaCl_3$ from other sources:

| $MCl_3$ | a | $b \cdot 10^3$ | t in °C | Ref. |
|---|---|---|---|---|
| $YCl_3$ | 2.9226 | 0.4594 | 809 to 1005 | [22] |
| $LaCl_3$ | 3.7407 | 0.6519 | 895 to 996 | [27] |
| | 3.8773 | 0.7774 | 873 to 973 | [28] |
| | 3.802 | 0.563 | 800 to 900 | [29] |
| $PrCl_3$ | 3.8095 | 0.7417 | 839 to 984 | [26] |
| $NdCl_3$ | 4.0101 | 0.9302 | 809 to 999 | [26] |
| $GdCl_3$ | 3.9668 | 0.6732 | 659 to 1007 | [26] |
| $DyCl_3$ | 4.2009 | 0.8136 | 689 to 996 | [26] |

The mean deviation of the values of [28], who used a buoyancy method, from those of the equation, using the constants above, is ±0.0001 g/cm$^3$. Constants derived by Nisel'son, Lyzlov [30] from measurements with a pycnometer-dilatometer for the temperature range between melting point and about 1100 °C are as follows:

| $MCl_3$ | a | $b \cdot 10^3$ | $MCl_3$ | a | $b \cdot 10^3$ |
|---|---|---|---|---|---|
| $LaCl_3$ | 3.664 | 0.671 | $TbCl_3$ | 4.002 | 0.694 |
| $CeCl_3$ | 3.648 | 0.642 | $DyCl_3$ | 4.02 | 0.66 |
| $PrCl_3$ | 3.684 | 0.658 | $HoCl_3$ | 4.029 | 0.621 |
| $NdCl_3$ | 3.751 | 0.663 | $ErCl_3$ | 3.97 | 0.51 |
| $SmCl_3$ | 3.919 | 0.731 | $TmCl_3$ | 4.133 | 0.664 |
| $EuCl_3$ [a)] | – | – | $YbCl_3$ | 4.228 | 0.663 |
| $GdCl_3$ | 4.000 | 0.706 | $LuCl_3$ | 4.256 | 0.682 |

[a)] $EuCl_3$ decomposes on heating.

The square error of the $D_l(t)$ equations is $3 \times 10^{-3}$ g/cm$^3$ [30]. The density of $ScCl_3$ is about 1.67 (estimated value) at the melting point (940 °C) and 1.63 at 1000 °C. The density of molten $YCl_3$ between 700 and 875 °C obeys the relation $D_l = 2.52 - 0.0005\,(t-700)$. The relation $D_l = 3.155 - 0.0005\,(t-860)$ holds for $LaCl_3$ between 860 and 1005 °C [3, 31]. A $D_l$ value of 2.595 is extrapolated for $YCl_3$ at the melting point of 714 °C from the dilatometric measurements of [22].

Fig. 12

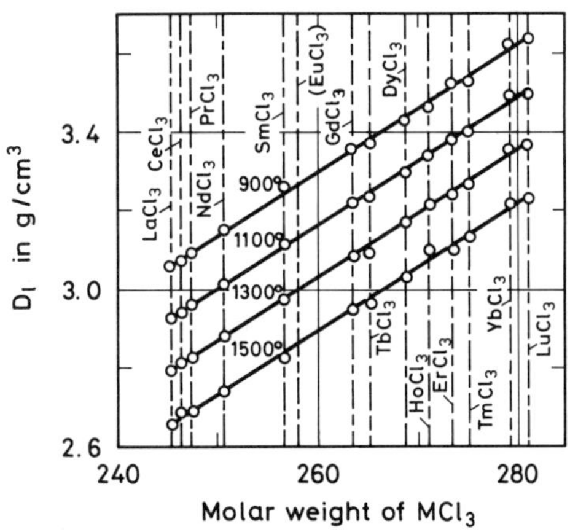


Density $D_l$ of molten rare earth chlorides $MCl_3$ as a function of the molar weight at 900 to 1500 °C.

**Fig. 12**, p. 103, gives the isotherms of the densities of a number of lanthanide chlorides calculated from the above $D_l(t)$ equations for temperatures of 900, 1100, 1300, and 1500 °C. It is thus established that the densities of the fused chlorides rise linearly with their molecular weights, so that a general first-order equation for the dependence of $D_l$ on temperature and molecular weight M can be derived: $D_l = -0.195 + 0.016\,M - 0.735 \times 10^{-3}\,t$. The square error in calculations by means of this equation does not exceed $8 \times 10^{-3}$ g/cm³ for temperatures between the melting points and ≈1100 °C [30].

### Molar Volume $V_{mol}$ in cm³/mol

A synoptic graph of older literature data is presented by Klemm et al. [32]. Molar volumes of $ScCl_3$, $YCl_3$, and $LaCl_3$ at 25 °C and at the respective melting points $t_m$ (940, 700, 860 °C) as derived from density measurements by [3] are:

| $MCl_3$ | $ScCl_3$ | $YCl_3$ | $LaCl_3$ |
|---|---|---|---|
| $V_{mol}$ at 25 °C | 63.4 | 73 | 64.2 |
| $V_{mol}$ at $t_m$ | ≈91 | 77.5 | 77.8 |

Additional molar volumes of fused chlorides between the melting points $t_m$ and 1000 °C from density determinations are:

| $MCl_3$ | $t_m$ | 800 °C | 900 °C | 1000 °C |
|---|---|---|---|---|
| $PrCl_3$ | 76.637 | 76.884 | 78.699 | 80.602 |
| $NdCl_3$ | 75.868 | 76.731 | 78.981 | 81.366 |
| $GdCl_3$ | 74.114 | 76.894 | 78.434 | 80.037 |
| $DyCl_3$ | 73.315 | 75.735 | 77.511 | 79.373 |

Interpolated values for $V_{mol}$ of other chlorides at $t_m$:

| $MCl_3$ | $PmCl_3$ | $SmCl_3$ | $EuCl_3$ | $TbCl_3$ |
|---|---|---|---|---|
| $V_{mol}$ | 75.6 | 75.0 | 74.5 | 73.2 |

The molar volume at $t_m$ is linearly dependent on the radius of the cations, Cho et al. [26]. The following relative changes of $V_{mol}$ occur on melting:

| $MCl_3$ | $LaCl_3$ | $PrCl_3$ | $NdCl_3$ | $GdCl_3$ | $DyCl_3$ |
|---|---|---|---|---|---|
| $\Delta V_{mol}/V_{mol}$ in % | 21.3 | 24.59 | 25.15 | 27.08 | 30.08 |
| Ref. | [27] | [26] | [26] | [26] | [26] |

The data are based on the authors' own density measurements for the fused chlorides and on literature values for the density of the crystalline substances at room temperature. Relationships between $V_{mol}$ in the molten state and the crystal structure are discussed [26].

### Surface Tension $\sigma$

According to measurements by the maximum bubble pressure method, the decrease of the surface tension of molten rare earth chlorides with temperature follows the linear relationship $\sigma = a - bt$ (t in °C). The following table lists the constants a and b of this equation and their standard error $\delta$. Values for the surface enthalpy H(surf) are also included; these were calculated by the formula H(surf) = $\sigma - T(d\sigma/dT)$ (T in K).

| $MCl_3$ | a in dyn/cm | b in $dyn \cdot cm^{-1} \cdot °C^{-1}$ | $\delta$ in dyn/cm | H(surf) in $erg/cm^2$ | temperature range in °C |
|---|---|---|---|---|---|
| $YCl_3$ | 106.3 | 0.0334 | 0.158 | 115.4 | 768 to 946 |
| $LaCl_3$ | 147.9 | 0.0423 | 0.121 | 159.5 | 892 to 950 |
| $PrCl_3$ | 135.9 | 0.0372 | 0.034 | 146.1 | 830 to 926 |
| $NdCl_3$ | 173.4 | 0.0855 | 0.166 | 196.8 | 800 to 865 |
| $GdCl_3$ | 110.1 | 0.0278 | 0.139 | 117.7 | 820 to 930 |
| $DyCl_3$ | 124.4 | 0.0465 | 0.069 | 137.1 | 760 to 901 |

The surface enthalpies of $YCl_3$ and $GdCl_3$ lie between those of the remaining rare earth chlorides (and of other molten salts) and those of covalent fluids such as benzene and methanol; H(surf) of $NdCl_3$ is striking due to its extraordinary high value, Igarashi et al. [33]. Measurements by the same method between 892 and 1027 °C by Smirnov et al. [34] for $LaCl_3$ (see "Rare Earth Elements" C4b, Section 11.2.11.3.23) obey the equation $\sigma = 272.2 - 0.132\ T \pm 0.1$ dyn/cm and thus lie somewhat higher than the corresponding values in the table. The same is true for the $\sigma$ values of $LaCl_3$ from the relationship $\sigma = 184.2 - 0.0541\ T \pm 1$ dyn/cm, which Smirnov, Stepanov [29] report for the temperature range from 800 to 900 °C; cf. also [35].

**Viscosity** $\eta$

The viscosity of molten $MCl_3$ (M = Pr, Nd, Gd, Dy) has been measured from 787 to 999 °C by Cho, Kuroda [36] using a capillary method based on the Ostwald principle, see **Fig. 13.** The apparent activation energies of viscous flow (in kcal/mol) are calculated as 7.29 for $PrCl_3$ at 855 to 999 °C, 7.37 for $NdCl_3$ at 886 to 967 °C, 7.68 for $GdCl_3$ at 787 to 960 °C, and 6.91 for $DyCl_3$ at 932 to 999 °C. The relationship found between fluidity ($1/\eta$) and specific volume of the molten chlorides is in good agreement with the rule discussed by Batschinski [37].

Fig. 13

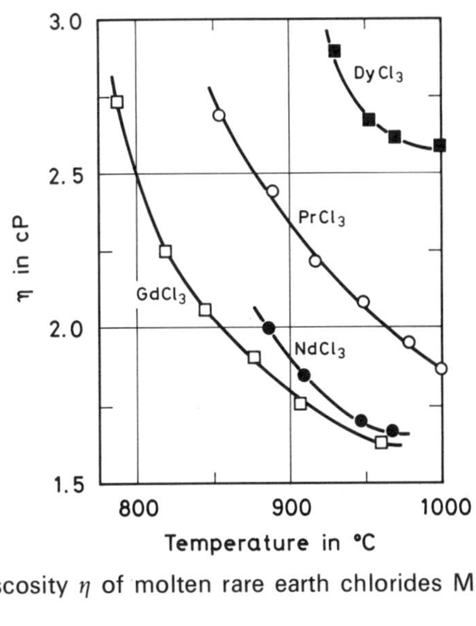


Viscosity $\eta$ of molten rare earth chlorides $MCl_3$.

The viscosity of molten $LaCl_3$ has been determined by the torsional pendulum method. Treatment of the results by the method of least squares yields the relation $\log \eta = -1.686 + 2852/T \pm 0.004$ ($\eta$ in cP, $T \approx 1180$ to 1280 K). The viscosity of the $LaCl_3$ melt is consistent with the viscosity data for other fused chlorides which are ionic liquids, Smirnov et al. [38].

References:

[1] W. Klemm, E. Krose (Z. Anorg. Allgem. Chem. **253** [1947] 218/25, 220). – [2] K.H. Schröder (Diss. Münster 1955, pp. 1/80, 20). – [3] W. Klemm, J. Rockstroh (Z. Anorg. Allgem. Chem. **152** [1926] 235/51, 250). – [4] H.E. Swanson, M.C. Morris, R.P. Stinchfield, E.H. Evans (NBS-MON-25-1 [1962] 1/56; C.A. **56** [1962] 13642). – [5] B. Morosin (J. Chem. Phys. **49** [1968) 3007/12).

[6] J.A. Sommers, E.F. Westrum (J. Chem. Thermodyn. **8** [1976] 1115/36, 1118). – [7] W.H. Zachariasen (TID-5212 [1955] 157/64; C.A. **1957** 35). – [8] J.H. Kleinheksel, H.C. Kremers (J. Am. Chem. Soc. **50** [1928] 959/67, 964). – [9] D.H. Templeton, C.H. Dauben (J. Am. Chem. Soc. **76** [1954] 5237/9). – [10] F. Ephraim, R. Bloch (Ber. Deut. Chem. Ges. **59** [1926] 2692/705).

[11] F. Weigel, V. Scherer (Radiochim. Acta **7** [1967] 40/6). – [12] V.F. Wishnevsky (BMWF-FB-K-68-42 [1968] 1/121; N.S.A. **23** [1969] No. 21678). – [13] J.A. Sommers, E.F. Westrum (J. Chem. Thermodyn. **9** [1977] 1/26, 3). – [14] V.F. Goryushkin, D.M. Laptev, I.S. Astakhova, E.S. Vorontsov (Zh. Neorgan. Khim. **21** [1976] 821/3; Russ. J. Inorg. Chem. **21** [1976] 445/7). – [15] D.H. Templeton, G.F. Carter (J. Phys. Chem. **58** [1954] 940/4).

[16] W. Klemm, J. Rockstroh (Z. Anorg. Allgem. Chem. **176** [1928] 181/99, 194). – [17] B.G. Korshunov, D.V. Drobot, V.V. Bukhtiyarov, Z.N. Shevtsova (Zh. Neorgan. Khim. **9** [1964] 1427/30; Russ. J. Inorg. Chem. **9** [1964] 773/5). – [18] N.M. Kulagin, D.M. Laptev, I.S. Astakhova (Zh. Neorgan. Khim. **21** [1976] 2549/51; Russ. J. Inorg. Chem. **21** [1976] 1401/3). – [19] J.D.H. Donnay, H.M. Ondik (Crystal Data, 3rd Ed., Vol. 2, Washington, D.C., 1973). – [20] A.L. Harris, C.R. Veale (J. Inorg. Nucl. Chem. **27** [1965] 1437/9).

[21] J.D. Forrester, A. Zalkin, D.H. Templeton, J.C. Wallmann (Inorg. Chem. **3** [1964] 185/8). – [22] J. Mochinaga, K. Irisawa (Bull. Chem. Soc. Japan **47** [1972] 364/7). – [23] F.H. Driggs, B.S. Hopkins (J. Am. Chem. Soc. **47** [1925] 363/9). – [24] D.M. Laptev, A.I. Poshevneva, I.S. Astakhova, N.M. Kulagin (Zh. Neorgan. Khim. **21** [1976] 2317/20; Russ. J. Inorg. Chem. **21** [1976] 1274/6). – [25] G.E. Revzin (Metody Poluch. Khim. Reaktiv. Prep. **1967** No. 16, pp. 124/9; C.A. **69** [1968] No. 48783).

[26] K. Cho, K. Irisawa, J. Mochinaga, T. Kuroda (Electrochim. Acta **17** [1972] 1821/7). – [27] K. Cho, T. Kuroda (Denki Kagaku **40** [1972] 45/9; C.A. **77** [1972] No. 93636). – [28] I.S. Jaffe, E.R. van Artsdalen (ORNL-2159 [1956] 77/80). – [29] M.V. Smirnov, V.P. Stepanov (Tr. Inst. Elektrokhim. Akad. Nauk SSSR Ural'sk. Filial No. 12 [1969] 9/16). – [30] L.A. Nisel'son, Yu.N. Lyzlov (Dokl. Akad. Nauk SSSR **220** [1975] 608/9; Dokl. Chem. Proc. Acad. Sci. USSR **220/225** [1975] 103/4).

[31] W. Biltz, W. Klemm (Z. Anorg. Allgem. Chem. **152** [1926] 267/94). – [32] W. Klemm, K. Meisel, H.U. v. Vogel (Z. Anorg. Allgem. Chem. **190** [1930] 123/44, 125). – [33] K. Igarashi, J. Mochinaga, S. Ueda (Bull. Chem. Soc. Japan **51** [1978] 1151/2). – [34] M.V. Smirnov, V.A. Khokhlov, V.P. Stepanov (Tr. Inst. Elektrokhim. Akad. Nauk

SSSR Ural'sk. Filial No. 10 [1967] 41/8; C.A. **69** [1968] No. 89873), A.N. Baraboshkin (Electrochemistry of Molten and Solid Electrolytes, Vol. 7, Consultants Bureau, New York 1969, pp. 31/6). – [35] M.V. Smirnov, V.P. Stepanov (Tr. Inst. Elektrokhim. Akad. Nauk SSSR Ural'sk. Filial No. 12 [1969] 3/8, No. 14 [1970] 58/61), V.P. Stepanov, M.V. Smirnov (Tr. Inst. Elektrokhim. Akad. Nauk SSSR Ural'sk. Filial No. 14 [1970] 62/5).

[36] K. Cho, T. Kuroda (Denki Kagaku **40** [1972] 878/81; C.A. **78** [1973] No. 88804). – [37] A.J. Batschinski (Z. Physik. Chem. **84** [1913] 643/706, 646). – [38] M.V. Smirnov, V.A. Khokhlov, V.P. Stepanov (Zh. Fiz. Khim. **40** [1966] 1248/51; Russ. J. Phys. Chem. **40** [1966] 671/3).

## 11.2.6.8 Thermal Properties

### 11.2.6.8.1 Thermal Expansion

Calculated values for the linear thermal expansion coefficients $\alpha$ of crystalline lanthanide chlorides $MCl_3$ are:

| M | La | Ce | Pr | Nd | Pm | Sm | Eu | Gd |
|---|---|---|---|---|---|---|---|---|
| $\alpha \cdot 10^6$ in $K^{-1}$ | 14.2 | 13.5 | 13.4 | 13.8 | 13.6 | 13.8 | 13.8 | 14.5 |

| M | Tb | Dy | Ho | Er | Tm | Yb | Lu |
|---|---|---|---|---|---|---|---|
| $\alpha \cdot 10^6$ in $K^{-1}$ | 13.7 | 13.7 | 13.2 | 13.3 | 12.8 | 12.6 | 13.1 |

The dependence of $\alpha$ on the number of electrons in the f-sublevel is presented graphically and discussed in terms of changes of the lattice energy by a crystal field effect, Krestov, Krestova [1].

The volume coefficients $\gamma$ of thermal expansion for molten chlorides were derived from density measurements (see p. 102). Data in $K^{-1}$ for $YCl_3$ from [2], for $LaCl_3$ from [3], and for all other chlorides from [4]:

$YCl_3$

| t in °C | 750 | 800 | 850 | 900 | 950 | 1000 |
|---|---|---|---|---|---|---|
| $\gamma \cdot 10^4$ | 1.782 | 1.798 | 1.814 | 1.831 | 1.844 | 1.865 |

$LaCl_3$

| t in °C | 872 | 902 | 932 | 962 | 992 |
|---|---|---|---|---|---|
| $\gamma \cdot 10^4$ | 2.055 | 2.068 | 2.081 | 2.094 | 2.107 |

$PrCl_3$

| t in °C | 786 | 800 | 900 | 1000 |
|---|---|---|---|---|
| $\gamma \cdot 10^4$ | 2.300 | 2.306 | 2.361 | 2.418 |

$NdCl_3$

| t in °C | 760 | 800 | 900 | 1000 |
|---|---|---|---|---|
| $\gamma \cdot 10^4$ | 2.816 | 2.848 | 2.932 | 3.020 |

$GdCl_3$

| t in °C | 609 | 800 | 900 | 1000 |
|---|---|---|---|---|
| $\gamma \cdot 10^4$ | 1.893 | 1.964 | 2.003 | 2.044 |

$DyCl_3$

| t in °C | 656 | 800 | 900 | 1000 |
|---|---|---|---|---|
| $\gamma \cdot 10^4$ | 2.219 | 2.292 | 2.346 | 2.402 |

The $\gamma$ values are very close to those of the fused alkaline and alkaline-earth chlorides, which are ionic liquids, and smaller than that of the nonpolar carbon tetrachloride [4]. See Klemm [5] for older data on the thermal expansion of molten $ScCl_3$, $YCl_3$, and $LaCl_3$. Cho, Kawamura [6] applied the (modified) significant structure theory of liquids to molten $LaCl_3$, $PrCl_3$, and $NdCl_3$ and calculated values for the expansivity which agree well with the experimental results of Cho et al. [3, 4].

References:

[1] G.A. Krestov, V.N. Krestova (Radiokhimiya **11** [1969] 62/74; Soviet Radiochem. **11** [1969] 55/66). – [2] J. Mochinaga, K. Irisawa (Bull. Chem. Soc. Japan **47** [1974] 364/7). – [3] K. Cho, T. Kuroda (Denki Kagaku **40** [1972] 45/9; C.A. **77** [1972] No. 93636). – [4] K. Cho, K. Irisawa, J. Mochinaga, T. Kuroda (Electrochim. Acta **17** [1972] 1821/7). – [5] W. Klemm (Z. Anorg. Allgem. Chem. **152** [1926] 295/313, 305).

[6] K. Cho, K. Kawamura (Denki Kagaku Oyobi Kogyo Butsuri Kagaku **47** [1979] 209/13; C.A. **91** [1979] No. 63573).

### 11.2.6.8.2 Melting Data

**Melting Points** $t_m$ in °C

| | |
|---|---|
| $ScCl_3$ | 940 [6], 953 [7], 957 [8], 960±2 [9 to 11], 967 [12, 13] |
| $YCl_3$ | 624 [14], 647±20 [15], 686 [16], 700 [17 to 24], 709 [8, 25, 26], 710±1 [27, 28], 714 to 717 [a] [104], 721 [12, 29 to 32] |
| $LaCl_3$ | 845 [33], 846 [34], 850 ±2 [35], 852 [19, 36 to 40], 854 [41], 858 [12, 42, 43], 859±2 [44], 860 [45 to 47], 862 [8, 25, 48, 49], 872 [14, 29] |
| $CeCl_3$ | 800 [29], 802 [33, 38, 50 to 52], 803 [41], 807 [34], 808.4 [53], 812 [54], 817 [8, 12, 25, 55], 820 [56], 821 to 822 [45], 822 [14, 101], 831 [39] |
| $PrCl_3$ | 765 [37, 57], 772 [49], 778 [39], 781 [34, 41], 785 [47], 786 [8, 12, 43], 790 [58], 823 [14] |
| $NdCl_3$ | 745 [28, 33, 59], 746 [41], 750 [34], 754 [39, 40], 758 [8, 25], 759 [12, 42, 43], 760 [38, 48], 761 [14], 765 [60], 766 [58, 61], 783 to 787 [45] |
| $PmCl_3$ | 720 [b] [39], 737 [62] |
| $SmCl_3$ | 665 [19, 63 to 65], 668 [39], 675 [66], 677 [31, 67, 68], 678±2 [22, 48],679±3 [69], 681±2 [70], 682 [8, 14], 747 [71] |
| $EuCl_3$ [c] | 623±2 [48, 72, 73], 624±1.5 [74, 75], 628 [68, 76], 660 [39] |
| $GdCl_3$ | 602 [8, 12, 77, 78], 605 [79], 609±2 [32, 65], 625 [28], 628 [39, 40], 670 [80] |
| $TbCl_3$ | 566 [81], 578 [73], 582 [8, 12], 588±2 [82], 595±2 [40, 83] |
| $DyCl_3$ | 640 [80], 647 [8, 25], 648±2 [40, 83], 651 [12, 84], 654±2 [32, 85], 655 [14] |
| $HoCl_3$ | 696 [14], 705±5 [86], 708±2 [83], 718 [32, 87], 720 [8, 12, 76, 77], 721.5 [102] |
| $ErCl_3$ | 760±5 [86], 764 [65, 76, 87], 773 [88, 89], 774 [32], 776 [8, 12, 77], 777±5 [30], 781±2 [40, 83] |

$TmCl_3$ 812±2 [83], 821±3 [90], 824 [8, 25], 828±1.5 [91], 830±5 [86], 866 [14]

$YbCl_3$ 843±2 [83], 854 [92], 854±3 [93], 860±5 [86], 865 d) [8], 866 [68], 875±2 [94]

$LuCl_3$ 892 [92, 93], 904±2 [40, 83], 905 [25], 925 [8]

a) The two values given are measured by differential thermal analyses on cooling and heating, respectively. – b) Derived from the linear change of $t_m$ with the atomic number. – c) Decomposes under 1 atm Ar. – d) According to thermographic studies by Laptev et al. [94], the value of [8] does not represent the melting point of pure (undecomposed) $YbCl_3$, but corresponds to the end of fusion of $YbCl_3$ in a $YbCl_3$-$YbCl_2$ specimen.

A critical summary of measured and estimated $t_m$ values from the literature is found in Thoma [95]. Older data are listed by [62, 96 to 99]; see also Revzin [1]. A nearly constant ratio $U/T_m \approx 1.04$ (U in kcal, $T_m$ in K) is found from literature data for the lattice energy U and melting point $T_m$. Thus hitherto unknown or questionable melting points of rare earth chlorides may be calculated and checked, using known values for the lattice energy; numerical values are given, Gopal, Husain [100].

**Enthalpy of Melting** $\Delta H_m$ in kcal/mol. **Entropy of Melting** $\Delta S_m$ in $cal \cdot mol^{-1} \cdot K^{-1}$

Values determined from calorimetric measurements for $\Delta H_m$ and $\Delta S_m$ (at the respective melting points) follow; the precision of the $\Delta H_m$ values is ±1 to 2%:

| $MCl_3$ | $\Delta H^\circ_m$ | $\Delta S^\circ_m$ | Ref. | $MCl_3$ | $\Delta H^\circ_m$ | $\Delta S^\circ_m$ | Ref. |
|---|---|---|---|---|---|---|---|
| $ScCl_3$ | 16.1 | 13.0 | [12] | $NdCl_3$ | 12.0 | 11.6 | [43] |
| $YCl_3$ | 7.5 | 7.6 | [12] | $GdCl_3$ | 9.7 | 11.1 | [12, 77] |
| $LaCl_3$ | 13.0 | 11.5 | [43] | $TbCl_3$ | 4.65 | 5.4 | [12] |
| | 7.8 | – | [2] | $DyCl_3$ | 6.1 | 6.6 | [12] |
| $CeCl_3$ | 12.8 | 11.7 | [3, 12, 53] | $HoCl_3$ | 7.3 | 7.3 | [12], cf. [77] |
| $PrCl_3$ | 12.1 | 11.4 | [43] | $ErCl_3$ | 7.8 | 7.4 | [77] |

The entropy of melting of the chlorides varies considerably with the crystal structure type, Dworkin, Bredig [12]. The chlorides $MCl_3$ with M=La, Ce, Pr, Nd, Gd, which possess the $UCl_3$ structure, show among themselves similar $\Delta H^\circ_m$ and $\Delta S^\circ_m$ values. Similarly the values for $DyCl_3$, $HoCl_3$, and $ErCl_3$, which have the $AlCl_3$ structure, differ little from each other [4, 12]. The chlorides with lower entropies of melting ($YCl_3$, $DyCl_3$, $HoCl_3$) have also a lower value for $S_{1300}$-$S_{298}$, which indicates either a higher degree of ordering in the melt (possibly increased ion association, as for example to $[MCl]^{2+}$) or a higher entropy in the solid state at room temperature [12], see also [77].

A value $\Delta H^\circ_m = 10 \pm 3$ for $YCl_3$ is inferred from changes in slope of the vapor pressure (ion current vs. 1/T) curves determined by mass-spectroscopic studies, McKinley [15]. Evaluations of thermographic results yield $\Delta H^\circ_m = 9.4$, 10.0, 8.9, and 9.9 for $LaCl_3$, $CeCl_3$, $PrCl_3$, and $NdCl_3$, respectively, Novikov et al. [41], $\Delta H^\circ_m = 12.2 \pm 0.5$ for $EuCl_3$, Kulagin, Laptev [74], and $\Delta H^\circ_m = 15.2 \pm 0.4$ for $YbCl_3$, Laptev et al. [94]. Lower values $\Delta H^\circ_m = 7.4 \pm 0.5$ and $6.9 \pm 0.5$ for $LaCl_3$ and $PrCl_3$ are determined by Savin et al. [49] using the same method. Applying Raoult's law to the Ce-$CeCl_3$ equilibrium state diagram, Kojima, Kikuchi [4] calculated $\Delta H^\circ_m = 9.2$ for $CeCl_3$. The following values were computed for

$TmCl_3$, $YbCl_3$, and $LuCl_3$ assuming a linear dependence of $\Delta H^\circ_m$ on the melting point for all chlorides of the same structure type: $\Delta H^\circ_m = 8.2$, 8.6, and 9.4, respectively. For $EuCl_3$, which is reported to decompose (see p. 143), a melting point $T_m$ of 1020 K has been estimated by interpolation of a plot of $T_m$ vs. atomic number for chlorides having the $UCl_3$-type structure. This value leads to an estimate of 11.8 for the enthalpy of melting, Myers, Graves [71]. See also Korshunov et al. [19] for estimated $\Delta H^\circ_m$ values for $YCl_3$ and $SmCl_3$.

Entropies of melting $\Delta S^\circ_m = 8.0$, 7.4, 7.6, and 7.8 were calculated for $LaCl_3$, $CeCl_3$, $PrCl_3$, and $NdCl_3$, respectively, using melting point data determined by themselves and $\Delta H^\circ_m$ values from the literature, Novikov, Baev [34].

Compilations of older $\Delta H^\circ_m$ and $\Delta S^\circ_m$ data are given by [5, 6, 62, 98].

References:

[1] G.E. Revzin (Metody Poluch. Khim. Reaktiv. Prep. **1967** No. 16, pp. 124/9; C.A. **69** [1968] No. 48783). – [2] V.D. Savin, N.P. Mikhailova (Zh. Fiz. Khim. **51** [1977] 1272; Russ. J. Phys. Chem. **51** [1977] 753/4). – [3] D.F. Smith (private communication from A.S. Dworkin, H.R. Bronstein, M.A. Bredig, Discussions Faraday Soc. No. 32 [1961] 188/96, 195). – [4] T. Kojima, K. Kikuchi (J. Electrochem. Soc. Japan **21** [1953] 177/80). – [5] C.E. Wicks, F.E. Block (U.S. Bur. Mines Bull. No. 605 [1963] 1/146; C.A. **59** [1963] 14656).

[6] W. Biltz, W. Klemm (Z. Anorg. Allgem. Chem. **131** [1923] 22/6). – [7] O.G. Polyachenok, G.I. Novikov (Zh. Neorgan. Khim. **8** [1963] 2819/21; Russ. J. Inorg. Chem. **8** [1963] 1479/80). – [8] F.H. Spedding, A.H. Daane (Met. Rev. **5** [1960] 297/348, 340). – [9] W. Fischer, R. Gewehr, H. Wingchen (Z. Anorg. Allgem. Chem. **242** [1939] 161/87, 172). – [10] W. Fischer, K. Brünger, H. Grieneisen (Z. Anorg. Allgem. Chem. **231** [1937] 54/62, 57).

[11] R. Gut, D.M. Gruen (J. Inorg. Nucl. Chem. **21** [1961] 259/61). – [12] A.S. Dworkin, M.A. Bredig (High Temp. Sci. **3** [1971] 81/90, 89). – [13] J.D. Corbett, B.N. Ramsey (Inorg. Chem. **4** [1965] 260/2). – [14] J.H. Kleinheksel, H.C. Kremers (J. Am. Chem. Soc. **50** [1928] 959/67, 962). – [15] J.D. McKinley (J. Chem. Phys. **42** [1965] 2245/6).

[16] H.C. Fogg, C. James (J. Am. Chem. Soc. **44** [1922] 307/16). – [17] D.V. Drobot, G.P. Anikina, L.V. Durinina, B.G. Korshunov (Zh. Neorgan. Khim. **10** [1965] 562/4; Russ. J. Inorg. Chem. **10** [1965] 306). – [18] B.G. Korshunov, D.V. Drobot, Z.N. Shevtsova (Zh. Neorgan. Khim. **10** [1965] 1901/5; Russ. J. Inorg. Chem. **10** [1965] 1035/7). – [19] B.G. Korshunov, D.V. Drobot, L.V. Durinina (Zh. Neorgan. Khim. **10** [1965] 2120/3; Russ. J. Inorg. Chem. **10** [1965] 1154/6). – [20] B.G. Korshunov, D.V. Drobot (Zh. Neorgan. Khim. **9** [1964] 222/3; Russ. J. Inorg. Chem. **9** [1964] 121/2).

[21] D.V. Drobot, B.G. Korshunov, Z.M. Shevtsova (Tr. 2nd Vses. Soveshch. Fiz. Khim. Rasplavlen. Solei, Kiev 1963 [1965], pp. 48/54; C.A. **65** [1966] 8082). – [22] A.S. Pashinkin, D.V. Drobot, Z.N. Shevtsova, B.G. Korshunov (Zh. Neorgan. Khim. **7** [1962] 2811/3; Russ. J. Inorg. Chem. **7** [1962] 1466/8). – [23] W. Klemm, J. Rockstroh (Z. Anorg. Allgem. Chem. **152** [1926] 235/51, 249, 250). – [24] W. Klemm, W. Biltz (Z. Anorg. Allgem. Chem. **152** [1926] 225/34, 230). – [25] A.H. Daane, F.H. Spedding (ISC-976 [1958] 21; N.S.A. **12** [1958] No. 13707).

[26] C.V. Banks, O.N. Carlson, A.H. Daane, V.A. Fassel, R.W. Fisher, E.H. Olson, J.E. Powell, F.H. Spedding (IS-1 [1959] 1/250, 46; C.A. **1960** 6446). – [27] A.T. Pustil'nik, B.A. Sakharov, T.S. Sedykh, R.V. Dubchak (Zh. Fiz. Khim. **44** [1970] 2947; Russ. J. Phys. Chem. **44** [1970] 1682/3). – [28] Fam Ngok T'en, I.S. Morozov (Zh. Neorgan. Khim. **14** [1969] 1361/4; Russ. J. Inorg. Chem. **14** [1969] 713/5). – [29] O.B. Troshkina (Spektrosk. At. Mol. **1969** 17/22, 18; C.A. **74** [1971] No. 16340). – [30] J.D. Corbett, D.L. Pollard, J.E. Mee (Inorg. Chem. **5** [1966] 761/6).

[31] O.G. Polyachenok, G.I. Novikov (Zh. Neorgan. Khim. **8** [1963] 2818/9; Russ. J. Inorg. Chem. **8** [1963] 1478/9). – [32] G. Jantsch, H. Jawurek, N. Skalla, H. Gawalowski (Z. Anorg. Allgem. Chem. **207** [1932] 353/67, 356). – [33] Sun In'-Chzhu, I.S. Morozov (Zh. Neorgan. Khim. **3** [1958] 1914/24; Russ. J. Inorg. Chem. **3** No. 8 [1958] 266/78). – [34] G.I. Novikov, A.K. Baev (Zh. Neorgan. Khim. **7** [1962] 1349/52; Russ. J. Inorg. Chem. **7** [1962] 694/6). – [35] M.V. Smirnov, P.M. Usov (Dokl. Akad. Nauk SSSR **151** [1963] 862/5; Dokl. Chem. Proc. Acad. Sci. USSR **148/153** [1963] 606/9).

[36] M.V. Smirnov, P.M. Usov, T.F. Khazemova (Dokl. Akad. Nauk SSSR **151** [1963] 591/4; Dokl. Chem. Proc. Acad. Sci. USSR **148/153** [1963] 583/6). – [37] D.V. Drobot, B.G. Korshunov, L.V. Durinina (Izv. Akad. Nauk SSSR Neorgan. Materialy **1** [1965] 2189/96; Inorg. Materials [USSR] **1** [1965] 1978/85, 1982). – [38] E. Shimazaki, K. Niwa (Z. Anorg. Allgem. Chem. **314** [1962] 21/34, 32). – [39] L.A. Nisel'son, Yu.N. Lyzlov (Zh. Neorgan. Khim. **21** [1976] 3344/7; Russ. J. Inorg. Chem. **21** [1976] 1842/4). – [40] Yu.N. Lyzlov, L.A. Nisel'son (Zh. Neorgan. Khim. **23** [1978] 857/9; Russ. J. Inorg. Chem. **23** [1978] 476/7).

[41] G.I. Novikov, A.K. Baev, O.G. Polyachenok (Khim. Redk. Elem. Leningr. Gos. Univ. **1964** 63/7; C.A. **61** [1964] 11389). – [42] A.S. Dworkin, H.R. Bronstein, M.A. Bredig (Discussions Faraday Soc. No. 32 [1961] 188/96, 190). – [43] A.S. Dworkin, M.A. Bredig (J. Phys. Chem. **67** [1963] 697/8). – [44] F.J. Keneshea, D.D. Cubicciotti (J. Chem. Eng. Data **6** [1961] 507/9). – [45] J. Shiokawa, T. Ishino (Denki Kagaku **32** [1964] 109/16 from C.A. **62** [1965] 6111).

[46] D.M. Gruen, J.G. Conway, R.D. McLaughlin (J. Chem. Phys. **25** [1956] 1102/11, 1103). – [47] A. Voigt, W. Biltz (Z. Anorg. Allgem. Chem. **133** [1924] 277/305, 287). – [48] G. Jantsch, H. Grubitsch, F. Hoffmann, H. Alber (Z. Anorg. Allgem. Chem. **185** [1930] 49/64, 63). – [49] V.D. Savin, N.P. Mihailova, V.A. Morozova (Zh. Fiz. Khim. **53** [1979] 1410/3; Russ. J. Phys. Chem. **53** [1979] 797/9). – [50] D. Cubicciotti (J. Am. Chem. Soc. **71** [1949] 4119/21).

[51] B.G. Korshunov, I.S. Morozov, V.I. Ionov (Izv. Vysshikh Uchebn. Zavedenii Khim. Khim. Tekhnol. **3** [1960] 402/4; C.A. **1960** 21969). – [52] V.I. Ionov, B.G. Korshunov, V.V. Kokorev, I.S. Morozov (Izv. Vysshikh Uchebn. Zavedenii Tsvetn. Met. **3** [1960] 102/8; C.A. **1961** 5101). – [53] G.E. Walden, D.F. Smith (U.S. Bur. Mines Rept. Invest. No. 5859 [1961] 1/4; C.A. **56** [1962] 9511). – [54] G.W. Mellors, S. Senderoff (J. Phys. Chem. **63** [1959] 1110/2). – [55] H.R. Bronstein, A.S. Dworkin, M.A. Bredig (J. Phys. Chem. **66** [1962] 44/8).

[56] G.L. Zverev (Dokl. Akad. Nauk SSSR **104** [1955] 242/5). – [57] Z.N. Shevtsova, E.N. Korzina, B.G. Korshunov (Zh. Neorgan. Khim. **7** [1962] 2596/9; Russ. J. Inorg. Chem. **7** [1962] 1348/50). – [58] G.I. Novikov, O.G. Polyachenok (Zh. Neorgan. Khim. **8** [1963] 1053/9; Russ. J. Inorg. Chem. **8** [1963] 545/50, 546). – [59] I.S. Morozov, V.I. Ionov, B.G. Korshunov (Zh. Neorgan. Khim. **4** [1959] 1457/8; Russ. J. Inorg. Chem.

4 [1959] 655/6). – [60] I.S. Morozov, Z.N. Shevtsova, L.V. Klyukina (Zh. Neorgan. Khim. **2** [1957] 1639/42; Russ. J. Inorg. Chem. **2** No. 7 [1957] 301/5).

[61] O.G. Polyachenok, G.I. Novikov (Zh. Neorgan. Khim. **8** [1963] 1567/73; Russ. J. Inorg. Chem. **8** [1963] 816/20). – [62] L. Brewer (Natl. Nucl. Energy Ser. Div. IVB **19** [1950] 193/275, 198). – [63] B.G. Korshunov, D.V. Drobot, V.V. Bukhtiyarov, Z.N. Shevtsova (Zh. Neorgan. Khim. **9** [1964] 1427/30; Russ. J. Inorg. Chem. **9** [1964] 773/5). – [64] B.G. Korshunov, D.V. Drobot, K.I. Petrov, V.V. Bukhtiyarov, M.V. Rubtsov (Zh. Neorgan. Khim. **10** [1965] 1675/80; Russ. J. Inorg. Chem. **10** [1965] 913/6). – [65] D.V. Drobot, B.G. Korshunov (Izv. Akad. Nauk SSSR Neorgan. Materialy **2** [1966] 1093/6; Inorg. Materials [USSR] **2** [1966] 931/4).

[66] Sung Yü-Lin, G.I. Novikov (Zh. Neorgan. Khim. **8** [1963] 700/1; Russ. J. Inorg. Chem. **8** [1963] 356/7). – [67] W. Klemm, J. Rockstroh (Z. Anorg. Allgem. Chem. **176** [1928] 181/99, 196). – [68] O.G. Polyachenok, G.I. Novikov (Zh. Neorgan. Khim. **9** [1964] 773/7; Russ. J. Inorg. Chem. **9** [1964] 429/31). – [69] V.F. Goryushkin, D.M. Laptev, I.S. Astakhova, E.S. Vorontsov (Zh. Neorgan. Khim. **21** [1976] 821/3; Russ. J. Inorg. Chem. **21** [1976] 445/7). – [70] D.M. Laptev, V.F. Goryushkin, I.S. Astakhova, G.G. Polyakova (Zh. Neorgan. Khim. **24** [1979] 1311/6; Russ. J. Inorg. Chem. **24** [1979] 730/3).

[71] C.E. Myers, D.T. Graves (J. Chem. Eng. Data **22** [1977] 440/4). – [72] S. Mroczkowski (J. Cryst. Growth **6** [1970] 147/50). – [73] B.G. Korshunov, D.V. Drobot, G.P. Borodulenko, I.E. Galchenko (Zh. Neorgan. Khim. **11** [1966] 1013/8; Russ. J. Inorg. Chem. **11** [1966] 547/50). – [74] N.M. Kulagin, D.M. Laptev (Zh. Fiz. Khim. **50** [1976] 810; Russ. J. Phys. Chem. **50** [1976] 483). – [75] N.M. Kulagin, D.M. Laptev, I.S. Astakhova (Zh. Neorgan. Khim. **21** [1976] 2549/51; Russ. J. Inorg. Chem. **21** [1976] 1401/3).

[76] B.G. Korshunov, D.V. Drobot (Zh. Neorgan. Khim. **10** [1965] 2310/5; Russ. J. Inorg. Chem. **10** [1965] 1256/9). – [77] A.S. Dworkin, M.A. Bredig (J. Phys. Chem. **67** [1963] 2499/500). – [78] A.L. Harris, C.R. Veale (J. Inorg. Nucl. Chem. **27** [1965] 1437/9). – [79] J.E. Mee, J.D. Corbett (Inorg. Chem. **4** [1965] 88/93, 89). – [80] K.V. Orlov, V.G. Kozlov, N.G. Pospelova (Met. Metalloved. Chist. Metal. Sb. Nauchn. Rabot No. 5 [1966] 73/6; C.A. **65** [1966] 9817).

[81] H. Mattausch, A. Simon, N. Holzer, R. Eger (Z. Anorg. Allgem. Chem. **466** [1980] 7/22, 12). – [82] G. Jantsch (Oesterr. Chemiker-Ztg. **40** [1937] 218/25). – [83] Yu.N. Lyzlov, L.A. Nisel'son (Zh. Neorgan. Khim. **22** [1977] 2245/7; Russ. J. Inorg. Chem. **22** [1977] 1215/6). – [84] J.D. Corbett, B.C. McCollum (Inorg. Chem. **5** [1966] 938/40). – [85] B.G. Korshunov, D.V. Drobot (Zh. Neorgan. Khim. **10** [1965] 939/42; Russ. J. Inorg. Chem. **10** [1965] 508/10).

[86] J.L. Moriarty (J. Chem. Eng. Data **8** [1963] 422/4). – [87] B.G. Korshunov, D.V. Drobot, I.E. Galchenko, Z.N. Shevtsova (Zh. Neorgan. Khim. **11** [1966] 411/4; Russ. J. Inorg. Chem. **11** [1966] 223/5). – [88] G.I. Novikov, F.G. Gavryuchenko (Zh. Neorgan. Khim. **10** [1965] 1668/74; Russ. J. Inorg. Chem. **10** [1965] 909/13). – [89] F.G. Gavryuchenko, G.I. Novikov (Zh. Neorgan. Khim. **11** [1966] 2172/4; Russ. J. Inorg. Chem. **11** [1966] 1165/6). – [90] G. Jantsch, N. Skalla, H. Grubitsch (Z. Anorg. Allgem. Chem. **212** [1933] 65/83).

[91] P.E. Caro, J.D. Corbett (J. Less-Common Metals **18** [1969] 1/10, 4). – [92] D.V. Drobot, B.G. Korshunov, G.P. Borodulenko (Zh. Neorgan. Khim. **13** [1968] 1635/41; Russ. J. Inorg. Chem. **13** [1968] 855/9). – [93] G. Jantsch, N. Skalla, H. Jawurek

(Z. Anorg. Allgem. Chem. **201** [1931] 207/20, 211). – [94] D.M. Laptev, A.I. Poshevneva, I.S. Astakhova, N.M. Kulagin (Zh. Neorgan. Khim. **21** [1976] 2317/20; Russ. J. Inorg. Chem. **21** [1976] 1274/6). – [95] R.E. Thoma (in: L. Eyring, Progress in the Science and Technology of the Rare Earths, Vol. 2, Pergamon, Oxford - London - Edinburgh - New York - Toronto - Sydney - Paris - Braunschweig 1966, pp. 90/122, 96/100).

[96] O. Kubaschewski, E.L. Evans (Metallurgical Thermochemistry, Pergamon, New York 1958, pp. 236/81). – [97] O. Kubaschewski, E.L. Evans, C.B. Alcock (Metallurgical Thermochemistry, Pergamon, New York 1967, pp. 314/63). – [98] R.C. Vickery (Chemistry of the Lanthanons, Butterworth, London 1953, pp. 228/9; The Chemistry of Yttrium and Scandium, Pergamon, New York - Oxford - London - Paris 1960, pp. 56/69, 56). – [99] F.D. Rossini, D.D. Wagman, W.H. Evans, S. Levine, I. Jaffe (NBS-Circ.-500 [1952] 1/1268, 740/60, 1131/9). – [100] R. Gopal, M.M. Husain (J. Indian Chem. Soc. **44** [1967] 726/7).

[101] T. Kojima (J. Electrochem. Soc. Japan **20** [1952] 173/6). – [102] U. Löchner, J.D. Corbett (Inorg. Chem. **14** [1975] 426/8). – [103] J.A. Gibson, J.F. Miller, P.S. Kennedy, G.W.P. Rengstorff (The Properties of the Rare Earth Metals and Compounds, Battelle Memorial Institute, Columbus, Ohio, 1959, p. 10, Section 1). – [104] J. Mochinaga, K. Irisawa (Bull. Chem. Soc. Japan **47** [1974] 364/7).

### 11.2.6.8.3 Vaporization Data

#### Composition of the Vapor Phase

From mass spectrometric studies of the vapor of $ScCl_3$ escaping from a diffusion cell at 943 K and <0.1 Torr, both dimeric and monomeric molecules are said to exist. The average molecular weight of the vapor amounts to 175 to 180 g/mol. Equations for the partial pressures of the monomer and dimer are given in "Rare Earth Elements" C 4b, Section 11.2.11.1.11.2, Patrikeev et al. [1]. Tensimetric studies at higher temperatures and pressures establish that above 1173 K probably trimeric and tetrameric molecules also exist in the vapor phase. For 1238 K and a 446 Torr total pressure two possible compositions have been calculated (mol %): 21.5 $ScCl_3$, 51.3 $Sc_2Cl_6$, and 27.2 $Sc_3Cl_9$, or 23.5 $ScCl_3$, 61.8 $Sc_2Cl_6$, and 14.6 $Sc_4Cl_{12}$, Polyachenok et al. [2], see also Polyachenok et al. [3]. According to Wagner, Schäfer [49], the saturated vapor over $ScCl_3$ at 850 K has an $ScCl_3$:$Sc_2Cl_6$ ratio of 11.3:1, and the fraction of $Sc_2Cl_6$ increases with increasing temperature. In contrast, Rat'kovskii et al. [50] report $ScCl_3$:$Sc_2Cl_6$:$Sc_3Cl_9$=0.01:1:0.001; however, their enthalpy of dimerization is in conflict with the results of other investigators [1, 49, 51].

Noticeable quantities of dimeric molecules have been identified in the $YCl_3$ vapor from a Knudsen cell at 900 to 1600 K by mass spectrometry, McKinley [4, 5]. On the contrary, only monomeric trichloride molecules could be detected in the vapor phase above $YCl_3$, $LaCl_3$, and $NdCl_3$ by electron diffraction studies, Akishin et al. [6]. Mass spectrometric studies indicate the existence of the dimers in only small quantities (1%) in the vapor above solid $LaCl_3$ and $EuCl_3$ as well as over a mixture of $EuCl_3$ with $LuCl_3$, at pressures <0.1 Torr. For temperatures near the respective boiling points, however, a higher proportion of dimeric species is to be expected, for $LaCl_3$ (boiling point 2085 K) 85%, for example, Hastie et al. [7]. In similar studies of the vapor above $SmCl_3$ and $YbCl_3$ about 2% dimer was determined. $EuCl_3$ decomposes on heating with evolution of $Cl_2$; at higher temperatures the $EuCl_2$ formed evaporates congruently, forming only monomeric molecules, Chervonnyi et al. [8]. The decomposition of liquid $EuCl_3$ is extensively impeded by addition of $LuCl_3$. For a mixture with a melting point around 1000 K the dichloride fraction in the vapor at 960 K amounts to less than 5% of the trichloride

[7]. For $NdCl_3$ and $GdCl_3$ at 1055 to 1093 or 963 to 1075 K, respectively, from mass spectrometric studies of the vapor phase only a very small dimerization occurs of the trichloride, which in these cases evaporates undecomposed, Clach et al. [9]. The same holds for $YCl_3$ at 770 to 1070 K, McKinley [5].

**Vapor Pressure** p

Introductory Remarks

Chlorides of the rare earth elements are difficultly volatile substances, whose vaporization normally becomes noticeable only above 1000 °C. The following table lists the experimentally determined temperatures necessary to yield vapor pressures of 0.1 to 4 Torr [10] along with estimated temperatures for 2 Torr and the boiling point [11, 12] based on data from the literature.

Temperatures in °C for vapor pressures:

| $MCl_3$ | 4 Torr | 2 Torr | 1 Torr | 0.1 Torr | 2 Torr (estim.) | 760 Torr (b.p.) |
|---|---|---|---|---|---|---|
| $YCl_3$ | 1050 | 975 | 909 | 735 | 950 | 1510 |
| $LaCl_3$ | 1027 | 997 | 969 | 886 | 1100 | 1750 |
| $CeCl_3$ | 1195 | 1125 | 1065 | 888 | 1090 | 1730 |
| $PrCl_3$ | 1144 | 1085 | 1031 | 878 | 1080 | 1710 |
| $NdCl_3$ | 1166 | 1106 | 1048 | 892 | 1060 | 1690 |
| $SmCl_3$ | 1400 | 1229 | 1087 | 764 | 1310 | 2030 |
| $EuCl_3$ | 998 | 930 | 869 | 703 | 940 | decomp. |
| $GdCl_3$ | 1048 | 995 | 947 | 808 | 980 | 1580 |
| $TbCl_3$ | 1121 | 1068 | 1010 | 854 | 960 | 1550 |
| $DyCl_3$ | 979 | 939 | 899 | 779 | 950 | 1530 |
| $HoCl_3$ | 986 | 953 | 919 | 827 | 950 | 1510 |
| $ErCl_3$ | 1155 | 1076 | 1000 | 809 | 950 | 1500 |
| $TmCl_3$ | 951 | 925 | 899 | 824 | 940 | 1490 |
| $YbCl_3$ | 1091 | 1039 | 994 | (856) [a] | 940 | decomp. |
| $LuCl_3$ | 998 | 959 | 926 | (819) [a] | 950 | 1480 |

[a] Extrapolated value below melting point.

The table has often been used to determine the appropriate temperatures for vacuum sublimations and distillations. All stable trichlorides have vapor pressures high enough to permit them to be purified by evaporation. Except for the chlorides of the lighter elements, all should be capable of being vacuum distilled at an acceptable rate at temperatures of approximately 1000 °C or less, Block, Campbell [13]. Approximate data on the vapor pressures in the range $10^{-1}$ to $10^{-3}$ Torr are also given by Harrison [14] on the basis of molecular effusion measurements.

Numerical values of vapor pressure measurements are presented and discussed in the sections on the individual rare earth chlorides in "Rare Earth elements" C4b. The log p−1/T equations are treated collectively in the following section.

### Vapor Pressure Equations for Solid Compounds

The temperature dependence of the vapor pressures is usually described by equations of the form log p=A−B/T. Values for the constants A and B are the following, for p in Torr and T in K:

| $MCl_3$ | temperature in K | A | B | Ref. |
|---|---|---|---|---|
| $ScCl_3$ | 873 to 1233 | 14.56 | 14580 | [1] |
| | 1065 to 1228 | 14.37 | 14223 | [15] |
| | 1074 to 1227 | 13.61±0.31 | 13470±360 | [2, 3, 16] |
| $YCl_3$ | 771 to 898 | 8.587 | 8796.6 | [17] |
| $LaCl_3$ | 1022 to 1110 | 11.828 | 15796 | [18] |
| $LaCl_3$ | 842 to 1101 | 11.9163 | 15284 | [19] |
| $CeCl_3$ | 1012 to 1071 | 12.035 | 15544 | [18] |
| $PrCl_3$ | 1002 to 1061 | 12.121 | 15439 | [18] |
| $NdCl_3$ | 973 to 1032 | 12.014 | 15145 | [18] |
| | 973 to 1073 | 13.49±0.84 | 16250±860 | [20] |
| $SmCl_3$ | 774 to 855 | 9.763 | 9686.6 | [17] |

Kubaschewski et al. [21, 22] reevaluate the result of [18] on the basis of the equation log p=A−B/T−C log T and give the following constants for the range between 298 K and the melting point, for p in Torr:

$LaCl_3$: A=36.20, B=19040, C=7.05
$CeCl_3$: A=36.38, B=18750, C=7.05
$PrCl_3$: A=36.31, B=18490, C=7.05
$NdCl_3$: A=36.27, B=18220, C=7.05

### Vapor Pressure Equations for Melts

Vapor pressures determined by the flow method [23], Knudsen effusion method [24], and the boiling point method in quartz ampules [other references] are represented by the equation log p=A−B/T. The following constants A and B are derived for p in Torr and T in K:

| $MCl_3$ | temperature in K | A | B | Ref. |
|---|---|---|---|---|
| $YCl_3$ | 988 to 1193 | 3.07±0.032 | 3750 | [23] |
| $LaCl_3$ | 1273 to 1673 | 10.19 a) | 13930 | [25] |
| | 1373 to 1573 | 10.679 a) | 14671 | [26] |
| | 1397 to 1493 | 11.535 a) | 15940 | [27] |
| $CeCl_3$ | 1366 to 1497 | 11.006 a) | 14860 | [27] |
| $PrCl_3$ | 1314 to 1465 | 10.444 a) | 13810 | [27] |
| | 1373 to 1573 | 9.202 a) | 12419 | [26] |
| $NdCl_3$ | 1235 to 1467 | 9.970 a) | 12930 | [27] |
| | 1373 to 1573 | 9.412 a) | 12420 | [26] |
| $GdCl_3$ | 933 to 1073 | 9.1±0.3 | 11600±300 | [24] |
| $ErCl_3$ | 1273 to 1473 | 9.5144 | 11579 | [28] |
| $LuCl_3$ | 1373 to 1573 | 9.521 a) | 11130 | [26] |

a) The original data are for p in atm.

The equation log p = A − B/T − 6.0 log T is preferred for the representation of p measurements from about 1273 to 1673 K by the boiling point method. The equation has been derived from the usual linear expression allowing for $\Delta C_p$ ($C_p$ = heat capacity) of the evaporation process. The coefficient 6.0 is $\Delta C_p/R$, when $\Delta C_p$ is taken as equal to $-12\ cal \cdot mol^{-1} \cdot K^{-1}$. The following numerical values for A and B are calculated by the method of least squares for p in Torr:

| $MCl_3$ | temperature in K | A | B | Ref. |
|---|---|---|---|---|
| $YCl_3$ | 1044 to 1376 | 31.244 | 15650 | [29] |
| $LaCl_3$ | 1000 to 1370 | 30.974 | 16463 | [30] a) |
| $CeCl_3$ | 1000 to 1370 | 31.084 | 16347 | [30] a) |
| $PrCl_3$ | 1101 to 1363 | 31.030 | 16100 | [29] |
| $NdCl_3$ | 1000 to 1370 | 30.845 | 15769 | [30] a) |
| $GdCl_3$ | 1096 to 1330 | 30.915 | 15715 | [29] |
| $TbCl_3$ | 1053 to 1380 | 30.660 | 14995 | [29] |
| $DyCl_3$ | 1052 to 1347 | 30.743 | 15064 | [29] |
| $HoCl_3$ | 1000 to 1370 | 30.735 | 14946 | [30] |
| $ErCl_3$ | 1000 to 1370 | 30.740 | 14816 | [30] a) |
| $TmCl_3$ | 1000 to 1370 | 30.571 | 14466 | [30] |
| $LuCl_3$ | 1000 to 1370 | 30.940 | 14718 | [30] |

a) The given values have been recalculated by [30] from earlier experimental data.

The p data for $YCl_3$ and $PrCl_3$ almost coincide with results obtained earlier by [31].

The equation log p = A − B/T − 9.061 log T has been used to describe vapor pressure measurements by [32, 33] from 1373 to 1673 K by a modified (thermographic) boiling point method. The equation is reported to hold for the range between melting points and boiling points of the respective chlorides. The coefficient 9.061 is $\Delta C_p/R$, when $\Delta C_p$ is assumed to be $-(18 \pm 2)\ cal \cdot mol^{-1} \cdot K^{-1}$. The constants A and B are the following:

| $MCl_3$ | A | B | Ref. |
|---|---|---|---|
| $YCl_3$ | 42.169 | 17470 | [32, 34] |
| $LaCl_3$ | 41.983 | 18392 | [32, 33] |
| $CeCl_3$ | 42.011 | 18153 | [32, 33] |
| $PrCl_3$ | 41.981 | 17946 | [32, 33] |
| $NdCl_3$ | 41.841 | 17691 | [32, 33] |
| $SmCl_3$ | 41.82 | 17450 | [33, 35] |
| $EuCl_3$ | 41.79 | 17300 | [33, 35] |
| $ErCl_3$ | 41.671 | 16624 | [32, 33] |
| $YbCl_3$ | 41.65 | 16380 | [33, 35] |

The equations for $SmCl_3$, $EuCl_3$, and $YbCl_3$ are obtained by interpolation and extrapolation from the results for the other chlorides, which show a monotonic change in the characteristics of the evaporation process across the series listed in the table.

See [36, 11] for data from the literature for temperatures at which the vapor pressure of the rare earth chlorides amounts to $10^{-4}$, $10^{-3}$, $10^{-2}$, and 1 atm, respectively.

Cho, Kawamura [37] applied the modified significant structure theory of liquids to molten $PrCl_3$ and $NdCl_3$ and calculated values for the vapor pressure which agree well with the experimental results of Moriarty [10].

## Boiling Point

The following normal boiling point temperatures $T_b$ have been determined from vapor pressure measurements in the temperature ranges given below the table:

| $MCl_3$ | $T_b$ in K | Ref. | $MCl_3$ | $T_b$ in K | Ref. | $MCl_3$ | $T_b$ in K | Ref. |
|---|---|---|---|---|---|---|---|---|
| $ScCl_3$ | 1240 | [10] | $NdCl_3$ | 1903 | [30] | $HoCl_3$ | 1796 | [30] |
| $YCl_3$ | 1761 | [29] | | 1921 | [32, 33] | | 1787 | [33] |
| | 1774 | [32, 34] | | 1947 | [18] | $ErCl_3$ | 1769 | [30] |
| $LaCl_3$ | 1981 | [30] | $PmCl_3$ | 1901 | [33] | | 1781 | [32, 33] |
| | 2002 | [32, 33] | $SmCl_3$ | 1883 | [33] | $TmCl_3$ | 1761 | [30] |
| | 2085 | [18] | $EuCl_3$ | 1867 | [33] | | 1759 | [33] |
| $CeCl_3$ | 1924 | [30] | $GdCl_3$ | 1870 | [29] | $YbCl_3$ | 1744 | [33] |
| | 1947 | [32, 33] | | 1849 | [33] | $LuCl_3$ | 1695 | [30] |
| | 2004 | [18] | $TbCl_3$ | 1827 | [29] | | 1729 | [33] |
| $PrCl_3$ | 1899 | [29] | | 1827 | [33] | | | |
| | 1918 | [32, 33] | $DyCl_3$ | 1812 | [29] | | | |
| | 1982 | [18] | | 1809 | [33] | | | |

Temperatures of vapor pressure measurements in K: $\leqq$1239 [10], 1317 to 1653 [29], 1373 to 1673 [32], 973 to 1110 [18], 1273 to 1643 [30]. The data of [30] for $LaCl_3$, $CeCl_3$, $NdCl_3$, and $ErCl_3$ are recalculated from earlier vapor pressure measurements. The values of [33] for $PmCl_3$ through $LuCl_3$ are calculated from estimated vapor pressure equations.

Lower boiling points of 1843, 1828, 1827, and 1824 K for $LaCl_3$, $CeCl_3$, $PrCl_3$, and $NdCl_3$, respectively, have been derived by Novikov, Baev [27] from their own vapor pressure determinations in the range from 1235 to 1497 K; lower values are also given by Harrison [14].

A nearly constant ratio $-U/T_b \approx 0.57\ kcal \cdot mol^{-1} \cdot K^{-1}$ follows from published values for the lattice energy U and the boiling point. Unknown or unreliable $T_b$ data may thus be estimated and checked using available data for the lattice energy; numerical values are given, Gopal, Husain [38]. Compilations of older $T_b$ data can be found in [11, 36, 39 to 41], see also [42].

## Thermodynamics of Vaporization

The following values for the enthalpy $\Delta H_v^\circ$ and entropy $\Delta S_v^\circ$ of vaporization at the boiling point have been calculated from the vapor pressure data presented on pp. 115/117. $\Delta H_v^\circ$ and $\Delta S_v^\circ$ are in the units kcal/mol and $cal \cdot mol^{-1} \cdot K^{-1}$, respectively:

| $MCl_3$ | $\Delta H_v^\circ$ | $\Delta S_v^\circ$ | Ref. |
|---|---|---|---|
| $ScCl_3$ | 46 | 37 | [11] |
| $YCl_3$ | 50.3 | 28.6 | [29] |
| | 48.0 | 27.1 | [32] |
| $LaCl_3$ | 51.5 | 26.0 | [30] |
| | 49.0 | 23.0 | [18] |
| | 48.1 | 24.0 | [32, 33] |
| $CeCl_3$ | 51.7 | 26.9 | [30] |
| | 49.0 | 24.0 | [18] |
| | 48.0 | 24.7 | [32, 33] |
| $PrCl_3$ | 50.8 | 26.7 | [29] |
| | 49.0 | 24.0 | [18] |
| | 47.6 | 24.8 | [32, 33] |
| $NdCl_3$ | 49.3 | 25.9 | [30] |
| | 48.0 | 24.0 | [18] |
| | 46.2 | 24.1 | [32, 33] |
| $PmCl_3$ | 46.3 | 24.4 | [33] |
| $SmCl_3$ | 45.9 | 24.4 | [33] |
| $EuCl_3$ | 45.5 | 24.4 | [33] |
| $GdCl_3$ | 49.4 | 26.4 | [29] |
| | 45.2 | 24.4 | [33] |
| $TbCl_3$ | 46.7 | 25.5 | [29] |
| | 44.9 | 24.6 | [33] |
| $DyCl_3$ | 47.2 | 26.0 | [29] |
| | 44.5 | 24.6 | [33] |
| $HoCl_3$ | 46.8 | 26.1 | [30] |
| | 44.2 | 24.7 | [33] |
| $ErCl_3$ | 46.6 | 26.3 | [30] |
| | 43.9 | 24.6 | [32, 33] |
| $TmCl_3$ | 45.1 | 25.6 | [30] |
| | 43.7 | 24.8 | [33] |
| $YbCl_3$ | 43.5 | 24.9 | [33] |
| $LuCl_3$ | 47.0 | 27.7 | [30] |
| | 43.4 | 25.1 | [33] |

The data of [18] are derived from measurements of the vapor pressure above the solid, assuming the enthalpy of fusion as 9.0 kcal/mol for $LaCl_3$ and as 8.0 kcal/mol for the other chlorides; the entropy of fusion has been uniformly taken as 8.0 $cal \cdot mol^{-1} \cdot K^{-1}$.

Evseeva, Zenkevich [24] report $\Delta H_v^\circ = 53.1 \pm 2.7$ and $\Delta S_v^\circ = 41.5 \pm 2.6$ for $GdCl_3$ at 1003 K. Data for $\Delta H_v^\circ$, which widely diverge from the tabulated values, have been computed by Moriarty [10] from his own rather scarce vapor pressure determinations. Deviating values are also given by Novikov, Baev [27] and Harrison [14].

Reduction to 0 K gives:

| $MCl_3$ | $YCl_3$ | $LaCl_3$ | $CeCl_3$ | $PrCl_3$ | $NdCl_3$ | $GdCl_3$ |
|---|---|---|---|---|---|---|
| $\Delta H_0^\circ$ | 71.6 | 75.3 [a)] | 74.8 [a)] | 73.6 | 72.1 [a)] | 71.9 |
| Ref. | [29] | [30] | [30] | [29] | [30] | [29] |

| $MCl_3$ | $TbCl_3$ | $DyCl_3$ | $HoCl_3$ | $ErCl_3$ | $TmCl_3$ | $LuCl_3$ |
|---|---|---|---|---|---|---|
| $\Delta H_0^\circ$ | 68.6 | 68.9 | 68.4 | 67.8 [a)] | 66.2 | 67.3 |
| Ref. | [29] | [29] | [30] | [30] | [30] | [30] |

[a)] Calculated from literature data.

This table supersedes the earlier compilation by [33].

The vapor pressures of $NdCl_3$ and $GdCl_3$ at 1055 to 1093 K and 963 to 1075 K, respectively, were determined by Clach et al. [9] using effusion mass spectrometry and used to calculate $\Delta H_v^\circ$ in kcal/mol at 298 K. The results are compared with recalculated values from the literature:

$NdCl_3$

| | | | | | |
|---|---|---|---|---|---|
| $\Delta H_v^\circ$ (298 K) | 69.3±4 | 64.4 | 71.1 | 75.4 | 77.5 |
| Ref. | [9] | [10] | [18] | [33] | [27] |

$GdCl_3$

| | | | |
|---|---|---|---|
| $\Delta H_v^\circ$ (298) . . . . | 65.5±4 | 62.7 | 68.3 |
| Ref. . . . . . . . | [9] | [10] | [29] |

Values for $\Delta H_v^\circ$ (298 K) and $\Delta H_v^\circ$ at the normal boiling point have also been tabulated by Gibson et al. [36] and Thoma [41]. Free energies of vaporization at 1000 K for several rare earth chlorides have been calculated from older literature data by Brewer [11].

Published data on the vaporization of the lanthanide trichlorides have been reviewed by Myers, Graves [43], and enthalpies of sublimation at 298.15 K, $\Delta H_s^\circ$ (298, II) and $\Delta H_s^\circ$ (298, III), have been calculated by the second and third law methods, respectively, using thermodynamic functions for gaseous and condensed states developed in an internally consistent manner. The results are given in the following table. Preferred values by the third law are marked with an asterisk; they were primarily selected on the basis of agreement between the second and third law results.

| $MCl_3$ | $\Delta H_s^\circ$ (298, II) | $\Delta H_s^\circ$ (298, III) | source |
|---|---|---|---|
| $LaCl_3$ | 77.9 | 81.9±3.0* | [18] |
| | 101.2 | 77.1 | [10] |
| | 83.6 | 80.4 | [30] |
| | 79.0 | – | [7] |
| $CeCl_3$ | 76.4 | 79.1±3.0* | [18] |
| | 62.5 | 76.2 | [10] |
| | 78.8 | 77.4 | [30] |
| $PrCl_3$ | 76.5 | 77.7±3.0* | [18] |
| | 64.1 | 75.6 | [10] |
| | 78.8 | 76.2 | [29] |
| $NdCl_3$ | 74.0 | 78.1±3.0* | [18] |
| | 59.5 | 76.4 | [10] |
| | 78.1 | 77.5 | [30] |
| | 78.1 | – | [9] |
| | 79.7 | 62.9 | [20] |
| $EuCl_3$ | 42.5 | 68.2(?) | [10] |
| $GdCl_3$ | 61.0 | 70.5 | [10] |
| | 75.7 | 73.9±4.0* | [30] |
| | 72.2 | – | [9] |
| $TbCl_3$ | 55.1 | 73.3 | [10] |
| | 71.1 | 72.6±4.0* | [29] |
| $DyCl_3$ | 58.4 | 71.1 | [10] |
| | 68.9 | 76.2±4.0* | [29] |
| $HoCl_3$ | 73.9 | 73.2 | [10] |
| | 69.3 | 76.4±4.0* | [30] |
| $ErCl_3$ | 47.7 | 75.3 | [10] |
| | 68.5 | 77.9±4.0* | [30] |
| $TmCl_3$ | 73.0 | 72.3 | [10] |
| | 67.0 | 76.7±4.0* | [30] |
| $YbCl_3$ | 42.7 | 76.7(?) | [10] |
| $LuCl_3$ | 58.2 | 75.8 | [10] |
| | 68.9 | 76.3±4.0* | [30] |

Experimental methods: Mass effusion [10, 18, 20], effusion mass spectroscopy [7, 9], boiling point techniques [29, 30]. Third law agreement is generally rather good within the limits of uncertainty of the selected values. No selected values are chosen for $EuCl_3$ and $YbCl_3$ since their vaporizations are probably incongruent and the results therefore uncertain. Satisfactory agreement between the second and third law results is achieved by Shimazaki, Niwa [18] and, for the early part of the series, by Dudchik et al. [29, 30], but the agreement for the results of the latter authors is not as good for compounds toward the end of the series. In their critical discussion Myers, Graves [43] also find that Moriarty [10] collected far too few vapor pressure data for a meaningful second law treatment, and that his experimental design raises some serious questions.

The results of Evseeva, Zenkevich [20] are said to be objectionable, because the Russian authors evidently were not able to overcome the problem of the hygroscopic behavior of their $NdCl_3$ sample.

Measurements of the vapor pressures of $PrCl_3$, $GdCl_3$, $TbCl_3$, and $DyCl_3$ by mass loss effusion from stainless steel cells at temperatures near 1000 K yielded the following second and third law enthalpies of sublimation:

| $MCl_3$ | $\bar{T}$ in K | $\Delta H_s^\circ$ ($\bar{T}$, II) | $\Delta H_s^\circ$ (298, II) | $\Delta H_s^\circ$ (298, III) | Ref. |
|---|---|---|---|---|---|
| $PrCl_3$ | 1000 | 73.3 | 78.9 | $78.4 \pm 3.0$ | [44] |
| $GdCl_3$ | 995 | 63.0 | 77.9 | $74.5 \pm 3.0$ | [45] |
| $TbCl_3$ | 980 | 60.0 | 73.3 | $72.7 \pm 3.0$ | [45] |
| $DyCl_3$ | 975 | 61.6 | 71.9 | $74.3 \pm 3.0$ | [45] |

$\bar{T}$ is the reciprocal of the mean value of 1/T. The agreement with the values calculated by [43] from the earlier data of [27, 29, 30] is good, and the agreement between the second and third law results is quite satisfactory.

Additional data for $\Delta H_s^\circ$ and the entropy of sublimation $\Delta S_s^\circ$ at various temperatures ($T_m$ = melting point):

$ScCl_3$: $\Delta H_s^\circ$ ($T_m$) = 65 (±5%), $\Delta S_s^\circ$ = 52, Fischer et al. [15]
$YCl_3$: $\Delta H_s^\circ$ (834.75) = $40.2 \pm 4.0$, Pashinkin et al. [17]
$LaCl_3$: $\Delta H_s^\circ$ (1030) = 74, $\Delta H_s^\circ$ (298) = 84.5, $\Delta S_s^\circ$ (1030) = 42.5, Hastie et al. [7], $\Delta H_s^\circ$ (842 to 1101) = 69.93 [19]
$SmCl_3$: $\Delta H_s^\circ$ (814.5) = $44.3 \pm 5.4$, Pashinkin et al. [17]

Enthalpies of sublimation at 298 K for all the rare earth trichlorides can also be obtained from the list of enthalpies of formation for solid and gaseous compounds by Dudchik et al. [47]. See also the compilation of $\Delta H_s^\circ$ (298) values for $CeCl_3$ through $LuCl_3$ by Feber [46] and the experimental data for $LaCl_3$ by Novikov, Tolmacheva [25], for $LaCl_3$, $CeCl_3$, $PrCl_3$, and $NdCl_3$ by Novikov, Baev [27], and for $GdCl_3$ by Novikov, Gavryuchenko [28, 48].

The sublimation entropy of the dimer $La_2Cl_6$ is $\Delta S_s^\circ$ (1030) = $55.0 \pm 3$ $cal \cdot mol^{-1} \cdot K^{-1}$ [7]; see also the pertinent discussion of Myers, Graves [43].

Free energies of sublimation for several rare earth chlorides at 1000 K have been calculated by Brewer [11] from older literature data.

The decrease of $\Delta H_s^\circ$ from $LaCl_3$ to $NdCl_3$ (and from the chlorides to the iodides) is explained as the result of ion deformation, Shimazaki, Niwa [18]. As shown in **Fig. 14** and as already noticed by Myers, Graves [43], the third law enthalpies of sublimation of the rare earth trichlorides at 298 K are not a smooth function of the lanthanide atomic number, especially in the region of $GdCl_3$, $TbCl_3$, and $DyCl_3$. This observation is discussed in relation to crystal structure, hard sphere ion and thermochemical lattice energies, and bonding models for the solids, Myers, Hannay [45]. Russian authors, see for example Dudchik et al. [47], earlier established that the enthalpies of evaporation of the chlorides decrease monotonically with the atomic number throughout the entire lanthanide series, while the entropies of evaporation and sublimation remain virtually constant.

Fig. 14

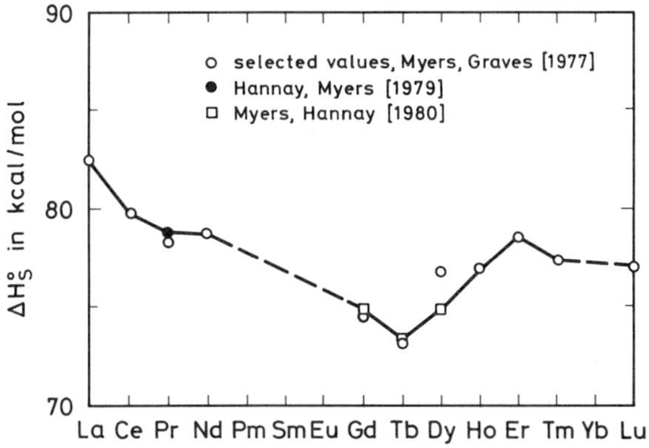


Third law enthalpies of sublimation at 298.15 K according to [45].

References:

[1] Yu.B. Patrikeev, V.A. Morozova, G.P. Dudchik, O.G. Polyachenok, G.I. Novikov (Zh. Fiz. Khim. **47** [1973] 266; Russ. J. Phys. Chem. **47** [1973] 152/3). – [2] L.D. Polyachenok, G.P. Dudchik, K. Nazarov, O.G. Polyachenok (Vesti Akad. Navuk Belarusk. SSR Ser. Khim. Navuk **1977** No. 3, pp. 113/5; C.A. **87** [1977] No. 58676). – [3] L.D. Polyachenok, K. Nazarov, G.P. Dudchik, O.G. Polyachenok (Rasshir. Tezisy Dokl. 7th Vses. Konf. Kalorim 1977, Vol. 2, pp. 309/13 from C.A. **92** [1980] No. 11942). – [4] J.D. McKinley (J. Chem. Phys. **41** [1964] 2814/7). – [5] J.D. McKinley (J. Chem. Phys. **42** [1965] 2245/6).

[6] P.A. Akishin, V.A. Naumov, V.M. Tatevskii (Vestn. Mosk. Univ. Ser. Mat. Mekhan. Astron. Fiz. Khim. **14** No. 1 [1959] 229/36; C.A. **1960** 31; Kristallografiya **4** [1959] 194/200; Soviet Phys.-Cryst. **4** [1959] 174/80). – [7] J.W. Hastie, P. Ficalora, J.L. Margrave (J. Less-Common Metals **14** [1968] 83/91). – [8] A.D. Chervonnyi, V.K. Il'in, V.A. Krenev (V Sb. Splavy Redk. Met. Osob. Fiz. Khim. Svoistv. **1975** 133/6; C.A. **85** [1976] No. 10960). – [9] S. Clach, A.J.C. Nicholson, D.L. Swingler, P.J. Thistlethwaite (Inorg. Chem. **12** [1973] 2072/4). – [10] J.L. Moriarty (J. Chem. Eng. Data **8** [1963] 422/4).

[11] L. Brewer (Natl. Nucl. Energy Ser. Div. IV B **19** [1950] 193/275). – [12] C.E. Wicks, F.E. Block (U.S. Bur. Mines Bull. No. 605 [1963] 1/146; C.A. **59** [1963] 14656). – [13] F.E. Block, T.T. Campbell (in: F.H. Spedding, A. Daane, The Rare Earths, Wiley, New York - London 1961, pp. 89/101, 98). – [14] E.R. Harrison (J. Appl. Chem. [London] **2** [1952] 601/2). – [15] W. Fischer, R. Gewehr, H. Wingchen (Z. Anorg. Allgem. Chem. **242** [1939] 161/87).

[16] L.D. Polyachenok, K. Nazarov, O.G. Polyachenok (Zh. Fiz. Khim. **52** [1978] 1767/8; Russ. J. Phys. Chem. **52** [1978] 1021/2). – [17] A.S. Pashinkin, D.V. Drobot, Z.N. Shevtsova, B.G. Korshunov (Zh. Neorgan. Khim. **7** [1962] 2811/3; Russ. J. Inorg. Chem. **7** [1962] 1466/8). – [18] E. Shimazaki, K. Niwa (Z. Anorg. Allgem. Chem. **314** [1962] 21/34, 30). – [19] A.N. Nesmeyanov, L.A. Sazonov (Zh. Neorgan. Khim. **4** [1959] 230; Russ. J. Inorg. Chem. **4** [1959] 97). – [20] G.V. Evseeva, L.V. Zenkevich (Vestn. Mosk. Univ. Khim. **31** No. 1 [1976] 110/1; Moscow Univ. Chem. Bull. **31** No. 1 [1976] 84/5).

[21] O. Kubaschewski, E.L. Evans, C.B. Alcock (Metallurgical Thermochemistry, 4th Ed., Oxford - London - Edinburgh - New York - Toronto - Sydney - Paris - Braunschweig 1967, pp. 409/20). – [22] O. Kubaschewski, C.B. Alcock (Metallurgical Thermochemistry, 5th Ed., Oxford - New York - Toronto - Sydney - Paris - Frankfurt 1979). – [23] A.I. Pustil'nik, B.A. Sakharov, T.S. Sedykh, R.V. Dubchak (Zh. Fiz. Khim. **44** [1970] 2947; Russ. J. Phys. Chem. **44** [1970] 1682/3). – [24] G.V. Evseeva, L.V. Zenkevich (Vestn. Mosk. Univ. Ser. II Khim. **19** No. 1 [1978] 89/91; Moscow Univ. Chem. Bull. **33** No. 1 [1978] 70/1). – [25] G.I. Novikov, V.D. Tolmacheva (Zh. Neorgan. Khim. **10** [1965] 2712/6; Russ. J. Inorg. Chem. **10** [1965] 1472/4).

[26] L.A. Nisel'son, Yu.N. Lyzlov, S.I. Slov'ev (Zh. Neorgan. Khim. **23** [1978] 787/9; Russ. J. Inorg. Chem. **23** [1978] 434/6). – [27] G.I. Novikov, A.K. Baev (Zh. Neorgan. Khim. **7** [1962] 1349/52; Russ. J. Inorg. Chem. **7** [1962] 694/96; Zh. Obshch. Khim. **32** [1962] 315/6; J. Gen. Chem. [USSR] **32** [1962] 309/10). – [28] G.I. Novikov, F.G. Gavryuchenko (Zh. Neorgan. Khim. **10** [1965] 1668/74, Russ. J. Inorg. Chem. **10** [1965] 909/13). – [29] G.P. Dudchik, O.G. Polyachenok, G.I. Novikov (Zh. Neorgan. Khim. **14** [1969] 3165/7; Russ. J. Inorg. Chem. **14** [1969] 1669/70). – [30] G.P. Dudchik, O.G. Polyachenok, G.I. Novikov (Zh. Fiz. Khim. **43** [1969] 2145; Russ. J. Phys. Chem. **43** [1969] 1203/4).

[31] G.I. Novikov, A.K. Baev (Vestn. Leningr. Univ. Fiz. Khim. **1962** No. 1, pp. 154/7). – [32] O.G. Polyachenok, G.I. Novikov (Vestn. Leningr. Univ. Fiz. Khim. **1963** No. 3, pp. 133/4; C.A. **60** [1964] 4875). – [33] O.G. Polyachenok, G.I. Novikov (Zh. Neorgan. Khim. **8** [1963] 1526/7; Russ. J. Inorg. Chem. **8** [1963] 793/4). – [34] O.G. Polyachenok, G.I. Novikov (Zh. Neorgan. Khim. **8** [1963] 2818/9; Russ. J. Inorg. Chem. **8** [1963] 1478/9). – [35] O.G. Polyachenok, G.I. Novikov (Zh. Neorgan. Khim. **9** [1964] 773/7; Russ. J. Inorg. Chem. **9** [1964] 429/31).

[36] J.A. Gibson, J.F. Miller, P.S. Kennedy, G.W.P. Rengstorff (The Properties of the Rare Earth Metals and Compounds, Battelle Memorial Institute, Columbus, Ohio, 1959, p. 10, Section 1). – [37] K. Cho, K. Kawamura (Denki Kagaku Oyobi Kogyo Butsuri Kagaku **47** [1979] 209/13; C.A. **91** [1979] No. 63573). – [38] R. Gopal, M.M. Husain (J. Indian Chem. Soc. **44** [1967] 726/7). – [39] R.C. Vickery (Chemistry of the Lanthanons, Butterworth, London 1953, pp. 228/9). – [40] R.C. Vickery (The Chemistry of Yttrium and Scandium, Pergamon, New York - Oxford - London - Paris 1960, pp. 56/69).

[41] R.E. Thoma (in: L. Eyring, Progress in the Science and Technology of the Rare Earths, Vol. 2, Pergamon Press, Oxford - London - Edinburgh - New York - Toronto - Sydney - Paris - Braunschweig 1966, pp. 90/122). – [42] G.E. Revzin (Metody Poluch. Khim. Reaktiv. Prep. **1967** No. 16, pp. 124/9; C.A. **69** [1968] No. 48783). – [43] C.E. Myers, D.T. Graves (J. Chem. Eng. Data **22** [1977] 440/5). – [44] M.H. Hannay, C.E. Myers (J. Less-Common Metals **66** [1979] 145/50). – [45] C.E. Myers, M.H. Hannay (J. Less-Common Metals **70** [1980] 15/24).

[46] R.C. Feber (LA-3164 [1965] 1/185, 175; N.S.A. **19** [1965] No. 30148). – [47] G.P. Dudchik, O.G. Polyachenok, G.I. Novikov (Zh. Fiz. Khim. **45** [1971] 728/91; Russ. J. Phys. Chem. **45** [1971] 409/10). – [48] G.I. Novikov, F.G. Gavryuchenko (Zh. Neorgan. Khim. **10** [1965] 2706/11; Russ. J. Inorg. Chem. **10** [1965] 1469/72). – [49] K. Wagner, H. Schäfer (Z. Anorg. Allgem. Chem. **430** [1977] 43/50). – [50] I.A. Rat'kovskii, L.N. Novikova, T.A. Pribytkova (Khim. Khim. Tekhnol. **9** [1975] 19/25; Zh. Fiz. Khim. **48** [1974] 2615; Russ. J. Phys. Chem. **48** [1974] 1551/2).

[51] H. Schäfer, M. Binnewies (Z. Anorg. Allgem. Chem. **410** [1974] 251/68).

### 11.2.6.8.4 Thermodynamic Data

#### Enthalpy of Formation

Table 11.2.6-5
Values for $\Delta H^{\circ}_{298}$ in kcal/mol for the Standard Enthalpy of Formation at 298 K.

| $MCl_3$ | $-\Delta H^{\circ}_{298}$ | method | Ref. |
|---|---|---|---|
| $ScCl_3$ | 220.7 | a | [1] |
| | 220.8 | b | [2] |
| | 221.0 | c | [3] |
| | 221.1 | c | [4] |
| | 225.44±0.34 | a | [5] |
| $YCl_3$ | 232.69±0.20 | a | [6] |
| | 232.69±0.30 | c | [7] |
| | 234.8 | a | [1] |
| | 235 | d | [8] |
| | 239.0 | c | [4, 9] |
| | 239.16±0.50 | a | [10] |
| $LaCl_3$ | 254.9 | a | [11] |
| | 255.68 | d | [8] |
| | 255.9 | c | [9] |
| | 255.91±0.20 | a | [6] |
| | 255.91±0.30 | c | [7] |
| | 256.0 | c | [12] |
| | 256.18±0.28 | a | [13] |
| | 256.3±0.3 | c | [14] |
| | 263.7 | a | [1] |
| $CeCl_3$ | 251.5 | c | [9] |
| | 251.8 | c | [12] |
| | 252.2±1 | c | [14] |
| | 252.84 | d | [8] |
| | 252.84 | a | [15] |
| | 252.98±0.30 | c | [7] |
| | 259.8 | a | [1] |
| $PrCl_3$ | 252.0 | c | [9, 14] |
| | 252.09±0.20 | a | [6] |
| | 252.09±0.30 | c | [7] |
| | 252.6 | c | [12] |
| | 253.02 | d | [8] |
| | 257.8 | a | [1] |
| $NdCl_3$ | 245.62 | d | [8] |
| | 245.62±0.02 | a | [15] |
| | 245.71±0.30 | c | [7] |
| | 246.5 | a | [16] |
| | 246.5±1.5 | a | [17] |
| | 248.7 | c | [9] |
| | 248.8 | c | [12] |
| | 248.97±0.26 | a | [18] |
| | 249.1±0.3 | c | [14] |
| $NdCl_3$ | 250 | e | [19] |
| | 254.3 | a | [1] |
| $PmCl_3$ | 245 | f | [20] |
| | 247 | g | [14, 47] |
| $SmCl_3$ | 243 | g | [21, 22] |
| | 244.1 | c | [9] |
| | 244.18±0.41 | a | [23] |
| | 245.0±3.0 | c | [7] |
| | 245.2 | c | [12] |
| | 248 | d | [8] |
| $EuCl_3$ | 218.74±2.0 | a | [24] |
| | 219.5 | c | [9] |
| | 220.0±2 | c | [14] |
| | 220.4 | c | [3] |
| | 223.7 | c | [12] |
| | 223.71±0.7 | a | [25] |
| | 241 | g | [21, 22] |
| $GdCl_3$ | 239.6 | c | [9] |
| | 240 | e | [19] |
| | 240±3 | c | [14] |
| | 240.09±0.20 | a | [6] |
| | 240.09±0.30 | c | [7] |
| | 241 | c | [12] |
| | 245.2 | a | [1] |
| $TbCl_3$ | 236 | g | [3] |
| | 238.3 | c | [12] |
| | 238.33±1.50 | a | [26] |
| | 238.4 | c | [9] |
| $DyCl_3$ | 233 | g | [3] |
| | 234.5 ($\gamma$ form) | a | [1] |
| | 236 ($\gamma$ form) | c | [12] |
| | 236.6 | c | [9] |
| | 237.5 ($\beta$ form) | a | [1] |
| | 238±4 | g | [14] |
| | 239 ($\beta$ form) | c | [12] |
| $HoCl_3$ | 231 | g | [3] |
| | 232.8 | a | [1] |
| | 237.8 | c | [9] |
| | 240.29±1.7 | a | [26] |
| | 240.3 | c | [12] |

Table 11.2.6-5 [continued]

| $MCl_3$ | $-\Delta H^\circ_{298}$ | method | Ref. | $MCl_3$ | $-\Delta H^\circ_{298}$ | method | Ref. |
|---|---|---|---|---|---|---|---|
| $ErCl_3$ | 227.40±0.46 | a | [27] | $TmCl_3$ | 235.82±0.40 | a | [29] |
| | 229.07±0.20 | a | [6] | | 236.3 | c | [9] |
| | 229.07±0.30 | c | [7] | | | | |
| | 231.8 | a | [1] | $YbCl_3$ | 215 | f | [3] |
| | 237.69±0.41 | a | [28] | | 223 | g | [21, 22] |
| | 237.94±0.56 | a | [27] | | 224±3 | f | [7] |
| | 238.0 | c | [9] | | 229.37±0.72 | a | [29] |
| | 238.7 | c | [12] | | 229.4 | c | [9, 12] |
| $TmCl_3$ | 226 | g | [3] | $LuCl_3$ | 221 | g | [3] |
| | 229.37±0.72 | a | [29] | | 226.0 | c | [12] |
| | 229.4 | a | [1] | | 227.8 | a | [1] |
| | 235.8 | c | [12] | | 234.5 | c | [9] |

Methods:

a) Solution calorimetry. – b) Experimental value, method not reported. – c) Critical evaluation of experimental data from literature. – d) Not reported. – e) Calculation from experimental heat of sublimation and auxiliary data from literature. – f) Estimated. – g) Linear interpolation and extrapolation of experimental $\Delta H^\circ_{298}$ data for chlorides of neighboring lanthanide elements.

The different $\Delta H^\circ_{298}$ values of [27] for $ErCl_3$ in the table have been calculated from irreconcilable auxiliary data on the heat of combustion and heat of solution for Er metal. Compilations of $\Delta H^\circ_{298}$ values based on literature data are also given by [20, 30 to 34]; estimated values of enthalpies of formation at 500, 1000, and 1500 K are also given in [30].

The temperature dependence of $\Delta H^\circ$ may be expressed by equations of the form

$$\Delta H^\circ_T = \Delta H^\circ_{298} + AT + {}^1/_2 B \cdot 10^{-3} T^2 + {}^1/_3 C \cdot 10^{-6} T^3 - {}^1/_2 D \cdot 10^5 T^{-1} + E$$

The following constants of the equation (except the nonused C) have been derived by Glassner [8] for the formation of rare earth chlorides from the elements under various conditions, when $\Delta H^\circ_T$ is in kcal/mol and $\Delta H^\circ_{298}$ takes the values given under [8] in table 11.2.6-5. The following table also includes values for an additional constant F from the equation given on p. 126 for the temperature dependence of the free energy and entropy of formation.

Table 11.2.6-6

| formation reaction | temperature range in K | A | B | C | D | E | F |
|---|---|---|---|---|---|---|---|
| $Y(s) + {}^3/_2 Cl_2(g) \rightleftharpoons YCl_3(s)$ | 298 to 973 | 1.5 | 11.6 | – | 1.0 | −0.619 | 64.1 |
| $Y(s) + {}^3/_2 Cl_2(g) \rightleftharpoons YCl_3(l)$ | 973 to 1725 | 14.3 | −2.6 | – | 1.0 | 2.650 | 129.1 |
| $La(s) + {}^3/_2 Cl_2(g) \rightleftharpoons LaCl_3(s)$ | 298 to 1125 | 3.1 | 7.4 | – | 1.0 | −0.918 | 75.9 |
| $La(l) + {}^3/_2 Cl_2(g) \rightleftharpoons LaCl_3(s)$ | 1125 to 1153 | 12.7 | −2.0 | – | 1.0 | 3.230 | 124.7 |
| $La(l) + {}^3/_2 Cl_2(g) \rightleftharpoons LaCl_3(l)$ | 1153 to 2020 | 11.6 | −0.4 | – | 1.0 | 1.169 | 120.6 |
| $Ce(s) + {}^3/_2 Cl_2(g) \rightleftharpoons CeCl_3(s)$ | 298 to 1048 | 4.5 | 2.4 | – | 1.0 | −1.104 | 84.8 |
| $Ce(l) + {}^3/_2 Cl_2(g) \rightleftharpoons CeCl_3(s)$ | 1048 to 1085 | 1.0 | 8.4 | – | 1.0 | −2.831 | 68.8 |
| $Ce(l) + {}^3/_2 Cl_2(g) \rightleftharpoons CeCl_3(l)$ | 1085 to 2000 | 11.0 | −0.4 | – | 1.0 | 0.189 | 121.1 |
| $Pr(s) + {}^3/_2 Cl_2(g) \rightleftharpoons PrCl_3(s)$ | 298 to 1049 | 3.7 | 3.4 | – | 1.0 | −0.900 | 80.2 |
| $Pr(s) + {}^3/_2 Cl_2(g) \rightleftharpoons PrCl_3(l)$ | 1049 to 1205 | 13.9 | −5.0 | – | 1.0 | 0.22 | 139.7 |
| $Pr(l) + {}^3/_2 Cl_2(g) \rightleftharpoons PrCl_3(l)$ | 1205 to 1980 | 10.9 | −0.4 | – | 1.0 | −1.202 | 121.1 |

Table 11.2.6-6 [continued]

| formation reaction | temperature range in K | A | B | C | D | E | F |
|---|---|---|---|---|---|---|---|
| | | | | constants | | | |
| $Nd(s)+{}^3/_2Cl_2(g) \rightleftharpoons NdCl_3(s)$ | 298 to 1033 | 4.1 | 1.7 | – | 1.0 | −0.949 | 82.2 |
| $Nd(s)+{}^3/_2Cl_2(g) \rightleftharpoons NdCl_3(l)$ | 1033 to 1297 | 12.2 | −5.7 | – | 1.0 | 2.528 | 123.7 |
| $Nd(l)+{}^3/_2Cl_2(g) \rightleftharpoons NdCl_3(l)$ | 1297 to 1960 | 8.8 | −0.4 | – | 1.0 | 0.012 | 107.6 |
| $Sm(s)+{}^3/_2Cl_2(g) \rightleftharpoons SmCl_3(s)$ | 298 to 951 | 2.5 | 6.6 | – | 1.0 | −0.694 | 69.4 |
| $Sm(s)+{}^3/_2Cl_2(g) \rightleftharpoons SmCl_3(l)$ | 951 to 1623 | 11.7 | −3.8 | – | 1.0 | 3.260 | 114.2 |
| $Sm(l)+{}^3/_2Cl_2(g) \rightleftharpoons SmCl_3(l)$ | 1623 to 1830 | 9.4 | −0.4 | – | 1.0 | −1.185 | 105.0 |

Emf measurements in the range from 1150 to 1250 K yield the values −241.7 [35] and −251.8 kcal/mol [36] for the enthalpies of formation of liquid $LaCl_3$ and $CeCl_3$ from liquid metal and gaseous chlorine. A somewhat more negative value of −245.0 kcal/mol at 1073 K for liquid $CeCl_3$ follows from the emf measurements of Senderoff et al. [37].

The enthalpy of formation changes nonmonotonically with the atomic number of the lanthanides, see the graphical representations by Bommer, Hohmann [1] and Krestov, Krestova [38]. The nature of the enthalpy change is associated with the lanthanide contraction, Gschneidner [39], and influences of a crystal field effect [38]; see also the recent discussion by Kim, Johnson [47].

**Gibbs Free Energy ΔG° and Entropy ΔS° of Formation**

Standard values for ΔG° in kcal/mol and ΔS° in $cal \cdot mol^{-1} \cdot K^{-1}$ at 298 K are:

| $MCl_3$ | $-\Delta G^\circ_{298}$ | $-\Delta S^\circ_{298}$ | Ref. | $MCl_3$ | $-\Delta G^\circ_{298}$ | $-\Delta S^\circ_{298}$ | Ref. |
|---|---|---|---|---|---|---|---|
| $ScCl_3$ | – | 78±2 | [2] | $NdCl_3$ | 227 | 63 | [41] |
| | | | | | 227.9 | – | [8] |
| $YCl_3$ | 215 | 59 | [41] | | – | 60.0 | [44] |
| | 215.2 | – | [6] | | 227.93 | 59.3 | [15] |
| | 219 | – | [8] | | 230.8 | 60.36 | [42] |
| | 222 | 59 | [10] | $PmCl_3$ | 227 | 60.2 | [42] |
| $LaCl_3$ | 237.9 | 60.65 | [42] | $SmCl_3$ | – | 60 | [21, 22] |
| | 238 | 59 | [41] | | 226 | 61 | [23] |
| | 238.3 | – | [6] | | 227 | 61 | [41] |
| | 238.8 | – | [8] | | 227.1 | 60.68 | [42] |
| | | | | | 232 | – | [8] |
| $CeCl_3$ | 233.6 | 61.14 | [42] | $EuCl_3$ | – | 60 | [21, 22] |
| | 233.7 | – | [12] | | 204.7 | 63.79 | [42] |
| | 234 | 62 | [41] | $GdCl_3$ | 222 | 61 | [41] |
| | 235.1 | – | [8] | | 222.5 | – | [6] |
| | 235.16 | 59.3 | [15] | | 223 | 60.01 | [42] |
| $PrCl_3$ | 233 | 63 | [41] | $ErCl_3$ | 210 | 62 | [41] |
| | 234.5 | 60.79 | [42] | | 211.4 | – | [6] |
| | 234.5 | – | [6] | $YbCl_3$ | – | 61 | [21, 22] |
| | 235.4 | – | [8] | | 206 | 62 | [41] |

The data of [6, 15] are based on their own $\Delta H^\circ_{298}$ values and entropies calculated by Latimer's method. Sommers, Westrum [42] used $\Delta H^\circ_{298}$ data of [12] and $\Delta S^\circ_{298}$ values which were partially recalculated on an empirical basis; for $PmCl_3$ they accepted the $\Delta H^\circ_{298}$ value of [20] and the $\Delta S^\circ_{298}$ value of [30]. The $\Delta S^\circ_{298}$ value of [44] for $NdCl_3$ is derived from tensimetric results. The rest of the tabulated data has been assessed by various methods.

Values for $\Delta G^\circ_{298}$ from the literature together with estimated values have also been compiled by Wicks, Block [30]; see also Karapet'yants [45], Emley [43].

The temperature dependence of $\Delta G^\circ$ and $\Delta S^\circ$ has been represented by the equations:

$$\Delta G^\circ_T = \Delta H^\circ_{298} - 2.303\,A\,T\log T - {}^1/_2 B\cdot 10^{-3}T^2 - {}^1/_6 C\cdot 10^{-6}T^3 - {}^1/_2 D\cdot 10^5 T^{-1} - (F-A)T - E$$

$$\Delta S^\circ_T = 2.303\,A\log T + B\cdot 10^{-3}T + {}^1/_2 C\cdot 10^{-6}T^2 - {}^1/_2 D\cdot 10^5 T^{-2} - F$$

The constants of the equations for the formation of $MCl_3$ (M=Y, La, Ce, Pr, Nd, Sm) from the elements under various conditions have been given in table 11.2.6-6, pp. 124/5, from Glassner [8]. Free energies of formation for solid $CeCl_3$ determined by Egan et al. [46] directly from emf measurements are somewhat lower than the values calculated from the pertinent equation by [8]:

| | | | |
|---|---|---|---|
| temperature in K . . . . . . . . | 673 | 773 | 873 |
| $-\Delta G^\circ_T$ in kcal/mol [47] . . . . . | 211.26 | 206.58 | 202.02 |
| $-\Delta G^\circ_T$ in kcal/mol [8] . . . . . | 214.2 | 208.8 | 203.4 |

Emf measurements in the range from 1150 to 1250 K yield the relations $\Delta G^\circ_T = -241\,700 + 62.3\,T$ cal/mol, Smirnov, Usov [35], and $\Delta G^\circ_T = -251\,800 + 58.8\,T$ cal/mol, Smirnov, Lbov [36], for the formation of liquid $LaCl_3$ and $CeCl_3$ from liquid metal and gaseous chlorine. From literature data $\Delta G^\circ_T = -249\,300 + 53.2\,T$ cal/mol is calculated for liquid $LaCl_3$ [35]. Values $\Delta G^\circ_{1073} = -180.45$, $\Delta G^\circ_{1198} = -172.85$ kcal/mol, and $\Delta S^\circ_{1073} = -60.2$ cal·mol$^{-1}$·K$^{-1}$ are derived for liquid $CeCl_3$ by Senderoff et al. [37], also from emf measurements. Free energies of formation for $MCl_3$ (M=La, Ce, Pr, Nd) have earlier been inferred tentatively from available data [40] on current potential measurements and the results given in the form of "Ellingham" diagrams in which $\Delta G^\circ_T$ is plotted against T; for $CeCl_3$ the results have also been represented by $\Delta G^\circ_T$ (T) equations, Emley [43]. Estimated values for $\Delta G^\circ$ and $(\Delta G^\circ_T - \Delta H^\circ_{298})/T$ at 500, 1000, and 1500 K have been compiled by Wicks, Block [30] and Brewer et al. [33], respectively.

References:

[1] H. Bommer, E. Hohmann (Z. Anorg. Allgem. Chem. **248** [1941] 373/82). – [2] Yu.B. Patrikeev, V.A. Morozova, G.P. Dudchik, O.G. Polyachenok, G.I. Novikov (Zh. Fiz. Khim. **47** [1973] 266; Russ. J. Phys. Chem. **47** [1973] 152/3). – [3] G.P. Dudchik, O.G. Polyachenok, G.I. Novikov (Zh. Fiz. Khim. **45** [1971] 728/9; Russ. J. Phys. Chem. **45** [1971] 409/10). – [4] D.D. Wagman, W.H. Evans, V.B. Parker, I. Halow, S.M. Bailey, R.H. Schumm, K.L. Churney (NBS-TN-270-5 [1971]). – [5] J.M. Stuve (U.S. Bur. Mines Rept. Invest. No. 6705 [1965] 1/9; C.A. **66** [1967] No. 41209).

[6] F.H. Spedding, J.P. Flynn (J. Am. Chem. Soc. **76** [1954] 1474/7). – [7] R.L. Montgomery (U.S. Bur. Mines Rept. Invest. No. 5468 [1959] 1/23, 11; C.A. **1959** 12815). – [8] A. Glassner (ANL-5750 [1957] 1/70) from J.A. Gibson, J.F. Miller, P.S. Kennedy, G.W.P. Rengstorff (The Properties of the Rare Earth Metals and Compounds,

Battelle Memorial Institute, Columbus, Ohio, 1959, Section 3, Y-7, La-13/14, Ce-14, Pr-10, Nd-10, Sm-7). – [9] L.R. Morss (J. Phys. Chem. **75** [1971] 392/9, 394). – [10] R.L. Montgomery, T.D. Hubert (USBM-RI-5659 [1960] 1/10, 5; C.A. **1961** 2264).

[11] A. Broido (Diss. Univ. California 1950; UCRL-666 [1950] 1/44 from N.S.A. **5** [1951] No. 1822). – [12] R.H. Schumm, D.D. Wagman, S. Bailey, W.H. Evans, V.B. Parker (NBS-TN-270-7 [1973]). – [13] R.L. Montgomery (U.S. Bur. Mines Rept. Invest. No. 5445 [1959] 1/12, 8; C.A. **1959** 7750). – [14] D.A. Johnson (J. Chem. Soc. A **1969** 2578/80). – [15] F.H. Spedding, C.F. Miller (J. Am. Chem. Soc. **74** [1952] 4195/8).

[16] O.G. Polyachenok, G.I. Novikov (Zh. Obshch. Khim. **33** [1963] 2797; J. Gen. Chem. [USSR] **33** [1963] 2725). – [17] O.G. Polyachenok, G.I. Novikov (Zh. Neorgan. Khim. **8** [1963] 1567/73; Russ. J. Inorg. Chem. **8** [1963] 816/20). – [18] J.M. Stuve (U.S. Bur. Mines Rept. Invest. No. 6697 [1965] 1/4; C.A. **64** [1966] 4344). – [19] S. Clach, A.J.C. Nicholson, D.L. Swingler, P.J. Thistlethwaite (Inorg. Chem. **12** [1973] 2072/4). – [20] R.C. Feber (LA-3164 [1965] 1/185, 175; N.S.A. **19** [1965] No. 30148).

[21] O.G. Polyachenok, G.I. Novikov (Zh. Neorgan. Khim. **9** [1964] 773/7; Russ. J. Inorg. Chem. **9** [1964] 429/31). – [22] O.G. Polyachenok, G.I. Novikov (Vestn. Leningr. Univ. Fiz. Khim. **1963** No. 16, pp. 133/4; C.A. **60** [1964] 4875). – [23] R.L. Montgomery, T.D. Hubert (BM-RI-5525 [1959] 1/11; N.S.A. **14** [1960] No. 1432). – [24] J.M. Stuve (U.S. Bur. Mines Rept. Invest. No. 6640 [1965] 1/9, 5; C.A. **63** [1965] 12407). – [25] L.R. Morss, H.O. Haug (J. Chem. Thermodyn. **5** [1973] 513/24).

[26] J.M. Stuve (U.S. Bur. Mines Rept. Invest. No. 7046 [1967] 1/7; C.A. **69** [1968] No. 13520). – [27] R.L. Montgomery, J.M. Stuve (U.S. Bur. Mines Rept. Invest. No. 5892 [1961] 1/10; C.A. **56** [1962] 5461). – [28] J. Fuger, L. Morss, D. Brown (J. Chem. Soc. Dalton Trans. **1980** 1076/8). – [29] J.M. Stuve (U.S. Bur. Mines Rept. Invest. No. 6902 [1967] 1/7; C.A. **67** [1967] No. 6282). – [30] C.E. Wicks, F.E. Block (U.S. Bur. Mines Bull. No. 605 [1963] 1/146; C.A. **59** [1963] 14656).

[31] R.E. Thoma (in: L. Eyring, Progress in the Science and Technology of the Rare Earths, Vol. 2, Pergamon Press, Oxford - London - Edinburgh - New York - Toronto - Sydney - Paris - Braunschweig 1966, pp. 90/122, 96/100). – [32] O. Kubaschewski, E.L. Evans (Metallurgical Thermochemistry, Pergamon Press, London - New York - Paris - Los Angeles 1958, pp. 236/81), O. Kubaschewski, E.L. Evans, C.B. Alcock (Metallurgical Thermochemistry, Pergamon Press, Oxford - London - Edinburgh - New York - Toronto - Sydney - Paris - Braunschweig 1967, pp. 314/60), O. Kubaschewski, C.B. Alcock (Metallurgical Thermochemistry, 5th Ed., Oxford - New York - Toronto - Sydney - Paris - Frankfurt 1979). – [33] L. Brewer, L.A. Bromley, P.W. Gilles, N.L. Lofgren (Natl. Nucl. Energy Ser. Div. IVB **19** [1950] 76/192, 105). – [34] G.E. Revzin (Metody Poluch. Khim. Reaktiv. Prep. **1967** No. 16, pp. 124/9; C.A. **69** [1968] No. 48783). – [35] M.V. Smirnov, P.M. Usov (Dokl. Akad. Nauk SSSR **151** [1963] 862/5; Dokl. Chem. Proc. Acad. Sci. **148/153** [1963] 606/9).

[36] M.V. Smirnov, V.S. Lbov (Tr. Inst. Elektrokhim. Akad. Nauk SSSR Ural'sk. Filial No. 8 [1966] 3/6; C.A. **66** [1967] No. 34233; Electrochemistry of Molten and Solid Electrolytes, Vol. 5, New York 1967, pp. 1/4). – [37] S. Senderoff, G.W. Mellors, R.I. Bretz (Ann. N.Y. Acad. Sci. **79** [1960] 878/96; N.S.A. **14** [1960] No. 11527). – [38] G.A. Krestov, V.N. Krestova (Radiokhimiya **11** [1969] 62/74; Soviet Radiochem. **11** [1969] 55/66). – [39] K.A. Gschneidner (J. Less-Common Metals **17** [1969] 1/12, 4). – [40] B. Neumann, H. Richter (Z. Elektrochem. **31** [1925] 296/304).

[41] R.L. Montgomery (U.S. Bur. Mines Rept. Invest. No. 5468 [1959] 1/23, 11; C.A. **1959** 12815). – [42] J.A. Sommers, E.F. Westrum (J. Chem. Thermodyn. **9** [1977]

1/26). – [43] E.F. Emley (Chem. Eng. Sci. **1** [1952] 131/44). – [44] G.I. Novikov, O.G. Polyachenok (Zh. Neorgan. Khim. **8** [1963] 1053/9; Russ. J. Inorg. Chem. **8** [1963] 545/50). – [45] M.K. Karapet'yants (Zh. Fiz. Khim. **28** [1954] 353/8; C.A. **1955** 5953).

[46] J.J. Egan, W. McCoy, J. Bracker (Thermodyn. Nucl. Mater. Proc. Symp., Vienna 1962 [1963], pp. 163/72, 169; C.A. **63** [1965] 1267). – [47]. K.-Y. Kim, C.E. Johnson (J. Chem. Thermodyn. **13** [1981] 13/25, 23).

### 11.2.6.8.5 Heat Capacity $C_p$. Enthalpy Increment $H^\circ_T - H^\circ_{298}$

The molar heat capacities $C_p$ determined experimentally between 5 and 350 K in an adiabatic calorimeter for $LaCl_3$, $PrCl_3$, and $NdCl_3$ are shown in **Fig. 15**, Sommers, Westrum [1], for $SmCl_3$, $EuCl_3$, and $GdCl_3$ in **Fig. 16**, Sommers, Westrum [5]. $C_p$ values in $cal \cdot mol^{-1} \cdot K^{-1}$ at 298.15 K (standard deviation 0.1% max.) are:

| $MCl_3$ . . . . | $LaCl_3$ | $PrCl_3$ | $NdCl_3$ | $SmCl_3$ | $EuCl_3$ | $GdCl_3$ |
|---|---|---|---|---|---|---|
| $C_p$ . . . . . | 23.43 | 23.65 | 23.72 | 23.79 | 25.57 | 23.37 |
| Ref. . . . . | [1] | [1] | [1] | [5] | [5] | [5] |

For the compounds $MCl_3$ with M = Pr, Nd, Sm, Eu, Gd, no cooperative thermal phenomena have been observed, in spite of the considerable magnetic contribution to the molar heat capacity, C(m), which these compounds exhibit (cooperative magnetic ordering occurs only at <5 K, cf. pp. 136/7). Rather, an important contribution of the non-coopera-

Fig. 15

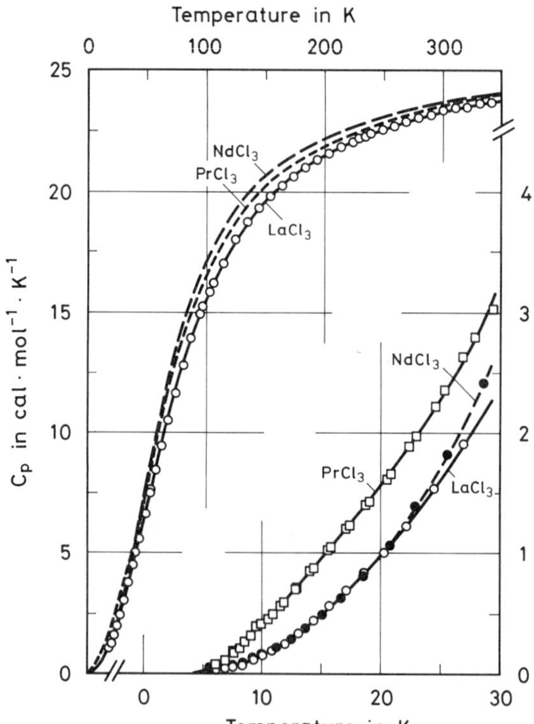


Molar heat capacity $C_p$ of $LaCl_3$, $PrCl_3$, and $NdCl_3$ as a function of temperature.

Fig. 16

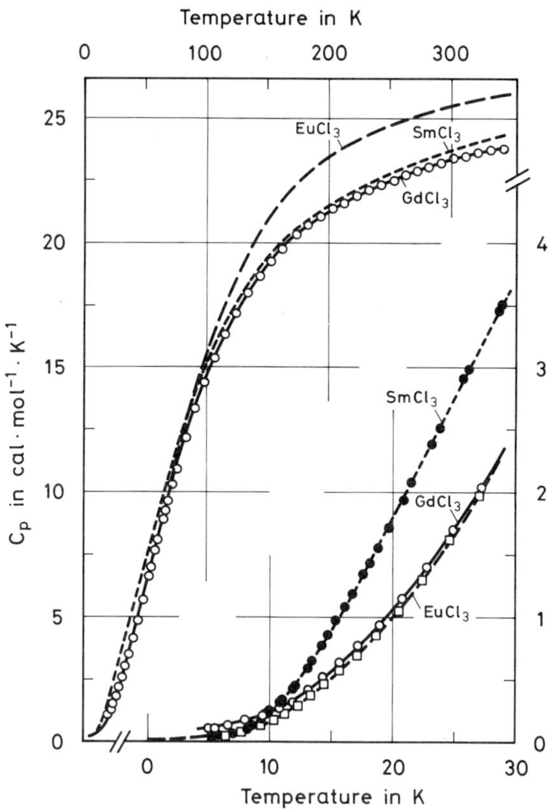


Molar heat capacity $C_p$ of $SmCl_3$, $EuCl_3$, and $GdCl_3$ as a function of temperature.

tive Schottky type, C(Sch), occurs as a result of the deep-lying crystal field levels of $Pr^{III}$, $Nd^{III}$, $Sm^{III}$, and $Eu^{III}$. In contrast $GdCl_3$ shows no Schottky anomaly, since the $^8S_{7/2}$ ground state of $Gd^{III}$ is only slightly perturbed by the crystal field (of the surrounding Cl ions) and the adjacent J state is too energetic for a Schottky contribution below 350 K. For obtaining of this C(Sch) contribution, the measured molar heat capacity of the isostructural and diamagnetic $LaCl_3$ is subtracted from that of the paramagnetic $MCl_3$ (M = Pr, Nd, Sm, Eu). These experimental C(Sch) values are compared with those calculated from the optical spectra in the literature (determined on $LaCl_3$ crystals doped with 0.1 to 2 mol % $M^{III}$); see for this purpose the figures in [1, 5]. The discrepancies between the experimental and calculated C(Sch) curves are probably primarily due to the differences in the lattice contributions to the molar heat capacity, C(lat), between $LaCl_3$ and $MCl_3$ (M = Pr, Nd, Sm, or Eu). See the papers [1, 5] for an exhaustive discussion.

Therefore, the volumetric lattice heat capacity approximation technique was applied to the homologous light $MCl_3$, M = La to Gd. The formulae are: $C_p(\text{lattice},MCl_3) = (1-f)\ C_p(LaCl_3) + fC_p(\text{lattice},GdCl_3)$, in which f may be expressed in terms of the molar volumes $V_i$ of the $MCl_3$ as $f = f(MCl_3) = [V(MCl_3) - V(LaCl_3)]/[V(GdCl_3) - V(LaCl_3)]$. For $PrCl_3$, $NdCl_3$, $SmCl_3$, and $EuCl_3$ f = 0.46, 0.57, 0.79 and 0.88, respectively. The heat capacity associated with the cooperative antiferromagnetic ordering of $GdCl_3$ has been

Fig. 17

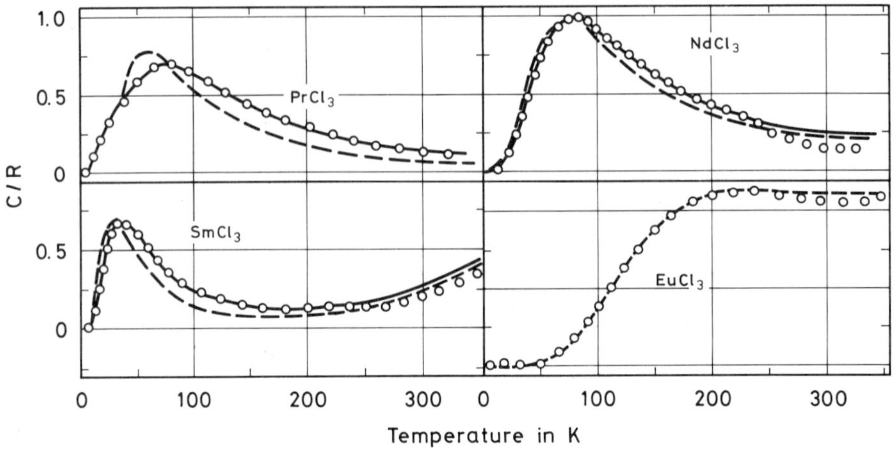


Schottky contribution to the molar heat capacity: calorimetric curves are shown by circles and the spectroscopic Schottky curves for doped $LaCl_3$ crystals by solid lines; those deduced from estimated energy levels for the pure chlorides by dashed lines.

subtracted. In this way good to excellent agreement between spectroscopic and calorimetric Schottky contribution is achieved, see **Fig. 17**, Westrum et al. [12].

The following $C_p^\circ$ values in $cal \cdot mol^{-1} \cdot K^{-1}$ for several trichlorides $MCl_3$ at 298.15 K are recommended by Schumm et al. [2]:

| $MCl_3$ . . | $LaCl_3$ | $CeCl_3$ | $PrCl_3$ | $NdCl_3$ | $GdCl_3$ | $HoCl_3$ | $ErCl_3$ |
|---|---|---|---|---|---|---|---|
| $C_p^\circ$ . . . | 26.0 | 20.9 | 24.0 | 27.0 | 21.0 | 21.0 | 24.0 |

With an increasing number of 4f electrons $C_p$ at first decreases, passes through a minimum at $4f^3$ and increases sharply to a maximum at $4f^7$. With further increase there is a substantial drop in $C_p$ to $4f^{14}$ with a minimum at $4f^{13}$. This is explained by a crystal field effect. A figure is given, Krestov, Krestova [16].

A linear temperature dependence of the heat capacity is assumed for $MCl_3$ (M = La, Pr, Nd, Eu, Ho, Er, Tm, Yb, and Lu): $C_p = a + bT$. The constants a and b have been estimated from literature values and the authors' own derived data for the electronic $C_p$ contribution: a = 21.5 for $EuCl_3$ and 20.1 $cal \cdot mol^{-1} \cdot K^{-1}$ for $TmCl_3$, $YbCl_3$, and $LuCl_3$; b = 0.012 or 0.0077 $cal \cdot mol^{-1} \cdot K^{-2}$, respectively, Myers, Graves [10]. A three term expression, $C_p = a + b \cdot 10^{-3} T + c \cdot 10^5 T^{-2}$, is used between 298 K and the melting point by Kubaschewski, Alcock [17] with the following constants:

| $MCl_3$ | a | $b \cdot 10^3$ | $c \cdot 10^5$ | page |
|---|---|---|---|---|
| $ScCl_3$ | 22.86 | 3.68 | −1.74 | 352 |
| $YCl_3$ | 25.03 | 0.77 | −2.90 | 355 |
| $PrCl_3$ | 20.6 | 11.4 | – | 351 |
| $NdCl_3$ | 18.65 | 14.6 | – | 350 |
| $SmCl_3$ [a)] | 19.66 | 11.4 | +0.18 | 352 |

| $MCl_3$ | a | $b \cdot 10^3$ | $c \cdot 10^5$ | page |
|---|---|---|---|---|
| $GdCl_3$ | 20.67 | 8.20 | +0.34 | 346 |
| $DyCl_3$ | 22.60 | 4.30 | −0.34 | 345 |
| $HoCl_3$ | 22.84 | 3.10 | −0.23 | 347 |
| $ErCl_3$ | 22.84 | 4.2 | −0.25 | 345 |
| $YbCl_3$ [a)] | 22.63 | 2.23 | −0.45 | 355 |

[a)] Estimated values.

$C_p$ values in the neighborhood of the melting points determined by calorimetric measurements to about ±5% precision are:

| $MCl_3$ . . . . | $LaCl_3$ | $PrCl_3$ | $NdCl_3$ | $GdCl_3$ | $HoCl_3$ | $EuCl_3$ |
|---|---|---|---|---|---|---|
| $C_p$ . . . . . | 34.7 | 32.3 | 35.4 | 29.1 | 29.0 | 32.0 |
| Ref. . . . . | [3] | [3] | [3] | [4] | [4] | [4] |

The molar heat capacity $C_p$ of $LaCl_3$ and $EuCl_3$ has been determined at temperatures <20 K in an adiabatic calorimeter. Comparison of the $C_p$ values obtained (see in this regard the $C_p/T$ vs. $T^2$ plot in a figure in [6]) shows that $LaCl_3$ and $EuCl_3$ in the defined temperature region possess nearly identical molar heat capacities. These consist only of (mainly acoustic) lattice contributions. This is consistent with the values reported by Varsanyi, Maita [6] for the characteristic Debye temperature $\theta_D = 118.7 \pm 1.0$ K and the lattice vibration frequencies of Stedman, Newman [7] for $LaCl_3$. The quoted value of $\theta_D$ agrees well with the authors' own value of 119±8 K. Since the heat capacities of $LaCl_3$ and $EuCl_3$ are almost identical and the optical contribution to $C_p$ is small (at 5, 10, 15, and 20 K about 0, 2, 13, and 25%, respectively, of the total value), it is believed that the acoustic parts of the heat capacities for all trichlorides of the series between M = La and Eu below 20 K are nearly the same, Sommers, Westrum [5].

Studies concerning the temperature dependence of the molar heat capacities at temperatures below 4.4 K exist for the trichlorides of La, Ce, Pr, Nd, Sm, Gd, Dy, and Ho, and are mainly concerned with the interpretation of the magnetic and electrical properties, see the sections dealing with the individual chlorides.

Values of $C_p$ determined calorimetrically for $MCl_3$ melts near the melting temperatures, with about ±5% precision, are:

| $MCl_3$ . . . . | $LaCl_3$ | $PrCl_3$ | $NdCl_3$ | $GdCl_3$ | $HoCl_3$ | $ErCl_3$ |
|---|---|---|---|---|---|---|
| $C_p$, liquid . . | 37.7 | 32.0 | 35.0 | 33.7 | 35.3 | 33.7 |
| Ref. . . . . | [3] | [3] | [3] | [4] | [4] | [4] |

According to newer determinations $C_p = 83.5\ cal \cdot mol^{-1} \cdot K^{-1}$ for especially carefully purified $LaCl_3$ at 862 to 910 °C, and 43.0 $cal \cdot mol^{-1} \cdot K^{-1}$ for similarly purified $PrCl_3$ at 770 to 920 °C. The anomalously high heat capacity of $LaCl_3$ in the vicinity of the melting point is attributed to "post-melting" processes. The extraordinarily large heat absorption on dissolution of $CaCl_2$ also speaks for such structural rearrangements in the melt, Savin et al. [8], Savin, Mikhailova [9]. Classical estimation methods yield for $NdCl_3$ a value of 32 $cal \cdot mol^{-1} \cdot K^{-1}$, Novikov, Polyachenok [11]. For the liquid state constant $C_p$ values have been derived of 36.4 for $EuCl_3$ and 34.0 $cal \cdot mol^{-1} \cdot K^{-1}$ for $TmCl_3$, $YbCl_3$, and $LuCl_3$ [10].

The enthalpy increment $H^\circ_{T_m} - H^\circ_{298}$ in cal/mol from calorimetric measurements with 0.1 to 0.2% precision is tabulated for the solid compounds at their melting points $T_m$:

| $MCl_3$ | $LaCl_3$ | $PrCl_3$ | $NdCl_3$ | $GdCl_3$ | $HoCl_3$ | $ErCl_3$ |
|---|---|---|---|---|---|---|
| $(H^\circ_{T_m}-H^\circ_{298})\cdot10^{-3}$ | 22.8 | 21.5 | 20.3 | 14.8 | 17.4 | 19.2 |
| Ref. | [3] | [3] | [3] | [4] | [4] | [4] |

Enthalpy increments between 298 and 1000 to 1500 K recalculated from literature data for $MCl_3$ with M=La to Nd, Eu to Lu are given by Myers, Graves [10].

The enthalpy increments can be expressed by the empirical relation $H^\circ_T-H^\circ_{298}=a+bT+cT^2+dT^{-1}$. Coefficients, temperature ranges $T_1$ to $T_2$ for which the equation is valid, and average deviations $\sigma$ are given for several chlorides:

| $MCl_3$ | $-a\cdot10^{-3}$ | b | $c\cdot10^3$ | $d\cdot10^{-4}$ | $\sigma$ in % | $T_1$ to $T_2$ in K |
|---|---|---|---|---|---|---|
| $ScCl_3$ | 7.565 | 22.864 | 1.840 | 17.430 | 0.4 | 298 to 1240 |
| | 2.675 | 34.284 | – | – | 0.05 | 1240 to 1300 |
| $YCl_3$ | 8.467 | 25.027 | 0.385 | 28.950 | 0.4 | 298 to 994 |
| | 7.622 | 32.436 | – | – | 0.05 | 994 to 1100 |
| $GdCl_3$ | 6.414 | 20.672 | 4.101 | 3.400 | 0.2 | 298 to 875 |
| | 4.717 | 33.346 | – | – | 0.05 | 875 to 1000 |
| $TbCl_3$ | 7.228 | 22.494 | 3.078 | 7.370 | 0.3 | 298 to 783 |
| | 7.432 | 29.620 | – | – | 0.1 | 783 to 855 |
| | 6.980 | 34.530 | – | – | 0.05 | 855 to 1000 |
| $DyCl_3$ | 7.043 | 22.601 | 2.149 | 3.400 | 0.1 | 298 to 924 |
| | 10.156 | 34.602 | – | – | 0.05 | 924 to 1000 |
| $HoCl_3$ | 7.024 | 22.840 | 1.550 | 2.310 | 0.4 | 298 to 993 |
| | 10.807 | 35.530 | – | – | 0.05 | 993 to 1100 |

Additional values for $H^\circ_T-H^\circ_{298}$ (T=400 to 1300 °C) which are used to calculate the $C_p$ data given above are represented in the table on p. 134 [13].

Estimated values for $H^\circ_T-H^\circ_{298}$ based on collected literature data for the trichlorides of Sc, Y, La to Lu at T=500, 1000, and 1500 K can also be found in Wicks, Block [14]. For older results for several chlorides at T=500, 1000, and 1500 K, cf. for example Brewer et al. [15].

References:

[1] J.A. Sommers, E.F. Westrum (J. Chem. Thermodyn. **8** [1976] 1115/36, 1122). – [2] R.H. Schumm, D.D. Wagman, S. Bailey, W.H. Evans, V.B. Parker (NBS-TN-270-7 [1973]). – [3] A.S. Dworkin, M.A. Bredig (J. Phys. Chem. **67** [1963] 697/8). – [4] A.S. Dworkin, M.A. Bredig (J. Phys. Chem. **67** [1963] 2499/500). – [5] J.A. Sommers, E.F. Westrum (J. Chem. Thermodyn. **9** [1977] 1/26, 6, 9, 11, 17, 21).

[6] F. Varsanyi, J.P. Maita (Bull. Am. Phys. Soc. **10** [1965] 609). – [7] G.E. Stedman, D.J. Newman (J. Phys. Chem. Solids **31** [1970] 1587/93). – [8] V.D. Savin, N.P. Mikhailova, V.A. Morozova (Zh. Fiz. Khim. **53** [1979] 1410/3; Russ. J. Phys. Chem. **53** [1979] 797/9). – [9] V.D. Savin, N.P. Mikhailova (Zh. Fiz. Khim. **51** [1977] 1272; Russ. J. Phys. Chem. **51** [1977] 753/4). – [10] C.E. Myers, D.T. Graves (J. Chem. Eng. Data **22** [1977] 440/4).

[11] G.I. Novikov, O.G. Polyachenok (Zh. Neorgan. Khim. **8** [1963] 1053/9; Russ. J. Inorg. Chem. **8** [1963] 545/50). – [12] E.F. Westrum, D.R. Chirico, J.B. Gruber (J. Chem. Thermodyn. **12** [1980] 717/36; Rare Earths Mod. Sci. Technol. **2** [1980] 387/92). – [13] A.S. Dworkin, M.A. Bredig (High Temp. Sci. **3** [1971] 81/90, 85). – [14] C.E. Wicks, F.E. Block (U.S. Bur. Mines Bull. No. 605 [1963] 1/146; C.A. **59** [1963] 14656). – [15] L. Brewer, L.A. Bromley, P.W. Gilles, N.L. Lofgren (Natl. Nucl. Energy Ser. Div. IV B **19** [1950] 76/192, 79).

[16] G.A. Krestov, N.V. Krestova (Radiokhimiya **11** [1969] 62/74; Soviet Radiochem. **11** [1969] 52/66, 60). – [17] O. Kubaschewski, C.B. Alcock (Metallurgical Thermochemistry, 5th Ed., Oxford – New York – Toronto –Sydney – Paris – Frankfurt 1979, pp. 340/56).

### 11.2.6.8.6 Entropy and Gibbs Free Energy Function

Standard entropy values $S^{\circ}_{298}$ in $cal \cdot mol^{-1} \cdot K^{-1}$ for the solid compounds at 298 K are:

| $MCl_3$ | $S^{\circ}_{298}$ | methods | Ref. | $MCl_3$ | $S^{\circ}_{298}$ | methods | Ref. |
|---|---|---|---|---|---|---|---|
| $ScCl_3$ | 22 | a) | [1] | $EuCl_3$ | 30 | a) | [1] |
| | 30.4 | b) | [16] | | 34 | c) | [4, 6] |
| $YCl_3$ | 27 | a) | [1] | $GdCl_3$ | 32 | a) | [1] |
| | 32.7 | b) | [2] | | 34.9 | b) | [2] |
| $LaCl_3$ | 27 | a) | [1] | $TbCl_3$ | 34 | a) | [1] |
| | 34.5 | b) | [2] | $DyCl_3$ | 35 | b) | [7] |
| $CeCl_3$ | 29 | a) | [1] | | 37 | a) | [1] |
| | 34.5 | b) | [3] | $HoCl_3$ | 38 | a) | [1] |
| | 36.0 | e) | [11] | | | | |
| | 41.0 | d) | [10] | $ErCl_3$ | 37 | a) | [1] |
| | | | | | 35.1 | b) | [2] |
| $PrCl_3$ | 29 | a) | [1] | $TmCl_3$ | 36 | a) | [1] |
| | 34.5 | b) | [2] | $YbCl_3$ | 30 | a) | [1] |
| $NdCl_3$ | 30 | a) | [1] | | 33 | c) | [4, 6] |
| | 34.6 | b) | [3] | $LuCl_3$ | 34 | a) | [1] |
| $SmCl_3$ | 27 | a) | [1] | | | | |
| | 34 | c) | [4, 6] | | | | |

Methods:

a) The values of [1] were obtained from experimental sublimation data with use of the $S^{\circ}_{298}$ values for the gaseous chlorides calculated by a comparison method. The precision is 3 to 4 $cal \cdot mol^{-1} \cdot K^{-1}$.

b) The sum of the atomic entropies determined according to Latimer [5] was used for the estimation by [2, 3, 7, 16].

c) The values of [4, 6] are calculated on the basis of their own dissociation pressure data for $MCl_3$ as well as auxiliary thermochemical data.

d) Reference [10] provides no data regarding the determination.

e) The results of [11] are based on literature data.

For $ScCl_3$ a value of $S^\circ_{298}=39.3$ cal·mol$^{-1}$·K$^{-1}$ was estimated by Kireev's method, Komissarova et al. [16]. Collected $S^\circ_{298}$ values (in part estimated) for all solid rare earth trichlorides can be found in Wicks, Block [6].

The significant structure theory of liquids has been applied to calculate the entropies of molten $LaCl_3$, $PrCl_3$, and $NdCl_3$. The agreement with the data of Glassner [14] is good, Cho, Kawamura [15]. The Debye temperatures are 188.463, 211.864 and 204.091 for $LaCl_3$, $PrCl_3$, and $NdCl_3$, Glassner [14]. Smoothed values of enthalpy and entropy increments for rounded temperatures from calorimetric measurements between 420 and 1273 K (experimental values are given in the paper) from Dworkin, Bredig [8] are:

| T in K | $H^\circ_T-H^\circ_{298}$ in cal/mol | $S^\circ_T-S^\circ_{298}$ in cal·mol$^{-1}$·K$^{-1}$ | T in K | $H^\circ_T-H^\circ_{298}$ in cal/mol | $S^\circ_T-S^\circ_{298}$ in cal·mol$^{-1}$·K$^{-1}$ |
|---|---|---|---|---|---|
| | $ScCl_3$ | | | $YCl_3$ | |
| 400 | 2310 | 6.66 | 400 | 2330 | 6.71 |
| 500 | 4675 | 11.93 | 500 | 4720 | 12.05 |
| 600 | 7105 | 16.36 | 600 | 7170 | 16.51 |
| 700 | 9590 | 20.19 | 700 | 9650 | 20.34 |
| 800 | 12120 | 23.57 | 800 | 12160 | 23.69 |
| 900 | 14700 | 26.60 | 900 | 14690 | 26.66 |
| 1000 | 17310 | 29.36 | 994 (s) | 17080 | 29.19 |
| 1100 | 19970 | 31.89 | 994 (l) | 24620 | 36.77 |
| 1200 | 22670 | 34.24 | 1000 | 24810 | 36.97 |
| 1240 (s) | 23760 | 35.13 | 1100 | 28060 | 40.06 |
| 1240 (l) | 39840 | 48.10 | 1200 | (31300) | (42.87) |
| 1300 | 41890 | 49.72 | 1300 | (34550) | (45.46) |
| | $GdCl_3$ | | | $TbCl_3$ | |
| 400 | 2425 | 7.00 | 400 | 2445 | 7.05 |
| 500 | 4880 | 12.47 | 500 | 4935 | 12.60 |
| 600 | 7410 | 17.08 | 600 | 7500 | 17.28 |
| 700 | 10020 | 21.10 | 700 | 10130 | 21.33 |
| 800 | 12710 | 24.68 | 783 (l-t) | 12370 | 24.35 |
| 875 (s) | 14780 | 27.16 | 783 (h-t) | 15760 | 28.68 |
| 875 (l) | 24460 | 38.23 | 800 | 16260 | 29.32 |
| 900 | 25290 | 39.16 | 855 (s) | 17890 | 31.29 |
| 1000 | 28630 | 42.68 | 855 (l) | 22540 | 36.72 |
| 1100 | (31970) | (45.85) | 900 | 24100 | 38.50 |
| 1200 | (35300) | (48.75) | 1000 | 27550 | 42.13 |
| 1300 | (38640) | (51.52) | 1100 | (31000) | (45.41) |
| | | | 1200 | (34460) | (48.41) |
| | | | 1300 | (37910) | (51.17) |
| | $DyCl_3$ | | | $HoCl_3$ | |
| 400 | 2425 | 6.99 | 400 | 2415 | 6.97 |
| 500 | 4860 | 12.43 | 500 | 4830 | 12.35 |
| 600 | 7350 | 16.96 | 600 | 7270 | 16.81 |
| 700 | 9880 | 20.86 | 700 | 9750 | 20.63 |
| 800 | 12460 | 24.30 | 800 | 12270 | 23.98 |

| T in K | $H^\circ_T - H^\circ_{298}$ in cal/mol | $S^\circ_T - S^\circ_{298}$ in $cal \cdot mol^{-1} \cdot K^{-1}$ | T in K | $H^\circ_T - H^\circ_{298}$ in cal/mol | $S^\circ_T - S^\circ_{298}$ in $cal \cdot mol^{-1} \cdot K^{-1}$ |
|---|---|---|---|---|---|
| | $DyCl_3$ | | | $HoCl_3$ | |
| 900 | 15080 | 27.39 | 900 | 14810 | 26.98 |
| 924 (s) | 15710 | 28.08 | 993 (s) | 17200 | 29.51 |
| 924 (l) | 21820 | 34.69 | 993 (l) | 24470 | 36.83 |
| 1000 | 24450 | 37.42 | 1000 | 24720 | 37.08 |
| 1100 | (27910) | (40.71) | 1100 | 28270 | 40.46 |
| 1200 | (31370) | (43.72) | 1200 | (31820) | (43.55) |
| 1300 | (34830) | (46.49) | 1300 | (35380) | (46.39) |

s = solid; l = liquid; l-t = low temperature structure; h-t = high temperature structure; values in parentheses are extrapolated.

Calorimetrically determined values (ignoring contributions from nuclear spin and isotopic mixing) for $H^\circ_{298} - H^\circ_0$ in $cal \cdot mol^{-1}$, and $-(G^\circ_{298} - H^\circ_0)/T$ and $S^\circ_{298} - S^\circ_0$, each in $cal \cdot mol^{-1} \cdot K^{-1}$, are:

| $MCl_3$ . . . . | $LaCl_3$ | $CeCl_3$ | $PrCl_3$ | $NdCl_3$ | $PmCl_3$ | $SmCl_3$ | $EuCl_3$ | $GdCl_3$ |
|---|---|---|---|---|---|---|---|---|
| $H^\circ_{298} - H^\circ_0$ . . . | 4826.5 | – | 5010 | 5080 | – | 4897.7 | 5163.0 | 4750.8 |
| $-(G^\circ_{298} - H^\circ_0)/T$ | 16.70 | – | 19.83 | 19.63 | – | 19.46 | 17.12 | 20.25 |
| $S^\circ_{298} - S^\circ_0$ . . . | 32.88 | (36.0) [a] | 36.64 [b] | 36.67 | (37.0) [a] | 35.88 | 34.43 | 36.19 |
| Ref. . . . . . | [12,13] | [12,13] | [12,13] | [12,13] | [12] | [12] | [12] | [12] |

[a] Values in parentheses include interpolated lattice contributions and a calculated Schottky contribution. – [b] Based on 0.294 K.

The authors' own $S^\circ_T - S^\circ_0$ values are compared with those obtained from various estimation schemes, see the paper, Sommers, Westrum [12]. Values newly calculated from literature data for the function $-(G^\circ_T - H^\circ_{298.15})/T$ are, from Myers, Graves [9]:

| $MCl_3$ . . . . . | $LaCl_3$ | | | $CeCl_3$ | | | $PrCl_3$ | | |
|---|---|---|---|---|---|---|---|---|---|
| $-(G^\circ_T - H^\circ_{298.15})/T$ | 49.95 | 48.99 | 44.75 | 49.54 | 47.45 | 43.56 | 49.15 | 47.03 | 43.32 |
| T in K . . . . . | 1070 | 1200 | 1500 | 1040 | 1180 | 1500 | 1030 | 1190 | 1500 |

| $MCl_3$ . . . . . | $NdCl_3$ | | | $EuCl_3$ | | $GdCl_3$ | |
|---|---|---|---|---|---|---|---|
| $-(G^\circ_T - H^\circ_{298.15})/T$ | 50.63 | 48.37 | 44.34 | 50.42 | 48.80 | 46.93 | 42.38 |
| T in K . . . . . | 1000 | 1170 | 1500 | 1020 | 1070 | 1100 | 1500 |

| $MCl_3$ . . . . . | $TbCl_3$ | | $DyCl_3$ | | $HoCl_3$ | |
|---|---|---|---|---|---|---|
| $-(G^\circ_T - H^\circ_{298.15})/T$ | 45.74 | 42.60 | 48.61 | 45.15 | 48.56 | 45.62 |
| T in K . . . . . | 1190 | 1500 | 1110 | 1500 | 1150 | 1500 |

| $MCl_3$ . . . . . | $ErCl_3$ | | $TmCl_3$ | | $YbCl_3$ | $LuCl_3$ | |
|---|---|---|---|---|---|---|---|
| $-(G^\circ_T - H^\circ_{298.15})/T$ | 50.35 | 47.00 | 50.09 | 46.51 | 49.73 | 50.05 | 47.16 |
| T in K . . . . . | 1140 | 1500 | 1120 | 1500 | 1190 | 1190 | 1500 |

These values are used to transform the enthalpies of vaporization into the enthalpies of sublimation [9]. Older data for the Gibbs free energy function $-(G^\circ_T - H^\circ_{298})/T$ at T = 298.1, 500, 1000, and 1500 K are given by Brewer et al. [17].

References:

[1] G.P. Dudchik, O.G. Polyachenok, G.I. Novikov (Zh. Fiz. Khim. **45** [1971] 728/9; Russ. J. Phys. Chem. **45** [1971] 409/10). – [2] F.H. Spedding, J.P. Flynn (J. Am. Chem. Soc. **76** [1954] 1474/7). – [3] F.H. Spedding, C.F. Miller (J. Am. Chem. Soc. **74** [1952] 4195/8). – [4] O.G. Polyachenok, G.I. Novikov (Zh. Neorgan. Khim. **9** [1964] 773/7; Russ. J. Inorg. Chem. **9** [1964] 429/31). – [5] W.M. Latimer (J. Am. Chem. Soc. **73** [1951] 1480/2).

[6] O.G. Polyachenok, G.I. Novikov (Vestn. Leningr. Univ. Fiz. Khim. **1963** No. 16, pp. 133/4; C.A. **60** [1964] 4875). – [7] F. Weigel, V. Wishnevsky (Ber. Deut. Chem. Ges. **102** [1969] 5/13). – [8] A.S. Dworkin, M.A. Bredig (High Temp. Sci. **3** [1971] 81/90, 86). – [9] C.E. Myers, D.T. Graves (J. Chem. Eng. Data **22** [1977] 440/5). – [10] K.K. Kelley, E.G. King (U.S. Bur. Mines Bull. No. 592 [1961] 1/149, 33).

[11] R.H. Schumm, D.D. Wagman, S.M. Bailey et al. (NBS-TN-270-7 [1973] 54). – [12] J.A. Sommers, E.F. Westrum (J. Chem. Thermodyn. **9** [1977] 1/26, 14). – [13] J.A. Sommers, E.F. Westrum (J. Chem. Thermodyn. **8** [1976] 1115/36, 1123). – [14] A. Glassner (ANL-5750 [1957] from [15]). – [15] K. Cho, K. Kawamura (Denki Kagaku Oyobi Kogyo Butsuri Kagaku **47** [1971] 209/13; C.A. **91** [1979] No. 63573).

[16] L.N. Komissarova, L.K. Tarasov (Zh. Neorgan. Khim. **14** [1969] 621/3; Russ. J. Inorg. Chem. **14** [1969] 324/6). – [17] L. Brewer, L.A. Bromley, P.W. Gilles, N.L. Lofgren (Natl. Nucl. Energy Ser. Div. IV B**19** [1950] 76/192, 79, 90).

### 11.2.6.9 Magnetic Properties

$ScCl_3$, $YCl_3$, and $LaCl_3$ must be diamagnetic like the trifluorides (see "Seltenerdelemente" C3, 1976, p. 97). The susceptibility $\chi$ of these chlorides has not yet been experimentally determined. Insofar as the trichlorides of the lanthanides have been investigated, they have proven to be paramagnetic as expected. $SmCl_3$ and $EuCl_3$ show only a weak Van Vleck paramagnetism.

Ferromagnetism (FM) or antiferromagnetism (AFM) at low temperatures (magnetic ordering temperatures $T_C$ or $T_N$) have up to now been determined for the following compounds:

| | $CeCl_3$ (hex) | $PrCl_3$ (hex) | $NdCl_3$ (hex) | $SmCl_3$ (hex) | $GdCl_3$ (hex) | $HoCl_3$ (monocl) | $ErCl_3$ (monocl) |
|---|---|---|---|---|---|---|---|
| $T_C$ or $T_N$ in K . | 0.1 | 0.4 | ≈0.4 | ≈0.4 | 2.21 | 2.15 | 0.307 |
| ordering . . . | AFM | AFE | AFM? | AFM | FM | AFM | FM or canted AFM |

The (except $GdCl_3$) still uncertain results refer to the dominant ordering force. In several compounds (e.g., $CeCl_3$, $NdCl_3$) there appears to be an initiation point of short-range ordering (or even linear ordering in chains) distinguishable from an initiation point lying at lower temperatures of long-range ordering (three dimensional ordering). Contrary to older opinion, in $PrCl_3$ the ordered state is of an antiferroelectric (AFE) nature. In the hexagonal crystals the easy magnetization direction lies parallel to the hexagonal c axis, in the monoclinic crystals in the a-b plane.

The origin of the magnetic order is the combined action of exchange interactions, which are isotropic in $GdCl_3$ and anisotropic in $CeCl_3$ and $NdCl_3$, and anisotropic magnetic dipole-dipole interactions. Thus the nondipolar exchange interactions between the nearest and those between the next-nearest neighbors can be opposed to each other, ferromagnetic and antiferromagnetic, as for $CeCl_3$ and $NdCl_3$. In $PrCl_3$, moreover, an electrical dipole-dipole interaction is added that is so large that it determines the ordered state. For disentangling the interaction forces, results for magnetically diluted crystals have occasionally been used (e.g., $Gd^{3+}$ in diamagnetic $LaCl_3$), which will when necessary also be used in the following sections. For details see the chapter on the individual compounds in "Rare Earth Elements" C4b. Additional data on magnetic properties are found in the Section 11.2.6.6, p. 87.

### 11.2.6.10 Electrical Properties

#### 11.2.6.10.1 Dielectric Properties

IR-"reststrahlen" reflection data have been fitted to a dielectric dispersion function of a general form using data on hexagonal $LaCl_3$ and $PrCl_3$ at 293 and 82 K taken in the range 50 to 300 $cm^{-1}$ of the three IR-active long-wavelength types of lattice vibration (one has $A_u$ symmetry, two $E_{1u}$). This process yielded the following dielectric constants $\varepsilon_{opt}$ and $\varepsilon_0$ (for static fields):

| | | $\varepsilon_{opt}$ | $\varepsilon_0$ (293 K) | $\varepsilon_0$ (82 K) |
|---|---|---|---|---|
| $LaCl_3$ | $E \parallel c$ | 3.71 | 9.66 | 9.56 |
| | $E \perp c$ | 3.71 | 8.45 | 8.02 |
| $PrCl_3$ | $E \parallel c$ | 3.184 | 10.26 | 9.45 |
| | $E \perp c$ | 2.814 | 8.688 | 8.665 |

Berreman, Unterwald [1].

The electronic polarizability of the molecules in the melts ($\alpha_\infty^m$) decreases in agreement with the radius r of the $M^{3+}$ ion ($r(Y^{3+}) < r(Dy^{3+})$). Values for $\alpha_\infty^m$ in $10^{-24}$ $cm^3$:

| | $YCl_3$ | | $LaCl_3$ | | $PrCl_3$ | | $DyCl_3$ | |
|---|---|---|---|---|---|---|---|---|
| T in K . . . | 1073 | 1193 | 1193 | 1253 | 1093 | 1193 | 993 | 1153 |
| $\alpha_\infty^m$ . . . . . | 10.59 | 10.50 | 11.91 | 12.05 | 11.52 | 11.51 | 10.92 | 10.85 |

Values for the electronic polarizability of the $M^{3+}$ ions in the melts are also given, Mochinaga et al. [2].

References:

[1] D.W. Berreman, F.C. Unterwald (Phys. Rev. [2] **174** [1968] 791/9). – [2] J. Mochinaga, K. Igarashi, T. Aoki, Y. Iwadate (Bull. Chem. Soc. Japan **51** [1978] 3107/10).

#### 11.2.6.10.2 Electrical Conductivity of the Melts

The specific conductance or conductivity $\varkappa$ is expressed in $\Omega^{-1} \cdot cm^{-1}$.

Within the lanthanide series the conductivity of the trichloride melts appears to decrease linearly with the atomic number, as shown by the measurement results of Förthmann, Schneider [1] in **Fig. 18**, p. 138. The decrease is explained by decreasing dissociation with decreasing ionic radius.

The conductivity increases with temperature, the usual measurement range being about 800 to 1000 °C. The temperature dependence is usually fitted by a formula of

Fig. 18

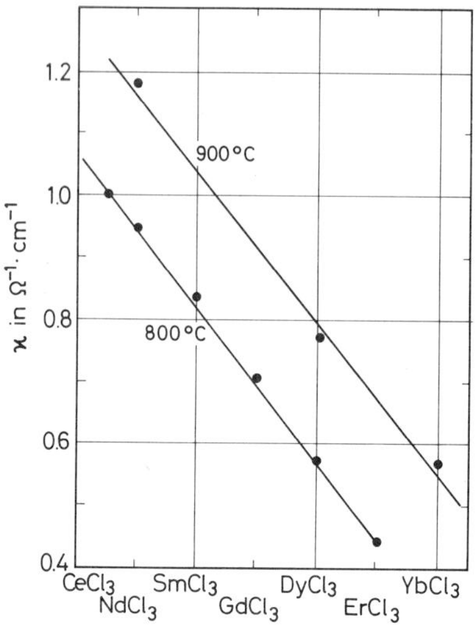


Specific electrical conductivity $\varkappa$ of rare earth trichloride melts at 800 and 900 °C.

Fig. 19

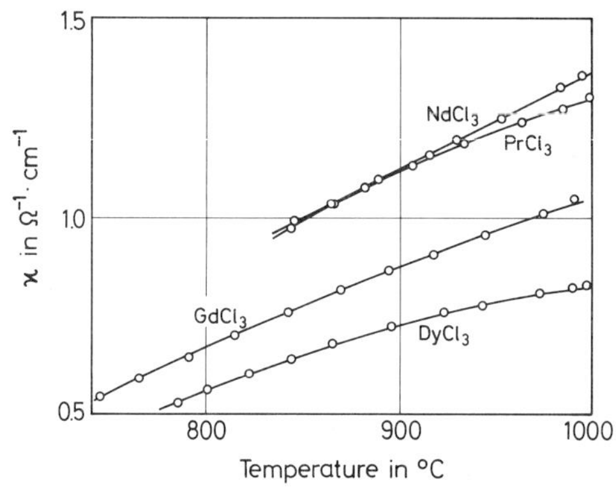


Specific electrical conductivity $\varkappa$ of rare earth trichloride melts as a function of temperature.

the type $\varkappa = a + bt + ct^2$. The measurement results for $LaCl_3$ in the temperature range from 892 to 1043 °C from Cho, Kuroda [2] as well as those for $PrCl_3$, $NdCl_3$, $GdCl_3$, and $DyCl_3$ which are shown in **Fig. 19** from Cho, Kuroda [3], yield for the parameters of this formula and for the activation energies $E_\Lambda$ in $kcal \cdot equiv^{-1} \cdot K^{-1}$:

| $MCl_3$ . . . . | $LaCl_3$ | $PrCl_3$ | $NdCl_3$ | $GdCl_3$ | $DyCl_3$ |
|---|---|---|---|---|---|
| $-a$ . . . . . | 2.1376 | 3.0331 | 1.4670 | 1.5036 | 2.8508 |
| $10^3 \cdot b$ . . . . | 4.9309 | 7.0871 | 3.2227 | 3.3316 | 6.6109 |
| $-10^6 \cdot c$ . . . | 1.3249 | 2.7592 | 0.3931 | 0.7796 | 2.9420 |
| $E_\Lambda$ . . . . . | – | 5.47 | 6.97 | 7.06 | 5.77 |

Values for the equivalent conductances can be found in [2, 3].

A selection of other measured conductivity values and constants a, b, and c of the temperature function $\varkappa = a + bt + ct^2$ for the melts of the anhydrous rare earth trihalides at the temperature t and the studied range (both in °C) is summarized in the following table:

| $MCl_3$ | $\varkappa$ | $-a$ | $10^3 \cdot b$ | $-10^6 \cdot c$ | t (range studied) | Ref. |
|---|---|---|---|---|---|---|
| $ScCl_3$ | 0.52 | – | – | – | 960 (960 to 990) | [4] |
| | 0.57 | – | – | – | 960 | [5] |
| $YCl_3$ | 0.73 | – | – | – | 875 (714 to 875) | [6] |
| $LaCl_3$ | 1.60 ± 0.02 | – | – | – | 920 | [7] |
| | 1.46 | 0.420 | 2.084 | – | 900 (900 to 1015) | [8] |
| | 1.23 | – | – | – | 895 (868 to 1005) | [6] |
| | 1.40 | 0.940 | 2.564 | – | 910 (880 to 920) | [9] |
| | 1.39 [a)] | – | – | – | 900 (950 to 1135) | [3, 10] |
| | – | 8.004 | 8.025 | 8.167 | 874 to 987 | [11] |
| $CeCl_3$ | 1,063 | 0.268 | 1.49 | – | 900 (840 to 1000) | [12] |
| | 0.960 | – | – | – | 850 | [13] |
| | 1.067 | – | – | – | 900 | [13] |
| | 1.1725 | – | – | – | 950 | [13] |
| | 1.12 | – | – | – | 828 | [14] |
| | 1.38 | – | – | – | 931 | [14] |
| $PrCl_3$ | 1.09 | 1.189 | 2.75 | – | 830 (800 to 860) | [15] |
| | 1.37 [a)] | – | – | – | 960 (825 to 965) | [5, 10] |
| $NdCl_3$ | 1.08 | 3.854 | 9.200 | 4.0 | 855 (800 to 900) | [9] |
| | 1.2 [a)] | – | – | – | 960 (775 to 900) | [5, 10] |
| | 0.94 | – | – | – | 800 | [7] |
| | 1.18 | – | – | – | 900 | [7] |
| $DyCl_3$ | 0.372 | – | – | – | 679 | [16] |
| | 0.466 | – | – | – | 730 | [16] |
| $HoCl_3$ | 0.431 | – | – | – | 747 | [16] |
| | 0.562 | – | – | – | 819 | [16] |
| $ErCl_3$ | 0.468 | – | – | – | 801 | [16] |
| | 0.531 | – | – | – | 839 | [16] |

[a)] Corrected value.

Additional conductivity values are usually given for the studied temperature range (in brackets) in the cited papers.

References:

[1] R. Förthmann, A. Schneider (Z. Anorg. Allgem. Chem. **367** [1969] 27/33). – [2] K. Cho, T. Kuroda (Denki Kagaku Oyobi Kogyo Butsuri Kagaku **40** [1972] 45/9; C.A. **77** [1972] No. 93636). – [3] K. Cho, T. Kuroda (Denki Kagaku Oyobi Kogyo Butsuri Kagaku **40** [1972] 837/9; C.A. **78** [1972] No. 89816). – [4] W. Biltz, W. Klemm (Z. Anorg. Allgem. Chem. **131** [1923] 22/6). – [5] W. Biltz, W. Klemm (Z. Physik. Chem. **110** [1924] 318/42, 323).

[6] W. Klemm, W. Biltz (Z. Anorg. Allgem. Chem. **152** [1926] 225/34, 231). – [7] R. Förthmann, G. Vogel, A. Schneider (Z. Anorg. Allgem. Chem. **367** [1969] 19/26). – [8] M.V. Smirnov, P.M. Usov, V.S. Lbov, O.M. Shabanov (Tr. Inst. Elektrokhim. Akad. Nauk SSSR Ural'sk. Filial No. 6 [1965] 57/64; C.A. **63** [1965] 14132; Electrochemistry of Molten and Solid Electrolytes, Vol. 3, New York 1966, pp. 49/56). – [9] A.S. Dworkin, H.R. Bronstein, M.A. Bredig (Discussions Faraday Soc. No. 32 [1961] 188/96). – [10] A. Voigt, W. Biltz (Z. Anorg. Allgem. Chem. **133** [1924] 277/305, 285/7), cf. also W. Biltz (Z. Anorg. Allgem. Chem. **133** [1924] 306/11).

[11] I.S. Yaffe, E.R. van Artsdalen (in: E.H. Taylor, ORNL-2159 [1956] 1/92, 77/80; N.S.A. **10** [1956] No. 11691). – [12] M.V. Smirnov, V.S. Lbov (Tr. Inst. Elektrokhim. Akad. Nauk SSSR Ural'sk. Filial No. 8 [1966] 7/12; C.A. **66** [1967] No. 41182; Electrochemistry of Molten and Solid Electrolytes, Vol. 5, New York 1967, pp. 5/10). – [13] G.W. Mellors, S. Senderoff (J. Phys. Chem. **64** [1960] 294/300). – [14] H.R. Bronstein, A.S. Dworkin, M.A. Bredig (J. Phys. Chem. **66** [1962] 44/8). – [15] A.S. Dworkin, H.R. Bronstein, M.A. Bredig (J. Phys. Chem. **66** [1962] 1201/2).

[16] A.S. Dworkin, H.R. Bronstein, M.A. Bredig (J. Phys. Chem. **67** [1963] 2715/7).

### 11.2.6.11 Optical Properties

Electronic spectra are treated in another Gmelin volume; for results of IR and Raman spectra see Section 11.2.6.3, p. 80.

#### 11.2.6.11.1 Color

Solid Compounds

The following colors have been reported for the transparent $MCl_3$ crystals or their powders:

| $MCl_3$ | color | Ref. | $MCl_3$ | color | Ref. |
|---|---|---|---|---|---|
| $ScCl_3$ | white | [1, 2] | $EuCl_3$ | bright yellow | [12, 13] |
| $YCl_3$ | colorless | [3] | | pale yellow | [8, 6, 14] |
| | white | [4] | | white | [17] |
| $LaCl_3$ | colorless | [5, 6] | $GdCl_3$ | colorless | [3] |
| | white | [4, 7, 8] | | white | [4, 9] |
| | cream | [9] | $TbCl_3$ | white | [4] |
| $CeCl_3$ | colorless | [6] | $DyCl_3$ | colorless | [3] |
| | white | [4, 8, 9] | | white | [9] |
| $PrCl_3$ | pale green | [6] | | green | [4] |
| | bright green | [7 to 9] | $HoCl_3$ | light yellow | [3] |
| $NdCl_3$ | violet-rose | [3, 4, 7] | | brown | [4] |
| | light blue-green | [9] | | light rose-brown | [9] |
| | colorless | [6] | $ErCl_3$ | light rose-violet | [3] |
| $PmCl_3$ | blue-violet | [10] | | rose | [4, 9] |
| $SmCl_3$ | pale yellow | [18] | $TmCl_3$ | green | [4] |
| | yellowish white | [11] | | white [b] | [15] |
| | light yellow-brown | [9] | $YbCl_3$ | white | [4, 8, 16] |
| | colorless [a] | [6] | $LuCl_3$ | colorless | [15] |
| | white | [8] | | white | [4] |

[a] According to [18] caused by contamination with oxide chloride. – [b] Contaminated with Er.

Melts

| $MCl_3$ | color | Ref. | $MCl_3$ | color | Ref. |
|---|---|---|---|---|---|
| $YCl_3$ | colorless | [3] | $GdCl_3$ | colorless | [3] |
| | white | [19] | $DyCl_3$ | colorless | [3] |
| $LaCl_3$ | white | [19] | | pale olive-green | [19] |
| $CeCl_3$ | white | [19] | $HoCl_3$ | orange-yellow | [3] |
| $PrCl_3$ | light green | [19] | | weakly yellowish | [19] |
| $NdCl_3$ | wine-red | [19] | $ErCl_3$ | rose | [3] |
| $SmCl_3$ | pale yellow | [19] | $TmCl_3$ | pale yellow-green | [19] |
| | wine-red [a] | [20] | | colorless [b] | [15] |
| $EuCl_3$ | dark green | [13] | $LuCl_3$ | yellowish | [15] |

[a] Cf. color of solid $SmCl_2$, p. 44. — [b] With Er contamination.

References:

[1] J.M. Stuve (U.S. Bur. Mines Rept. Invest. No. 6705 [1965] 1/5; C.A. **66** [1967] No. 41209). — [2] R.W. Stotz, G.A. Melson (Inorg. Chem. **11** [1972] 1720/1). — [3] G. Jantsch, H. Jawurek, N. Skalla, H. Gawalowski (Z. Anorg. Allgem. Chem. **207** [1932] 353/67, 354/5, 361). — [4] J.L. Moriarty (J. Chem. Eng. Data **8** [1963] 422/4). — [5] E.V. Sayre, K.M. Sancier, S. Freed (J. Chem. Phys. **23** [1955] 2060/5).

[6] H.E. Swanson, M.C. Morris, R.P. Stinchfield, E.H. Evans (Natl. Bur. Std. [U.S.] Monograph No. 25 [1962] 1/56). — [7] A. Voigt, W. Biltz (Z. Anorg. Allgem. Chem. **133** [1924] 277/305, 284). — [8] M.D. Taylor, C.P. Carter (J. Inorg. Nucl. Chem. **24** [1962] 387/91). — [9] F.L. Carter, J.F. Murray (Mater. Res. Bull. **7** [1972] 519/23). — [10] F. Weigel, V. Scherer (Radiochim. Acta **7** [1967] 40/6, 41).

[11] F. Ephraim, P. Rây (Ber. Deut. Chem. Ges. **62** [1929] 1639/50, 1646). — [12] C.T. Stubblefield, L. Eyring (J. Am. Chem. Soc. **77** [1955] 3004/5). — [13] O.G. Polyachenok, G.I. Novikov (Zh. Neorgan. Khim. **9** [1964] 773/7; Russ. J. Inorg. Chem. **9** [1964] 429/31). — [14] W. Kapfenberger (Z. Anorg. Allgem. Chem. **238** [1938] 273/86, 281/2). — [15] G. Jantsch, N. Skalla, H. Grubitsch (Z. Anorg. Allgem. Chem. **212** [1933] 65/83, 71/4, 78/81).

[16] G. Jantsch, N. Skalla, H. Jawurek (Z. Anorg. Allgem. Chem. **201** [1931] 207/20, 210/11). — [17] J.W. Hastie, P. Ficalora, J.L. Margrave (J. Less-Common Metals **14** [1968] 83/91, 88). — [18] V.F. Goryushkin, D.M. Laptev, I.S. Astakhova, E.S. Vorontsev (Zh. Neorgan. Khim. **21** [1976] 821/3; Russ. J. Inorg. Chem. **21** [1976] 445/7). — [19] J.H. Kleinheksel, H.C. Kremers (J. Am. Chem. Soc. **50** [1928] 959/67). — [20] W. Klemm, J. Rockstroh (Z. Anorg. Allgem. Chem. **176** [1928] 181/99, 196).

### 11.2.6.11.2 Optical Refraction

**Refractive Index** n

Solid Compounds

The refractive indices of the trichlorides are in view of the hygroscopic nature of these substances difficult to determine and thus also little studied. Measurements on $SmCl_3$ crystals give the result that $n \parallel c > n \perp c$, i.e., $n_e > n_o$ (optically positive); $n_o = 1.895 \pm 0.005$ and $n_e = 1.96$ for $\lambda = 589.3$ nm, Staritzky, Walker [1]. In $PrCl_3$, by the method of minimum deviation, $(n \perp c) = 1.866 \pm 0.007$ for $\lambda = 632.8$ nm (He-Ne laser), Chivian et al. [2]. In

$LaCl_3$ and $PrCl_3$ the difference in the refractive indices for visible light polarized perpendicular and parallel to the c axis of the crystal lies below the measurement precision, Berreman, Unterwald [3].

For mixed crystals of $LaCl_3$ with 0.26 mol % $PrCl_3$ n=1.6 to 1.7 is reported by Dorman [4], with 15 mol % $PrCl_3$ n=1.7 by Wright et al. [5], without discussion.

Melts

The refractive indices of $MCl_3$ melts (M=Y, La, Pr, Nd, Dy), determined by the cylindrical lens method at the wavelengths $\lambda$=434.1, 486.2, 589.3, and 656.3 nm (486.2 nm omitted for $PrCl_3$), change as a function of wavelength at 1073 and 1173 K as portrayed in **Fig. 20** [6]. In a limited study of the temperature region between 935 and 1266 K, n decreases linearly with increasing temperature, see **Fig. 21** [6] for $MCl_3$ with M=Y, La, Nd, and Dy at 486.2 nm. Values for the constants a and b in the corresponding expression $n=a+bt\times10^{-3}$ (t in °C) at the given wavelengths are listed in the paper. The following refractive indices $n_\infty$ at $\lambda=\infty$ have been calculated by use of Cauchy's relation for dispersion, at the given temperatures, Mochinaga et al. [6]:

| $MCl_3$ . . . . | $YCl_3$ | | $LaCl_3$ | | $PrCl_3$ | | $DyCl_3$ | |
|---|---|---|---|---|---|---|---|---|
| T in K . . . | 1073 | 1193 | 1193 | 1253 | 1093 | 1193 | 993 | 1153 |
| $n_\infty$ . . . . . . | 1.617 | 1.594 | 1.696 | 1.695 | 1.702 | 1.672 | 1.663 | 1.628 |

**Molar Refractivity**

The average values of the molar refractivity of $MCl_3$ melts (in $cm^3$) have been calculated in the given temperature regions for various wavelengths [6]:

| $\lambda$ in nm | 434.1 | 486.2 | 589.3 | 656.3 | temperature range in K |
|---|---|---|---|---|---|
| $LaCl_3$ | 31.7 | 31.2 | 30.9 | 30.9 | 1173 to 1253 |
| $PrCl_3$ | 30.8 | – | 30.2 | 29.7 | 1073 to 1193 |
| $NdCl_3$ | – | 30.0 | – | – | 1073 to 1193 |
| $DyCl_3$ | 27.7[1)] | 28.6 | 28.2 | 28.0 | 973 to 1153 |
| $YCl_3$ | 28.3 | 27.6 | 27.3 | 27.3 | 1013 to 1053 |

[1)] 993 to 1053 K.

Data for the electronic polarizability are found on p. 137, Mochinaga et al. [6].

Fig. 20

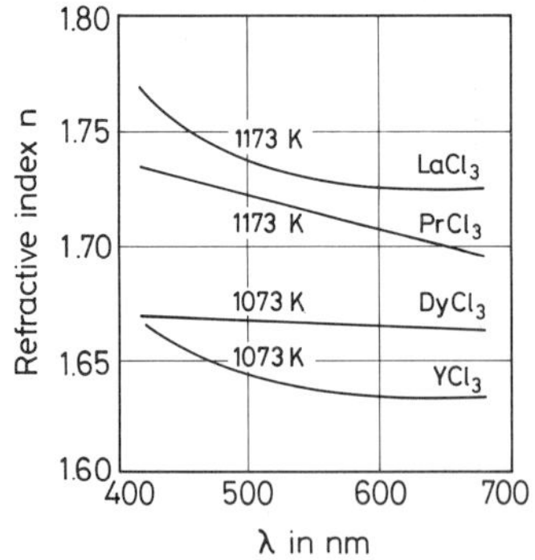


Fig. 21

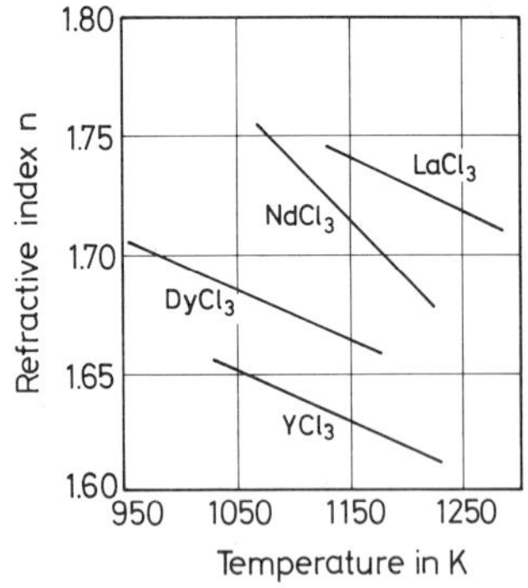


Refractive index n of $MCl_3$ melts vs. wavelength $\lambda$ at various temperatures (Fig. 20); refractive index n of $MCl_3$ melts at $\lambda$=486.2 nm vs. temperature (Fig. 21).

References:

[1] E. Staritzky, D.I. Walker (LA-1439 [1955] 1/43, 10; N.S.A. **10** [1956] No. 2391). – [2] J.S. Chivian, B.R. King, J.R. Trimmier (J. Opt. Soc. Am. **65** [1975] 357). – [3] D.W. Berreman, F.C. Unterwald (Phys. Rev. [2] **174** [1968] 791/9, 796). – [4] E. Dorman (J. Chem. Phys. **44** [1966] 2910/6, 2913/4). – [5] J.C. Wright, D.J. Zalucha, H.V. Lauer, D.E. Cox, F.K. Fong (J. Appl. Phys. **44** [1973] 781/6).

[6] J. Mochinaga, K. Igarashi, T. Aoki, Y. Iwadate (Bull. Chem. Soc. Japan **51** [1978] 3107/10).

## 11.2.6.12 Chemical Reactions

### 11.2.6.12.1 Reactions on Heating

Rare earth trichlorides, with the exception of $SmCl_3$, $EuCl_3$, and $YbCl_3$, evaporate congruently, see Sections 11.1.3, p. 10, 11.2.6.8.3, p. 113. The following equilibrium constants K and thermodynamic data, ΔH and ΔS, for the reaction $MCl_3(l) \rightleftharpoons MCl_2(l) + {}^1/_2 Cl_2(g)$ (M = Sm, Eu, Yb) have been determined by static vapor pressure measurements with a quartz-membrane null-manometer in the temperature ranges given; K is expressed by $\log K = A - BT - C \log T$:

| $MCl_3$ | temperature range in K | A | B | C | $\Delta H_{1173}$ in kcal/mol | $\Delta S_{1173}$ in $cal \cdot mol^{-1} \cdot K^{-1}$ |
|---|---|---|---|---|---|---|
| $SmCl_3$ | 988 to 1249 | 22.908 | 8856 | 6.0 | 26.4 | 8.5 |
| $SmCl_3$ | 796 to 1241 | 23.142 | 9050 | 6.0 | 27.3 | 9.6 |
| $EuCl_3$ | 875 to 1223 | 22.902 | 5625 | 6.0 | 11.6 | 8.5 |
| $YbCl_3$ | 1149 to 1234 | 23.176 | 8337 | 6.0 | 24.0 | 9.7 |

At 620 °C the equilibrium chlorine pressure above the solid trichlorides has the approximate values $5 \times 10^{-7}$, 5, and $2 \times 10^{-5}$ Torr for $SmCl_3$, $EuCl_3$, and $YbCl_3$, respectively. Thus solid $SmCl_3$ and $YbCl_3$ dissociate with difficulty, the equilibrium pressure of saturated $SmCl_3$ and $YbCl_3$ being greater by several orders of magnitude than the equilibrium chlorine pressures. $EuCl_3$ decomposes in a vacuum, and this may provide a basis for the preparation of $EuCl_2$. In the fused state, $SmCl_3$ and $YbCl_3$ also dissociate to the dichlorides to a considerable extent in a vacuum, Polyachenok, Novikov [1, 2]; see also Novikov, Polyachenok [3]. According to Jantsch et al. [4, p. 209] $SmCl_3$ starts to give off $Cl_2$ in high vacuum above 900 °C, $YbCl_3$ at 870 °C. Standard free energies at 298 K in kcal/mol for the reaction $MCl_3(s) \rightarrow MCl_2(s) + {}^1/_2 Cl_2(g)$ have been calculated by Johnson [5]:

| | | | | | | | |
|---|---|---|---|---|---|---|---|
| $MCl_3$ . . . . | $LaCl_3$ | $CeCl_3$ | $PrCl_3$ | $NdCl_3$ | $PmCl_3$ | $SmCl_3$ | $EuCl_3$ |
| $\Delta G^\circ_{298}$ . . . | 153±5 | 128±7 | 90±7 | 77±4 | 68±10 | 43±5 | 21±4 |
| $MCl_3$ . . . . | $GdCl_3$ | $TbCl_3$ | $DyCl_3$ | $HoCl_3$ | $ErCl_3$ | $TmCl_3$ | $YbCl_3$ |
| $\Delta G^\circ_{298}$ . . . | 136±15 | 95±7 | 71±7 | 75±7 | 85±7 | 62±7 | 32±4 |

The very low value for $EuCl_3$ again reflects the tendency for dissociation. A general qualitative description of the thermal decomposition of chlorides in the rare earth elements series is given by Abalduev et al. [6]. The dissociation of gaseous chlorides is treated in Section 11.1, p. 1.

References:

[1] O.G. Polyachenok, G.I. Novikov (Zh. Neorgan. Khim. **9** [1964] 773/7; Russ. J. Inorg. Chem. **9** [1964] 429/31). – [2] O.G. Polyachenok, G.I. Novikov (Vestn. Leningr. Univ. Fiz. Khim. **1963** No. 16, pp. 133/4; C.A. **60** [1964] 4875). – [3] G.I. Novikov, O.G. Polyachenok (Usp. Khim. **33** [1964] 732/47; Russ. Chem. Rev. **33** [1964] 342/50). – [4] G. Jantsch, N. Skalla, H. Jawurek (Z. Anorg. Allgem. Chem. **201** [1931] 207/20). – [5] D.A. Johnson (J. Chem. Soc. A **1969** 2578/80).

[6] B.V. Abalduev, A.F. Bol'shakov, L.G. Sokolova, A.V. Lapitskaya, S.B. Pirkes (Issled. Obl. Khim. Redkozem. Elem. **1975** 47/9; C.A. **84** [1976] No. 98768).

### 11.2.6.12.2 Reactions with Nonmetallic Elements and Their Compounds

#### Reactions with Hydrogen

The onset of appreciable reaction between hydrogen and the trichlorides $MCl_3$ (M = Sm, Eu, Yb) has been studied thermogravimetrically at a constant 1 atm $H_2$ pressure and a heating rate of 2 to 3 K/min. A change in the specimen weight was observed at 525 °C for $SmCl_3$, at 275 °C for $EuCl_3$, and at 460 °C for $YbCl_3$. The heterogeneous processes pass through a number of stages. The initial stage involves the adsorption of $H_2$ on the surface of the chlorides, which was investigated gravimetrically in the temperature ranges 20 to 480 °C for $SmCl_3$, 20 to 240 °C for $EuCl_3$, and 20 to 350 °C for $YbCl_3$ at $H_2$ pressures between 0.05 and 0.3 atm. The results are presented in **Fig. 22**. Physical adsorption of $H_2$ was observed up to 260 °C for $SmCl_3$, 240 °C for $EuCl_3$, and 160 °C for $YbCl_3$. The nature of the physical adsorption was the same for all three chlorides, which was reflected by the virtually identical heats of adsorption: 4.2, 4.1, and 4.0 kcal/mol for $SmCl_3$, $EuCl_3$, and $YbCl_3$, respectively. The adsorption equilibrium is described satisfactorily by the Langmuir equation and is established almost instantaneously under conditions of moderate surface coverage. The average hydrogen adsorption coefficients are of the same order of magnitude, viz. 23.2 to 42.5, 36.2 to 48.2, and 36.2 to 47.4 $atm^{-1}$, respectively, on $SmCl_3$, $EuCl_3$, and $YbCl_3$. An increase of temperature induces the desorption of the physically adsorbed hydrogen and in the later stages leads to chemisorption. In the chemisorption range the amount of adsorbed hydrogen increases, until around 480 °C ($SmCl_3$), 240 °C ($EuCl_3$), and 350 °C ($YbCl_3$) "active"

Fig. 22

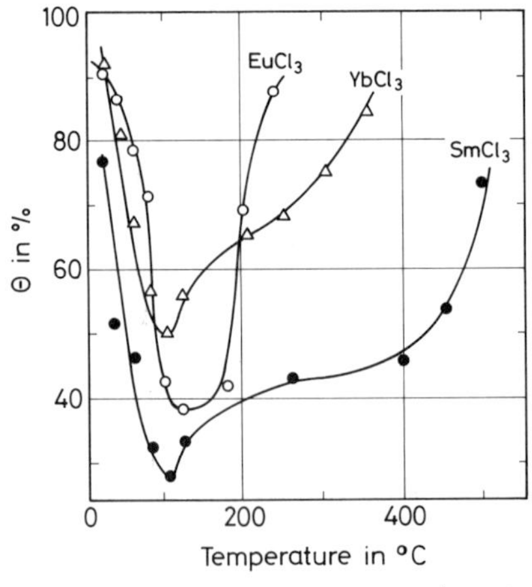


Temperature variation of the degree of surface coverage $\Theta$ of $MCl_3$ with hydrogen.

reduction of the trichlorides sets in, accompanied by the formation and desorption of hydrogen chloride. X-ray electron spectroscopic studies of the surface of specimens equilibrated in the chemisorption range revealed inclusions of the dichlorides in layers up to 30 Å thick. Chemisorbed compounds of the trichloride-hydrogen type were not detected. The change in the oxidation state of the metal during the chemisorption of hydrogen was confirmed by electron paramagnetic resonance measurements on specimens of $EuCl_3$ which had been heated in an $H_2$ atmosphere at 270 to 300 °C, Il'in et al. [7, 8].

The kinetics of the reduction of the above three trichlorides by $H_2$ have been investigated gravimetrically under static conditions in the ranges 570 to 710 °C for $SmCl_3$, 360 to 480 °C for $EuCl_3$, and 570 to 660 °C for $YbCl_3$ at $H_2$ pressures of 0.02 to 1 atm. The specific surfaces of the samples were $5 \pm 0.5\ m^2/g$. The reaction rates V, expressed in mol converted chloride per $m^2$ and min, can be described by the following equations:

$$V_{SmCl_3} = 8.66 \times 10^{12} \cdot p_{H_2}^{0.35} \cdot \exp(-18\,900/RT)$$

$$V_{EuCl_3} = 1.0 \times 10^{13} \cdot p_{H_2}^{0.41} \cdot \exp(-28\,900/RT)$$

$$V_{YbCl_3} = 8.84 \times 10^{12} \cdot p_{H_2}^{0.40} \cdot \exp(-25\,700/RT)$$

The reduction rates and rate constants were determined with an accuracy of $\pm 10\%$. The apparent activation energies have an error limit of $\pm 5$ kcal/mol. The most probable limiting stage of the reduction process is the surface reaction between adsorbed hydrogen and $MCl_3$, producing hydrogen chloride adsorbed on the formed $MCl_2$, Il'in et al. [7, 9]. See pp. 46/7 for the kinetic parameters of the reverse reaction, i.e., the oxidation of the dichlorides by hydrogen chloride.

Reduction of the trichlorides with hydrogen is the usual method for the preparation of $SmCl_2$, $EuCl_2$, and $YbCl_2$, see p. 30. The free energy change for the reduction of the solid trichlorides at 620 °C is respectively −5, −20, −9 kcal/mol. It is doubtful whether solid $SmCl_3$ can be completely reduced, all the more so since the formation of intermediate compounds $SmCl_3 \cdot n\,SmCl_2$ cannot be ruled out. Reduction in the melt should theoretically go to completion, but in the case of $SmCl_3$ the rate of reaction is apparently too low. In most cases reduction of the trichlorides of the other rare earth elements can be ignored. Calculation using the approximate heats of formation of the dichlorides shows that marked reduction can only be expected in the case of the chlorides of Nd, Pm, Er, Tm, and Sc, Novikov, Polyachenok [1, 3]. According to thermodynamic estimates, the reduction equilibrium in the cases of $SmCl_3$, $EuCl_3$, and $YbCl_3$ at 1100 K is almost completely on the side of the dichlorides (degree of reduction $>0.9$). Reduction to the metals is possible only above 3500 K. Stable hydrides are not formed under usual conditions, Il'in et al. [9]. Calculations of the equilibrium constants K for the reactions $MCl_3(l) + {}^1/_2\,H_2(g) \rightarrow MCl_2(l) + HCl(g)$ (M = lanthanides) at 1273 K also indicate almost complete reduction of $SmCl_3$, $EuCl_3$, and $YbCl_3$ and partial reduction of Nd, Pm, Dy, Ho, Er, and Tm trichlorides; numerical values for K, ΔH, and ΔG are given, Polyachenok, Novikov [10].

Free energy changes for the reactions of gaseous rare earth chlorides with $H_2$ to HCl and liquid metal at 1000 and 1900 K have been calculated by Novikov, Baev [11], see p. 155.

The different reducibility of the rare earth chlorides provides a welcome purifying effect in the preparation of $SmCl_2$, $EuCl_2$, and $YbCl_2$. During the reduction and a subsequent vacuum treatment at 700 to 750 °C, the amount of lanthanide impurities in these dichlorides decreases by more than an order of magnitude, Il'in et al. [7]. Initiation of reduction of $SmCl_3$, $EuCl_3$, and $YbCl_3$ to the dichlorides was earlier observed at about 400, 270, and 560 °C, respectively. In the case of the other rare earth trichlorides the metals were formed, which alloyed with the gold boats used for the experiments. The latter reactions started within 10 h at the following temperatures t:

| $MCl_3$ . . . . . | $YCl_3$ | $LaCl_3$ | $NdCl_3$ | $GdCl_3$ | $TmCl_3$ | $LuCl_3$ |
|---|---|---|---|---|---|---|
| t in °C . . . . | 880 | 880 | 840 | 820 | 650 | 750 |

Jantsch et al. [12 to 14]. Analogous experiments with mixtures of the chlorides were performed at 800 °C by Holleck [15]. A 5.5 h treatment of $DyCl_3$ and $ErCl_3$ with $H_2$ at 600 to 700 °C did not result in any definite indication of reduction, Klemm, Schüth [2]. Pure $EuCl_3$ is readily reduced to $EuCl_2$ by 1:1 $H_2$-HCl mixtures at 400 °C, but $EuCl_3$ in mixed crystals with $SrCl_2$ is not so reduced, Brauer, Knausenberger [16].

At 650 °C, $GdCl_3$ is reduced in a vacuum apparatus by proton bombardment. At a certain exposure the irradiated product was found to contain 93% Gd and 4% Cl. X-ray analyses showed GdCl and Gd. The formation of Gd was also observed during proton irradiation of eutectic ACl-$GdCl_3$ mixtures (A=K, Na), Bochin et al. [17].

## Reactions with Air and Oxygen

The rare earth chlorides are highly moisture-sensitive substances, which rapidly absorb water on exposure to air, forming hydrates; later they deliquesce. Quantitative data on the hydration rate of $YCl_3$, $LaCl_3$, $EuCl_3$, and $YbCl_3$ in moist air under precisely defined conditions are presented in the sections dealing with the chemical reactions of the individual chlorides in "Rare Earth Elements" C 4b.

On heating in air to moderately high temperatures, the oxide chlorides MOCl are formed by the reactions $MCl_3 + {}^1/_2\,O_2 \rightarrow MOCl + Cl_2$ and $MCl_3 + H_2O \rightarrow MOCl + 2\,HCl$, respectively; see for example Haeseler et al. [18], Matignon [19 to 21], Baev, Novikov [22], Bourion [23], Frerichs, Smith [24]. At higher temperatures the reaction $4\,MCl_3 + 3\,O_2 \rightarrow 2\,M_2O_3 + 6\,Cl_2$ must also be considered, T'en, Morozov [4]. $CeCl_3$ is converted into $CeO_2$, Haeseler, Matthes [25].

The temperatures at which the oxide chlorides are formed decrease noticeably with increasing atomic number of the rare earth elements. Thermogravimetric analyses at a heating rate of 6 K/min in an air stream indicate the following starting temperatures t for MOCl formation:

| M . . | Y | La | Ce | Pr | Nd | Sm | Eu | Gd | Tb | Dy | Ho | Er | Tm | Yb | Lu |
|---|---|---|---|---|---|---|---|---|---|---|---|---|---|---|---|
| t in °C . | 320 | 410 | 350 | 370 | 360 | 350 | 350 | 330 | 320 | 310 | 310 | 310 | 300 | 300 | 290 |

The air flow rate was 15 to 20 mL/min; the $H_2O$ vapor pressure was in the range 16 to 20 Torr. The anhydrous chlorides were formed by preceding thermal decomposition of the trichloroacetates. On further heating to higher temperatures the oxide chlorides are slowly converted into the oxides. $CeO_2$ begins to form even at 350 °C. In the case of the lighter rare earths, oxide formation is incomplete even at 1000 °C, Mathur, Bhat [26]. According to Sommers, Westrum [27], the ignition of $LaCl_3$, $CeCl_3$, and $PrCl_3$ in air at 1000 °C gives phases approximating the composition MOCl. Quantitative conversion into the oxides for analytical purposes is attained, however, at this temperature after a preceding hydrolysis treatment. Nearly complete transformation of the chlorides into oxide chlorides was observed under the following conditions:

| $MCl_3$ | $YCl_3$ | $PrCl_3$ | $SmCl_3$ | $EuCl_3$ | $GdCl_3$ |
|---|---|---|---|---|---|
| temp./time | 700 °C/10 min | 800 °C/– | 900 °C/30 min | 700 °C/5 min | 500 °C/– |
| Ref. | [28, 29] | [30] | [28] | [28] | [28] |

The formation of oxide chlorides from intermediately formed anhydrous chlorides during thermal decomposition of the hydrated chlorides in air is discussed in Section 11.2.7.11.1, p. 213.

The equilibria of the reactions of solid anhydrous chlorides with oxygen

$$MCl_3(s) + {}^1/_2\,O_2(g) \rightleftharpoons MOCl(s) + Cl_2(g)$$

(M = La, Pr, Sm, Gd, Er) have been studied by a circulation method. The dependence of the equilibrium constants on temperature can be represented by the following equations, when the partial pressures entering into $K_p$ are expressed in Torr:

$$\log K_p(LaCl_3) = 5.70 - 5800/T \quad (T = 1013 \text{ to } 1123\text{ K})\ [31]$$

$$\log K_p(PrCl_3) = 6.06 - 5450/T \quad (T = 901 \text{ to } 1035\text{ K})\ [31]$$

$$\log K_p(SmCl_3) = 5.96 - 4780/T \quad (T = 829 \text{ to } 935\text{ K})\ [32]$$

$$\log K_p(GdCl_3) = 5.87 - 4650/T \quad (T = 823 \text{ to } 880\text{ K})\ [32]$$

$$\log K_p(ErCl_3) = 5.75 - 4900/T \quad (T = 904 \text{ to } 1003\text{ K})\ [32]$$

The change in free energy $\Delta G^\circ_T$ is generally described by

$$\Delta G^\circ_T = \Delta H^\circ_0 + 7.04\ T \log T + 0.72 \times 10^{-3}\ T^2 + 24000\ T^{-1} + I\,T.$$

The constants $\Delta H^\circ_0$ and I and $\Delta H^\circ$ values for 298 and 875 K are the following:

| $MCl_3$ | $\Delta H^\circ_0$ | $\Delta H^\circ_{298}$ | $\Delta H^\circ_{875}$ | $-I$ |
|---|---|---|---|---|
| $LaCl_3$ | 30.7 | 29.6 | 27.4 | 45.2 |
| $PrCl_3$ | 28.6 | 27.8 | 25.4 | 46.7 |
| $SmCl_3$ | 25.0 | 24.2 | 21.8 | 45.6 |
| $GdCl_3$ | 24.4 | 23.7 | 21.3 | 45.2 |
| $ErCl_3$ | 25.9 | 25.1 | 22.7 | 45.1 |

The $\Delta H^\circ$ values are in kcal/mol, Drobot et al. [31, 32].

Morozov, Korshunov [33] assumed reactions

$$4\,MCl_3(s) + 3\,O_2(g) \leftrightarrows 2\,M_2O_3(s) + 6\,Cl_2(g)$$

(M = Sc, La, Nd, Sm) even below 600 °C and determined the following equilibrium constants and thermodynamic characteristics:

$4\,ScCl_3(s) + 3\,O_2(g) \rightleftharpoons 2\,Sc_2O_3(s) + 6\,Cl_2(g)$

| temp. in °C | $K_p$ | $\Delta G^\circ_T$ cal | $\Delta H^\circ$ cal |
|---|---|---|---|
| 250 | $3.762 \times 10^{-4}$ | 8190 | 24580 |
| 295 | $1.051 \times 10^{-3}$ | 7740 | |
| 340 | $8.747 \times 10^{-3}$ | 6310 | |
| 364 | $2.136 \times 10^{-2}$ | 4860 | |

$4\,LaCl_3(s) + 3\,O_2(g) \rightleftharpoons 2\,La_2O_3(s) + 6\,Cl_2(g)$

| temp. in °C | $K_p$ | $\Delta G^\circ_T$ cal | $\Delta H^\circ$ cal |
|---|---|---|---|
| 274 | $1.478 \times 10^{-4}$ | 8450 | 25230 |
| 330 | $9.198 \times 10^{-4}$ | 8380 | |
| 415 | $1.264 \times 10^{-2}$ | 5980 | |
| 425 | $1.972 \times 10^{-2}$ | 5440 | |
| 505 | $1.393 \times 10^{-1}$ | 3050 | |
| 533 | $2.733 \times 10^{-1}$ | 2070 | |

| $4NdCl_3(s)+3O_2(g) \rightleftharpoons 2Nd_2O_3(s)+6Cl_2(g)$ | | | |
|---|---|---|---|
| temp. in °C | $K_p$ | $\Delta G^\circ_T$ cal | $\Delta H^\circ$ cal |
| 270 | $8.857\times10^{-5}$ | 10070 | 25700 |
| 368 | $1.504\times10^{-3}$ | 8270 | |
| 440 | $1.344\times10^{-2}$ | 6100 | |
| 575 | $3.438\times10^{-1}$ | 1800 | |
| 600 | $7.962\times10^{-1}$ | 395 | |
| 615 | $9.319\times10^{-1}$ | 120 | |

| $4SmCl_3(s)+3O_2(g) \rightleftharpoons 2Sm_2O_3(s)+6Cl_2(g)$ | | | |
|---|---|---|---|
| temp. in °C | $K_p$ | $\Delta G^\circ_T$ cal | $\Delta H^\circ$ cal |
| 290 | $7.406\times10^{-5}$ | 10630 | 25520 |
| 340 | $3.762\times10^{-4}$ | 9600 | |
| 372 | $4.597\times10^{-4}$ | 9750 | |
| 480 | $1.069\times10^{-2}$ | 6785 | |
| 585 | $1.393\times10^{-1}$ | 3360 | |

The equilibria were studied statically by analyzing gas samples after equilibrium was reached, which from both sides of the reaction equation took 3 to 5 days (the gas phase compositions are given in vol %) [33], see also Morozov, Korshunov [34].

Oxide chlorides are also formed by the action of air or oxygen on the molten chlorides, see for example Bourion [23], or on solutions of the chlorides in alkali chloride melts, see the chapters on the formation of the oxide chlorides in "Seltenerdelemente" C 5. $CeCl_3$ is transformed to $CeO_2$ under these conditions. Newer studies on the interaction of rare earth chloride melts with oxygen have been reported by Buturlov et al. [35].

**Reactions with Graphite and Arsenic**

Reactions of rare earth chlorides with graphite were first studied by Croft [36, 37] and Vickery, Campbell [38]. These authors attempted to prepare intercalation compounds by heating the hydrated chlorides with graphite in sealed glass tubes at 550 to 560 °C. The results were contradictory. Vickery, Campbell [38] found that only $YCl_3$ and $GdCl_3$ intercalated consistently and in appreciable quantities. According to Stumpp, Nietfeld [39, 40] intercalation compounds are not likely to be formed under these conditions. However, by using the anhydrous salts in a $Cl_2$ atmosphere and by adding small amounts of $AlCl_3$, the latter authors succeeded in obtaining intercalation compounds of graphite with the trichlorides of Sc, Y, Sm, Eu, Gd, Tb, Dy, Ho, Er, Tm, Yb, and Lu. The chlorides of La, Ce, Pr, and Nd did not intercalate. The reaction starts only when the vapor pressure of the chlorides has surpassed a certain, individual threshold value. Addition of $AlCl_3$ serves to increase the chloride concentration in the gas phase by complex formation; this was found to be unnecessary in the cases of $TbCl_3$ and $DyCl_3$, but $YCl_3$ only reacted in the presence of $AlCl_3$, even at 600 °C. At higher temperatures the graphite compounds are decomposed to products poorer in chloride. Composition and reaction conditions of representative preparations from mixtures of 2 g $MCl_3$, 500 mg graphite, and 50 mg $AlCl_3$ are:

| $MCl_3$ | time in d | temp. in °C | atomic ratio C:M:Cl | stage |
|---|---|---|---|---|
| $ScCl_3$ | 10 | 520 | 20.8:1:3.06 | 3 |
| $YCl_3$ | 3 | 550 | 19.9:1:3.11 | 2 |
| $SmCl_3$ | 6 | 520 | 40.0:1:3.21 | 3+4 |
| $EuCl_3$ | 3 | 500 | 31.2:1:3.4 | 3 |
| $EuCl_3$ | 3 | 550 | 37.1:1:3.2 | 4 |
| $GdCl_3$ | 2 | 550 | 20.4:1:3.1 | 3 |
| $TbCl_3$ | 3 | 500 | 18.3:1:3.24 | 2 |

| $MCl_3$ | time in d | temp. in °C | atomic ratio C:M:Cl | stage |
|---|---|---|---|---|
| $TbCl_3$ [a)] | 3 | 500 | 19.3:1:3.35 | 2 |
| $DyCl_3$ | 3.5 | 500 | 19.3:1:3.0 | 2 |
| $DyCl_3$ | 3 | 600 | 18.8:1:3.12 | 2 |
| $DyCl_3$ [a)] | 3 | 550 | 18.4:1:3.24 | 2 |
| $HoCl_3$ | 3 | 500 | 18.9:1:3.12 | 2 |
| $ErCl_3$ | 3 | 550 | 20.6:1:3.04 | 3 |
| $TmCl_3$ | 3 | 500 | 23.0:1:3.02 | 3+2 |
| $YbCl_3$ | 3 | 550 | 25.4:1:3.16 | 3 |
| $LuCl_3$ | 1 | 600 | 32.7:1:3.14 | 4+3 |

[a)] Without $AlCl_3$ addition.

The samples were characterized by chemical and X-ray analyses. The chlorides are intercalated in the form of a chloride layer sandwich similar to the sheets in $YCl_3$, as indicated by the observed amount of c-axis increase. Intercalation compounds can be classified according to the repeat frequency of occupied interlamellar spaces. A compound in which every space is occupied, thus containing the maximum amount of metal chloride, is referred to as "first stage"; if every second interlamellar space is occupied it is called "second stage", and so on. These stages are specified in the fifth column of the table. The chlorides which were found not to intercalate crystallize in the $UCl_3$ structure having 3-dimensional arrangements of ions; these chlorides cannot form sheets between C layers, Stumpp, Nietfeld [40].

Reactions of Sc, Y, La, Ce, Sm, Eu, and Tm chlorides with arsenic were investigated in quartz ampules under a He atmosphere at 800 to 1000 °C. $ScCl_3$, $YCl_3$, and $LaCl_3$ did not react at 800 °C, the observed Ce concentration in As was 3 times that of Eu or Tm, Avdeeva et al. [41].

### Reactions with Water Vapor

At higher temperatures anhydrous rare earth chlorides $MCl_3$ react with water vapor according to

$$MCl_3(s) + H_2O(g) \rightarrow MOCl(s) + 2HCl(g)$$

The equilibrium constants of the reaction were determined in the range 700 to 900 K by a flow method. A stream of $N_2$ containing HCl and $H_2O(g)$ in known ratios was passed over a mixture of $MCl_3$ and MOCl; weight changes were indicated by a quartz-fiber balance. The following values for the equilibrium constants $K_p$ were found:

| $MCl_3$ | $YCl_3$ | $LaCl_3$ | $PrCl_3$ | $NdCl_3$ | $PmCl_3$ |
|---|---|---|---|---|---|
| T in K | 704 to 884 | 733 to 892 | 703 to 865 | 730 to 897 | 732 to 872 |
| $K_p$ in atm | 40.77 to 260.8 | 0.12 to 2.78 | 0.732 to 13.26 | 2.532 to 36.48 | 9.085 to 72.44 |
| Ref. | [42] | [43 to 46] | [43, 45, 47] | [43, 47] | [42, 48] |

| $MCl_3$ | $SmCl_3$ | $GdCl_3$ | $\gamma$-$DyCl_3$ | $\beta$-$DyCl_3$ |
|---|---|---|---|---|
| T in K | 703 to 893 | 701 to 895 | 709 to 877 | 708 to 761 |
| $K_p$ in atm | 7.53 to 140 | 14.0 to 192 | 72.94 to 400.9 | 32.36 to 64.86 |
| Ref. | [43, 45, 46, 49] | [43, 46, 49] | [42, 50] | [42, 50] |

| $MCl_3$ . . | $HoCl_3$ | $ErCl_3$ | $TmCl_3$ | $YbCl_3$ | $LuCl_3$ |
|---|---|---|---|---|---|
| T in K . . | 712 to 833 | 696 to 869 | 700 to 880 | 699 to 875 | 702 to 877 |
| $K_p$ in atm . | 55.95 to 224.94 | 52.58 to 333.5 | 82.30 to 483.9 | 113.1 to 609.7 | 136.4 to 727.9 |
| Ref. . . . | [51] | [42, 52] | [42, 52] | [53] | [53] |

For the highly radioactive $PmCl_3$ the determination method was adapted to closed system operation [42, 48].

From the measured data the Gibbs free energies $\Delta G^\circ_T$ of reaction were calculated and reduced to the general equation

$$\Delta G^\circ_T = \Delta H_0 + 6.4 \times 10^{-3}\ T \log T + 1.8 \times 10^{-7}\ T^2 - 22\ T^{-1} + IT \text{ kcal},$$

using an estimated $\Delta C_p$ function

$$\Delta C_p = -2.8 - 0.36 \times 10^{-3}\ T + 0.44 \times 10^5\ T^{-2} \text{ cal/K}.$$

The heat of reaction is then represented by

$$\Delta H^\circ_T = \Delta H_0 - 2.8 \times 10^{-3}\ T - 1.8 \times 10^{-7}\ T^2 - 44\ T^{-1} \text{ kcal}.$$

The following constants of the equations and numerical values of the thermodynamic functions at 785 and 298 K are given:

| $MCl_3$ . . | $\Delta G^\circ_{785}$ | $\Delta G^\circ_{298}$ | $\Delta H^\circ_{785}$ | $\Delta H^\circ_{298}$ | $\Delta S^\circ_{785}$ | $\Delta S^\circ_{298}$ | $\Delta H_0$ | I | Ref. |
|---|---|---|---|---|---|---|---|---|---|
| $YCl_3$ . . | −7.3 | 5.7 | 12.8 | 14.2 | 25.6 | 28.5 | 15.09 | −47.13 | [42] |
| $LaCl_3$ . . | +1.4 | 17.0 | 25.6 | 27.0 | 30.8 | 33.5 | 27.9 | −52.4 | [43 to 46] |
| $PrCl_3$ . . | −2.07 | 13.22 | 21.53 | 22.98 | 30.07 | 32.75 | 24.01 | −51.86 | [43, 45, 47] |
| $NdCl_3$ . | −2.88 | 12.41 | 20.91 | 22.24 | 30.31 | 32.99 | 23.20 | −52.10 | [43, 47] |
| $PmCl_3$ . | −4.9 | 10.4 | 18.9 | 20.3 | 30.3 | 33.2 | 21 | −51.8 | [42, 48] |
| $SmCl_3$ . | −5.4 | 10.4 | 19.2 | 20.6 | 31.4 | 34.0 | 21.6 | −53.0 | [43, 45, 46, 49] |
| $GdCl_3$ . | −6.2 | 9.2 | 17.6 | 19.0 | 30.2 | 32.9 | 20.0 | −51.9 | [43, 46, 49] |
| $\gamma$-$DyCl_3$ . | −8.0 | 5.2 | 12.4 | 13.8 | 26.0 | 28.9 | 14.72 | −47.61 | [42, 50] |
| $\beta$-$DyCl_3$ . | −7.0 | 7.2 | 15.0 | 16.4 | 28.0 | 30.9 | 17.28 | −49.55 | [42, 50] |
| $HoCl_3$ . | −7.7 | 6.0 | 13.4 | 14.8 | 26.7 | 29.7 | 15.8 | −48.5 | [51] |
| $ErCl_3$ . . | −7.8 | 5.6 | 13.0 | 14.4 | 26.5 | 29.5 | 15.28 | −48.03 | [42, 52] |
| $TmCl_3$ . | −8.4 | 4.8 | 12.1 | 13.5 | 26.1 | 29.2 | 14.41 | −47.58 | [42, 52] |
| $YbCl_3$ . . | −8.8 | 4.5 | 11.8 | 13.2 | 26.3 | 29.1 | 14.13 | −47.89 | [53] |
| $LuCl_3$ . . | −9.1 | 4.4 | 11.7 | 13.15 | 26.5 | 29.4 | 14.07 | −48.11 | [53] |

The $\Delta G$ and $\Delta H$ values are in kcal, the $\Delta S$ values in cal/K. In the case of $DyCl_3$, two different sets of K values were obtained. They were explained by the existence of two polymorphic forms ($\beta$- and $\gamma$-$DyCl_3$), which differ in their heats of formation by 2.6 kcal/mol [42, 50]. New measurements on $NdCl_3$ by [42] yielded $K_p$ values ranging from 4.43 at 761 K to 26.08 at 872 K that were in agreement with the data of [43, 47]. Data on the vapor phase hydrolysis of $TbCl_3$ have been reported by Koch [43]. The hydrolysis of $LaCl_3$, $CeCl_3$, $PrCl_3$, and $NdCl_3$ has also been studied by a static method with a membrane null-manometer. The results are in satisfactory agreement with the findings of [44, 47] (numerical values for $K_p$, $\Delta H$, and $\Delta S$ are presented in the sections on chemical reactions of the individual chlorides), Novikov, Baev [54]. Studies of the kinetics of hydrolysis of $CeCl_3$ have been reported by Hollabaugh et al. [55]. The heats of reaction of the trichlorides of La through Gd at 785 K show an approximately linear dependence on the reciprocal of the cation radius, Koch, Cunningham [47].

**Reactions with Ammonia and Hydrazine**

If $NH_3$ is condensed onto $ScCl_3$, $LaCl_3$, and $EuCl_3$ in a sealed container at −80 °C and the excess $NH_3$ is then evaporated off at room temperature, solid phases remain of the compositions $ScCl_3 \cdot 6NH_3$, $LaCl_3 \cdot 8NH_3$, and $EuCl_3 \cdot 10NH_3$, respectively, Pistorius [56]. At room temperature and 0 °C gaseous $NH_3$ reacts with $LaCl_3$, $CeCl_3$, $PrCl_3$, and $NdCl_3$ with great swelling to yield octammines, from which on thermal decomposition compounds with 5, 3, 2, and 1 mol $NH_3$ are formed, Ephraim, Block [57]. The following addition compounds have been reported by other authors:

$ScCl_3 \cdot nNH_3$ with n=5, 4, and 2 [58, 59];

$LaCl_3 \cdot nNH_3$ with n=8, 6, 4, and 2 [60];

$CeCl_3 \cdot nNH_3$ with n=20, 12, 8, 4, and 2 [61], 12, 8, 5, and 3 [86];

$NdCl_3 \cdot nNH_3$ with n=12, 11, 8, 5, 4, 2, and 1 [62];

$SmCl_3 \cdot nNH_3$ with n=11.5, 9.5, 8, 5, 4, 3, 2, and 1 [63],
10, 8, 5, 4, 3, 2.5, 2, 1.5, and 1 [64].

See the chapters on the chemical reactions of the individual chlorides in "Rare Earth Elements" C4b for more detailed information. Physicochemical studies of the synthesis and decomposition processes of $LaCl_3$ ammoniates have also been published by Torgonskaya et al. [65]. The $LaCl_3$-$NH_3$-$H_2O$ system was investigated at 20 and 40 °C by Mikaelyan et al. [66].

$LaCl_3$, $PrCl_3$, and $NdCl_3$ are not dissolved by anhydrous hydrazine, Welsh, Broderson [67]. A complex $[La(N_2H_4)_3Cl_3] \cdot 2H_2O$ is obtained by reaction of hydrazine hydrate with weakly acidic aqueous solutions of $LaCl_3$, Aliev et al. [68].

At elevated temperatures $NH_3$ and $N_2H_4$ reduce $SmCl_3$, $EuCl_3$, and $YbCl_3$ to the dichlorides; see the sections on the preparation of the individual dichlorides in "Rare Earth Elements" C4b.

**Reactions with Hydrogen Halides**

The exchange of $^{36}Cl$ between HCl vapor and solid rare earth chlorides has been investigated in the range from 30 to 300 °C. The following table gives the time in min required to achieve the percent exchange indicated at head of column:

| $MCl_3$ | % of activity transferred at 200 °C | | % of activity transferred at 300 °C | | |
|---|---|---|---|---|---|
| | 20 | 50 | 20 | 50 | 80 |
| $LaCl_3$ | 120 | – | 2 | 5 | 30 |
| $PrCl_3$ | 10 | – | *) | 2 | 13 |
| $NdCl_3$ | – | – | 1 | 5 | 40 |
| $GdCl_3$ | 5 | – | *) | 2 | 5 |
| $DyCl_3$ | 3 | 15 | – | – | 10 |
| $YbCl_3$ | – | – | 1 | 6 | – |

The asterisk marks tests in which the exchange was too fast to be measured. All of the measurements were made with $1.5 \times 10^{-4}$ mol HCl in a 30 mL reaction vessel at a solid-to-gas Cl ratio of 38:1. Estimated particle sizes were in the 5 μm range. With $ScCl_3$ significantly faster exchange of 20% after 4 min at 30 °C and 50% after 25 min at 100 °C was observed. $YCl_3$ showed a 20% exchange after 15 min at 100 °C. The mechanism of the exchange is discussed, Wilson, Willard [69], Wilson [70].

Hydrogen bromide and hydrogen iodide convert the rare earth chlorides into the respective bromides and iodides [13, 14, 19 to 21, 71 to 77]. In order to prepare the bromides, 1 to 2 g chloride is slowly heated over a period of 7 to 10 h from about 400 °C to a temperature close to the melting point in a stream of pure dry HBr. The transformation of the anhydrous chlorides to the iodides occurs at 400 to 600 °C and may take as long as 40 h. Instead of pure HI, a more readily obtainable mixture of $H_2$ and HI is commonly used, but pure $EuI_3$ cannot be obtained in this manner because of its easy reducibility, Taylor [78], Wetzel [79].

### Reactions with Hydrogen Chalcogenides

An $H_2S$ stream at 600 to 1000 °C converts the chlorides into the sulfides [79 to 83]; $H_2Se$ gives the selenides [79, 83].

### Reactions with $SiO_2$ and $As_2O_3$

Quartz and silicate glasses react with liquid and solid rare earth chlorides at elevated temperatures according to the equation

$$2\,MCl_3(s, l) + SiO_2(s) \rightarrow 2\,MOCl(s) + SiCl_4(g) \quad (M = \text{rare earth metal}).$$

The reaction enthalpies range from about 25 kcal ($LaCl_3$) to about 50 kcal ($LuCl_3$), Corbett [84, 85]. According to Haschke [5], a variety of oxide silicates and chloride silicates such as the $Yb_3(SiO_4)_2Cl$ phase characterized by Ayasse, Eick [6] are also possible products. The equilibrium of the reaction $^4/_3\,ScCl_3(s) + {}^7/_3\,SiO_2(s) \rightleftharpoons {}^2/_3\,Sc_2Si_2O_7(s) + SiCl_4(g)$, assumed by Polyachenok et al. [88], is discussed in "Rare Earth Elements" C 4b, Section 11.2.11.1.11.3.

The reaction $3\,MCl_3 + As_2O_3 \rightarrow 3\,MOCl + 2\,AsCl_3$ was found to occur to a small extent, when mixtures of rare earth chlorides and $As_2O_3$ were heated for 48 h at 350 to 400 °C in a bomb tube, Haeseler et al. [18].

References:

[1] O.G. Polyachenok, G.I. Novikov (Zh. Neorgan. Khim. **9** [1964] 773/7; Russ. J. Inorg. Chem. **9** [1964] 429/31). – [2] W. Klemm, W. Schüth (Z. Anorg. Allgem. Chem. **184** [1929] 352/8). – [3] G.I. Novikov, O.G. Polyachenok (Usp. Khim. **33** [1964] 732/47; Russ. Chem. Rev. **33** [1964] 342/50). – [4] Fam Ngok T'en, I.S. Morozov (Zh. Neorgan. Khim. **14** [1969] 2246/52; Russ. J. Inorg. Chem. **14** [1969] 1179/83). – [5] J.M. Haschke (in: K.A. Gschneidner, L.R. Eyring, Handbook on the Physics and Chemistry of Rare Earths, Vol. 4, Amsterdam - New York - Oxford 1979, pp. 89/151, 95).

[6] C. Ayasse, H.A. Eick (Inorg. Chem. **12** [1973] 1140/3). – [7] V.K. Il'in, A.D. Chervonnyi, V.A. Krenev (Zh. Neorgan. Khim. **21** [1976] 614/8; Russ. J. Inorg. Chem. **21** [1976] 332/4). – [8] V.K. Il'in, V.A. Krenev, V.I. Evdokimov (Zh. Neorgan. Khim. **19** [1974] 2583/6; Russ. J. Inorg. Chem. **19** [1974] 1411/2). – [9] V.K. Il'in, V.A. Krenev, V.I. Evdokimov (Zh. Neorgan. Khim. **18** [1973] 2336/40; Russ. J. Inorg. Chem. **18** [1973] 1236/8). – [10] O.G. Polyachenok, G.I. Novikov (Zh. Neorgan. Khim. **8** [1963] 1567/73; Russ. J. Inorg. Chem. **8** [1963] 816/20).

[11] G.I. Novikov, A.K. Baev (Izv. Vysshikh Uchebn. Zavedenii Khim. Khim. Tekhnol. **9** No. 2 [1966] 180/4; C.A. **65** [1966] 9824). – [12] G. Jantsch, N. Skalla, H. Grubitsch (Z. Anorg. Allgem. Chem. **216** [1933] 75/9). – [13] G. Jantsch, N. Skalla, H. Grubitsch (Z. Anorg. Allgem. Chem. **212** [1933] 65/83). – [14] G. Jantsch (Österr. Chemiker-Ztg. **40** [1937] 218/25). – [15] L. Holleck (Angew. Chem. **51** [1938] 243/51).

[16] H.G. Brauer, M. Knausenberger (Z. Anorg. Allgem. Chem. **320** [1963] 54/8). – [17] V.P. Bochin, E.I. Gutsevich, G.I. Novikov (Khim. Khim. Tekhnol. [Minsk] **14** [1979] 65/7 from C.A. **92** [1980] No. 9392). – [18] G. Haeseler, H. Lusky, F. Matthes (Z. Chem. [Leipzig] **5** [1965] 64/5). – [19] C. Matignon (Compt. Rend. **140** [1905] 1637/9). – [20] C. Matignon (Ann. Chim. Phys. [8] **8** [1906] 243/83).

[21] C. Matignon (Ann. Chim. Phys. [8] **8** [1906] 386/401). – [22] A.K. Baev, G.I. Novikov (Zh. Neorgan. Khim. **10** [1965] 2457/64; Russ. J. Inorg. Chem. **10** [1965] 1337/41). – [23] F. Bourion (Compt. Rend. **145** [1907] 62/4). – [24] F. Frerichs, F. Smith (Liebigs Ann. Chem. **191** [1878] 331/66, 356). – [25] G. Haeseler, F. Matthes (J. Less-Common Metals **9** [1965] 133/51).

[26] B.S. Mathur, T.R. Bhat (Z. Anorg. Allgem. Chem. **343** [1966] 220/4). – [27] J.A. Sommers, E.F. Westrum (J. Chem. Thermodyn. **8** [1976] 1115/36, 1117). – [28] H.E. Swanson, M.C. Morris, R.P. Stinchfield, E.H. Evans (NBS-MON-25-Pt. 1 [1962] 1/56). – [29] V.P. Glybin, R.B. Dobrotin, G.V. Akulova (Zh. Neorgan. Khim. **16** [1971] 2640/4; Russ. J. Inorg. Chem. **16** [1971] 1408/10). – [30] H.E. Swanson, M.I. Cook, T. Isaacs, E.H. Evans (NBS-Circ.-539-9 [1960] 1/64, 47).

[31] D.V. Drobot, B.G. Korschunov, L.V. Durinina (Izv. Akad. Nauk SSSR Neorgan. Materialy **1** [1965] 2189/96; Inorg. Materials [USSR] **1** [1965] 1978/85). – [32] D.V. Drobot, B.G. Korshunov (Izv. Akad. Nauk SSSR Neorgan. Materialy **2** [1966] 1093/6; Inorg. Materials [USSR] **2** [1966] 931/4). – [33] I.S. Morozov, B.G. Korshunov (Dokl. Akad. Nauk SSSR **119** [1958] 523/5; C.A. **1959** 4873). – [34] I.S. Morozov, B.G. Korshunov (Zh. Neorgan. Khim. **1** [1956] 2606/12; Russ. J. Inorg. Chem. **1** No. 11 [1956] 180/6). – [35] A.S. Buturlov, N.A. Mal'tsev, V.A. Krokhin, V.N. Zaikanov (Nauchn. Tr. Gos. Nauchn. Issled Proektn. Inst. Redkomet. Promsti **1980** 109/14 from C.A. **93** [1980] No. 223691).

[36] R.C. Croft (Nature **172** [1953] 725/6). – [37] R.C. Croft (Australian J. Chem. **9** [1956] 184/93). – [38] R.C. Vickery, N.L. Campbell (J. Am. Chem. Soc. **79** [1957] 5897/9). – [39] E. Stumpp (Mater. Sci. Eng. **31** [1977] 53/9). – [40] E. Stumpp, G. Nietfeld (Z. Anorg. Allgem. Chem. **456** [1979] 261/8).

[41] A.K. Avdeeva, B.S. Azikov, E.I. Perov, V.V. Serebrennikov, T.M. Shkol'nikova (Khim. Svoistva Soedin. Redkozemel. Elem. Dokl. 6th Vses. Soveshch. Fiz. Khim. Primen. Redkozemel. Elem. Ikh Soedin. Splavov, Moscow 1969 [1973], pp. 12/3 from C.A. **79** [1973] No. 99906; Tr. Tomsk. Gos. Univ. **237** [1973] 88/95 from C.A. **80** [1974] No. 74674). – [42] V. Wishnevsky (Diss. Univ. München 1967, pp. 1/121; BMWF-FB-K-68-42 [1968] 1/121; N.S.A. **23** [1969] No. 21678). – [43] C.W. Koch (UCRL-2286 [1953] 1/101 from N.S.A. **8** [1954] No. 772). – [44] C.W. Koch, A. Broido, B.B. Cunningham (J. Am. Chem. Soc. **74** [1952] 2349/51). – [45] A. Broido (UCRL-666 [1950] 1/44 from N.S.A. **5** [1951] No. 1822).

[46] C.W. Koch, B.B. Cunningham (J. Am. Chem. Soc. **75** [1953] 796/7). – [47] C.W. Koch, B.B. Cunningham (J. Am. Chem. Soc. **76** [1954] 1471/4). – [48] F. Weigel, V. Wishnevsky, V. Scherer, M. van Ghemen (Radiochim. Acta **11** [1969] 38/43, 41/2). – [49] C.W. Koch, B.B. Cunningham (UCRL-1918 [1952] 1/9; N.S.A. **6** [1952] No. 6330). – [50] F. Weigel, V. Wishnevsky (Ber. Deut. Chem. Ges. **102** [1969] 5/13, 6/7).

[51] F. Weigel, H. Haug (Ber. Deut. Chem. Ges. **94** [1961] 1548/54). – [52] F. Weigel, V. Wishnevsky (Ber. Deut. Chem. Ges. **103** [1970] 193/9). – [53] F. Weigel, V. Wishnevsky (Ber. Deut. Chem. Ges. **105** [1972] 95/102). – [54] G.I. Novikov, A.K. Baev (Vestn. Leningr. Univ. **17** No. 4 Ser. Fiz. Khim. No. 1 [1962] 154/60; C.A. **57**

[1962] 7988). – [55] C.H. Hollabaugh, E.I. Onstott, T.C. Wallace, M.G. Bowman (Advan. Hydrogen Energy **1** [1979] 809/28; C.A. **92** [1980] No. 149251).

[56] A. Pistorius according to [60]. – [57] F. Ephraim, R. Bloch (Ber. Deut. Chem. Ges. **59** [1926] 2692/705). – [58] I.V. Tananaev, V.P. Orlovskii (Zh. Neorgan. Khim. **7** [1962] 2299/303; Russ. J. Inorg. Chem. **7** [1962] 1192/4). – [59] Yu.Ya. Kharitonov, V.P. Orlovskii, I.V. Tananaev (Zh. Neorgan. Khim. **8** [1963] 1093/103; Russ. J. Inorg. Chem. **8** [1963] 567/72). – [60] G.F. Hüttig, W. Dauschan (Monatsh. Chem. **82** [1951] 742/7).

[61] Barre (Compt. Rend. **156** [1913] 1017/9). – [62] C. Matignon, R. Trannoy (Compt. Rend. **142** [1906] 1042/5; Ann. Chim. Phys. [8] **8** [1906] 284/8). – [63] C. Matignon, R. Trannoy (Compt. Rend. **140** [1905] 141/3). – [64] F. Ephraim, P. Rây (Ber. Deut. Chem. Ges. **62** [1929] 1639/50). – [65] T.I. Torgonskaya, N.D. Severinova, A.A. Vecher, G.I. Kurnevich (Vestn. Beloruss. Gos. Univ. II **1975** 8/13 from C.A. **84** [1976] No. 129803).

[66] A.G. Mikaelyan, S.M. Manukova, R.S. Mkhitaryan, G.G. Babayan (Mater. 5th Resp. Soveshch. Neorgan. Khim., Yerevan 1976 [1977], pp. 73/7 from C.A. **91** [1979] No. 113172). – [67] T.W.B. Welsh, H.J. Broderson (J. Am. Chem. Soc. **37** [1915] 816/22). – [68] R.Ya. Aliev, M.N. Guseinov, A.D. Kuliev, D.B. Musaev (Zh. Neorgan. Khim. **18** [1973] 382/4; Russ. J. Inorg. Chem. **18** [1973] 199/201). – [69] J.R. Wilson, J.E. Willard (J. Phys. Chem. **70** [1966] 1665/7). – [70] J.R. Wilson (Diss. Univ. Wisconsin 1965, pp. 1/155; Diss. Abstr. **25** [1965] 5593/4).

[71] W. Klemm, R. Rockstroh (Z. Anorg. Allgem. Chem. **176** [1928] 181/99). – [72] G. Jantsch, H. Grubitsch, F. Hoffmann, H. Alber (Z. Anorg. Allgem. Chem. **185** [1930] 49/64). – [73] G. Jantsch, N. Skalla, H. Jawurek (Z. Anorg. Allgem. Chem. **201** [1931] 207/20). – [74] G. Jantsch, H.Jawurek, N. Skalla, G. Gawalowski (Z. Anorg. Allgem. Chem. **207** [1932] 353/67). – [75] W. Klemm, W. Döll (Z. Anorg. Allgem. Chem. **241** [1939] 233/7).

[76] E. Hohmann, H. Bommer (Z. Anorg. Allgem. Chem. **248** [1941] 383/96). – [77] E.R. Harrison (J. Appl. Chem. [London] **2** [1952] 601/2). – [78] M.D. Taylor (Chem. Rev. **62** [1962] 503/11). – [79] K. Wetzel (in: G. Brauer, Handbuch der Präparativen Anorganischen Chemie, 2nd Ed., Vol. 2, Stuttgart 1962, pp. 982/1013). – [80] W. Muthmann, L. Stützel (Ber. Deut. Chem. Ges. **32** [1899] 3413/9).

[81] W. Klemm, K. Meisel, H.U. v. Vogel (Z. Anorg. Allgem. Chem. **190** [1930] 123/44). – [82] G. Brauer (Handbuch der Präparativen Anorganischen Chemie, Stuttgart 1954, pp. 841/64). – [83] W. Klemm, A. Koczy (Z. Anorg. Allgem. Chem. **233** [1937] 84/96). – [84] J.D. Corbett (Inorg. Nucl. Chem. Letters **8** [1972] 337/40). – [85] J.D. Corbett (Rev. Chim. Minerale **10** [1973] 239/57).

[86] T.I. Torgonskaya, N.P. Masherova, G.N. Rzhetskaya (Izv. Vysshikh Uchebn. Zavedenii Khim. Khim. Tekhnol. **22** [1979] 906/10 from C.A. **91** [1979] No. 217538). – [87] L.D. Polyachenok, K. Nazarov, O.G. Polyachenok (Zh. Fiz. Khim. **52** [1978] 1767/8; Russ. J. Phys. Chem. **52** [1978] 1021/2).

### 11.2.6.12.3 Reactions with Metals and Their Compounds

As long ago as 1827, Mosander [1] prepared metallic cerium by reducing $CeCl_3$ with potassium. Lanthanum, yttrium, neodymium, and didymium (Nd-Pr mixture) were subsequently obtained by similar metallothermic reactions when their halides (usually chlorides) were reacted with various active metals (usually alkali metals), see for example [2 to 6].

Klemm and Bommer [7] prepared all rare earth metals except promethium in the form of powder mixtures with KCl by reduction of the chlorides with potassium in glass ampules at low temperatures and used these powders for X-ray and magnetic studies. The reaction with excess K commenced at 200 °C and yielded a loose product, which did not show any tendency to sinter even on later heating to 350 to 400 °C. In the cases of $SmCl_3$, $EuCl_3$, and $YbCl_3$ the reaction was found to be retrograde above 250 °C in that the dichlorides were formed. Sodium also gives incomplete reduction of these chlorides, not exceeding the dichloride stage [7]; see also Brauer [8], Wetzel [9]. The mixtures of rare earth metals and KCl obtained with potassium in high vacuum were extensively used by Bommer, Hohmann [10, 11] for studies of heats of solution. Reduction attempts by Cueilleron [12] yielded no satisfactory results. A technique for preparing small quantities of rare earth metals that are pure with respect to most nonmetallic impurities consists of reducing $GdCl_3$ and other rare earth chlorides by K or Na vapors in a tantalum apparatus, Daane [13]. A method for the reduction of $LaCl_3$ with vaporized K is described by Power et al. [14]. The following changes in free energy in kcal have been calculated from literature data for the reaction $MCl_3(g)+3A(g)\rightarrow 3ACl(g)+M(l)$ (M = rare earth elements, $A={}^1/_2H_2$ and alkali metals) at 1000 and 1900 K:

| $MCl_3$ | $^1/_2(H_2)$ | | Li | | Na | | K | |
|---|---|---|---|---|---|---|---|---|
| | 1000 K | 1900 K | 1000 K | 1900 K | 1000 K | 1900 K | 1000 K | 1900 K |
| $YCl_3$ | 46 | 16 | −107 | −103 | −65 | −64 | −76 | −78 |
| $LaCl_3$ | 63 | 35 | − 90 | − 86 | −48 | −47 | −59 | −61 |
| $CeCl_3$ | 62 | 34 | − 91 | − 87 | −49 | −48 | −60 | −62 |
| $PrCl_3$ | 62 | 34 | − 91 | − 87 | −49 | −48 | −60 | −62 |
| $NdCl_3$ | 58 | 30 | − 95 | − 91 | −53 | −52 | −64 | −66 |
| $PmCl_3$ | 58 | 30 | − 85 | − 91 | −53 | −52 | −64 | −66 |
| $SmCl_3$ | 56 | 28 | − 97 | − 93 | −55 | −54 | −66 | −68 |
| $EuCl_3$ | 54 | 25 | − 99 | − 95 | −57 | −56 | −68 | −70 |
| $GdCl_3$ | 56 | 27 | − 97 | − 93 | −55 | −54 | −66 | −68 |
| $TbCl_3$ | 55 | 26 | − 98 | − 94 | −56 | −55 | −67 | −69 |
| $DyCl_3$ | 51 | 21 | −101 | − 97 | −60 | −59 | −71 | −73 |
| $HoCl_3$ | 49 | 18 | −102 | − 98 | −62 | −61 | −73 | −75 |
| $ErCl_3$ | 46 | 16 | −104 | −100 | −65 | −64 | −76 | −78 |
| $TmCl_3$ | 44 | 14 | −109 | −105 | −67 | −66 | −78 | −80 |
| $YbCl_3$ | 41 | 11 | −112 | −108 | −70 | −69 | −81 | −83 |
| $LuCl_3$ | 38 | 9 | −115 | −111 | −73 | −72 | −84 | −86 |

From these results, lithium should be the most suitable of all the alkali metals as a reducing agent for rare earth chlorides from the energetic point of view, Novikov, Baev [15]. This holds true also for lower temperatures, all the more since a high heat of reaction goes together with a number of other virtues, not the least of which is a low melting point of the chloride. Practical experiments with $CeCl_3$, however, met with poor success, Gray [16]. Better results were obtained with $YCl_3$. Reduction studies with sodium in this case also showed promise, Campbell et al. [17, 18], Carlson, Schmidt [19].

Molten magnesium reduces the chlorides of Ce, Nd, and Gd with formation of rare earth alloys in 50% yields. The Mg may be distilled off from the alloys to give metals of 99% purity, Trombe, Mahn [20]. Judging from estimated values for the free energy change, the reduction of $YCl_3$ by magnesium does not seem to be feasible. Calcium reduction should be possible over the entire temperature range from 298 to 1500 K. However, operating temperatures higher than 1055 K are required to prevent the solidification of $CaCl_2$, which is expected to impede the reaction, Wicks, Block [21]. The following free energy changes in kcal have been calculated by Novikov, Baev [15] for the reaction $MCl_3(g)+{}^3/_2Me(g)\rightarrow{}^3/_2MeCl_2(g)+M(l)$ (M = rare earth elements, Me = Mg, Ca, Zn) at 1000 and 1900 K:

| $MCl_3$ | Ca | | Mg | | Zn | |
|---|---|---|---|---|---|---|
| | 1000 K | 1900 K | 1000 K | 1900 K | 1000 K | 1900 K |
| $YCl_3$ | −116 | −117 | −56 | −31 | +18 | +22 |
| $LaCl_3$ | − 99 | − 89 | −39 | − 5 | +35 | +40 |
| $CeCl_3$ | −100 | − 90 | −40 | − 6 | +34 | +39 |
| $PrCl_3$ | −100 | − 90 | −40 | − 6 | +34 | +39 |
| $NdCl_3$ | −104 | − 94 | −44 | −10 | +30 | +35 |
| $PmCl_3$ | −104 | − 94 | −44 | −10 | +30 | +35 |
| $SmCl_3$ | −106 | − 96 | −46 | −12 | +28 | +33 |
| $EuCl_3$ | −108 | − 99 | −48 | −15 | +26 | +30 |
| $GdCl_3$ | −106 | − 97 | −46 | −15 | +28 | +32 |
| $TbCl_3$ | −107 | − 98 | −47 | −13 | +27 | +31 |
| $DyCl_3$ | −111 | −103 | −51 | −14 | +23 | +26 |
| $HoCl_3$ | −113 | −106 | −53 | −19 | +21 | +23 |
| $ErCl_3$ | −116 | −108 | −56 | −22 | +18 | +21 |
| $TmCl_3$ | −118 | −110 | −58 | −24 | +16 | +19 |
| $YbCl_3$ | −121 | −113 | −61 | −26 | +13 | +16 |
| $LuCl_3$ | −124 | −115 | −64 | −29 | +10 | +14 |

The heats of reaction of Ca with molten chlorides $MCl_3$ (M = La, Ce, Pr, Nd) and with solutions of these chlorides in KCl melts have been determined directly by a thermographic calorimetry method, Savin et al. [22, 23]; the kinetics of these reactions were also studied, Savin et al. [24, 25]; see Sections 11.2.11.3.2.6, 11.2.11.4.3.6, 11.2.11.5.4.5, and 11.2.11.6.3.5 in "Rare Earth Elements" C 4 b.

In order to prepare massive highly pure rare earth metals a technique was devised whereby rare earth chlorides are reduced by calcium in inductively heated tantalum crucibles in an inert atmosphere; the resulting metals are vacuum cast in Ta molds to remove volatile impurities. The yields are generally high (>95%). Efforts to prepare pure rare earth metals with calcium by bomb-reduction techniques using iodine and other auxiliary oxidants as boosters yielded no completely satisfactory results. Sm, Eu, and Yb chlorides are reduced only to their stable divalent state and can thus be efficiently removed from rare earth mixtures by such metallurgical processing, see for example [13, 17, 26 to 33]. Instructions for the preparation of Ce (and other rare earth metals) by calciothermic reduction of the chloride are given by Brauer [8], Wetzel [9].

Reactions of rare earth chlorides with their basis metals are treated in connection with the relevant phase diagrams. A general survey of the lower chloride phases formed in the M-$MCl_3$ systems and of the extent to which the metals react with their molten trichlorides is presented in Section 11.2.1.1, p. 19, of the present volume.

Metallic gallium takes up about 0.01 wt % of rare earth metals when it is heated for 2 h with rare earth chlorides in a helium atmosphere at 800 to 1200 °C. Thermodynamic parameters for the reaction $LaCl_3 + Ga \rightarrow GaCl_3 + La$ have been calculated for 298 to 2000 K, Avdeeva et al. [34].

Reduction of Sm, Tm, and Yb chlorides to the divalent state in hexamethylphosphoramide solutions has been effected by the action of metallic sodium, Kamenskaya et al. [35, 36]. The interaction of $Sm^{3+}$ with magnesium metal in solutions of $SmCl_3$ in acidic aqueous ethanol also produces $Sm^{2+}$, which is precipitated as the scantily soluble $SmCl_2$. Microquantities of Eu and Yb and several actinide elements (Np, Pu, Am, Cm, Fm) are likewise reduced under these conditions and by cocrystallization with $SmCl_2$ efficiently removed from these solutions. Cocrystallization with $EuCl_2$ also takes place when $EuCl_3$ is the predominant solute and the solution is treated in a similar manner, Mikheev et al. [38, 39]. The Eu in 6 N aqueous HCl solutions is reduced to the divalent state by boiling with zinc amalgam, Wilkinson, Hicks [40]. For iodometric titrations, weakly acidic (0.05 N HCl) solutions of Eu are passed through a Jones reductor; this procedure may also be used for the quantitative determination of Eu in rare earth mixtures, McCoy [41 to 43].

Sodium amalgam reacts with the rare earth chlorides in alcoholic solution to give rare earth amalgams according to the equation

$$3\,NaHg_x + MCl_3 \rightarrow 3\,NaCl + MHg_{3x}$$

The sodium chloride is precipitated as is also some rare earth chloride. Although the reaction goes nearly to completion, there is a trace of sodium left in the amalgam. After a time there is a tendency for the amalgam to decompose, West, Hopkins [44]. The yield of rare earth amalgams in absolute ethanol is acceptable, formation of sodium alcoholate is also observed. Strontium amalgam reduces $SmCl_3$ to insoluble red-brown $SmCl_2$ and thus provides a path for the separation of Sm from rare earth mixtures. Calcium amalgam shows a similar behavior, Brukl [45].

The reaction of the rare earth chlorides with $Sb_2O_3$ provides a convenient method for preparing the rare earth oxide chlorides, especially those of the heavy rare earth elements, which have low thermal stability. The reaction proceeds according to the equation

$$3\,MCl_3 + Sb_2O_3 = 3\,MOCl + 2\,SbCl_3 \quad (M = \text{rare earth element})$$

and is usually conducted at 250 to 450 °C. The by-product $SbCl_3$ is oxidized in a $Cl_2$ stream to $SbCl_5$, which is more volatile than $SbCl_3$ and is thus more rapidly removed from the reaction mixture. This displaces the equilibrium towards the right and increases the rate of preparation of the oxide chlorides. CeOCl does not form even under these conditions, because of its very low decomposition temperature.

Oxide chlorides of the lighter rare earth elements form by the reaction of the chlorides with the corresponding rare earth oxides at the melting point of the chlorides, T'en, Morozov [46]. Reactions of this kind and equilibrium diagrams of $MCl_3$-$M_2O_3$ and $MCl_3$-MOCl systems, where M stands for the same as well as different rare earth elements, have been treated in "Seltenerdelemente" C5, 1977, pp. 3/41; see also "Rare Earth Elements" C4b for the formation of oxide chlorides on introduction of oxygen into $CeCl_3$-$YCl_3$, $NdCl_3$-$YCl_3$, $NdCl_3$-$CeCl_3$, and $GdCl_3$-$NdCl_3$ melts in the absence and presence of alkali chlorides.

Since rare earth fluorides are insoluble in most chloride mixtures, any soluble fluoride will be precipitated from molten salt mixtures as $MF_3$ (M = rare earth element), if rare earth chlorides are present, Emley [47]. The formation of double compounds by the reaction of rare earth chlorides with alkali chlorides and fluorides under various conditions has been treated in "Seltenerdelemente" C5, 1977, pp. 120/259. $MCl_3$-$M'Cl_3$ and $MCl_3$-$M'Cl_3$-$H_2O$ systems (M, M'=different rare earth elements) are discussed in Sections 11.2.9.1, p. 259 and 11.2.9.2, p. 260 of the present volume. Data on the formation of gaseous complexes with $AlCl_3$ and $FeCl_3$, which greatly enhances the vaporization of rare earth chlorides, are presented in Section 11.1.3, p. 10.

The reaction of anhydrous $MCl_3$ (M=Y, Sm, Gd, Tb, Dy, Er, Yb) with Li borohydride in tetrahydrofuran proceeds according to the equation

$$MCl_3 + 2\,LiBH_4 \rightarrow MCl(BH_4)_2 + 2\,LiCl$$

Attempts to obtain the chloride diborohydrides of La, Ce, Pr, and Nd failed. The Eu compound decomposes even in the cold; in heated mixtures reduction of $EuCl_3$ to $EuCl_2$ takes place:

$$EuCl_3 + LiBH_4 \rightarrow EuCl_2 + LiCl + {}^1/_2\,B_2H_6 + {}^1/_2\,H_2$$

Rossmanith, Muckenhuber [48], Brukl, Rossmanith [37].

The reaction of rare earth chlorides with GaAs was studied in quartz ampules at 800 to 1200 °C under He. The concentration of rare earth metals in GaAs drops with increasing temperature, Avdeeva et al. [34].

References:


[1] C.G. Mosander (Ann. Physik Chem. [2] **11** [1827] 406/16). – [2] F. Wöhler (Liebigs Ann. Chem. [2] **144** [1867] 251/5). – [3] O. Popp (Liebigs Ann. Chem. **131** [1864] 179/201, 186). – [4] C. Marignac (Ann. Chim. Phys. [3] **38** [1853] 148/77, 154). – [5] E. Zintl, S. Neumayr (Z. Elektrochem. **39** [1933] 84/6).

[6] H.C. Kremers (Trans. Am. Electrochem. Soc. **47** [1925] 365/71). – [7] W. Klemm, H. Bommer (Z. Anorg. Allgem. Chem. **231** [1937] 138/71, 141, 143). – [8] G. Brauer (Handbuch der Präparativen Anorganischen Chemie, Stuttgart 1954, pp. 841/64). – [9] K. Wetzel (in: G. Brauer, Handbuch der Präparativen Anorganischen Chemie, 2nd Ed., Vol. 2, Stuttgart 1962, pp. 982/1013). – [10] H. Bommer, E. Hohmann (Z. Anorg. Allgem. Chem. **248** [1941] 357/72).

[11] H. Bommer, E. Hohmann (Z. Anorg. Allgem. Chem. **248** [1941] 373/82). – [12] J. Cueilleron (Bull. Soc. Chim. France [5] **12** [1945] 88/9). – [13] A.H. Daane (in: F.H. Spedding, A.H. Daane, The Rare Earths, New York - London 1961, pp. 102/12). – [14] W.H. Power, G.L. Fox, J.B. Otto (MLM-995 [1954] 1/38 from N.S.A. **10** [1956] No. 2979). – [15] G.I. Novikov, A.K. Baev (Izv. Vysshikh Uchebn. Zavedenii Khim. Khim. Tekhnol. **9** No. 2 [1966] 180/4; C.A. **65** [1966] 9824).

[16] P.M.J. Gray (Metal Ind. [London] **83** [1952] 43/6). – [17] T.T. Campbell, F.E. Block, R.E. Mussler, G.B. Robidart (U.S. Bur. Mines Rept. Invest. No. 5880 [1961] 1/29; C.A. **56** [1962] 5613). – [18] T.T. Campbell, R. Mussler, G. Robidart, G. Engel (USBM-U-550 [1958] 1/36, 3/9; N.S.A. **14** [1960] No. 551). – [19] O.N. Carlson, F.A. Schmidt (in: F.H. Spedding, A.H. Daane, The Rare Earths, New York - London 1961, pp. 113/25). – [20] F. Trombe, F. Mahn (Ann. Chim. [Paris] [11] **19** [1944] 345/61).

[21] C.E. Wicks, F.E. Block (U.S. Bur. Mines Bull. No. 605 [1963] 1/146, 8, 9). – [22] V.D. Savin, N.P. Mikhailova (Zh. Fiz. Khim. **51** [1977] 1272/3, **54** [1980] 2246/50; Russ. J. Phys. Chem. **51** [1977] 754, **54** [1980] 1282/4). – [23] V.D. Savin, N.P.

Mikhailova, V.A. Morozova (Zh. Fiz. Khim. **53** [1979] 1142/6; Russ. J. Phys. Chem. **53** [1979] 645/8). – [24] V.D. Savin, N.P. Mikhailova (Zh. Fiz. Khim. **51** [1977] 1273, **54** [1980] 2241/5; Russ. J. Phys. Chem. **51** [1977] 754, **54** [1980] 1279/81). – [25] V.D. Savin, N.P. Mikhailova, V.N. Morozova (Zh. Fiz. Khim. **53** [1979] 1218/21; Russ. J. Phys. Chem. **53** [1979] 686/9).

[26] W.H. Keller, R.P. Ericson, C. Hach (TID-5212 [1945] 32/6; C.A. **1956** 7698). – [27] D.H. Ahmann (ISC-68 [1951] 1/79 from N.S.A. **5** [1951] No. 4401). – [28] F.H. Spedding, H.A. Wilhelm, W.H. Keller et al. (ISC-75 [1951] 1/20 from N.S.A. **5** [1951] No. 4402). – [29] A.H. Daane (ISC-121 [1951] 1/94 from N.S.A. **5** [1951] No. 5112). – [30] F.H. Spedding, A.H. Daane (ISC-205 [1952] 1/9 from N.S.A. **6** [1952] No. 1464).

[31] F.H. Spedding, H.A. Wilhelm, W.H. Keller et al. (Ind. Eng. Chem. **44** [1952] 553/6). – [32] F.H. Spedding, A.H. Daane (J. Am. Chem. Soc. **74** [1952] 2783/5). – [33] A.H. Daane, F.H. Spedding (J. Electrochem. Soc. **100** [1953] 442/4). – [34] A.K. Avdeeva, B.S. Azikov, E.I. Perov et al. (Khim. Svoistva Soedin. Redkozemel. Elem. Dokl. 6th Vses. Soveshch. Fiz. Khim. Primen. Redkozemel. Elem. Ikh. Soedin. Splavov, Moscow 1969 [1973], pp. 12/3 from C.A. **79** [1973] No. 99906; Tr. Tomsk. Gos. Univ. **237** [1973] 88/95 from C.A. **80** [1974] No. 74674). – [35] A.N. Kamenskaya, N.B. Mikheev, N.A. Konovalova (Zh. Neorgan. Khim. **22** [1977] 2130/4; Russ. J. Inorg. Chem. **22** [1977] 1152/5).

[36] A.N. Kamenskaya, K. Bukietynska, N.B. Mikheev et al. (Zh. Neorgan. Khim. **24** [1979] 1139/46; Russ. J. Inorg. Chem. **24** [1979] 633/7). – [37] A. Brukl, K. Rossmanith (Monatsh. Chem. **90** [1959] 481/7). – [38] N.B. Mikheev, A.N. Kamenskaya, R.A. Dyachkova et al. (Inorg. Nucl. Chem. Letters **8** [1972] 523/8). – [39] N.B. Mikheev, V.I. Spitsyn, A.N. Kamenskaya et al. (Inorg. Nucl. Chem. Letters **8** [1972] 929/36). – [40] G. Wilkinson, H.G. Hicks (Phys. Rev. [2] **75** [1949] 1370/8).

[41] H.N. McCoy (J. Am. Chem. Soc. **57** [1935] 1756). – [42] H.N. McCoy (J. Am. Chem. Soc. **58** [1936] 1577/80). – [43] H.N. McCoy, L. Pauling, G.B. Baxter, F.D. Tuemmler (J. Am. Chem. Soc. **59** [1937] 1131/4). – [44] D.H. West, B.S. Hopkins (J. Am. Chem. Soc. **57** [1935] 2185/6). – [45] A. Brukl (Angew. Chem. **52** [1939] 151/4).

[46] Fam Ngok T'en (Pham Ngok Thien), I.S. Morozov (Zh. Neorgan. Khim. **16** [1971] 1683/8; Russ. J. Inorg. Chem. **16** [1971] 890/3). – [47] E.F. Emley (Chem. Eng. Sci. **1** [1952] 131/44, 141). – [48] K. Rossmanith, E. Muckenhuber (Monatsh. Chem. **92** [1961] 600/4).

### 11.2.6.12.4 Reactions with Water, Acids, and Bases

Reactions with water vapor and atmospheric moisture are discussed in Section 11.2.6.12.2.

Anhydrous rare earth trichlorides dissolve in water easily and with great evolution of heat. The solutions are clear, provided that O-containing impurities are absent, but mostly slightly hydrolyzed as for example pH measurements indicate, see Section 11.2.8.1.1 on the structure of aqueous solutions.

$AOH-MCl_3-H_2O$ systems (A = alkali metal, M = rare earth element) and hydroxide chlorides existing in them are treated in "Seltenerdelemente" C 5, 1977, pp. 42/77.

**Solubility in Water**

The following solubilities S in water were determined by Nikolaev et al. [1] at low temperatures:

| MCl₃ | $LaCl_3$ | $CeCl_3$ | $PrCl_3$ | $NdCl_3$ | $SmCl_3$ | $EuCl_3$ | $GdCl_3$ | $TbCl_3$ | $DyCl_3$ | $HoCl_3$ |
|---|---|---|---|---|---|---|---|---|---|---|
| S in wt% at −15 °C | 46.8 | 47.0 | 47.8 | 49.2 | 48.0 | 47.5 | 48.0 | 48.4 | 49.0 | 50.2 |
| S in wt% at −10 °C | 47.4 | 47.7 | 47.9 | 49.1 | 48.1 | 48.1 | 48.4 | 48.9 | 49.3 | 50.9 |
| S in wt% at − 5 °C | 47.3 | 47.3 | 47.7 | – | 47.8 | 47.5 | 48.2 | 48.3 | 49.0 | – |

Values at 0 to 50 °C are [2]:

| $MCl_3$ | $YCl_3$ | $LaCl_3$ | $CeCl_3$ | $PrCl_3$ | $NdCl_3$ | $SmCl_3$ | $EuCl_3$ | $GdCl_3$ |
|---|---|---|---|---|---|---|---|---|
| S in wt% at 0 °C | 43.3 | 47.2 | 47.6 | 47.4 | 49.2 | 47.9 | 47.7 | 48.2 |
| S in wt% at 10 °C | 43.4 | 48.2 | 48.1 | 48.2 | 49.3 | 47.8 | 47.6 | 48.3 |
| S in wt% at 25 °C | 43.5 | 48.8 | 48.6 | 49.0 | 49.7 | 48.3 | 48.3 | 48.7 |
| S in wt% at 30 °C | 44.0 | 49.8 | 49.3 | 50.0 | 50.5 | 49.0 | 48.5 | 49.3 |
| S in wt% at 50 °C | 45.2 | 51.6 | 52.2 | 52.2 | 52.2 | 50.5 | 50.4 | 50.8 |

| $MCl_3$ | $TbCl_3$ | $DyCl_3$ | $HoCl_3$ | $ErCl_3$ | $TmCl_3$ | $YbCl_3$ | $LuCl_3$ |
|---|---|---|---|---|---|---|---|
| S in wt% at 0 °C | 48.5 | 49.4 | 50.2 | 50.7 | 51.5 | 52.7 | 52.6 |
| S in wt% at 10 °C | 48.8 | 49.4 | 49.9 | 51.0 | 51.6 | 53.3 | 54.2 |
| S in wt% at 25 °C | 49.1 | 49.9 | 50.7 | 51.3 | 52.2 | 53.3 | 54.2 |
| S in wt% at 30 °C | 49.7 | 50.2 | 50.9 | 51.6 | – | 53.4 | 54.6 |
| S in wt% at 50 °C | 50.7 | 51.1 | 51.8 | 52.5 | 53.1 | 54.8 | 55.1 |

The ratio mol $H_2O$/mol $MCl_3$ in the saturated solutions is shown in **Fig. 23** as a function of the atomic number, Nikolaev et al. [2]. The solubility of $YbCl_3 \cdot 9H_2O$ at −10 °C is 51.9 wt%, Sokolova et al. [3]. Earlier data [4 to 6] for $LaCl_3$, $PrCl_3$, and $NdCl_3$ at 0 to 100 °C are shown in **Fig. 24** from Friend, Hale [6]. Data from Fischer et al. [7] for 0 °C are in general agreement with those from [2], except for $GdCl_3$ and $ErCl_3$:

| $MCl_3$ | $YCl_3$ | $LaCl_3$ | $PrCl_3$ | $NdCl_3$ | $SmCl_3$ | $GdCl_3$ | $ErCl_3$ |
|---|---|---|---|---|---|---|---|
| S in wt% | 43.0 | 47.0 | 47.5 | 48.8 | 48.0 | 44.9 | 49.8 |

The value for $NdCl_3$ is extrapolated from measurements in hydrochloric acid solutions at various HCl concentrations. Zhuravlev et al. [8, 9] report the solubility of $NdCl_3$ at 0 °C as 48.5 to 48.8 wt%.

The following compilation of solubilities in mol $MCl_3$ per 1000 g $H_2O$ at 25 °C from literature data has been given by Hinchey, Cobble [10]:

| $MCl_3$ | $YCl_3$ | $LaCl_3$ | $CeCl_3$ | $PrCl_3$ | $NdCl_3$ | $SmCl_3$ | $EuCl_3$ |
|---|---|---|---|---|---|---|---|
| S in m | 3.948 | 3.8944 | 3.748 | 3.975 | 3.9307 | 3.6414 | 3.619 |
| Ref. | [11] | [12] | [10, 13] | [6] | [12] | [12] | [14] |

| $MCl_3$ | $GdCl_3$ | $TbCl_3$ | $DyCl_3$ | $HoCl_3$ | $ErCl_3$ | $YbCl_3$ | $LuCl_3$ |
|---|---|---|---|---|---|---|---|
| S in m | 3.5898 | 3.5795 | 3.6302 | 3.739 | 3.7840 | 4.0028 | 4.136 |
| Ref. | [12] | [14] | [15] | [14] | [11] | [12] | [10] |

A solubility of 4.02 mol $PrCl_3$ per 1000 g $H_2O$ at 25 °C is reported by Grigorovich, Samoilov [16].

Fig. 23

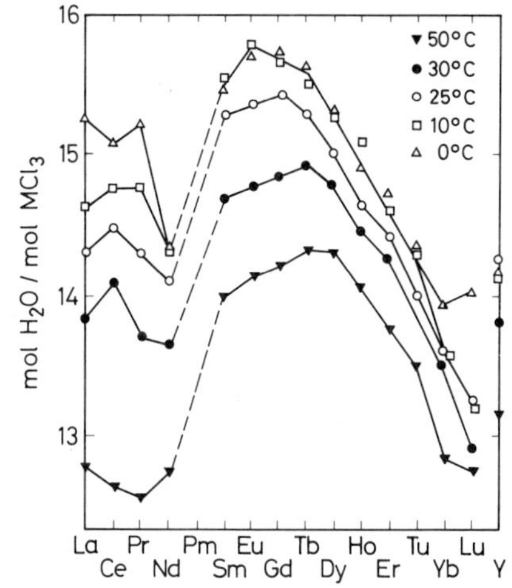


Molar ratio $H_2O:MCl_3$ in saturated aqueous solutions of the rare earth chlorides $MCl_3$ at temperatures between 0 and 50 °C.

Fig. 24

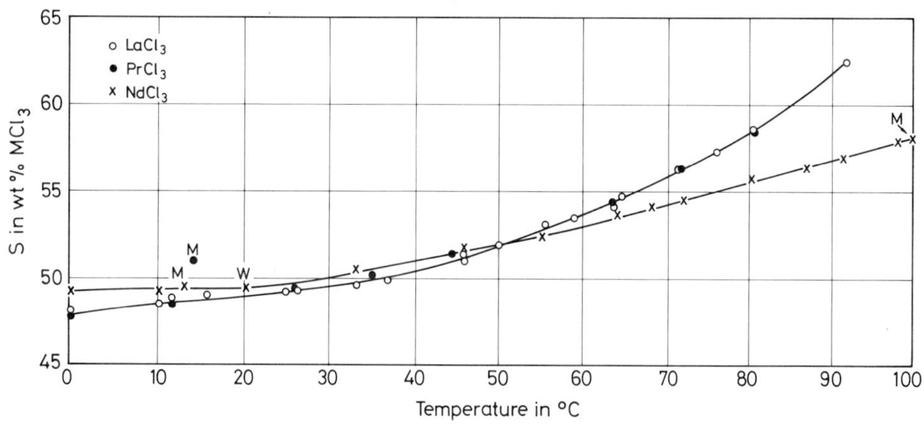


Solubilities S of $LaCl_3$, $PrCl_3$, and $NdCl_3$ in water at 0 to 100 °C. The data points denoted by M and W are from Matignon [4] and Williams et al. [5], respectively, the rest are from Friend and Hale [6]. Values refer to the anhydrous salts.

Values in wt% at 25 °C from various sources are:

| $MCl_3$ . . | $YCl_3$ | | $LaCl_3$ | | $CeCl_3$ | | | | $PrCl_3$ | |
|---|---|---|---|---|---|---|---|---|---|---|
| S . . . | 43.3 | 44.14 | 49.00 | 49.1 | 48.40 | 48.6 | 49.3 | 50.2 | 48.0 | 50.03 |
| Ref. . . | [27] | [26] | [26] | [27] | [26] | [27] | [17 to 24] | [25] | [27] | [26] |

| $MCl_3$ . . | $NdCl_3$ | | $SmCl_3$ | | $EuCl_3$ | $GdCl_3$ | $HoCl_3$ | $ErCl_3$ | $YbCl_3$ | |
|---|---|---|---|---|---|---|---|---|---|---|
| S . . . | 49.01 | 49.7 | 48.1 | 48.30 | 47.9 | 48.5 | 49.7 | 51.3 | 52.8 | 53.3 |
| Ref. . . | [26] | [27] | [27] | [26] | [27] | [27] | [27] | [28] | [27] | [28] |

Additional data for the chlorides of the lighter rare earth elements:

| $MCl_3$ | temperatures in °C 20 | 30 | 40 | 45 | 50 | 75 |
|---|---|---|---|---|---|---|
| $YCl_3$ | – | 43.6 [31] | – | – | 43.6 [42]<br>45.27 [43] | 44.3 [42] |
| $LaCl_3$ | 48.0 [29, 30]<br>48.2 [32]<br>48.5 [33 to 35] | 49.0 [29] | 50.2 [29]<br>50.5 [32 to 35] | – | 50.0 [42]<br>51.5 [29, 30] | 55.6 [42] |
| $CeCl_3$ | 47.5 [37]<br>48.5 [36, 38] | 49.2 [36] | 50 [37]<br>50.5 [36] | – | 51.7 [17 to 24, 36, 38]<br>52.8 [25] | 58.5 [a)] [37] |
| $PrCl_3$ | 49.0 [29, 30] | 50.0 [29] | 51.2 [29] | 49.89 [39] | 52.0 [29, 30] | – |
| $NdCl_3$ | 49.0 [30]<br>49.5 [29]<br>49.8 [8, 9] | 48.52 [39] | 50.6 [8, 9] | 48.89 [39] | 50.0 [42]<br>50.8 [29, 30] | 52.5 [42]<br>55 [a)] [37] |
| $SmCl_3$ | 48.0 [40]<br>48.3 [41] | 47.16 [39] | 49.0 [40]<br>49.2 [41] | 48.09 [39] | 50.01 [43] | – |

a) at 80 °C.

Solubilities at 10 °C in wt% are: 47.2 for $LaCl_3$ [29], 48.0 for $CeCl_3$ [36], and 48.2 for $PrCl_3$ [29].

The solubility of $YCl_3$ and $NdCl_3$ at 100 and 90 °C is 45.1 and 54.5 wt%, respectively, Zelikman et al. [42].

Data for the hexahydrated chlorides $MCl_3 \cdot 6H_2O$ in g hydrate per 100 g $H_2O$ from Powell [14]:

| M . . . | Y | La | Nd | Sm | Eu | Gd | Tb | Dy | Ho | Er | Tm | Yb | Lu |
|---|---|---|---|---|---|---|---|---|---|---|---|---|---|
| 0 °C . . | 199 | – | 225 | 210 | 207 | 210 | 208 | 215 | 225 | 238 | 248 | 260 | 267 (?) |
| 25 °C . . | 217 | – | 247 | 224 | 218 | 219 | 218 | 226 | 238 | 245 | 257 | 274 | 295 |
| 60 °C . . | 233 | 850 | 298 | 260 | 250 | 252 | 246 | 254 | 262 | 270 | 279 | 300 | 319 |

Corresponding values for the hexahydrates of $YCl_3$, $NdCl_3$, and $SmCl_3$ at 10 to 50 °C from Williams et al. [5] are:

| t in °C . . . . . . . | 10 | 20 | 30 | 40 | 50 |
|---|---|---|---|---|---|
| $YCl_3 \cdot 6H_2O$ . . . . | 213.5 | 217.0 | 221.0 | 227.0 | 233.2 |
| $NdCl_3 \cdot 6H_2O$ . . . | 237.5 | 243.0 | 250.2 | 259.2 | 272.6 |
| $SmCl_3 \cdot 6H_2O$ . . . | 214.7 | 218.4 | 222.8 | 232.5 | 244.7 |

The solubility of $YCl_3$ expressed in g anhydrous salt per 100 g $H_2O$ increases from 77.08 at 25 to 78.7 at 40 °C, that of $ErCl_3$ from 103.55 to 105.7 in the same temperature range, Petheram, Spedding [11]. See Crew et al. [44] for data for $YCl_3$ at 0 to 80 °C.

**Solubility in Hydrochloric Acid**

While the solubilities of the rare earth chlorides in water at 0 °C differ by only a few percent from each other, the solubilities in concentrated hydrochloric acid decrease from $LaCl_3$ to $ErCl_3$ by about two orders of magnitude. In detail the solubility (in g of anhydrous salt per 100 g solution) changes with the HCl concentration at 0 °C as follows (selected values):

| wt% HCl | 5 | 10 | 15 | 20 | 26 | 30 | 36 | 40 | 42 | 45 |
|---|---|---|---|---|---|---|---|---|---|---|
| $YCl_3 \cdot 6H_2O$ | 35.7 | 28.5 | 21.4 | 14.6 | 7.1 | 3.3 | 0.43 | 0.115 | 0.060 | 0.034 [a] |
| $LaCl_3 \cdot 7H_2O$ | 38.2 | 29.5 | 20.9 | 12.9 | 5.8 | 2.8 | 1.0 | 1.8 | 3.4 | – [b] |
| $PrCl_3 \cdot 6H_2O$ | 38.9 | 30.2 | 21.5 | 13.6 | 6.1 | 2.5 | 0.78 | 0.56 | 0.64 | 0.86 [a] |
| $NdCl_3 \cdot 6H_2O$ | 40.3 | 31.8 | 23.4 | 15.1 | 6.4 | 2.9 | 0.54 | 0.27 | 0.26 | 0.31 [a] |
| $SmCl_3 \cdot 6H_2O$ | 38.9 | 29.9 | 21.1 | 12.9 | 5.0 | 1.7 | 0.27 | 0.12 | 0.09 | 0.08 [a] |
| $GdCl_3 \cdot 6H_2O$ | 37.0 | 29.1 | 21.2 | 13.4 | 4.7 | 1.9 | 0.153 | 0.063 | 0.045 | 0.034 [a] |
| $ErCl_3 \cdot 6H_2O$ | 41.3 | 32.8 | 24.4 | 16.0 | 7.0 | 2.8 | 0.36 | 0.11 | 0.050 | 0.025 [a] |

[a] Saturated with HCl gas at 1 atm. – [b] The solution saturated with $LaCl_3$ absorbs at 1 atm HCl pressure only 42.5 wt% HCl.

Recalculation of the results into equivalents of rare earth chloride or moles of hydrogen chloride, respectively, which are dissolved in 1000 g $H_2O$, yields the solubility isotherms displayed in **Fig. 25a** and **b** in comparison with those of $ScCl_3$ and $AlCl_3$ (see the authors' discussion of the curves, especially in view of possible separation effects on fractional crystallization) [7]. Several older values from [5, 6, 44, 45] for the solubility in water or dilute hydrochloric acid, in part after extrapolation to 0 °C, are said to agree in general within a few tenths of a gram per 100 g solution with the values given in the table, Fischer et al. [7, 46]; cf. also Fischer et al. [47], in which additional results can be found concerning the solubilities and changes of solubility in ether/water saturated with HCl at 0 °C. Solubilities of $MCl_3$ hydrates (M=Y, La, Ce, Pr, Nd, Sm) in hydrochloric

Fig. 25

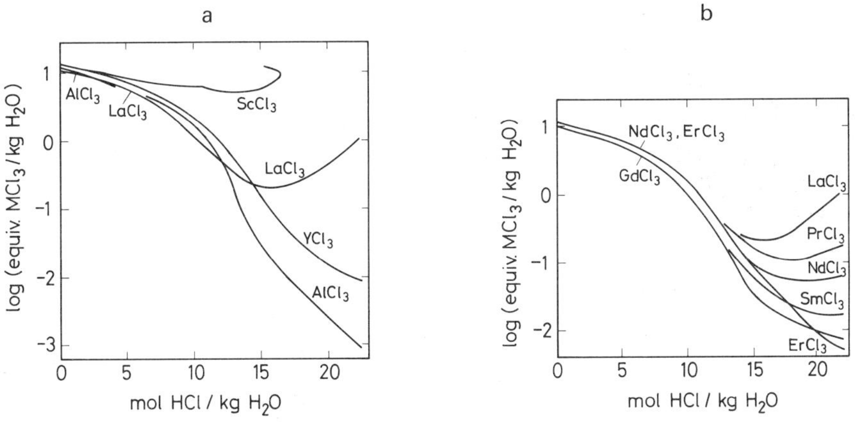


Solubility of rare earth chlorides $MCl_3$ in hydrochloric acid at 0 °C: Fig. 25a for $MCl_3$ with M=Sc, Y, and La compared with that of $AlCl_3$, and Fig. 25b for $MCl_3$ with M=La, Pr, Nd, Sm, Gd, and Er.

acid of varying concentrations at 25 °C; data in wt% of HCl and anhydrous salt, respectively, from Shevtsova et al. [26]:

| | | | | | | | | |
|---|---|---|---|---|---|---|---|---|
| HCl . . . . . . . | 2.57 | 4.10 | 5.02 | 9.75 | 17.13 | 17.89 | 19.23 | 23.00 |
| $YCl_3 \cdot 6H_2O$ . . . | 39.89 | 37.54 | 36.84 | 30.25 | 19.70 | 18.25 | 16.64 | 12.80 |
| HCl . . . . . . . | 2.55 | 5.08 | 9.54 | 13.32 | 20.05 | 23.53 | 28.68 | – |
| $LaCl_3 \cdot 7H_2O$ . . | 43.25 | 38.31 | 29.10 | 23.85 | 14.23 | 9.50 | 4.13 | – |
| HCl . . . . . . . | 2.78 | 5.97 | 12.70 | 15.08 | 24.16 | 26.80 | – | – |
| $CeCl_3 \cdot 7H_2O$ . . | 45.58 | 39.35 | 27.15 | 23.97 | 10.89 | 5.91 | – | – |
| HCl . . . . . . . | 1.80 | 5.47 | 8.34 | 13.45 | 22.02 | 26.50 | – | – |
| $PrCl_3 \cdot 7H_2O$ . . | 47.22 | 40.54 | 37.17 | 28.53 | 15.65 | 8.54 | – | – |
| HCl . . . . . . . | 0.55 | 1.99 | 3.75 | 5.62 | 12.50 | 16.08 | 23.12 | – |
| $NdCl_3 \cdot 6H_2O$ . . | 48.07 | 46.39 | 42.39 | 37.82 | 27.50 | 20.23 | 9.94 | – |
| HCl . . . . . . . | 1.30 | 5.43 | 10.45 | 18.71 | 21.45 | 26.50 | – | – |
| $SmCl_3 \cdot 6H_2O$ . . | 46.01 | 38.16 | 28.91 | 15.83 | 10.13 | 4.37 | – | – |

Solubilities of $YCl_3 \cdot 6H_2O$, $LaCl_3 \cdot 7H_2O$, and $NdCl_3 \cdot 6H_2O$ in hydrochloric acid of varying concentrations at t=50, 75, 90, and 100 °C, selected values from Zelikman et al. [42]; data in wt% HCl and anhydrous salt:

| | | | | | | | | | | |
|---|---|---|---|---|---|---|---|---|---|---|
| t . . . . . . . . | 50 | 50 | 50 | 50 | 75 | 75 | 75 | 100 | 100 | 100 |
| HCl . . . . . . . | 3.4 | 11.2 | 23.8 | 34.2 | 6.7 | 15.3 | 24.2 | 2.3 | 3.8 | 5.9 |
| $YCl_3 \cdot 6H_2O$ . . . | 39.1 | 27.6 | 11.2 | 2.5 | 35.3 | 24.0 | 13.8 | 42.2 | 40.2 | 37.6 |
| t . . . . . . . . | 50 | 50 | 50 | 50 | 75 | 75 | 75 | 90 | 90 | 90 |
| HCl . . . . . . . | 3.3 | 11.7 | 25.5 | 30.0 | 1.5 | 8.0 | 11.4 | 1.0 | 2.9 | 3.7 |
| $LaCl_3 \cdot 7H_2O$ . . | 45.0 | 31.4 | 12.5 | 12.0 | 53.4 | 43.7 | 41.0 | 63.1 | 61.6 | 61.4 |
| t . . . . . . . . | 50 | 50 | 50 | 50 | 75 | 75 | 75 | 90 | 90 | 90 |
| HCl . . . . . . . | 7.5 | 20.4 | 28.3 | 35.0 | 3.5 | 9.9 | 21.4 | 2.3 | 4.1 | 6.7 |
| $NdCl_3 \cdot 6H_2O$ . . | 37.2 | 16.8 | 7.8 | 3.5 | 46.8 | 36.4 | 21.8 | 50.9 | 47.9 | 44.4 |

The solubilities of $CeCl_3$ and $NdCl_3$ in aqueous hydrochloric acid between 20 and 80 °C decrease linearly with increasing HCl concentration; at 40 °C the solubility of $CeCl_3$, for instance, drops from 50 wt % at 0 wt % HCl to 27 wt % at 15 wt % HCl. At 20, 40, or 60 to 80 °C the solubility of $CeCl_3$ is lower than, equal to, or higher than that of $NdCl_3$, respectively, Kreingauz [37]. Older results concerning the solubility of $MCl_3 \cdot 6H_2O$ (M=Y, Nd, Sm, Yb) in HCl of density 1.1051 g/cm$^3$ at 10, 20, 30, 40, and 50 °C can be found in Williams et al. [5].

In the total range from 0 to 100 °C and at all investigated HCl concentrations, the hydrates identified by the formulas given constitute the stable phases in contact with the liquid in the $MCl_3$-HCl-$H_2O$ systems, with the exception of the $ScCl_3$-HCl-$H_2O$ system at high HCl concentrations (see "Rare Earth Elements" C 4b, Section 11.2.11.1.12). Apparently at these hydration numbers there is a particularly high bonding energy between water molecules and the $M^{3+}$ ions. The minima in the solubility isotherms for $LaCl_3$ at 0 and at 50 °C and high HCl concentrations speak in contrast for a change of the hydration state in the solution. These minima occur at 0 °C according to Fig. 25b, p.163 up to $SmCl_3$ and disappear at $GdCl_3$; they shift with increasing atomic number to higher HCl contents.

**Thermodynamics of the Dissolution Process**

A compilation of enthalpies of solution (in kJ/mol) in water at room temperature and infinite dilution has been given by Blachnik, Selle [48] from the authors' own measurements in an isoperibolic calorimeter and data from the post-1940 literature:

| | | | | | | |
|---|---|---|---|---|---|---|
| $LaCl_3$ | – | 132.21 | 137.70 | 137.78 | – | 133.72 |
| $CeCl_3$ | 140.67 | 136.82 | 144.01 | 142.30 | – | 137.19 |
| $PrCl_3$ | 147.65 | 141.84 | 149.24 | 149.24 | – | 144.31 |
| $NdCl_3$ | 153.26 | 150.21 | 156.86 | 155.23 | – | 147.70 |
| $SmCl_3$ | 165.94 | 163.18 | 167.07 | 167.07 | – | – |
| $EuCl_3$ | – | 171.96 | 170.29 | – | 174.89 | – |
| $GdCl_3$ | 180.79 | 177.40 | 181.59 | 179.87 | – | – |
| $TbCl_3$ | – | 191.63 | 192.38 | – | 186.61 | – |
| $DyCl_3$ | 198.11 | 197.90 | (209.20) | – | 193.30 | – |
| $HoCl_3$ | 210.25 | 209.20 | 213.43 | – | 199.58 | – |
| $ErCl_3$ | 214.97 | 207.11 | 215.06 | 208.32 | – | – |
| $YbCl_3$ | 215.43 | 212.97 | 216.06 | 216.06 | – | – |
| $LuCl_3$ | – | 214.64 | 218.53 | – | 222.59 | – |
| Source | [48] | [53] | [54] | [55, 56] | [57] | [58] |

Early data of Matignon [49] for the heats of solution of $LaCl_3$, $PrCl_3$, and $NdCl_3$ at 16 °C agree well with those of [53], the values for $YCl_3$ and $SmCl_3$ are significantly lower; see also Morss [50].

Krestov et al. [51, 52] determined calorimetrically the integral changes in enthalpy during the dissolution of anhydrous rare earth chlorides in water over the temperature range from 15 to 93 °C at concentrations from $5\times10^{-3}$ to $4\times10^{-2}$ m. The values obtained were extrapolated to infinite dilution using the extended Debye-Hückel equation. The following table gives the results at rounded temperatures and values calculated therefrom for the changes in free energy, entropy, and heat capacity on dissolution; ΔH and ΔG are in kcal/mol, ΔS and $\Delta C_p$ are in $cal\cdot mol^{-1}\cdot K^{-1}$:

| $MCl_3$ | function | temperature in K 273.15 | 283 | 293 | 298 | 313 | 323 | 343 | 363 | 373 |
|---|---|---|---|---|---|---|---|---|---|---|
| $YCl_3$ | $-\Delta H^\circ$ | 42.39 | 44.14 | 45.76 | 46.48 | 48.59 | 49.87 | 52.26 | 54.71 | 56.01 |
| | $-\Delta G^\circ$ | 33.8 | 33.3 | 32.9 | 32.6 | 32.0 | 31.4 | 30.2 | 28.8 | 25.1 |
| | $-\Delta S^\circ$ | 31.3 | 38.1 | 43.8 | 46.4 | 53.2 | 57.2 | 64.4 | 71.3 | 74.8 |
| | $-\Delta C_p^\circ$ | 204 | 177 | 156 | 148 | 131 | 124 | 120 | 126 | 133 |
| $LaCl_3$ | $-\Delta H^\circ$ | 27.93 | 29.74 | 31.42 | 32.19 | 34.37 | 35.75 | 37.45 | 41.31 | 42.79 |
| | $-\Delta G^\circ$ | 22.5 | 22.2 | 22.0 | 21.8 | 21.2 | 20.8 | 19.8 | 18.5 | 17.8 |
| | $-\Delta S^\circ$ | 19.9 | 26.5 | 32.3 | 34.9 | 42.1 | 46.4 | 54.6 | 62.7 | 66.9 |
| | $-\Delta C_p^\circ$ | 194 | 174 | 159 | 153 | 141 | 137 | 138 | 149 | 161 |
| $PrCl_3$ | $-\Delta H^\circ$ | 31.67 | 33.35 | 34.91 | 35.63 | 37.65 | 38.89 | 41.37 | 43.91 | 45.14 |
| | $-\Delta G^\circ$ | 25.2 | 24.9 | 24.6 | 24.4 | 23.8 | 23.3 | 22.3 | 21.1 | 20.3 |
| | $-\Delta S^\circ$ | 23.6 | 29.7 | 35.1 | 37.5 | 44.1 | 48.1 | 55.6 | 62.8 | 66.6 |
| | $-\Delta C_p^\circ$ | 179 | 168 | 147 | 141 | 129 | 125 | 125 | 133 | 140 |
| $NdCl_3$ | $-\Delta H^\circ$ | 33.37 | 94.96 | 36.43 | 37.11 | 39.05 | 40.29 | 42.58 | 45.44 | 46.86 |
| | $-\Delta G^\circ$ | 26.6 | 26.3 | 26.0 | 25.8 | 25.2 | 24.7 | 23.5 | 22.5 | 21.8 |
| | $-\Delta S^\circ$ | 24.9 | 30.5 | 35.5 | 37.8 | 44.2 | 48.1 | 55.6 | 63.1 | 67.1 |
| | $-\Delta C_p^\circ$ | 164 | 150 | 140 | 135 | 127 | 124 | 128 | 142 | 153 |

| $MCl_3$ | function | temperature in K | | | | | | | | |
|---|---|---|---|---|---|---|---|---|---|---|
| | | 273.15 | 283 | 293 | 298 | 313 | 323 | 343 | 363 | 373 |
| $SmCl_3$ | $-\Delta H^\circ$ | 36.28 | 37.80 | 39.23 | 39.89 | 41.82 | 43.05 | 45.83 | 48.22 | 49.63 |
| | $-\Delta G^\circ$ | 28.6 | 28.3 | 27.9 | 27.7 | 27.0 | 26.5 | 25.4 | 24.2 | 23.5 |
| | $-\Delta S^\circ$ | 28.1 | 33.7 | 38.6 | 40.9 | 47.2 | 51.1 | 58.6 | 66.2 | 70.1 |
| | $-\Delta C_p^\circ$ | 164 | 148 | 137 | 133 | 125 | 124 | 128 | 142 | 153 |
| $TbCl_3$ | $-\Delta H^\circ$ | 41.03 | 42.50 | 43.98 | 44.67 | 46.64 | 47.93 | 50.54 | 53.07 | 54.35 |
| | $-\Delta G^\circ$ | 32.9 | 32.5 | 32.2 | 32.0 | 31.3 | 30.7 | 29.7 | 28.4 | 27.6 |
| | $-\Delta S^\circ$ | 29.8 | 35.3 | 40.3 | 42.6 | 49.1 | 53.2 | 60.8 | 58.0 | 71.6 |
| | $-\Delta C_p^\circ$ | 158 | 148 | 140 | 137 | 130 | 128 | 127 | 130 | 134 |
| $ErCl_3$ | $-\Delta H^\circ$ | 46.85 | 48.24 | 49.85 | 50.19 | 52.04 | 53.24 | 55.70 | 58.38 | 59.82 |
| | $-\Delta G^\circ$ | 38.2 | 37.8 | 37.7 | 37.2 | 36.5 | 36.0 | 34.9 | 33.6 | 32.9 |
| | $-\Delta S^\circ$ | 31.8 | 36.8 | 41.4 | 43.5 | 49.5 | 53.3 | 60.7 | 68.2 | 72.2 |
| | $-\Delta C_p^\circ$ | 145 | 135 | 128 | 125 | 121 | 121 | 127 | 141 | 151 |

The data for $YCl_3$, $TbCl_3$, and $ErCl_3$ are from [52], the rest from [51]. The error in $\Delta H$ probably does not exceed 0.3 to 0.4%.

The $\Delta H$ values of [51, 52] at 298 K are in satisfactory agreement with those of Spedding et al. [55, 56]. The two sets of data lie on a single straight line when plotted against the atomic numbers of the rare earth elements, if $\Delta H(YCl_3)$ is put in the series between $DyCl_3$ and $HoCl_3$ which is in accordance with the crystallographic radius of the $Y^{3+}$ ion. Finogenov [57] demonstrated that the $\Delta H$ values of Spedding et al. [55, 56] are inversely proportional to the crystallographic radii of the cations $r_c$. The standard deviation of the experimental values from the $\Delta H$ vs. $1/r_c$ straight line fitted by least squares is $\pm 0.3$ kcal/mol, i.e., is within the experimental error limits. Bommer, Hohmann [53, 59] earlier found a non-linear dependence of the heats of solution on the atomic number of the rare earth elements and a strong influence of the crystal structure.

For all the salts investigated $\Delta H$ becomes more exothermic as the temperature is raised, because the endothermic component of $\Delta H$ associated with the breakdown of the structure of water by the ions is decreased. The temperature dependence of the change in heat capacity has an extremum. In the range up to 40 °C there is an appreciable increase in $\Delta C_p$, which then passes through a maximum, and from 50 to 70 °C onwards decreases, becoming more negative. This temperature characteristic of $\Delta C_p$ is evidently due to the anomalous behavior of the heat capacity of water itself with rising temperature [51]. The $\Delta C_p$ characteristics are in satisfactory agreement with calculations, as is shown by a figure in [52].

The concentration dependence of the enthalpies of solution can be represented by an equation of the general form $\Delta H$ (in cal/mol) $= A + b\, m^{1/2} + c\, m$ (m = molality). The constant b is taken as the theoretical limiting slope given by the Debye-Hückel theory for a 3-1 electrolyte. The extrapolated constant a is the value of the enthalpy of solution at infinite dilution. The following values of the constants in the above equation are listed by Spedding, Flynn [56] for the trichlorides studied in their work:

| $MCl_3$ | $-a$ | b | $-c$ | $\sigma$ | $m_{max}$ |
|---|---|---|---|---|---|
| $LaCl_3$ | 32933 | 6925 | 5640 | 35 | 0.08145 |
| $PrCl_3$ | 35672 | 6925 | 5774 | 21 | 0.09351 |
| $SmCl_3$ | 39932 | 6925 | 8949 | 15 | 0.09163 |

| $MCl_3$ | $-a$ | $b$ | $-c$ | $\sigma$ | $m_{max}$ |
|---|---|---|---|---|---|
| $GdCl_3$ . . | 42992 | 6925 | 4027 | 39 | 0.06350 |
| $ErCl_3$ . . . | 49792 | 6925 | 7873 | 40 | 0.06812 |
| $YbCl_3$ . . | 51638 | 6925 | 5433 | 40 | 0.08815 |
| $YCl_3$ . . . | 48690 | 6925 | 5223 | 40 | 0.08451 |

$\sigma$ is the average deviation in cal/mol of calculated from experimental values, $m_{max}$ is the highest molality studied.

Similar equations with an additional $m^{3/2}$ term were earlier used by Spedding, Miller [55] for the representation of ΔH data in cal/mol for $CeCl_3$ and $NdCl_3$, respectively, at 25 °C:

$$\Delta H = -34423 + 6925\ m^{1/2} - 9294\ m + 6147\ m^{3/2}$$

$$\Delta H = -37489 + 6925\ m^{1/2} - 10402\ m + 7096\ m^{3/2}$$

The average difference between the observed values and those calculated by the given equations was 12 cal/mol for $CeCl_3$ and 14 cal/mol for $NdCl_3$.

The very slight hydrolysis accompanying the dissolution of rare earth chlorides in water should not influence the calorimetric results beyond the limits of experimental error. Measurements made for comparison in weakly acidified water (pH $\approx$ 3 to 5) confirm this assumption, Krestov et al. [51], Spedding, Flynn [56]. Even in 0.05 N HCl no noticeable difference of the ΔH values was observed from those in pure water, Bommer, Hohmann [53]. At higher HCl concentrations, however, ΔH decreases somewhat with increasing HCl molality [56]. The heats of solution of $CeCl_3$ and $NdCl_3$ are reduced to a greater amount than can be accounted for on the basis of ionic strength alone, Spedding, Miller [60]. The following Table 11.2.6-7 lists ΔH data of various authors for different HCl concentrations at 25 °C (if not otherwise noted):

Table 11.2.6-7
Heats of Solution of Rare Earth Chlorides.

| $MCl_3$ | m(HCl) | m($MCl_3$) | -ΔH in kcal/mol | -ΔH in kJ/mol | Ref. |
|---|---|---|---|---|---|
| $YCl_3$ | 0 | 0 | 53.7 | 224.7 | [54] |
| | $\approx$0.05 | $\approx$0.010 | 52.3 (20 °C) | 218.8 | [53] |
| | 0 | 0.014 | 52.93 ± 0.30 | 221.5 | [61] |
| | 0.5 | 0.014 | 51.48 ± 0.16 | 215.4 | [61] |
| | 0 | 0.014 | 47.94 | 200.6 | [56] |
| | 1.472 | 0.008 | 46.78 | 195.7 | [62] |
| | 1.472 | 0.019 | 46.83 | 195.9 | [62] |
| | 1.472 | 0.032 | 46.24 | 193.5 | [62] |
| $LaCl_3$ | a) | $3 \times 10^{-4}$ | 32.5 ± 0.5 | 136.0 | [64] |
| | 0 | 0.013 | 32.17 ± 0.09 | 134.6 | [63] |
| | 0.543 | 0.015 | 31.00 ± 0.09 | 129.7 | [63] |
| | 0.251 | 0.015 | 31.28 | 130.9 | [62] |
| | 0.251 | 0.031 | 30.99 | 129.7 | [62] |
| $CeCl_3$ | 0.241 | 0.018 | 32.85 | 137.4 | [60] |
| | 0.241 | 0.023 | 32.66 | 136.6 | [60] |
| | 0.371 | 0.018 | 32.26 | 135.0 | [60] |
| | 0.371 | 0.040 | 32.04 | 134.1 | [60] |
| | 0.371 | 0.061 | 31.89 | 133.4 | [60] |

Table 11.2.6-7 [continued]

| $MCl_3$ | m(HCl) | m($MCl_3$) | -ΔH in kcal/mol | -ΔH in kJ/mol | Ref. |
|---|---|---|---|---|---|
| $CeCl_3$ | 1.462 | 0.002 | 31.00 | 129.7 | [65] |
| | 1.462 | 0.005 | 30.84 | 129.0 | [65] |
| $PrCl_3$ | 0.251 | 0.010 | 34.13 | 142.8 | [62] |
| | 0.251 | 0.026 | 33.86 | 141.7 | [62] |
| | 1.469 | 0.012 | 30.01 [b] | 125.6 | [62] |
| $NdCl_3$ | [a] | $3\times10^{-4}$ | 36.0±0.7 | 150.6 | [64] |
| | 0.241 | 0.021 | 35.64 | 149.1 | [60] |
| | 0.241 | 0.036 | 35.43 | 148.2 | [60] |
| | 4.360 | ? | 30.42±0.06 | 127.2 | [66] |
| $PmCl_3$ | 0 | 0 | 39.0 [c] | 163.2 | [57] |
| | 0 | 0 | 38.7 [d] | 161.9 | [54] |
| $SmCl_3$ | 0 | dilute | 40.8±0.2 | 170.7 | [67] |
| | [a] | $3\times10^{-4}$ | 38.4±0.2 | 160.7 | [64] |
| | 0.476 | 0.030 | 37.89±0.24 | 158.5 | [68] |
| $EuCl_3$ | 4.360 | ≈0.003 | 34.32±0.10 | 143.6 | [69] |
| $GdCl_3$ | [a] | $3\times10^{-4}$ | 40.1±1.0 | 167.8 | [64] |
| | 0.251 | 0.006 | 41.72 | 174.6 | [62] |
| | 0.251 | 0.015 | 41.64 | 174.2 | [62] |
| | 0.251 | 0.023 | 41.45 | 173.4 | [62] |
| $TbCl_3$ | 4.360 | ≈0.003 | 40.13±0.19 | 167.9 | [70] |
| $DyCl_3$ | [a] | $3\times10^{-4}$ | 49.3±1.0 | 206.2 | [64] |
| $HoCl_3$ | [a] | $3\times10^{-4}$ | 47.6±1.1 | 199.2 | [64] |
| | 4.360 | ≈0.002 | 43.14±0.08 | 180.5 | [70] |
| $ErCl_3$ | [a] | $3\times10^{-4}$ | 47.0±0.6 | 196.6 | [64] |
| | 0 | 0 | 51.3 | 214.8±0.3 | [71] |
| | 0.001 | 0.003 | 51.0 | 213.5±0.3 | [71] |
| | 1.43 | 0.003 | 48.2 | 201.7±0.3 | [71] |
| | 1.472 | 0.008 | 46.30 | 193.7 | [62] |
| | 1.472 | 0.017 | 46.12 | 193.0 | [62] |
| | 0 | 0.010 | 50.84±0.18 | 212.7 | [72] |
| | 1.43 | 0.011 | 48.25±0.21 | 201.9 | [72] |
| $TmCl_3$ | 4.360 | ≈0.002 | 44.60±0.21 | 186.6 | [73] |
| $YbCl_3$ | [a] | $3\times10^{-4}$ | 49.0±0.9 | 205.0 | [64] |
| | 4.360 | ≈0.004 | 44.73±0.20 | 187.0 | [73] |

[a] A few drops of concentrated perchloric acid was added to water to block any hydrolysis of the lanthanide ions. – [b] Mean value. – [c] Derived by graphical interpolation. – [d] Estimated value.

References:

[1] A.V. Nikolaev, A.A. Sorokina, N.P. Sokolova et al. (Izv. Sibirsk. Otd. Akad. Nauk SSSR Ser. Khim. Nauk **1978** No. 1, pp. 46/7; C.A. **88** [1978] No. 159295). – [2] A.V.

Nikolaev, A.A. Sorokina, N.P. Sokolova et al. (Izv. Sibirsk. Otd. Akad. Nauk SSSR Ser. Khim. Nauk **1977** No. 4, pp. 84/90; C.A. **87** [1977] No. 173507). – [3] N.P. Sokolova, I.I. Yakovlev, G.N. Tschechova (Izv. Sibirsk. Otd. Akad. Nauk SSSR Ser. Khim. Nauk **4** No. 9 [1979] 77/8). – [4] C. Matignon (Ann. Chim. Phys. [8] **8** [1906] 243/83, 386/401). – [5] M.D. Williams, H.C. Fogg, C. James (J. Am. Chem. Soc. **47** [1925] 297/301).

[6] J.N. Friend, R.W. Hale (J. Chem. Soc. **1940** 670/2). – [7] W. Fischer, H. Bauer, I. Dillo et al. (Z. Anorg. Allgem. Chem. **357** [1968] 177/83). – [8] E.F. Zhuravlev, A.G. Samotov, L.F. Shlentova (Zh. Neorgan. Khim. **16** [1971] 2804/7; Russ. J. Inorg. Chem. **16** [1971] 1493/5). – [9] E.F. Zhuravlev, R.V. Yurkevich, A.G. Samotov, L.F. Shlentova (Zh. Neorgan. Khim. **18** [1973] 2247/51; Russ. J. Inorg. Chem. **18** [1973] 1189/91). – [10] R.J. Hinchey, J.W. Cobble (Inorg. Chem. **9** [1970] 917/21).

[11] H.H. Petheram, F.H. Spedding (ISC-770 [1963] 1/69; N.S.A. **18** [1964] No. 13722). – [12] V.W. Saeger, F.H. Spedding (IS-338 [1960] 1/207, 36; N.S.A. **16** [1962] No. 88). – [13] C.W. DeKock, F.H. Spedding (IS-1356 [1965] 1/162, 44; C.A. **66** [1967] No. 59520). – [14] P.E. Powell (IS-15 [1959] 11/5; N.S.A. **14** [1960] No. 11488). – [15] D.A. Csejka, F.H. Spedding (IS-417 [1961] 1/143; N.S.A. **16** [1962] No. 31584).

[16] Z.I. Grigorovich, O.Ya. Samoilov (Zh. Strukt. Khim. **3** [1962] 464/6; J. Struct. Chem. [USSR] **3** [1962] 446/8). – [17] M.I. Kadantseva, E.F. Zhuravlev (Zh. Neorgan. Khim. **15** [1970] 211/4; Russ. J. Inorg. Chem. **15** [1970] 107/9). – [18] M.I. Kadantseva, E.F. Zhuravlev (Zh. Neorgan. Khim. **16** [1971] 273/5; Russ. J. Inorg. Chem. **16** [1971] 143/5). – [19] M.I. Kadantseva, E.F. Zhuravlev (Zh. Neorgan. Khim. **17** [1972] 3066/8; Russ. J. Inorg. Chem. **17** [1972] 1612/3). – [20] M.I. Kadantseva, E.F. Zhuravlev (Izv. Vysshikh Uchebn. Zavedenii Khim. Khim. Tekhnol. **15** [1972] 481/3; C.A. **77** [1972] No. 80131).

[21] M.I. Kadantseva, E.F. Zhuravlev (Zh. Neorgan. Khim. **18** [1973] 537/43; Russ. J. Inorg. Chem. **18** [1973] 281/4). – [22] M.I. Kadantseva, E.F. Zhuravlev (Zh. Neorgan. Khim. **16** [1971] 2836/9; Russ. J. Inorg. Chem. **16** [1971] 1510/1). – [23] E.F. Zhuravlev, M.I. Kadantseva (Zh. Neorgan. Khim. **14** [1969] 2854/6; Russ. J. Inorg. Chem. **14** [1969] 1504/5). – [24] E.F. Zhuravlev, M.I. Kadantseva (Zh. Neorgan. Khim. **16** [1971] 2867/9; Russ. J. Inorg. Chem. **16** [1971] 1524/6). – [25] E.F. Zhuravlev, M.I. Kadantseva (Zh. Neorgan. Khim. **15** [1970] 770/3; Russ. J. Inorg. Chem. **15** [1970] 395/7).

[26] Z.N. Shevtsova, L.S. Nam, B.G. Korshunov (Zh. Neorgan. Khim. **13** [1968] 1682/5; Russ. J. Inorg. Chem. **13** [1968] 877/9). – [27] A.V. Nikolaev, A.A. Sorokina, A.Ya. Vilenskaya et al. (Izv. Sibirsk. Otd. Akad. Nauk SSSR **1967** No. 6, pp. 5/21; C.A. **69** [1968] No. 90501). – [28] A.V. Nikolaev, L.M. Kosobudskaya, A.A. Sorokina (Izv. Sibirsk. Otd. Akad. Nauk SSSR Ser. Khim. Nauk **1975** No. 5, pp. 59/64; C.A. **84** [1976] No. 35887). – [29] E.N. Khutorskoi, A.D. Sheveleva (Uch. Zap. Permsk. Gos. Univ. No. 178 [1968] 57/63; C.A. **73** [1970] No. 124112). – [30] E.N. Khutorskoi, A.D. Sheveleva (Uch. Zap. Permsk. Gos. Univ. No. 178 [1968] 49/56; C.A. **73** [1970] No. 124111).

[31] G.A. Ashimkulova, K. Nogoev, K. Sulaimankulov (Zh. Neorgan. Khim. **19** [1974] 2588/90; Russ. J. Inorg. Chem. **19** [1974] 1414/5). – [32] E.F. Zhuravlev, R.V. Yurkevich, N.K. Kalegina (Zh. Neorgan. Khim. **17** [1972] 2536/40; Russ. J. Inorg. Chem. **17** [1972] 1327/9). – [33] E.F. Zhuravlev, R.V. Yurkevich, N.G. Kalegina (Zh. Neorgan. Khim. **18** [1973] 1109/12; Russ. J. Inorg. Chem. **18** [1973] 585/7). – [34] E.F. Zhuravlev, R.V. Yurkevich, N.A. Rasskazova (Izv. Vysshikh Uchebn. Zavedenii Khim. Khim. Tekhnol. **17**

[1974] 1123/6; C.A. **82** [1975] No. 22318). – [35] E.F. Zhuravlev, R.V. Yurkevich, N.A. Rasskazova (Zh. Neorgan. Khim. **19** [1974] 1977/8; Russ. J. Inorg. Chem. **19** [1974] 1083/4).

[36] A.D. Sheveleva, R.L. Bogdanovskaya, E.F. Zhuravlev (Zh. Neorgan. Khim. **9** [1964] 1435/40; Russ. J. Inorg. Chem. **9** [1964] 778/81). – [37] B.P. Kreingauz, E.A. Subbotina, B.N. Rabinovich (Met. Tsvetn. Redkikh Metal. Akad. Nauk SSSR Inst. Met. **1967** 210/5; C.A. **68** [1968] No. 52184). – [38] A.D. Sheveleva (Uch. Zap. Permsk. Gos. Univ. No. 159 [1966] 3/14; C.A. **69** [1968] No. 39117). – [39] A. Dilebaeva. K. Sulaimankulov (Zh. Neorgan. Khim. **18** [1973] 2257/61; Russ. J. Inorg. Chem. **18** [1973] 1194/6). – [40] E.F. Zhuravlev, N.A. Rasskazova, N. Prudenko (Zh. Neorgan. Khim. **23** [1978] 841/3; Russ. J. Inorg. Chem. **23** [1978] 466/7).

[41] E.F. Zhuravlev, N.A. Rasskazova, G.B. Khakimova (Zh. Neorgan. Khim. **23** [1978] 800/3; Russ. J. Inorg. Chem. **23** [1978] 442/4). – [42] A.N. Zelikman, N.V. Baryshnikov, A.I. Khokhlov (Zh. Neorgan. Khim. **16** [1971] 2023/4; Russ. J. Inorg. Chem. **16** [1971] 1080). – [43] Z.N. Shevtsova, N. Wei-Chüan (Zh. Neorgan. Khim. **8** [1963] 1949/52; Russ. J. Inorg. Chem. **8** [1963] 908/9). – [44] M.C. Crew, H.E. Steinert, B.S. Hopkins (J. Phys. Chem. **29** [1925] 34/8). – [45] J.E. Powell, H.R. Burkholder (J. Inorg. Nucl. Chem. **14** [1960] 65/70).

[46] W. Fischer (Angew. Chem. **75** [1963] 173). – [47] W. Fischer, J. Wernet, M. Zumbusch-Pfisterer (Z. Anorg. Allgem. Chem. **258** [1949] 157/61). – [48] R. Blachnik, D. Selle (Thermochim. Acta **33** [1979] 301/10). – [49] C. Matignon (Ann. Chim. Phys. [8] **8** [1906] 386/401, 402/16, 426/32, 433/9). – [50] L.R. Morss (Diss. Univ. California 1969, pp. 1/273 from Diss. Abstr. Intern. B **31** [1970] 605; UCRL-18951 [1969] 1/262 from C.A. **73** [1970] No. 49600).

[51] G.A. Krestov, V.A. Kobenin, S.V. Semenovskii (Zh. Neorgan. Khim. **17** [1972] 805/8; Russ. J. Inorg. Chem. **17** [1972] 421/3). – [52] G.A. Krestov, V.A. Kobenin, S.V. Semenovskii (Zh. Neorgan. Khim. **18** [1973] 3/6; Russ. J. Inorg. Chem. **18** [1973] 1/2). – [53] H. Bommer, E. Hohmann (Z. Anorg. Allgem. Chem. **248** [1941] 373/82). – [54] L.R. Morss (J. Phys. Chem. **75** [1971] 392/9). – [55] F.H. Spedding, C.F. Miller (J. Am. Chem. Soc. **74** [1952] 3158/62).

[56] F.H. Spedding, J.P. Flynn (J. Am. Chem. Soc. **76** [1954] 1477/80). – [57] A.D. Finogenov (Zh. Fiz. Khim. **45** [1971] 723/4; Russ. J. Phys. Chem. **45** [1971] 405/6). – [58] G.I. Novikov, A.K. Baev (Vestn. Leningr. Univ. **16** No. 22 Fiz. Khim. **1961** No. 4, pp. 116/26; C.A. **56** [1962] 9504). – [59] H. Bommer, E. Hohmann (Naturwissenschaften **27** [1939] 583). – [60] F.H. Spedding, C.F. Miller (J. Am. Chem. Soc. **74** [1952] 4195/8).

[61] R.L. Montgomery, T.D. Hubert (USBM-RI-5659 [1960] 1/10, 4; N.S.A. **15** [1961] No. 194). – [62] F.H. Spedding, J.P. Flynn (J. Am. Chem. Soc. **76** [1954] 1474/7). – [63] R.L. Montgomery (USBM-RI-5445 [1959] 1/12; C.A. **1959** 7750). – [64] M.E. Clark, J.L. Bear (J. Inorg. Nucl. Chem. **31** [1969] 2619/20). – [65] R.L. Montgomery (USBM-RI-6146 [1962] 1/9; C.A. **58** [1963] 5102).

[66] J.M. Stuve (U.S. Bur. Mines Rept. Invest. No. 6697 [1965] 1/4; C.A. **64** [1966] 4344). – [67] Yu-Lin Sung, G.I. Novikov (Zh. Neorgan. Khim. **8** [1963] 700/3; Russ. J. Inorg. Chem. **8** [1963] 356/7). – [68] R.L. Montgomery, T.D. Hubert (BM-RI-5525 [1959] 1/11; N.S.A. **14** [1960] No. 1432). – [69] J.M. Stuve (BM-RI-6640 [1964] 1/11; N.S.A. **19** [1965] No. 34135). – [70] J.M. Stuve (U.S. Bur. Mines Rept. Invest. No. 7046 [1967] 1/7; C.A. **69** [1968] No. 13520).

[71] J. Fuger, L. Morss, D. Brown (J. Chem. Soc. Dalton Trans. **1980** 1076/8). – [72] R.L. Montgomery, J.M. Stuve (U.S. Bur. Mines Rept. Invest. No. 5892 [1961] 1/10; C.A. **56** [1962] 5461). – [73] J.M. Stuve (U.S. Bur. Mines Rept. Invest. No. 6902 [1967] 1/7; C.A. **67** [1967] No. 6282).

### 11.2.6.12.5 Chemical Reactions of Rare Earth Chlorides with Organic Compounds

Preliminary Remarks

In this chapter some reactions of rare earth chlorides or their hydrates with organic compounds are presented. The reactions described here are analogous for all or most of the elements Sc, Y, La, and the lanthanide series. However, completeness is not intended. For reactions concerning only individual elements, see the respective chapters in "Rare Earth Elements" C4b.

#### 11.2.6.12.5.1 Reactions with Benzene

While Hopkins, Audrieth [1] state that $LaCl_3$ and $NdCl_3$ are insoluble in benzene, Matignon [2] found a very slight solubility of about $3\times10^{-6}$ g $LaCl_3$ and $CeCl_3$ per 100 g benzene at 20 °C.

References:

[1] B.S. Hopkins, L.F. Audrieth (Trans. Electrochem. Soc. **66** [1934] 135/42, 139). – [2] C. Matignon (Ann. Chim. Phys. [8] **8** [1906] 433/9).

#### 11.2.6.12.5.2 Reactions with Chlorohydrocarbons

$LaCl_3$ and $CeCl_3$ are insoluble in $CCl_4$, Hopkins, Audrieth [1].

Rare earth chloride hydrates $MCl_3\cdot nH_2O$ (M = La, Ce, Pr, Nd, or Sm, n = 6 or 7) dissolve in hexachloro-1,3-butadiene ($CCl_2$=CCl-CCl=$CCl_2$) with partial dehydration. Solubilities in wt% (calculated for anhydrous $MCl_3$) and residual water content n′ (mol $H_2O$ per mol $MCl_3$) are:

| $MCl_3$ | 25 °C | | 50 °C | | 75 °C | |
|---|---|---|---|---|---|---|
| | wt% $MCl_3$ | n′ | wt% $MCl_3$ | n′ | wt% $MCl_3$ | n′ |
| $LaCl_3$ | 0.040 | 4 | 0.042 | 4 | 0.057 | 2 |
| $CeCl_3$ | 0.036 | 4 | 0.043 | 4 | 0.062 | 2 |
| $PrCl_3$ | 0.039 | 4.5 | 0.041 | 4.5 | 0.062 | 2 |
| $NdCl_3$ | 0.029 | 3.5 | 0.037 | 3.5 | 0.055 | 3 |
| $SmCl_3$ | 0.024 | 3.5 | 0.035 | 3.5 | 0.054 | 3 |

With increased temperature the water content declines further; however, no anhydrous product is obtained because at about 120 °C the residual hydrate water hydrolyzes the chlorides to the oxide chlorides, Shevtsova et al. [2].

References:

[1] B.S. Hopkins, L.F. Audrieth (Trans. Electrochem. Soc. **66** [1934] 135/42, 139). – [2] Z.N. Shevtsova, B.G. Korshunov, V.V. Safonov, L.M. Kogan, V.I. Gudkova (Zh. Neorgan. Khim. **13** [1968] 3096/9; Russ. J. Inorg. Chem. **13** [1968] 1596/8).

### 11.2.6.12.5.3 Reactions with Ethers

Solubilities S (in wt%) for $MCl_3$ (M=La, Ce, Nd, Sm, Gd, Dy, Tm, and Yb) and compositions of the solid phases $MCl_3 \cdot n\,(C_2H_5)_2O$ in diethyl ether at 20 °C are the following:

| $MCl_3$ . . . . | $LaCl_3$ | $CeCl_3$ | $NdCl_3$ | $SmCl_3$ | $GdCl_3$ | $DyCl_3$ | $TmCl_3$ | $YbCl_3$ |
|---|---|---|---|---|---|---|---|---|
| S in wt% $\times 10^3$ | 45 | 7.3 | 5.8 | 7.4 | 28 | 32 | 53 | 85 |
| n . . . . . | – | 0.1 | 0.3 | 0.5 | 0.5 | 1 | 1 | 1 |

$LaCl_3$ does not form an etherate. On heating to 500 °C, solvolysis of the etherates occurs with formation of MOCl (M=Nd, Gd, Yb), Dzhuraev et al. [2].

The solubility of Y, La, and lanthanide chlorides in ethers increases with increasing atomic numbers of the rare earth elements. The following S values at 25 °C have been determined by Kirmse et al. [1] in methyl-n-amyl ether (I), ethyl-n-butyl ether (II), di-n-propyl ether (III), 1,4-dioxane (IV), 1,3-dioxolane (V), and ethyleneglycol-methyl-ethyl ether (VI):

| $MCl_3$ | S in g/100 g solution | | | | | |
|---|---|---|---|---|---|---|
| | I | II | III | IV | V | VI |
| $YCl_3$ | 3.0 | 0.05 | – | 0.1 | – | 0.4 |
| $LaCl_3$ | – | – | – | 0.02 | 0.5 | 0.02 |
| $CeCl_3$ | – | – | – | – | 1.0 | – |
| $PrCl_3$ | – | – | – | 0.04 | 0.9 | 0.04 |
| $NdCl_3$ | 0.07 | – | – | 0.1 | 0.8 | 0.04 |
| $SmCl_3$ | – | – | – | 0.07 | 2.6 | 0.45 |
| $EuCl_3$ | – | – | – | 0.07 | 3.5 | 0.6 |
| $GdCl_3$ | 0.08 | – | – | 0.1 | – | 0.55 |
| $TbCl_3$ | – | – | – | 0.3 | – | 0.6 |
| $DyCl_3$ | 0.08 | – | – | 0.4 | – | 0.6 |
| $HoCl_3$ | 2.4 | 0.01 | 0.1 | 0.55 | – | 0.65 |
| $ErCl_3$ | 33.5 | 0.5 | 0.2 | 0.7 | – | 0.9 |
| $YbCl_3$ | 43.5 | 12.0 | 13.7 | 0.8 | – | 1.45 |

In isomeric ethers, S decreases with increasing symmetry of the ether molecules. The solubility in V could only be determined for the lighter lanthanide chlorides; the chlorides of the heavier elements react with resinification [1].

In ethyleneglycol diethyl ether $C_2H_5\text{-}O\text{-}C_2H_4\text{-}O\text{-}C_2H_5$ the chlorides of the Ce group do not dissolve, the chlorides of the Y group dissolve sparingly, the solubility (S in wt%) at 25 °C generally increasing with increasing atomic number of M. The undissolved residue is in all cases an etherate $MCl_3 \cdot nE$ (E=ether); n varies between 0.7 and 2:

| $MCl_3$ . . . | $YCl_3$ | $LaCl_3$ | $PrCl_3$ | $NdCl_3$ | $SmCl_3$ | $EuCl_3$ | $GdCl_3$ |
|---|---|---|---|---|---|---|---|
| S . . . . | – | – | – | – | – | 0.25 | 0.33 |
| n . . . . . | 1.37 | 0.70 | 0.89 | 0.92 | 0.96 | 1.10 | 0.99 |

| $MCl_3$ . . . | $TbCl_3$ | $DyCl_3$ | $HoCl_3$ | $ErCl_3$ | $TmCl_3$ | $YbCl_3$ |
|---|---|---|---|---|---|---|
| S . . . . | 0.22 | 0.25 | 0.37 | 0.67 | 0.88 | 1.15 |
| n . . . . . | 1.37 | 1.7 | 1.82 | 1.86 | 2.00 | 0.94 |

Kirmse, Zwietasch [3].

Solubility data in g per L solution for $MCl_3$ (M = La, Nd, Sm, Gd) in glycol-methyl ether (GME) and glycol ethyl ether (GEE) at 20 °C from Hopkins, Audrieth [4]:

| $MCl_3$ . . . . | $LaCl_3$ | $NdCl_3$ | $SmCl_3$ | $GdCl_3$ |
|---|---|---|---|---|
| GME . . . . | 66.4 | 44.5 | 42.6 | 59.7 |
| GEE . . . . | 44.2 | 37.9 | 33.4 | 36.1 |

Addition compounds $MCl_3 \cdot nTHF$ with n = 1.5 for La and Ce, n = 2 for Pr and Nd, n = 3.5 for Y, Sm to Tm, and n = 3 for Yb and Lu, crystallize at room temperature from solutions of $MCl_3$ in tetrahydrofuran (THF), which had been obtained in an extraction apparatus. The saturated solutions contain the following amounts S in g/100 mL THF at 20 °C [5] and 22 °C [6]:

| $MCl_3$ . . . | $YCl_3$ | $LaCl_3$ | $CeCl_3$ | $PrCl_3$ | $NdCl_3$ | $SmCl_3$ | $EuCl_3$ | $GdCl_3$ |
|---|---|---|---|---|---|---|---|---|
| S at 20 °C . | 0.930 | 0.126 | 0.593 | 0.590 | 1.16 | 1.49 | – | 1.91 |
| n . . . . . | 3.59 | 1.36 | 1.46 | 2.00 | 1.99 | 1.98 | – | 2.07 |
| S at 22 °C . | 0.252 | – | – | – | 0.812 | 1.47 | 1.26 | 1.25 |
| n . . . . . | 3.5 | 1.5 | – | – | 2 | 3.5 | 3.5 | 3.5 |

| $MCl_3$ . . . | $TbCl_3$ | $DyCl_3$ | $HoCl_3$ | $ErCl_3$ | $TmCl_3$ | $YbCl_3$ | $LuCl_3$ |
|---|---|---|---|---|---|---|---|
| S at 20 °C . | – | 0.645 | 0.698 | 0.786 | – | 1.98 | – |
| n . . . . . | – | 3.45 | 3.33 | 3.47 | – | 2.92 | – |
| S at 22 °C . | 0.685 | 0.512 | 0.307 | 0.397 | 0.486 | 1.08 | 0.880 |
| n . . . . . | 3.5 | 3.5 | 3.55 | 3.5 | 3.5 | 3 | 3 |

The addition compounds formed have colors somewhat more pale than those of the respective trichlorides. The THF adducts of $SmCl_3$ and $GdCl_3$ having high vapor pressures easily pass over into adducts $MCl_3 \cdot 2THF$ with lower THF content. Treatment by ether or benzene generally converts the adducts $MCl_3 \cdot 3.5THF$ (M = Sm, Eu, Tb, Dy) into the di-solvates [5]. Thermal decomposition yields lower adducts $MCl_3 \cdot THF$ and $MCl_3 \cdot 0.5THF$, Rossmanith, Auer-Welsbach [6].

Crystalline complexes $ErCl_3 \cdot 3.5THF$ and $YbCl_3 \cdot 3THF$ are isolated by evaporation from solutions of $HgCl_2$ in THF which have been reacted with the rare earth metals. The solvates loose THF on heating. The process can be used for the preparation of anhydrous chlorides from the metals (see p. 52), Deacon, Koplick [7]. For the preparation of $YbCl_3 \cdot 3THF$ see also Rossmanith [8]; for the preparation of adducts $YCl_3 \cdot 3.2THF$ and $LaCl_3 \cdot 1.5THF$ see also Herzog et al. [9].

References:

[1] E.M. Kirmse, K.-J. Zwietasch, J. Tirschmann, L. Oelsner, U. Niedergesäss (Z. Chem. [Leipzig] **8** [1968] 472/3). – [2] Kh. Dzhuraev, U. Mirsaidov, A. Kurbanbekov, A. Rakhimova (Dokl. Akad. Nauk Tadzh. SSR **19** No. 9 [1976] 32/4; C.A. **86** [1977] No. 178197). – [3] E.M. Kirmse, K.-J. Zwietasch (Z. Chem. [Leipzig] **7** [1967] 281). – [4] B.S. Hopkins, L.F. Audrieth (Trans. Electrochem. Soc. **66** [1934] 135/42, 139). – [5] K. Rossmanith, C. Auer-Welsbach (Monatsh. Chem. **96** [1965] 602/5).

[6] K. Rossmanith (Monatsh. Chem. **100** [1969] 1484/8). – [7] G.B. Deacon, A.J. Koplick (Inorg. Nucl. Chem. Letters **15** [1979] 263/5). – [8] K. Rossmanith (Monatsh. Chem. **110** [1979] 109/14). – [9] S. Herzog, K. Gustav, E. Krüger, H. Oberender, R. Schuster (Z. Chem. [Leipzig] **3** [1963] 428/9).

### 11.2.6.12.5.4 Reactions with Alcohols

Anhydrous rare earth chlorides $MCl_3$ dissolve in alcohols easily with liberation of heat to form alcoholates $MCl_3 \cdot nROH$ (R=alkyl, aryl). The solubility of the chlorides and the amount of heat evolved generally decrease with increasing molecular weight of the alcohol. In isomeric alcohols, the solubility is higher for the primary than for the secondary or tertiary alcohols, Mehrotra et al. [1], Grigorovich [2], Merbach et al. [3], Misra et al. [4], Sankhla, Kapoor [5].

The solubilities S in mol $MCl_3$ per 1000 g alcohol (M=Y, La, Nd, Sm, Gd, Dy, Er, Yb) in $CH_3OH$, $C_2H_5OH$, and $i\text{-}C_3H_7OH$ and molar ratios alcohol/$MCl_3$ (=n) in the solid phase at 25 °C are:

| $MCl_3$ | $YCl_3$ | $LaCl_3$ | $NdCl_3$ | $SmCl_3$ | $GdCl_3$ | $DyCl_3$ | $ErCl_3$ | $YbCl_3$ |
|---|---|---|---|---|---|---|---|---|
| $S(CH_3OH)$ | 4.38 | 2.45 | 2.75 | 3.33 | 4.21 | 4.00 | 4.53 | 4.90 |
| $n(CH_3OH)$ | 4.1 | 4.0 | 4.1 | 4.1 | 4.1 | 5.1 | 4.0 | 3.8 |
| $S(C_2H_5OH)$ | 2.92 | 1.26 | 1.35 | 1.97 | 2.43 | 2.85 | 3.41 | 4.26 |
| $n(C_2H_5OH)$ | 4.1 | 3.6 | 3.3 | 3.4 | 4.1 | 4.0 | 4.1 | 3.6 |
| $S(i\text{-}C_3H_7OH)$ | 0.91 | 0.004 | 0.04 | 0.23 | 0.32 | 1.14 | 0.84 | 0.80 |
| $n(i\text{-}C_3H_7OH)$ | 3.3 | 4.6 to 5.1 | 3.6 to 4.4 | 3.7 | 3.0 to 3.1 | 3.4 to 3.9 | 3.7 | 3.5 to 3.6 |

In these determinations, for convenience, alcoholates were used which had been prepared from the hydrated chlorides by reaction with orthoformates (see p.175). Solubility results with unsolvated $MCl_3$ generally agreed satisfactorily with the data in the table except for $LaCl_3$ and $NdCl_3$ in $i\text{-}C_3H_7OH$. The somewhat higher S values found for the unsolvated chlorides were attributed to the formation of small amounts of chloroalcoholates [3].

The following solubilities of $LaCl_3$ and $NdCl_3$ in g/L solution in various alcohols at 20 °C are given by Hopkins, Audrieth [7]:

| $MCl_3$ | methanol | ethanol | 1-propanol | 2-propanol | butanol | amyl alcohol |
|---|---|---|---|---|---|---|
| $LaCl_3$ | – | 425.99 | 302.8 | 4.92 | – | – |
| $NdCl_3$ | 557.5 | 299.11 | 273.1 | 2.70 | 318 | 377.7 |

| $MCl_3$ | methyl cellosolve | ethyl cellosolve | 1,2-ethane-diol | 1,2,3-pro-panetriol |
|---|---|---|---|---|
| $LaCl_3$ | 66.4 | 44.2 | 372.64 | 162.08 |
| $NdCl_3$ | 44.5 | 37.9 | 417.97 | 162.43 |

See also Hopkins, Quill [8].

"Mischmetall" chloride containing (in wt%) 25.6 La, 45.2 Ce, 5.6 Pr, 19.9 Nd, and 3.75 Sm showed the following solubilities S in g/100 g solution at 25 °C:

| alcohol | $CH_3OH$ | $C_2H_5OH$ | $n\text{-}C_3H_7OH$ | $n\text{-}C_4H_9OH$ | $n\text{-}C_5H_{11}OH$ |
|---|---|---|---|---|---|
| S | 38.4±0.2 | 28.8±0.3 | 26.7±0.3 | 33.0±0.5 | 29.9±0.3 |

| alcohol | $n\text{-}C_6H_{13}OH$ | $n\text{-}C_8H_{17}OH$ | $n\text{-}C_9H_{19}OH$ | $n\text{-}C_{10}H_{21}OH$ |
|---|---|---|---|---|
| S | 25.3±0.4 | 15.8±0.4 | 8.5±0.2 | 5.3±0.1 |

Hartley, Wylie [6].

The solubilities of the hexahydrated chlorides $MCl_3 \cdot 6H_2O$ in 96.8% ethanol at 20 to 60 °C were measured for M = La, Ce, Pr, Nd, Sm, Eu, and Gd by Sakharova et al. [12], for M = Y, Tb, Dy, Ho, Er, Tm, Yb, and Lu by Sakharova, Ezhova [13]. Mean values in g per 100 g $C_2H_5OH$ are:

| temp. in °C | $YCl_3$ | $LaCl_3$ | $CeCl_3$ | $PrCl_3$ | $NdCl_3$ | $SmCl_3$ | $EuCl_3$ | $GdCl_3$ |
|---|---|---|---|---|---|---|---|---|
| 20.0 | 35.34 | 40.08 | 44.15 | 43.32 | 35.47 | 29.00 | 29.23 | 30.56 |
| 30.0 | 34.73 | 44.23 | 48.9 | 45.03 | 36.80 | 29.59 | 29.27 | 29.79 |
| 40.0 | 34.52 | 48.69 | 54.99 | 48.29 | 39.05 | 30.56 | 29.71 | 29.94 |
| 50.0 | 34.83 | 56.50 | 68.50 | 54.37 | 42.89 | 31.90 | 31.05 | 30.45 |
| 60.0 | 35.54 | 71.23 | 84.53 | 63.04 | 47.84 | 34.73 | 33.01 | 31.69 |

| temp. in °C | $TbCl_3$ | $DyCl_3$ | $HoCl_3$ | $ErCl_3$ | $TmCl_3$ | $YbCl_3$ | $LuCl_3$ |
|---|---|---|---|---|---|---|---|
| 20.0 | 30.63 | 33.43 | 34.79 | 37.90 | 42.11 | 47.37 | 53.26 |
| 30.0 | 30.08 | 32.38 | 33.87 | 37.53 | 41.11 | 47.71 | 54.78 |
| 40.0 | 30.27 | 32.09 | 33.97 | 37.51 | 41.83 | 49.60 | 57.00 |
| 50.0 | 30.83 | 33.43 | 34.49 | 38.70 | 44.33 | 51.95 | 61.51 |
| 60.0 | 32.80 | 34.91 | 35.93 | 40.27 | 47.29 | 58.31 | 68.28 |

The addition compounds $MCl_3 \cdot nC_2H_5OH$ crystallizing from saturated solutions in ethanol differ in the degree of solvation: n = 1 for M = Y, n = 3 and 2 for M = Sc, La, n = 3 for M = Ce and "mischmetall", and n = 3 and 1 for M = Nd [6, reviewing literature]. Adducts $LaCl_3 \cdot 2C_2H_5OH$ and $DiCl_3 \cdot 3C_2H_5OH$ (Di = Pr + Nd mixture with low Pr content) are reported to deposit from ethanolic solutions saturated with HCl. The alcoholate of $PrCl_3$ could not be identified, that of $CeCl_3$ is very instable. The presence of small amounts of water suppresses the formation of alcoholates in favor of that of hydrates, Meyer, Koss [9]. The formation of trialcoholates has been established for various alcohols and $MCl_3$ [1, 4, 5]. For details see "Rare Earth Elements" D 3, 1981, pp. 15/36.

When $PrCl_3$ and $NdCl_3$ were refluxed with sec-butyl alcohol, products with an appreciably lower M:Cl ratio of about 1:2.2 were obtained, indicating the occurrence of some side reactions. Refluxing of $PrCl_3$ with tert-butyl alcohol produced an insoluble solid with an M:Cl ratio as low as about 1:0.7, Merbach et al. [1].

Upon treating $SmCl_3 \cdot 6H_2O$ with excess 2-propanol in the presence of benzene and distilling off the benzene – 2-propanol – water azeotrope, the mixed hydrate – 2-propanolate $SmCl_3 \cdot H_2O \cdot 2(2\text{-}C_3H_7OH)$ forms. Analogous compounds are obtained when the monohydrate $SmCl_3 \cdot H_2O$ is allowed to react with $CH_3OH$, $C_2H_5OH$, and $2\text{-}C_3H_7OH$ [5]. In the same manner Mehrotra [1] earlier prepared hydrated dialcoholates $MCl_3 \cdot H_2O \cdot 2ROH$ with M = Pr, Nd and R = methyl, ethyl, and 2-propyl. The corresponding 1-butanol derivatives were won by alcohol interchange techniques. Azeotropic drying with a 2-propanol/benzene mixture converted the di-2-propanolates into the water-free trialcoholates [1].

Alcohol interchange reactions are generally used to obtain solvates of higher alcohols from more easily prepared lower alcoholates. Several examples are described by Mehrotra et al. [1], Merbach et al. [3], El-Aggan et al. [14].

Alcoholates are also produced by reaction of hydrated chlorides with orthoformates:

$$MCl_3 \cdot xH_2O + xHC(OR)_3 \rightarrow MCl_3 \cdot yROH + (2x-y)\ ROH + xHCOOR$$
$$(R = CH_3 \text{ or } C_2H_5)$$

The following adducts were obtained in this manner: $MCl_3 \cdot 4\,CH_3OH$ and $MCl_3 \cdot 3\,C_2H_5OH$ with M = La, Nd, Sm, Gd, and Yb, $MCl_3 \cdot 4\,C_2H_5OH$ with M = Y, Dy, Er. Adducts $MCl_3 \cdot 3\,i\text{-}C_3H_7OH$ were prepared by successive action of methyl orthoformate and of 2-propanol [3]. Tetramethanolated chlorides are also prepared by the reaction of the hydrated chlorides with 2,2-dimethoxypropane:

$$6(CH_3)_2C(OCH_3)_2 + MCl_3 \cdot 6\,H_2O \rightarrow MCl_3 \cdot 4\,CH_3OH + 8\,CH_3OH + 6\,(CH_3)_2CO$$
$$(M = La,\ Pr,\ Nd,\ Gd)$$

Vacuum drying converts the products into the trimethanolated salts, Quill, Clink [15].

References:

[1] R.C. Mehrotra, T.N. Misra, S.N. Misra (J. Indian Chem. Soc. **42** [1965] 351/8, 351/2). – [2] Z.I. Grigorovich (Zh. Neorgan. Khim. **8** [1963] 986/9; Russ. J. Inorg. Chem. **8** [1963] 507/9). – [3] A. Merbach, M.-N. Pitteloud, P. Jaccard (Helv. Chim. Acta **55** [1972] 44/52, 47). – [4] S.N. Misra, T.N. Misra, R.C. Mehrotra (J. Inorg. Nucl. Chem. **27** [1965] 105/13). – [5] B.S. Sankhla, R.N. Kapoor (Australian J. Chem. **20** [1967] 383/6).

[6] F.R. Hartley, A.W. Wylie (J. Chem. Soc. **1962** 679/81). – [7] B.S. Hopkins, L.F. Audrieth (Trans. Electrochem. Soc. **66** [1934] 135/42, 139). – [8] B.S. Hopkins, L.L. Quill (Proc. Natl. Acad. Sci. U.S. **19** [1933] 64/8). – [9] R.J. Meyer, M. Koss (Ber. Deut. Chem. Ges. **35** [1902] 2622/6). – [10] T. Moeller, D.S. Smith (AFOSR-TN-58-559 [1958]; AD-158377 [1958] 1/112, 25; N.S.A. **14** [1960] No. 8389).

[11] Z.A. Sheka, E.E. Kriss (Zh. Neorgan. Khim. **4** [1959] 1809/13; Russ. J. Inorg. Chem. **4** [1959] 816/8). – [12] N.N. Sakharova, Yu.G. Sakharova, T.A. Ezhova, A.A. Izmailova (Zh. Neorgan. Khim. **20** [1975] 1479/83; Russ. J. Inorg. Chem. **20** [1975] 830/2). – [13] Yu.G. Sakharova, T.A. Ezhova (Zh. Neorgan. Khim. **21** [1976] 551/4; Russ. J. Inorg. Chem. **21** [1976] 296/8). – [14] A.M. El-Aggan, D.C. Bradley, W. Wardlaw (J. Chem. Soc. **1958** 2092/9). – [15] L.L. Quill, G.L. Clink (Inorg. Chem. **6** [1967] 1433/5).

### 11.2.6.12.5.5 Reactions with Ketones

According to Hopkins, Audrieth [1], $LaCl_3$ and $NdCl_3$ are scarcely soluble in acetone. Barnebey [2] reports near insolubility in acetone for $YCl_3$, $LaCl_3$, $CeCl_3$, and $NdCl_3$. The presence of a number of common chlorides, e.g., $CuCl_2$, $BiCl_3$, $SnCl_2$, $FeCl_3$, greatly enhanced the solubility due to the formation of double chlorides.

Solubilities S in g/100 g solvent in various ketones at 20 °C and in benzophenone at 50 °C for $LaCl_3$ and $CeCl_3$ as given by Sheka, Kriss [3] are:

| $MCl_3$ | acetone | acetophenone | acetylacetone | cyclohexanone | benzophenone |
|---|---|---|---|---|---|
| $LaCl_3$ | $1.2 \times 10^{-3}$ | $2.2 \times 10^{-2}$ | $4.0 \times 10^{-1}$ | $1.8 \times 10^{-2}$ | $2.6 \times 10^{-4}$ |
| $CeCl_3$ | $4.3 \times 10^{-3}$ | $1.0 \times 10^{-1}$ | $1.0 \times 10^{-1}$ | $3.8 \times 10^{-1}$ | – |

For mixed ketones the same authors [3] found: 0.52 and 1.5 g $LaCl_3$ in 100 g acetophenone + benzoin (2:1) and acetylacetone + benzoin (2:1), respectively, 0.33 and 1.1 g $CeCl_3$ in 100 g acetophenone + benzoin (2:1) and acetylacetophenone + cyclohexanone (1:1, mole ratios), respectively.

References:

[1] B.S. Hopkins, L.F. Audrieth (Trans. Electrochem. Soc. **66** [1934] 135/42, 139). – [2] O.L. Barnebey (J. Am. Chem. Soc. **34** [1912] 1174/89, 1175). – [3] Z.A. Sheka, E.E. Kriss (Zh. Neorgan. Khim. **4** [1959] 1809/13; Russ. J. Inorg. Chem. **4** [1959] 816/8).

### 11.2.6.12.5.6 Reactions with Esters

Upon shaking anhydrous $MCl_3$ with an excess of diethyl esters of dicarboxylic acids, addition compounds $MCl_3 \cdot E$ and $2\,MCl_3 \cdot E$ (E=ester) form. In this way the compounds $2\,LaCl_3 \cdot C_2H_5OOC\text{-}COOC_2H_5$, $2\,LaCl_3 \cdot C_2H_5OOC\text{-}CH_2\text{-}COOC_2H_5$, and $LaCl_3 \cdot C_2H_5OOC\text{-}CH_2\text{-}CH_2\text{-}COOC_2H_5$ (white) and $NdCl_3 \cdot C_2H_5OOC\text{-}CH_2\text{-}COOC_2H_5$ and $NdCl_3 \cdot C_2H_5OOC\text{-}CH_2\text{-}CH_2\text{-}COOC_2H_5$ (lilac) have been obtained, Osipov, Romova [1]. By analogous reactions additional complex compounds with diethylesters of oxalic, malonic, and succinic acids have been made:

| $MCl_3$ | $MCl_3$:mol ligand | ligand | color |
|---|---|---|---|
| $LaCl_3$ | 1:1 | $C_2H_5OOC\text{-}[CH_2]_4\text{-}COOC_2H_5$ | white |
| $PrCl_3$ | 2:1 | $C_2H_5OOC\text{-}CH_2\text{-}COOC_2H_5$ | pale green |
| $PrCl_3$ | 1:1 | $C_2H_5OOC\text{-}CH_2\text{-}CH_2\text{-}COOC_2H_5$ | pale green |
| $NdCl_3$ | 1:1 | $C_2H_5OOC\text{-}[CH_2]_4\text{-}COOC_2H_5$ | lilac |
| $SmCl_3$ | 2:1 | $C_2H_5OOC\text{-}COOC_2H_5$ | yellowish |
| $SmCl_3$ | 2:1 | $C_2H_5OOC\text{-}CH_2\text{-}COOC_2H_5$ | yellowish |
| $SmCl_3$ | 1:1 | $C_2H_5OOC\text{-}CH_2\text{-}CH_2\text{-}COOC_2H_5$ | yellowish |
| $GdCl_3$ | 2:1 | $C_2H_5OOC\text{-}COOC_2H_5$ | white |
| $GdCl_3$ | 1:2 | $C_2H_5OOC\text{-}CH_2\text{-}COOC_2H_5$ | white |
| $GdCl_3$ | 1:1 | $C_2H_5OOC\text{-}CH_2\text{-}CH_2\text{-}COOC_2H_5$ | white |
| $DyCl_3$ | 2:1 | $C_2H_5OOC\text{-}COOC_2H_5$ | white |
| $DyCl_3$ | 2:1 | $C_2H_5OOC\text{-}CH_2\text{-}COOC_2H_5$ | white |
| $DyCl_3$ | 1:1 | $C_2H_5OOC\text{-}CH_2\text{-}CH_2\text{-}COOC_2H_5$ | white |

For discussions of the O–C bonds and the structures of the complexes based on spectroscopic data, see the source paper, Romova, Osipov [2].

The simultaneous action of ethylacetoacetate (EAA) and phenanthroline (phen) on $MCl_3$ produces mixed ligand complexes $MCl_3$:EAA:phen=1:3:1 (M=Nd, Sm, Eu, Tb, Dy, Ho, Er). The compounds with M=Sm, Eu, Tb, and Dy luminesce, Bel'tyukova et al. [3].

In solutions of $MCl_3$ (M=La, Pr, Nd, Eu) in triethyl or tributyl phosphate evidence for complexes $MCl_3 \cdot 3\,R_3PO_4$ and possibly $MCl_3 \cdot 6\,R_3PO_4$ has been found by proton NMR measurements, Galaktionova et al. [4].

References:

[1] O.A. Osipov, M.G. Romova (Zh. Neorgan. Khim. **9** [1964] 2126/8; Russ. J. Inorg. Chem. **9** [1964] 1149/50). – [2] M.G. Romova, O.A. Osipov, L.K. Isaeva (Zh. Neorgan. Khim. **11** [1966] 536/9; Russ. J. Inorg. Chem. **11** [1966] 291/3). – [3] S.V. Bel'tyukova, N.S. Poluektov, L.I. Kononenko, V.N. Drobyazko (Zh. Neorgan. Khim. **22** [1977] 1756/60; Russ. J. Inorg. Chem. **22** [1977] 953/5). – [4] O.V. Galaktionova, E.N. Lebedeva, V.V. Yastrebov, S.S. Korovin (Zh. Neorgan. Khim. **25** [1980] 2660/5; Russ. J. Inorg. Chem. **25** [1980] 1467/70).

### 11.2.6.12.5.7 Reactions with Amines or Their Salts

For additional data see "Seltenerdelemente" C 5, 1977, p. 212 and "Rare Earth Elements" D 1, 1980.

Monoalkylamines

The rare earth chlorides combine readily with $CH_3NH_2$ between room temperature and 300 °C at 700 Torr to form complexes $MCl_3 \cdot nCH_3NH_2$ with M = La, Ce, Pr, Nd, Sm, and Gd and $0.75 \leqq n \leqq 5$:

| $MCl_3$ | n | $MCl_3$ | n |
|---|---|---|---|
| $LaCl_3$ | 5, 4, 3, 2, 1.25, 1 | $NdCl_3$ | 5, 4, 3, 2.25, 2, 1.5, 1 |
| $CeCl_3$ | 5, 4, 3, 2, 1.5, 0.75 | $SmCl_3$ | 5, 4, 3, 2.25, 2, 1.5 |
| $PrCl_3$ | 5, 4, 3.33, 3, 2.33, 2, 1.5 | $GdCl_3$ | 5, 4.5, 4, 3, 2, 1.25 |

The decomposition temperature of these complexes increases from below 50 °C for n = 5 to above 200 °C for $n \leqq 1.5$ at the given pressure. At 305 Torr the same compounds were formed, Popov, Wendlandt [1]; see also Moeller et al. [2]. Analogous experiments with $YCl_3$ established the existence of complexes $YCl_3 \cdot nCH_3NH_2$ with n = 4, 3, 2, and 1, Wendlandt [6].

Thermogravimetric studies of the gas-solid reactions of $i\text{-}C_3H_7NH_2$ with $MCl_3$ (M = Y, La, Pr, Nd, Sm, Er) have been made by Nesbitt [3].

Dialkylamines

With dimethylammonium chloride in 96% alcohol solutions, $CeCl_3$ forms the compound $CeCl_3 \cdot n[(CH_3)_2NH \cdot HCl] \cdot 2H_2O$ with n = 2, 4, 6, and 7. With $SmCl_3$ and $SmCl_3$:amine ratios ranging from 1:5 to 1:8, the double salts $SmCl_3 \cdot n(CH_3)_2NH \cdot HCl \cdot 2H_2O$ with n = 1, 2, 4, 6, and 7 are obtained. With $SmCl_3$:amine = 1:8, the compound with n = 7 forms, Sakharova, Avdeev [4].

Trialkylamines

With trimethylammoniumchloride in 96% alcohol solutions, the double salts $MCl_3 \cdot 4[(CH_3)_3N \cdot HCl]$ form with M = La, Ce, Pr, and Nd, Sakharova, Zemskov [5]. With $SmCl_3$, the compounds $SmCl_3 \cdot n[(CH_3)_3N \cdot HCl]$ with n = 1, 2, 3, and 4 form; at a ratio $SmCl_3$:amine = 1:5 to 1:8, the value of n is always four [4].

Diamines

Coordination compounds with ethylenediamine, 1,2-propanediamine, and benzidine are surveyed by [2, pp. 183/4]. On pouring a saturated anhydrous alcoholic solution of $MCl_3$ into a large volume of ethylenediamine (= en) in anhydrous $CCl_4$ with constant agitation, the amount of ethylenediamine being at least 10times that of $MCl_3$ (molar ratio), complexes $M(en)_4Cl_3$ are obtained. With smaller amounts of ethylenediamine complexes of ill-defined composition form, containing 5 to 10% tightly bound alcohol, depending on the working conditions, Castellani-Bisi [7]. The same $M(en)_4Cl_3$ complexes form by reaction of ethylenediamine with $MCl_3$ in $CH_3CN$ under strictly anhydrous conditions, Forsberg, Moeller [8, 9]. Complexes $M(pn)_4Cl_3$ with M = La, Nd form on dropwise addition of a solution of 1,2-propanediamine (= pn) in completely anhydrous $CH_3CN$ to an also anhydrous solution of $MCl_3$ in the same solvent, Charpentier, Moeller [10, pp. 3577/8].

Complexes $MCl_3 \cdot$(benzidine)$\cdot$2(THF) are obtained by heating combined solutions of benzidine (4,4′-diaminobiphenyl) in THF (tetrahydrofuran) and $MCl_3$ (M=Y, La, Ce, Pr, Nd, Sm) in alcohol for a short time on a water bath, Rahman et al. [11].

### Hexamethylenetetramine

Urotropine ($C_6H_{12}N_4$) reacts with aqueous $MCl_3$ solutions to form adducts $M(C_6H_{12}N_4)_2Cl_3 \cdot 14H_2O$ (M=La, Ce, Nd) or $M(C_6H_{12}N_4)_2Cl_3 \cdot 11H_2O$ (M=Y, Er) [2, citing older work].

### Ethanolamines

In the aqueous system of $PrCl_3$ with the chloride of monoethanolamine, $PrCl_3$ forms a double salt $PrCl_3(HOCH_2\text{-}CH_2\text{-}NH_2 \cdot HCl)_2$. The chlorides of di- or triethanolamine do not form double salts with $PrCl_3$, Nasyrov et al. [12].

### Aromatic Monoamines or Monohydrazines

In aniline and phenylhydrazine, $LaCl_3$ or $NdCl_3$ dissolve only slightly, Hopkins, Audrieth [13]. Mixing solutions of $MCl_3 \cdot nH_2O$ in ethanol with solutions of excess phenylhydrazine in diethyl ether results in complexes $M(C_6H_5NHNH_2)_2(H_2O)Cl_3$ (M=Y, La, Ce, Pr, Nd, Sm), Rahman et al. [14]. Spectroscopic studies on these compounds are briefly discussed by [2]. See "Rare Earth Elements" D1, 1980, pp. 34ff. for formation of various complexes.

### Heterocyclic Amines

In pyridine, $LaCl_3$ and $NdCl_3$ are only scarcely soluble [13]. At 15 °C 100 g pyridine dissolves about 6 g $YCl_3$, Matignon [15], and about 1.8 g $NdCl_3$, Matignon [16]. Complexes $M(bpy)_2Cl_3 \cdot nH_2O$ with 2,2′-bipyridine (=bpy) and M=La through Ho are obtained by treating the hydrated $MCl_3$ with the ligand in an ethanol-diethyl ether medium. The chlorides of Er through Lu form complexes $M(bpy)Cl_3 \cdot nH_2O$. Comparable compounds (M=Sm, Eu, n=2; M=Pr, Gd, n=4) precipitate from absolute ethanol solutions after warming to 60 °C. Compounds $M(bpy)_2Cl_3 \cdot H_2O$ (M=La through Nd) and $M(bpy)_2Cl_3$ (M=Eu through Lu) have been isolated from absolute ethanol at room temperature. Complexes $M(bpy)_2Cl_3 \cdot C_2H_5OH$ (M=Sm, Tb) and $M(bpy)Cl_3 \cdot H_2O$ (M=La, Ce) were also prepared [2, citing earlier publications].

Reaction products of hydrated $MCl_3$ with 1,10-phenanthroline (=phen) are also reviewed by [2]. In diethyl ether/95% ethanol mixtures all $MCl_3$ hydrates react with 1,10-phenanthroline to form $MCl_3(phen)_2 \cdot nH_2O$. For products obtained from absolute alcohol, in some cases the value of n is smaller. The following compounds have been obtained:

| | | | | | | | |
|---|---|---|---|---|---|---|---|
| $MCl_3$ . . . | $LaCl_3$ | $CeCl_3$ | $PrCl_3$ | $NdCl_3$ | $SmCl_3$ | $EuCl_3$ | $TbCl_3$ |
| n . . . . . | 1 | 2.5, 6 | 2 | 6.5 | 4 | 2.5 | 4.5 |
| color . . . | white | yellow | pale green | pale lilac | white | white | white |

| | | | | | | |
|---|---|---|---|---|---|---|
| $MCl_3$ . . . | $DyCl_3$ | $HoCl_3$ | $ErCl_3$ | $TmCl_3$ | $YbCl_3$ | $LuCl_3$ |
| n . . . . . | 4.5 | 5 | 6 | 5 | 6 | 6 |
| color . . . | white | white | pink | white | white | white |

Lobanov, Smirnova [17]. For mixed ligand complexes containing 1,10-phenanthroline plus dibenzoylmethane or ethylacetoacetate see p. 177.

In appropriate solvents, 1,2-bis(pyridine-α-aldimino)ethane (=pale) and hydrated $MCl_3$ yield complexes $M(pale)Cl_3 \cdot nH_2O$ with n=1 for M=La through Gd and n=0 for M=Y, Tb through Lu. Similarly, 2,2′,2″-terpyridyl ($=C_{15}H_{11}N_3$) in ethanol gives $M(C_{15}H_{11}N_3)Cl_3 \cdot nH_2O$ with M=Ce through Gd, excepting Pm, and n=1 through 2.5. $EuCl_3$ also forms $Eu(C_{15}H_{11}N_3)_2Cl_3$. The reaction of 2,4,6-tri(α-pyridyl)-1,3,5-triazine (TPT) with $MCl_3 \cdot nH_2O$ results in compounds $M(TPT)Cl_3 \cdot 3H_2O$ with M=Ce and Eu, and $M(TPT)Cl_3 \cdot 2H_2O$ with M=Y, Ho, Lu. There are further complexes $M(TPT)Cl_3 \cdot nH_2O$ reported with:

| | | | | | | | | |
|---|---|---|---|---|---|---|---|---|
| M . . . . | Nd | Sm | Eu | Gd | Dy | Ho | Tb | Yb |
| n . . . . . | 4 | 5 | 5 | 4 | 4 | 3 | 4 | 3 |

[2, discussing literature data].

References:

[1] A.I. Popov, W.W. Wendlandt (J. Am. Chem. Soc. **77** [1955] 857/9). – [2] T. Moeller, R.L. Dieck, J.E. McDonald (Rev. Chim. Minerale **10** [1973] 177/98, 178/9). – [3] L.E. Nesbitt (Diss. Texas Technological College 1964, pp. 1/124 from Diss. Abstr. **26** [1966] 4200). – [4] N.N. Sakharova, V.P. Avdeev (Zh. Obshch. Khim. **35** [1965] 1171/3; J. Gen. Chem. [USSR] **35** [1965] 1175/7). – [5] N.N. Sakharova, S.V. Zemskov (Dokl. Akad. Nauk SSSR **120** [1958] 539/40).

[6] W.W. Wendlandt (Science **122** [1955] 197/8). – [7] C. Castellani-Bisi (Inorg. Chim. Acta **3** [1969] 660/2). – [8] J.H. Forsberg, T. Moeller (Inorg. Chem. **8** [1969] 883/8). – [9] J.H. Forsberg, T. Moeller (J. Am. Chem. Soc. **90** [1968] 1932). – [10] L.J. Charpentier, T. Moeller (J. Inorg. Nucl. Chem. **32** [1970] 3575/84).

[11] S.M.F. Rahman, J. Ahmad, M.M. Haq (J. Inorg. Nucl. Chem. **34** [1972] 1460/2). – [12] Ya.N. Nasyrov, E.F. Zhuravlev, N.I. Rakhmangulova (Izv. Vysshikh Uchebn. Zavedenii Khim. Khim. Tekhnol. **16** [1973] 1338/42 from C.A. **80** [1974] No. 7611). – [13] B.S. Hopkins, L.F. Audrieth (Trans. Electrochem. Soc. **66** [1934] 135/42, 139). – [14] S.M.F. Rahman, J. Ahmad, M.M. Haq (J. Inorg. Nucl. Chem. **33** [1971] 4351/4). – [15] C. Matignon (Ann. Chim. Phys. [8] **8** [1906] 433/9).

[16] C. Matignon (Ann. Chim. Phys. [8] **8** [1906] 270/2). – [17] N.I. Lobanov, V.A. Smirnova (Zh. Neorgan. Khim. **10** [1965] 1593/7; Russ. J. Inorg. Chem. **10** [1965] 868/71).

### 11.2.6.12.5.8 Reactions with Amides

The trichlorides $LaCl_3$ and $NdCl_3$ are rather soluble in acetamide, Hopkins, Audrieth [1]. With acetamide, $YCl_3$ in aqueous solution forms the hydrated complex $YCl_3 \cdot 4CH_3CONH_2 \cdot 5H_2O$ dissolving congruently in water and forming eutonics with both $YCl_3 \cdot 6H_2O$ and $CH_3CONH_2$, Ashimkulova et al. [2]. $ErCl_3$ forms a similar complex $ErCl_3 \cdot 5CH_3CONH_2 \cdot H_2O$, Aitimbekov et al. [3].

Crystalline solvates with dimethyl formamide could not be obtained, Moeller, Smith [4, p. 83], Smith [5].

Urea forms with $YCl_3$ in aqueous systems three double salts: $YCl_3 \cdot 6CO(NH_2)_2$, $YCl_3 \cdot 4CO(NH_2)_2$, and $2YCl_3 \cdot 3CO(NH_2)_2 \cdot 4H_2O$. From these, the 1:4 and the 2:3:4 compounds dissolve congruently while the 1:6 compound decomposes in water, Ashimkulova et al. [6]. Corresponding compounds with Pr, Nd, and Sm are $MCl_3 \cdot nCO(NH_2)_2$ with n=4 and 6 for all three rare earths. All compounds dissolve congruently in water, Dilebaeva, Sulaimankulov [7].

If $MCl_3 \cdot 6H_2O$ is ground together with thiourea in a molar ratio of $MCl_3 \cdot 6H_2O : CS(NH_2)_2 = 4:5$ for two hours, then dissolved in a little 94 to 96% ethanol at 65 to 70 °C, and the solution evaporated to half its initial volume, complexes $4MCl_3 \cdot 5CS(NH_2)_2 \cdot 20H_2O$ (M=Y, Tb, Dy, Ho, Er, Tm, Yb, or Lu) crystallize, Sakharova et al. [8]. With M=La through Gd, complexes $4MCl_3 \cdot 5CS(NH_2)_2 \cdot nH_2O$ with varying n are obtained from aqueous or ethanol solutions:

| | | | | | | | |
|---|---|---|---|---|---|---|---|
| M . . . . | La | Ce | Pr | Nd | Sm | Eu | Gd |
| n . . . . . | 24 | 24 | 20 | 20 | 22 | 22 | 22 |

The products are stable in air, but unstable in water, Sakharova et al. [9].

Upon dissolving anhydrous $MCl_3$ in hot absolute ethanol and adding slowly hexamethylphosphortriamide (HMPA) (mole ratio $MCl_3$ : HMPA = 1:5) complexes $MCl_3(HMPA)_3$ form which crystallize on cooling with an ice bath and addition of absolute diethyl ether. The compounds have been prepared with all rare earth chlorides, M=Sc, Y, and La through Lu including Pm, Donoghue, Peters [10]. The solubility S of unsolvated anhydrous $MCl_3$ in HMPA is at about 25 °C:

| | | | | |
|---|---|---|---|---|
| $MCl_3$ . . . . . . . | $LaCl_3$ | $CeCl_3$ | $PrCl_3$ | $NdCl_3$ |
| S in mmol/L . . . | 106±2 | 109±3 | 113±3 | 119±5 |

The chlorides of Sm, Tm, and Yb dissolve in HMPA exothermically and the residual solids stick together; the chlorides of La, Ce, Pr, and Nd dissolve well only on external heating and the remaining solid is freely mobile. In all cases the undissolved residue transforms to the solvate $MCl_3(HMPA)_3$, with M=La, Ce, Pr, and Nd more difficultly than with the heavier rare earths. The solubility S of these solvates in HMPA is:

| | | | | | | | |
|---|---|---|---|---|---|---|---|
| $MCl_3$ . . . . . . . | $LaCl_3$ | $CeCl_3$ | $PrCl_3$ | $NdCl_3$ | $SmCl_3$ | $EuCl_3$ | $GdCl_3$ |
| S in mmol/L . . . | 107±2 | 107±3 | 128±1 | 125±1 | 126±1 | 121±1 | 125±2 |
| $MCl_3$ . . . . . . . | $TbCl_3$ | $DyCl_3$ | $HoCl_3$ | $ErCl_3$ | $TmCl_3$ | $YbCl_3$ | $LuCl_3$ |
| S in mmol/L . . . | 128±7 | 109±2 | 108±1 | 98±2 | 93.5±1 | 85±10 | 73±1 |

Mikheev et al. [11].

References:

[1] B.S. Hopkins, L.F. Audrieth (Trans. Electrochem. Soc. **66** [1934] 135/42, 139). – [2] G.A. Ashimkulova, K. Nogoev, K. Sulaimankulov (Zh. Neorgan. Khim. **19** [1974] 2588/90; Russ. J. Inorg. Chem. **19** [1974] 1414/5). – [3] K. Aitimbekov, K. Sulaimankulov, A.G. Batyur, M. Ismailov (Zh. Neorgan. Khim. **19** [1974] 3153/5; Russ. J. Inorg. Chem. **19** [1974] 1724/6). – [4] T. Moeller, D.S. Smith (AFOSR-TN-58-559 [1958]; AD-158377 [1958] 1/112; N.S.A. **14** [1960] No. 8389). – [5] D.S. Smith (Diss. Univ. Illinois 1959, pp. 1/123 from Diss. Abstr. **19** [1959] 2456/7).

[6] G.A. Ashimkulova, K. Sulaimankulov, S.T. Turdaliev, K. Nogoev (Zh. Neorgan. Khim. **18** [1973] 2011/3; Russ. J. Inorg. Chem. **18** [1973] 1069/70). – [7] A. Dilebaeva, K. Sulaimankulov (Zh. Neorgan. Khim. **18** [1973] 2257/61; Russ. J. Inorg. Chem. **18** [1973] 1194/6). – [8] Yu.G. Sakharova, V.N. Perov, V.I. Loginov (Zh. Neorgan. Khim. **23** [1978] 632/6; Russ. J. Inorg. Chem. **23** [1978] 350/4). – [9] Yu.G. Sakharova, T.A. Ezhova, V.I. Loginov, A.G. Krylova (Zh. Neorgan. Khim. **21** [1976] 1727/32; Russ. J. Inorg. Chem. **21** [1976] 946/9). – [10] J.T. Donoghue, D.A. Peters (J. Inorg. Nucl. Chem. **31** [1969] 467/70).

[11] N.B. Mikheev, A.N. Kamenskaya, N.A. Konovalova, T.A. Zhilina (Zh. Neorgan. Khim. **22** [1977] 1761/6; Russ. J. Inorg. Chem. **22** [1977] 955/8).

### 11.2.6.12.5.9 Reactions with Nitriles

Solvates with acetonitrile crystallize from solutions of $MCl_3$ in excess $CH_3CN$. The color of these complexes is a paler version of that of the parent rare earth chlorides, Carter, O'Hara [1]. Heating $MCl_3$ with a large excess of $CH_3CN$ at 80 °C for 3 or 4 days in a sealed tube results in the following adducts:

| | | | | | | | |
|---|---|---|---|---|---|---|---|
| $MCl_3$ . . . . . . . | $ScCl_3$ | $YCl_3$ | $LaCl_3$ | $PrCl_3$ | $NdCl_3$ | $SmCl_3$ | $EuCl_3$ |
| $MCl_3:CH_3CN$ . . | 1:3 | 1:3 | 1:1 | 1:2 | 1:2 | 1:2 | 1:2 |

The 1:1 compounds of $PrCl_3$, $NdCl_3$, and $SmCl_3$ form by thermal decomposition of the 1:2 compounds at 177, 185, and 170 to 180 °C, respectively, MacCordick, Brun [2], see also Moeller et al. [3].

Heating o-phthalonitrile with $ErCl_3 \cdot 5H_2O$ in a molar ratio 4:1 for 1 hour at 270 to 280 °C, gives a dark green, microcrystalline phthalocyanine $C_{32}H_{16}N_8 \cdot ErCl \cdot 2H_2O$, Plyushchev, Shklover [4]. See also "Rare Earth Elements" D1, 1980, p. 97.

References:

[1] F.L. Carter, M. O'Hara (Proc. 7th Rare Earth Res. Conf., Coronado, Calif., 1968 [1969], Vol. 1, p. 147). — [2] J. MacCordick, C. Brun (Chemiker-Ztg. **94** [1970] 848/9). — [3] T. Moeller, R.L. Dieck, J.E. McDonald (Rev. Chim. Minerale **10** [1973] 177/98). — [4] V.E. Plyushchev, L.P. Shklover (Zh. Neorgan. Khim. **9** [1964] 335/9; Russ. J. Inorg. Chem. **9** [1964] 183/6).

### 11.2.6.12.5.10 Reactions with Other Organic Compounds

With antipyrine (phenazone, 2,3-dimethyl-1-phenyl-5-pyrazolone) some rare earth chlorides $MCl_3$ (M=Y, La, Ce, Pr, Nd, Er) form complexes $[M(ap)_6]Cl_3$. With pyramidone (amidopyrine, 4-dimethylamino antipyrine), complexes $[M(pd)_3]Cl_3$ form with the same rare earth ions; the pyramidone behaves as a bidentate ligand, Ryabchikov, Terent'eva [1].

In an aqueous medium $M_2O_3$ and an equimolar amount of iminodiacetic acid react with $MCl_3$ to give complexes $[M(Z)]Cl \cdot nH_2O$ (Z=chelating agent). In the case of the heavier lanthanides the hydroxides are treated with hydrochloric acid and iminodiacetic acid. The solutions are evaporated to a syrup and then treated with ethanol. The following compounds were prepared and spectroscopically studied:

| | | | | | | | | |
|---|---|---|---|---|---|---|---|---|
| M . . . . | Sc | Y | La | Pr | Nd | Sm | Eu | Gd |
| n . . . . . | 0, 2 | 2, 3 | 0, 2 | 0, 2 | 0, 2 | 0, 2 | 0, 2 | 0, 2, 3 |
| M . . . . | Tb | Dy | Ho | Er | Tm | Yb | Lu | |
| n . . . . . | 0, 3 | 0, 3 | 0, 3 | 0, 3 | 0, 3 | 0, 3 | 0, 3 | |

Prutkova et al. [2].

Enthalpies of solution $\Delta H_S^\circ$ (in kcal/mol) in dimethylsulfoxide (DMSO) at room temperature for some rare earth chlorides and final concentrations of about $3 \times 10^{-4}$ M are:

| | | | | | | | | |
|---|---|---|---|---|---|---|---|---|
| $MCl_3$ . . . | $LaCl_3$ | $NdCl_3$ | $SmCl_3$ | $GdCl_3$ | $DyCl_3$ | $HoCl_3$ | $ErCl_3$ | $YbCl_3$ |
| $-\Delta H_S^\circ$ . . . | 39.0 | 43.7 | 49.2 | 49.8 | 51.8 | 52.3 | 52.8 | 54.0 |

The data show DMSO to be a much better solvent medium for the rare earth ions than water, Clark, Bear [3].

Cyclopentadienyl sodium in THF (tetrahydrofuran) reacts with $MCl_3$ (M = Sc, Y, La, Ce, Pr, Nd, Sm, and Gd) to form tris-pentadienyl compounds $M(C_5H_5)_3$, Wilkinson, Birmingham [4], see also Wetzel [5]. Compounds $M(C_5H_5)_3$ with M = La, Pr, and Nd are obtained under pure Ar or in vacuo with an anhydrous solution of 25 to 30 wt% Na cyclopentadienyl in THF at 60 °C within 7 to 8 hours. The products are sublimed at 240 to 250 °C and 0.133 $N/m^2$, Krasnova et al. [6]. By the same method $M(C_5H_5)_3$ with M = Ce, Sm, and Gd are also obtained, Borisov et al. [7].

References:

[1] D.I. Ryabchikov, E.A. Terent'eva (Izv. Akad. Nauk SSSR Otd. Khim. Nauk **1949** 44/55 from C.A. **1949** 4969). – [2] N.M. Prutkova, L.I. Martynenko, A.I. Grigor'ev, N.D. Mitrofanova (Zh. Neorgan. Khim. **11** [1966] 1277/81; Russ. J. Inorg. Chem. **11** [1966] 682/5). – [3] M.E. Clark, J.L. Bear (J. Inorg. Nucl. Chem. **31** [1969] 2019/20). – [4] G. Wilkinson, J.M. Birmingham (J. Am. Chem. Soc. **76** [1954] 6210). – [5] K. Wetzel (in: G. Brauer, Handbuch der Präparativen Anorganischen Chemie, 2nd Ed., Vol. 2, Stuttgart 1962, pp. 982/1013, 1013).

[6] S.G. Krasnova, G.K. Borisov, G.G. Devyatykh (Zh. Neorgan. Khim. **16** [1971] 1733/5; Russ. J. Inorg. Chem. **16** [1971] 918/9). – [7] G.K. Borisov, S.G. Krasnova, G.G. Devyatykh (Zh. Neorgan. Khim. **18** [1973] 663/6; Russ. J. Inorg. Chem. **18** [1973] 346/8).

### 11.2.7 Comparative Data on $MCl_3 \cdot nH_2O$

#### 11.2.7.1 General Remarks

Numerous $MCl_3 \cdot nH_2O$ compounds with M = Sc, Y, La to Lu, n = 0.5 to 9, are mentioned in the literature. The hydrates $MCl_3 \cdot 7H_2O$, M = La to Pr, Haeseler, Matthes [1], and $MCl_3 \cdot 6H_2O$, M = Sc, Y, La, Nd to Lu, [1], Arkhangel'skii et al. [2], Powell, Burkholder [3], crystallize from aqueous solutions. Their thermal decomposition yields $MCl_3 \cdot nH_2O$ with n = 3.5, 3, 2.5, 2, or 1 [1], Su, Li [4], Ashcroft, Mortimer [7]. Other hydrates mentioned include $ScCl_3 \cdot 9H_2O$, Fischer, Buhler [5], $HoCl_3 \cdot 9H_2O$, Morrow, Palm [10], $ScCl_3 \cdot 5H_2O$ [2], $YCl_3 \cdot 5.5$ and $5H_2O$, Markovskii et al. [6], $MCl_3 \cdot 4H_2O$, M = Sm, Tb, Tm, Yb, Lu [7], and $EuCl_3 \cdot 0.5H_2O$, Wendlandt [8]. Sometimes the compounds contain some occluded water [7]. For a discussion of the water content see Lembert [9].

References:

[1] G. Haeseler, F. Matthes (J. Less-Common Metals **9** [1965] 133/51). – [2] I.V. Arkhangel'skii, L.N. Komissarova, V.M. Shatskii, N.P. Shepelev (Zh. Neorgan. Khim. **17** [1972] 310/4; Russ. J. Inorg. Chem. **17** [1972] 163/5). – [3] J.E. Powell, H.R. Burkholder (J. Inorg. Nucl. Chem. **14** [1960] 65/70). – [4] M.-Z. Su, G.-P. Li (Hua Hsueh Tung Pao **1979** 322/6; C.A. **91** [1979] No. 146463). – [5] W. Fischer, G. Buhler (Z. Anorg. Allgem. Chem. **285** [1956] 156/9).

[6] L.Ya. Markovskii, E.Ya. Pesina, Yu.A. Omel'chenko, Yu.D. Kondrashev (Zh. Neorgan. Khim. **14** [1969] 14/8; Russ. J. Inorg. Chem. **14** [1969] 7/9). – [7] S.J. Ashcroft, C.T. Mortimer (J. Less-Common Metals **14** [1968] 403/6). – [8] W.W. Wendlandt (J. Inorg. Nucl. Chem. **9** [1959] 136/9). – [9] M.E. Lembert (J. Phys. Chem. **104** [1923] 101/46, 126). – [10] R.J. Morrow, A. Palm (J. Inorg. Nucl. Chem. **24** [1962] 239/44).

### 11.2.7.2 Formation and Preparation

Various hydrates of the rare earth trichlorides were early prepared by Matignon [1] (who also reported data from older publications), namely $YCl_3 \cdot 6H_2O$ and $YCl_3 \cdot H_2O$ [1, pp. 435/9], $PrCl_3 \cdot 7H_2O$ and $PrCl_3 \cdot H_2O$ [1, pp. 364/401], $NdCl_3 \cdot 6H_2O$ and $NdCl_3 \cdot H_2O$ [1, pp. 243/83], $SmCl_3 \cdot 6H_2O$, $SmCl_3 \cdot 4H_2O$, and $SmCl_3 \cdot H_2O$ [1, pp. 402/16], $YbCl_3 \cdot 6H_2O$ and $YbCl_3 \cdot H_2O$ [1, pp. 440/3].

The hydrated $MCl_3$ are prepared from 99.9% pure $M_2O_3$ or $Ce(OH)_3$ by solution in a slight excess of hot 0.1 N HCl (M=La to Lu), Ashcroft, Mortimer [3]; in diluted (1:3) HCl (M=La to Gd, all are designated as hexahydrates), Sakharova et al. [16], (M=Y, Tb to Lu), Sakharova, Ezhova [7]; in a 10% excess of 1:1 HCl (M=Y, La, Pr to Lu), Vinarov, Kornelli [8]; in excess 2N HCl (M=Y, La to Lu, not Pm and Tm), Hinchey, Cobble [18]; in 5 N HCl (M=Sc, Y, La to Lu), Wendlandt [9], Ivanov [10]; in concentrated HCl, Merbach et al. [2], Graeber et al. [12], Morrow, Palm [14]. Excess $H_2O$ and acid are removed on a water bath [2]; the crystals are washed with small amounts of conductivity water [18] or alcohol and dried at room temperature [10]. Or after several evaporations the solutions are allowed to crystallize over 60% $H_2SO_4$ in a desiccator [3]. Crystals may be dried in a desiccator over $CaCl_2$, $P_2O_5$, and alkali [16, 17]; or stored in an evacuated desiccator for several days. In terms of cation concentration, deviations from stoichiometric ratios vary between −3.2 per cent for ytterbium and −0.43 per cent for dysprosium [14]. $MCl_3 \cdot 6H_2O$ (M=Gd, Tb, Ho, Lu) were crystallized over a period of weeks by evaporation of the saturated solutions, which were maintained at $t \leqq 45$ °C and $pH \cong 1$. The concentration of the water of hydration in the crystals was maintained at the stoichiometric six per mol of rare earth by means of the following procedure: Crystals were brought into equilibrium with $CaCl_2$ solutions of concentrations chosen such that the equilibrium partial pressure of water above the crystals was consistent with the hexahydrates being the only rare earth chloride species present at room temperature. It was found experimentally that the appropriate value for the partial pressure of $H_2O(g)$ above the solutions was $\cong 4 \pm 3$ Torr. During the equilibration process the crystals were periodically crushed to allow occluded solution to escape, Spedding et al. [5], see also Rulf [4].

Hexahydrated chlorides of Nd, Er, and Eu were obtained by Davidenko, Lugina [6] by slow evaporation of saturated solutions of the chlorides in air at normal temperatures. However, Repin [7] had previously established that the hexahydrates thus formed contain less Cl than that corresponding to the formula $MCl_3 \cdot 6H_2O$, due to the formation of oxide chlorides. Precipitation of the hepta- or hexahydrates with M=La, Ce, Pr, Nd, Sm, Eu, Gd, Tb, Dy, Ho, Er, Tm, Yb, and Lu from the hydrochloric acid solution with HCl gas followed by drying over KOH was employed by Haeseler, Matthes [11]. Preparation of $MCl_3 \cdot 6H_2O$ (M=Y, Sm, Eu, Gd, Tb, Dy, Ho, Er, Tm, Yb, Lu) can also be accomplished by reaction of the corresponding oxalates with concentrated HCl at 25 °C according to $M(C_2O_4)_3 \cdot H_2O + 6HCl = 2MCl_3 \cdot 6H_2O + 3H_2C_2O_4$, Möbius, Matthes [13].

Partially deuterated chloride hydrates of Nd, Sm, and Ho are obtained by a one hour treatment of the trichloride with dry HCl and Ar at 300 °C, followed by standing for 1 h at this temperature. The substances are then slowly cooled under a flow of gas, and some $D_2O$ is added to the crystalline material, Morrow, Palm [14]. $EuCl_3 \cdot 6D_2O$ and $TbCl_3 \cdot 6D_2O$ are obtained when the oxide is dissolved in DCl and the solution evaporated at room temperature. In order to avoid a contamination of the crystals by $H_2O$, the evaporation is undertaken in a desiccator with anhydrous $CaCl_2$, Freeman, Crosby [15].

For the formation of $MCl_3 \cdot n\,H_2O$ with $n<6$, see the discussion of the thermal decomposition of hydrated chlorides with higher water content.

References:

[1] C. Matignon (Ann. Chim. Phys. [8] **8** [1906] 243/88, 364/443). – [2] A. Merbach, M.-N. Pitteloud, P. Jaccard (Helv. Chim. Acta **55** [1972] 44/52). – [3] S.J. Ashcroft, C.T. Mortimer (J. Less-Common Metals **14** [1968] 403/6). – [4] D.C. Rulf (IS-T-378 [1970] 1/124, 24). – [5] F.H. Spedding, D.C. Rulf, B.C. Gerstein (J. Chem. Phys. **56** [1972] 1498/506, 1499).

[6] N.K. Davidenko, L.N. Lugina (Zh. Neorgan. Khim. **13** [1968] 980/7; Russ. J. Inorg. Chem. **13** [1968] 512/6). – [7] S.A. Repin (Redkozem. Elem. Akad. Nauk SSSR Inst. Geokhim. Analit. Khim. **1963** 71/4; C.A. **61** [1964] 3892). – [8] I.V. Vinarov, M.E. Kornelli (Khim. Prom. Nauk. Tekhn. Zb. **1962** No. 1, pp. 28/30; C.A. **59** [1963] 1257). – [9] W.W. Wendlandt (J. Inorg. Nucl. Chem. **5** [1958] 118/22, **9** [1959] 136/9). – [10] V.I. Ivanov (Kristallografiya **13** [1968] 905/7; Soviet Phys.-Cryst. **13** [1968] 786/8).

[11] G. Haeseler, F. Matthes (J. Less-Common Metals **9** [1965] 133/51). – [12] E.J. Graeber, G.H. Conrad, S.F. Duliere (Acta Cryst. **21** [1966] 1012/3). – [13] R. Möbius, F. Matthes (Z. Chem. [Leipzig] **4** [1964] 234/5). – [14] R.J. Morrow, A. Palm (J. Inorg. Nucl. Chem. **24** [1962] 239/44). – [15] J.J. Freeman, G.A. Crosby (J. Mol. Spectrosc. **13** [1964] 399/406, 400).

[16] N.N. Sakharova, Yu.G. Sakharova, T.A. Ezhova, A.A. Izmanilova (Zh. Neorgan. Khim. **20** [1975] 1479/83; Russ. J. Inorg. Chem. **20** [1975] 830/2). – [17] Yu.G. Sakharova, T.A. Ezhova (Zh. Neorgan. Khim. **21** [1976] 551/4; Russ. J. Inorg. Chem. **21** [1976] 296/8). – [18] R.J. Hinchey, J.W. Cobble (Inorg. Chem. **9** [1970] 917/21).

## 11.2.7.3 Crystallographic Properties

### 11.2.7.3.1 Isotypism

#### $MCl_3 \cdot 7\,H_2O$

The heptahydrates of $LaCl_3$, $CeCl_3$, and $PrCl_3$ crystallize triclinically and are isotypic, Burns, Peterson [1]. A comparison of the X-ray data with the goniometric measurements assembled from the older literature by Groth [2] for triclinic "$LaCl_3 \cdot 6\,H_2O$" and "$CeCl_3 \cdot 6\,H_2O$" single crystals shows that these compounds should be treated as the heptahydrates. This is also true for the "$LaCl_3 \cdot n\,H_2O$" and "$PrCl_3 \cdot n\,H_2O$" samples which were recognized as isotypic by Iveronova et al. [3] from X-rays and by Hellwege et al. [4] from comparison of the absorption spectrum of $La_{0.90}Pr_{0.10}Cl_3 \cdot n\,H_2O$ with that of $PrCl_3 \cdot n\,H_2O$.

It is questionable whether $CeCl_3 \cdot 7\,H_2O$ also occurs with orthorhombic symmetry, as is reported by Kojima et al. [5] from X-ray powder patterns and Dennis, Magee [6] from goniometric measurements. See "Rare Earth Elements" C4b, Section 11.2.11.4.4.1, for relevant data.

#### $MCl_3 \cdot 6\,H_2O$

The hexahydrates of the trichlorides of Y and Nd to Lu crystallize monoclinically and are isotypic, from crystallographic studies on the compounds with Nd to Tm and Lu ($PmCl_3 \cdot 6\,H_2O$ excluded), see for example Pabst [7], Iveronova et al. [3], Graeber et al. [8], and Habenschuss, Spedding [13 to 16]. Isotypism of $YCl_3 \cdot 6\,H_2O$, $HoCl_3 \cdot 6\,H_2O$, and $LuCl_3 \cdot 6\,H_2O$ is indicated by comparison of the absorption spectra of $HoCl_3 \cdot 6\,H_2O$ and the mixed crystal $M_{0.90}Ho_{0.10}Cl_3 \cdot 6\,H_2O$ for M = Y by Grohmann et al. [9] and for M = Lu by D. Kuse, from Pfeffer [10].

$YbCl_3 \cdot 6H_2O$ is assigned to this isotypic series on the basis of its observed crystal habit, Dieke, Crosswhite [12]. Criteria for the structure of $ScCl_3 \cdot 6H_2O$ yield only the Raman and IR spectra, see p. 196. For $GdCl_3 \cdot 6H_2O$ tetragonal symmetry has been reported by Benedicks [11], presumably on the basis of its crystal habit.

For the most of these hexahydrates instead of the monoclinic symmetry a hexagonal (pseudo)symmetry is derived from the absorption spectra in the visible and from the number of infrared lattice vibrations, see pp. 192/3.

References:

[1] J.H. Burns, J.R. Peterson (Inorg. Chem. **10** [1971] 147/51). – [2] P. Groth (Chemische Krystallographie, Pt. 1, Leipzig 1906, pp. 249/52). – [3] V.I. Iveronova, V.P. Tarasova, M.M. Umanskii (Izv. Akad. Nauk SSSR Ser. Fiz. **15** [1951] 164/9; Structure Reports, Vol. 15, 1951, p. 248). – [4] K.H. Hellwege, U. Johnsen, W. Pfeffer (Z. Physik **154** [1959] 301/9, 304). – [5] T. Kojima, T. Inoue, T. Ishiyama (J. Electrochem. Soc. Japan **19** [1951] 285/9; C.A. **1952** 1328).

[6] L.M. Dennis, W.H. Magee (Z. Anorg. Allgem. Chem. **7** [1894] 250/64, 259/61; J. Am. Chem. Soc. **16** [1894] 649/64, 659/61). – [7] A. Pabst (Am. J. Sci. [5] **22** [1931] 426/30). – [8] E.J. Graeber, G.H. Conrad, S.F. Duliere (Acta Cryst. **21** [1966] 1012/3). – [9] I. Grohmann, K.H. Hellwege, H.G. Kahle (Z. Physik **164** [1961] 243/56). – [10] W. Pfeffer (Z. Physik **168** [1962] 305/15, 306).

[11] C. Benedicks (Z. Anorg. Allgem. Chem. **22** [1900] 393/421, 403). – [12] G.H. Dieke, H.M. Crosswhite (J. Opt. Soc. Am. **46** [1956] 885/9). – [13] A. Habenschuss, F.H. Spedding (Cryst. Struct. Commun. **9** [1980] 71/6). – [14] A. Habenschuss, F.H. Spedding (Cryst. Struct. Commun. **9** [1980] 157/60). – [15] A. Habenschuss, F.H. Spedding (Cryst. Struct. Commun. **9** [1980] 207/11).

[16] A. Habenschuss, F.H. Spedding (Cryst. Struct. Commun. **9** [1980] 213/8).

## 11.2.7.3.2 Crystal Form, Cleavage

In the older literature compiled by Groth [1], the triclinic pinacoidal "$LaCl_3 \cdot 6H_2O$" and "$CeCl_3 \cdot 6H_2O$" crystals, which must be regarded as heptahydrate crystals (see 11.2.7.3.1, p. 185), are described as tabular on (001) (axis assignment $a>b>c$). In the La-containing crystals the following faces occur in sequence of decreasing significance: (001), (111), $(1\bar{1}1)$, (110), $(1\bar{1}0)$, $(11\bar{1})$, $(1\bar{1}\bar{1})$, (100), $(20\bar{1})$, (201), and (010). For the Ce-containing crystals (201) is not mentioned; (100) follows $(20\bar{1})$ [1]. For $PrCl_3 \cdot 7H_2O$ crystals with the same habit the faces (110), $(1\bar{1}0)$, (100), (010), (001), and (111) are observed, Söderström [2]. The (twinned) crystals, which form on evaporation on a water bath of a solution of La or Pr oxide or Ce carbonate in 5 N HCl, are described by Ivanov [3] as needle shaped. Goniometric measurements of the La- and Ce-containing crystals can be found in Groth [1].

The compounds $MCl_3 \cdot 6H_2O$ with M = Sm, Eu, Gd, Tb, Dy, Ho, Er, and Tm crystallize tabularly along (010), as shown by Pabst [4] for M = Sm, Gd, and Er (see figure in paper) and confirmed by Graeber et al. [5] for M = Sm to Tm. The crystals are elongated along the c axis [4], Bel'skii [6]. The following faces are observed on $SmCl_3 \cdot 6H_2O$ crystals in sequence of diminishing significance: (010), (110), (011), $(\bar{1}01)$, (101), and (100). The two last-named faces occur more rarely with $GdCl_3 \cdot 6H_2O$ and $ErCl_3 \cdot 6H_2O$. In addition the Sm-containing crystals exhibit (111) faces; see [4] for goniometric measurements. Occasionally the crystal form is denoted as pillow or tablet shaped, see for example Kahle [7] and Lagendijk, Huiskamp [8].

The monoclinic hexahydrates show perfect cleavage on (100), as demonstrated by Pabst [4] for the Sm and Gd compounds. The triclinic heptahydrates in contrast show no clear cleavage according to the data of Groth [1] for the supposed $CeCl_3 \cdot 6H_2O$ see Section 11.2.7.3.1, p. 185.

References:

[1] P. Groth (Chemische Krystallographie, Pt. 1, Leipzig 1906, pp. 249/52). – [2] C. Söderström (Z. Krist. **36** [1902] 194). – [3] V.I. Ivanov (Kristallografiya **13** [1968] 905/7; Soviet Phys.-Cryst. **13** [1968] 786/8). – [4] A. Pabst (Am. J. Sci. [5] **22** [1931] 426/30). – [5] E.J. Graeber, G.H. Conrad, S.F. Duliere (Acta Cryst. **21** [1961] 1012/3).

[6] N.K. Bel'skii (Opt. Spektrosk. **15** [1963] 245/8; Opt. Spectrosc. [USSR] **15** [1963] 131/3). – [7] H.G. Kahle (Z. Physik **145** [1956] 347/60, 348/9). – [8] E. Lagendijk, W.H. Huiskamp (Physica **65** [1973] 118/55, 123).

### 11.2.7.3.3 Twinning

$LaCl_3 \cdot 7H_2O$ according to Brouty, Herpin [1] and $CeCl_3 \cdot 7H_2O$ according to Hellwege et al. [2] crystallize in a twinned form from aqueous solutions. Monoclinic symmetry was thus simulated in certain X-ray patterns. According to the Weissenberg and precession patterns of [1] the pseudomonoclinic b axis is the twin axis. The triclinic individual crystals are inclined with respect to each other by around 6°. However, it cannot be decided by X-rays whether a purely penetration twinning is involved or whether the plane of the so-called orthorhombic section is the intergrowth plane [1].

References:

[1] C. Brouty, P. Herpin (Compt. Rend. C **272** [1971] 2049/52). – [2] K.H. Hellwege, R. v. Klot, G. Weber (Physik Kondensierten Materie **2** [1964] 397/407, 398).

### 11.2.7.3.4 Crystal Structure

#### 11.2.7.3.4.1 $MCl_3 \cdot 7H_2O$ with M = La, Ce, Pr

These compounds crystallize triclinically (pseudomonoclinically), Iveronova et al. [5], Burns, Peterson [1]. The space group has been determined for $LaCl_3 \cdot 7H_2O$ from Weissenberg and precession patterns as $P1-C_1^1$ (No. 1) with Z=2, by Brouty, Herpin [2] and Bakakin et al. [3]. Also, for $CeCl_3 \cdot 7H_2O$ comparison of the structural refinement with and without symmetry centers shows that the space group is acentric (R=0.0323; X-ray diffractometer data, counter tube technique, correction for absorption, H atoms incorporated in the refinement), Peterson et al. [8]. In contrast, Habenschuss, Spedding hold on the basis of their less precise structural refinement of $LaCl_3 \cdot 7H_2O$ to R=0.066 [9] and of $PrCl_3 \cdot 7H_2O$ to R=0.059 [10] (both X-ray diffractometer data, counter tube technique, without correction for absorption, location of the H atoms not determined) to the results of Iveronova et al. [5], that these two compounds crystallize in the space group $P\bar{1}-C_i^1$ (No. 2). The fact that the crystals of all three compounds show distinct piezoelectricity speaks against the existence of a symmetry center [4]. The absence of a symmetry center in $LaCl_3 \cdot 7H_2O$ has also been established by a statistical method [2]. The space group $Cm-C_s^3$ (No. 8) or $C2-C_2^3$ (No. 5) determined by Ivanov [4] for the three heptahydrates does not take into account the twinning.

In the structure each $M^{3+}$ ion is surrounded by $7H_2O$ molecules and 2 quasi-equivalent Cl ions, Bakakin et al. [3]. The coordination polyhedron is a simple quadratic antiprism capped by one $H_2O$ molecule [8 to 10]. Two such polyhedra are linked by a common edge formed by the Cl ions into a dimeric complex, whose structure can be written as $[(H_2O)_7MCl_2M(H_2O)_7]^{4+}$. These complexes are connected with each other via the remaining $^2/_3$ of the Cl ions [3, 8 to 10]; see the projection of the structure of $LaCl_3 \cdot 7\,H_2O$ along [010] in **Fig. 26** and on the (104) plane of the pseudomonoclinic A-cell in **Fig. 27** [3]. The coordination polyhedron in $LaCl_3 \cdot 7\,H_2O$ was in the first structure determination of Bakakin et al. [3] (R=0.107) described as a base-centered strongly distorted tetragonal antiprism. The normalized edge lengths and angles derived from this result by Drew [11], however, show that a capped quadratic antiprism is present.

In the acentric structure of $CeCl_3 \cdot 7\,H_2O$ (see the paper for atomic parameters and also for the H atoms determined from the Fourier synthesis), the following atomic distances occur within the two capped antiprisms of the dimeric complex: Ce↔Cl 2.895, 2.954 Å and 2.916, 2.949 Å; Ce↔O 2.482 to 2.601 Å and 2.464 to 2.574 Å, Cl↔Cl 3.509 Å. The observed edge lengths and angles of the capped antiprisms and their normalized values are listed in the paper [8]. The normalized values agree with those of Robertson [12], which are derived by means of Coulomb repulsion between the ligands for this type of polyhedron. In the centrosymmetric refined structure the following interatomic

Fig. 26

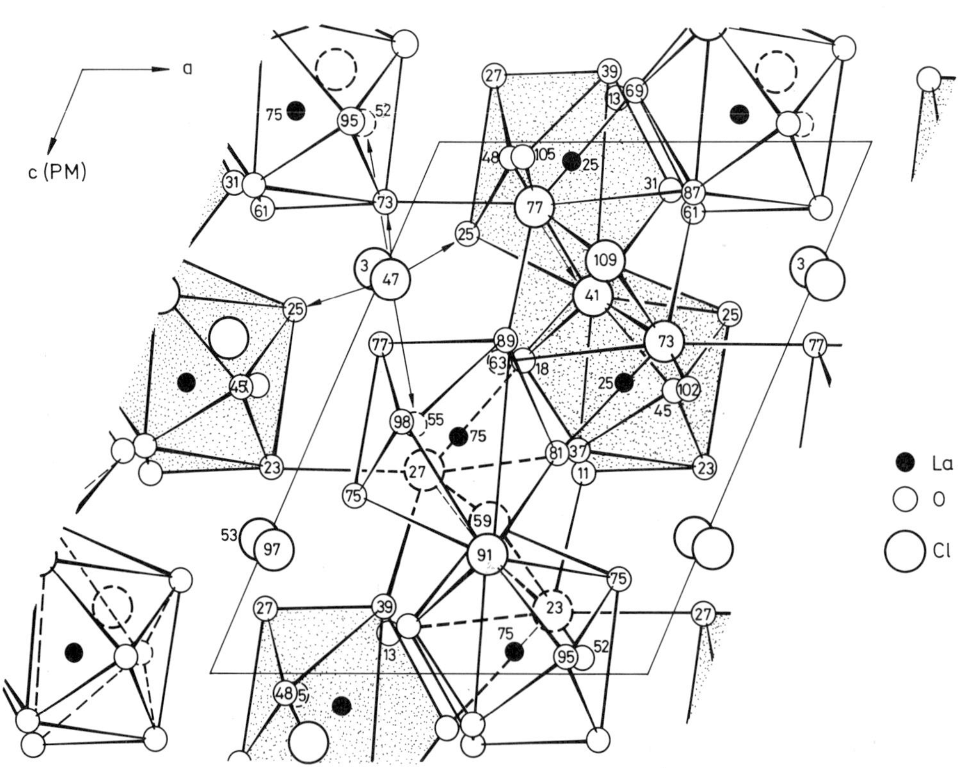


Structure of $LaCl_3 \cdot 7\,H_2O$ projected along [010] of the pseudomonoclinic A-cell (c(PM) = 12.40 Å, see p. 190); the numbers designate the y-parameters (×100) in this cell.

Fig. 27

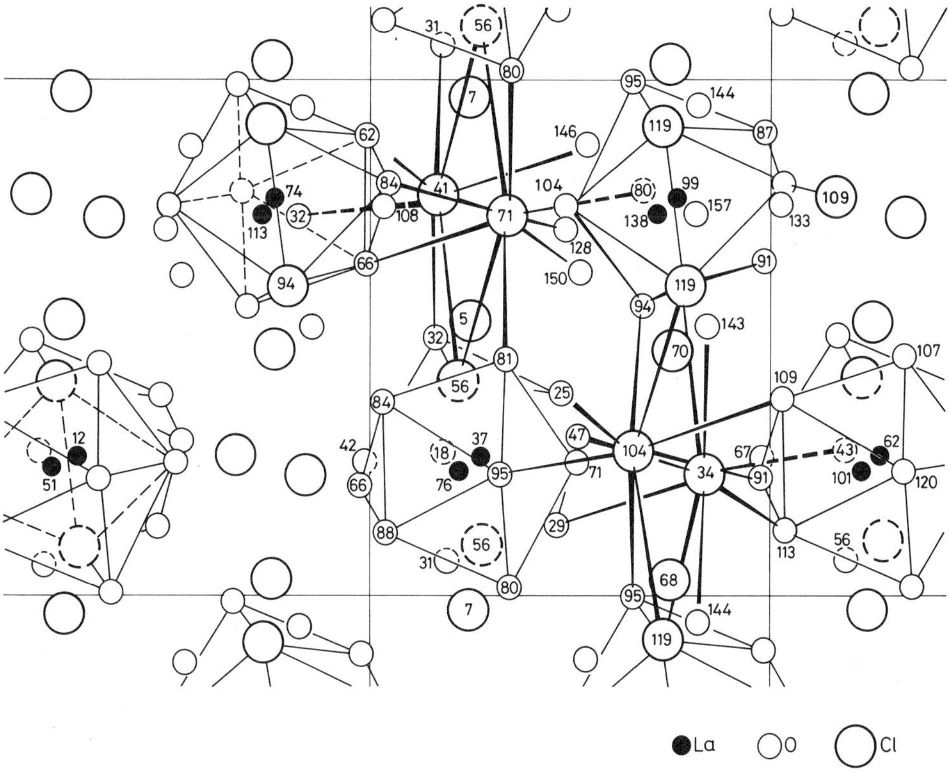


Section of the structure of $LaCl_3 \cdot 7H_2O$ projected on the (104) plane of the pseudomonoclinic A-cell; the relative heights of the atoms are given in hundredth of the diagonal (a+c(PM)).

distances in the inner complex have been determined from the atomic parameters given in the papers: In $LaCl_3 \cdot 7H_2O$ La↔Cl 2.924 and 2.958 Å, La↔O 2.529 to 2.587 Å, and Cl↔Cl 3.509 Å [9]; in $PrCl_3 \cdot 7H_2O$ Pr↔Cl 2.885 and 2.933 Å, Pr↔O 2.487 to 2.547 Å, and Cl↔Cl 3.470 Å [10]. For further atomic distances, including those for the Cl ions which link the dimers, and bond angles, see the papers [9, 10].

In connection with the lattice spectroscopic studies the question arises whether the Cl ions in the dimeric complex in fact do belong to the coordination spheres of the lanthanide ions, see p. 194. See Section 11.2.9.3, p. 263, for the structural relationship of these heptahydrates to the hexahydrates.

## Lattice Constants

Values at room temperature from single crystal patterns of the triclinic reduced Niggli (N) and Delauny (D) cells with Z=2 and the pseudomonoclinic A-cell (PM) with Z=4 are:

| cell | a in Å | b in Å | c in Å | α | β | γ | Ref. |
|---|---|---|---|---|---|---|---|
| $LaCl_3 \cdot 7H_2O$ | | | | | | | |
| N | 8.282 | 9.237 | 8.058 | 107.81° | 98.63° | 71.52° | [9] |
| N | 8.1 | 9.1 | 8.0 | 107°30′ | 99°00′ | 72°00′ | [6] |
| D | 8.18 | 10.44 | 8.03 | 122.1° | 104.1° | 91.8° | [1] |
| D | 9.30 | 10.70 | 8.04 | 129.5° | 107° | 91° | [3] |
| PM | 9.1 | 10.6 | 12.3 | 89° | 113°10′ | 91°20′ | [5] |
| PM | 9.30 | 10.70 | 12.40 | 88° | 114° | 91° | [3] |
| PM | 9.32 | 10.69 | 12.17 | 86°45′ | 113°35′ | 91°10′ | [2] |
| $CeCl_3 \cdot 7H_2O$ | | | | | | | |
| N | 8.276 | 9.212 | 8.014 | 107.57° | 98.39° | 71.51° | [8] |
| D | 8.20 | 10.43 | 7.98 | 122.1° | 104.0° | 91.9° | [1] |
| $PrCl_3 \cdot 7H_2O$ | | | | | | | |
| N | 8.232 | 9.164 | 7.984 | 107.45° | 98.32° | 71.44° | [10] |
| N | 8.2 | 9.0 | 8.0 | 107°00′ | 98°40′ | 72°00′ | [6] |
| D | 8.22 | 10.29 | 7.90 | 121.3° | 105.1° | 91.9° | [1] |
| PM | 9.0 | 10.5 | 12.4 | 88°40′ | 113° | 91°10′ | [5] |

Burns, Peterson [1] use for the Delauny cell the axis assignment a>b>c, likewise Brouty, Herpin [2] and Iveronova et al. [5] for the pseudomonoclinic cell. The constants of Donnay, Onduk [6] are calculated using data from Iveronova et al. [5] for the unreduced cell. In the older literature tabulated by Groth [7] axis ratios and angles are reported for $LaCl_3 \cdot 6H_2O$ and $CeCl_3 \cdot 6H_2O$ which agree with the pseudomonoclinic cell constants for the heptahydrates from [5].

References:

[1] J.H. Burns, J.R. Peterson (Inorg. Chem. **10** [1971] 147/51, 151). – [2] C. Brouty, P. Herpin (Compt. Rend. C **272** [1971] 2049/52). – [3] V.V. Bakakin, R.F. Klevstova, L.P. Solov'eva (Zh. Strukt. Khim. **15** [1974] 820/30; J. Struct. Chem. [USSR] **15** [1974] 723/32). – [4] V.I. Ivanov (Kristallografiya **13** [1968] 905/7; Soviet Phys.-Cryst. **13** [1968] 786/8). – [5] V.I. Iveronova, V.P. Tarasova, M.M. Umanskii (Izv. Akad. Nauk SSSR Ser. Fiz. **15** [1951] 164/9; Structure Reports, Vol. 15, 1951, p. 248).

[6] J.H.D. Donnay, H.M. Ondik (Crystal Data, 3rd Ed., Vol. 2, Washington, D.C., 1973, pp. A 26/A 28). – [7] P. Groth (Chemische Krystallographie, Pt. 1, Leipzig 1906, pp. 249/52). – [8] E.J. Peterson, E.I. Onstott, R.B. v. Dreele (Acta Cryst. B **35** [1979] 805/9). – [9] A. Habenschuss, F.H. Spedding (Cryst. Struct. Commun. **8** [1979] 511/6). – [10] A. Habenschuss, F.H. Spedding (Cryst. Struct. Commun. **7** [1978] 535/41).

[11] M.G.B. Drew (Coord. Chem. Rev. **24** [1977] 179/275, 239/41). – [12] B.E. Robertson (Inorg. Chem. **16** [1977] 2735/42).

### 11.2.7.3.4.2 $MCl_3 \cdot 6H_2O$ with M=Y, Nd to Lu

The compounds $MCl_3 \cdot 6H_2O$ with M=Y [1], Nd [2], Pm (if it exists) [3], Sm [2], Eu to Tm [4], Yb (see Section 11.2.7.3.1, p. 185) and Lu [18], crystallize monoclinically in their own characteristic $GdCl_3 \cdot 6H_2O$ structure type. The space group P2/n(P2/c) – $C_{2h}^4$ (No. 13) has been established by structural studies on single crystals with M = Nd [15], Sm [16], Eu [5], Gd [6, 17], Dy, Ho [7], and Lu [18]. No piezoelectric effect occurs, Bel'skii, Struchkov [5]. The unit cell contains two formula units. The atomic positions are listed in the papers [5, 7, 15 to 18]. Specific data for $GdCl_3 \cdot 6H_2O$ from Marezio et al. [6] are:

2 M in (e): $\pm(^1/_4, y, ^1/_4)$ with y = 0.1521
2 Cl(1) in (f): $\pm(^3/_4, y, ^1/_4)$ with y = 0.3769
4 Cl(2) in (g): $\pm(x, y, z; ^1/_2-x, y, ^1/_2-z)$ with x = 0.0587, y = 0.8370, z = 0.2601

4 O(1) in (g) with x = 0.2813, y = 0.0471, z = 0.5432
4 O(2) in (g) with x = 0.1432, y = 0.4254, z = 0.0888
4 O(3) in (g) with x = 0.4406, y = 0.2988, z = 0.1058

For provisional sites for the H atoms, see the paper [6].

The structure is portrayed in projection along [010] in **Fig. 28** and along [100] in **Fig. 29**, p. 192, from Bakakin et al. [8]. The atom heights presented (in hundredth of the y or x parameters) are calculated from the data of [5] for $EuCl_3 \cdot 6H_2O$. The structure is built up from complex ions $[M(H_2O)_6Cl_2]^+$ and "isolated" Cl ions. The coordination polyhedron around the $M^{3+}$ ion (coordination number = 8) is a distorted dodecahedron with triangular faces [5, 6, 7] or a severely deformed tetragonal [8] or quadratic [15 to 18] antiprism, whose sole symmetry element is one twofold axis [6]. The isolated Cl(1) ion (coordination number 6) is surrounded by a distorted octahedron of $H_2O$ molecules, see Fig. 28.

Fig. 28

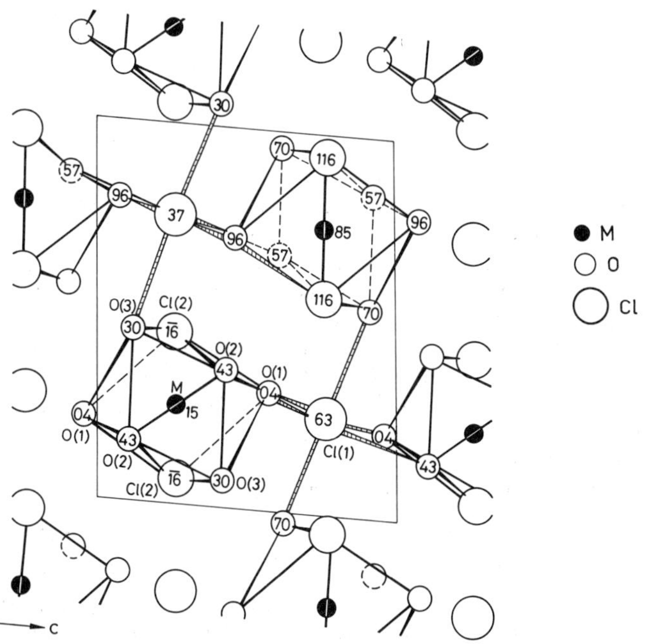


Structure of $MCl_3 \cdot 6H_2O$ projected along [010]. Numbers correspond to the y-parameters (×100) in $EuCl_3 \cdot 6H_2O$.

Fig. 29

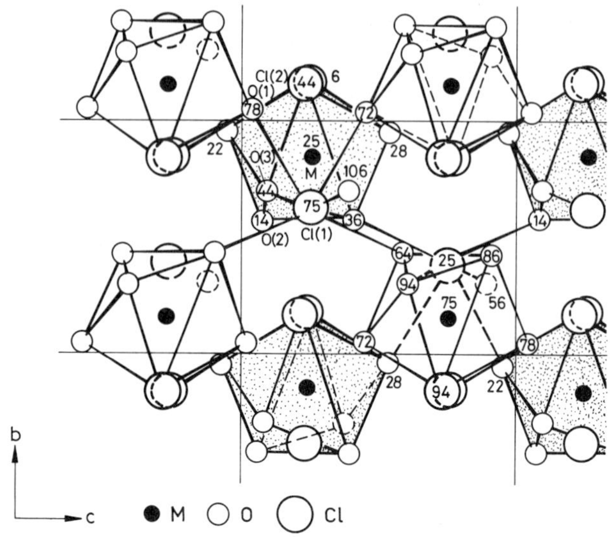


Structure of $MCl_3 \cdot 6H_2O$ projected along [100]. Numbers correspond to the x-parameters ($\times$100) in $EuCl_3 \cdot 6H_2O$.

The atomic distances in $EuCl_3 \cdot 6H_2O$ from the Fourier synthesis (R=0.15) by [5] and in $GdCl_3 \cdot 6H_2O$ from refinement with isotropic temperature factors (R=0.06) by [6] are compared with those of $DyCl_3 \cdot 6H_2O$ and $HoCl_3 \cdot 6H_2O$ after refinement with anisotropic temperature factors (R=0.029 and 0.043, respectively) by Ugro [7]. A nearly linear contraction of the coordination sphere of the $M^{3+}$ ion occurs from $EuCl_3 \cdot 6H_2O$ to $HoCl_3 \cdot 6H_2O$. The interatomic distance changes are $M \leftrightarrow Cl(2)$ from 2.76 to 2.739 Å, $M \leftrightarrow H_2O(1)$ from 2.52 to 2.375 Å, $M \leftrightarrow H_2O(2)$ from 2.37 to 2.336 Å, and $M \leftrightarrow H_2O(3)$ from 2.42 to 2.368 Å [7]. Structure refinements for $NdCl_3 \cdot 6H_2O$ (R=0.053) [15], $SmCl_3 \cdot 6H_2O$ (R=0.063) [16], $GdCl_3 \cdot 6H_2O$ (R=0.060) [17], and $LuCl_3 \cdot 6H_2O$ (R=0.078) [18] with anisotropic temperature factors yield the following changes in interatomic distances between the Nd and Lu compounds: $M \leftrightarrow Cl(2)$ from 2.815 to 2.712 Å, $M \leftrightarrow H_2O(1)$ from 2.472 to 2.335 Å, $M \leftrightarrow H_2O(2)$ from 2.449 to 2.312 Å, and $M \leftrightarrow H_2O(3)$ from 2.472 to 2.338 Å. The coordination sphere of the Cl(2) includes in addition to an $M^{3+}$ ion only $3H_2O$ molecules ($Cl(2) \leftrightarrow H_2O$ from 3.14 to 3.18 Å in $GdCl_3 \cdot 6H_2O$) according to Marezio et al. [6]. On the other hand Bel'skii, Struchkov [5] and Ugro [7] include three further removed $H_2O$ molecules so that the coordination number is 7 ($Cl(2) \leftrightarrow H_2O$ 3.12 to 3.33 Å for the Eu [5], 3.14 to 3.28 Å for the Dy [7], 3.143 to 3.382 Å in the Nd [15], and 3.116 to 3.209 Å in the Lu compounds [18]). Additional interatomic distances and bond angles are given in the papers. Different opinions exist regarding the location of the hydrogen bridging bonds [6, 7]. For the structural relationship of these hexahydrates with the heptahydrates, see Section 11.2.9.3, p. 263.

Lattice spectroscopic studies raise the question of whether the Cl(2) ions in fact belong to the coordination sphere of the lanthanide ion, see p. 194.

While crystallographic studies give $C_2$ for the point symmetry of the $M^{3+}$ ion site (see above), the strong dependence of the Zeeman effect in the visible on the direction of the external magnetic field and the Stark field splitting are explained in good approximation by the point symmetry $D_6$ [11, 13], for which the pseudohexagonal axis is perpendicular to the monoclinic axis, see for example [1, 9 to 13] for M = Sm, Eu, Gd, Dy, Tm, and Yb. Studies of the paramagnetic resonance and relaxation of various trivalent rare earth ions in single crystals of $YCl_3 \cdot 6H_2O$ yield similar results, Schulz, Jeffries [14].

When the correlation method is applied to the hexagonal (pseudo-)symmetry, the pseudo space group $P6_322-D_6^6$ (No. 182) results. In this two $M^{3+}$ ions occupy $D_3$ sites, six $Cl^-$ ions occupy $C_2$ sites and twelve $H_2O$ molecules are on two distinct $C_2$ sites with six equivalent units per site. The hexagonal pseudosymmetry is so pronounced that the number of the observed infrared vibrations (for example, 14 for $TmCl_3 \cdot 6H_2O$, see p. 197) agrees much better with the theoretically predicted 15 for the hexagonal space group than with the theoretically predicted 29 for the monoclinic space group, Berringer [19]. Nevertheless, these compounds are optically biaxial, in disagreement with hexagonal symmetry, see p. 211.

No pseudohexagonal axis was proposed for $Ho^{3+}$ and $Er^{3+}$ from the absorption spectra of their trichloride hydrates. Comparison of the sign and magnitude of the electric $4f^n$ multipole moments $Q_n$ (n = 2, 4, 6) for free ion ground states of various heavy rare earths in a table by Stöhr et al. [13] shows that $Ho^{3+}$ and $Er^{3+}$ both have comparatively large $Q_6$. Thus both these ions are expected to be much more sensitive to perturbations of the higher (hexagonal) symmetry approximation by monoclinic symmetry components [13].

Lattice constants at room temperature:

| $MCl_3 \cdot 6H_2O$ | a in Å | b in Å | c in Å | β | Ref. |
|---|---|---|---|---|---|
| $NdCl_3 \cdot 6H_2O$ | 9.761(4) | 6.579(2) | 8.033(3) | 93°45′ (2′) | [3] |
| | 9.72 | 6.60 | 7.90 | 93°00′ | [2] |
| $PmCl_3 \cdot 6H_2O$ | 9.73(1) | 6.56(1) | 8.00(1) | 93°43′ (5′) | [3] |
| $SmCl_3 \cdot 6H_2O$ | 9.670(15) | 6.550(10) | 7.960(12) | 93°40′ (5′) | [4] |
| | 9.60 | 6.61 | 8.00 | 93°40′ | [2] |
| $EuCl_3 \cdot 6H_2O$ | 9.680(15) | 6.530(10) | 7.960(12) | 93°40′ (5′) | [4] |
| | 9.67(3) | 6.52(1) | 7.99(2) | 94°36′ (30′) | [5] |
| $GdCl_3 \cdot 6H_2O$ | 9.640(15) | 6.530(10) | 7.930(12) | 93°40′ (5′) | [4] |
| | 9.651(1) | 6.523(1) | 7.923 | 93.65(2)° | [6] |
| $TbCl_3 \cdot 6H_2O$ | 9.630(15) | 6.510(10) | 7.890(12) | 93°40′ (5′) | [4] |
| | 9.62(3) | 6.52(3) | 7.88(3) | 93.66(12)° | [7] |
| $DyCl_3 \cdot 6H_2O$ | 9.610(15) | 6.490(10) | 7.870(12) | 93°40′ (5′) | [4] |
| | 9.596(4) | 6.491(3) | 7.868(3) | 93.69(2)° | [7] |
| $HoCl_3 \cdot 6H_2O$ | 9.580(15) | 6.470(10) | 7.840(12) | 93°40′ (5′) | [4] |
| | 9.581(5) | 6.480(4) | 7.852(4) | 93.68(2)° | [7] |
| $ErCl_3 \cdot 6H_2O$ | 9.570(15) | 6.470(10) | 7.840(12) | 93°40′ (5′) | [4] |
| | 9.582 | 6.447 | 7.834 | 93°43′ | [3] |
| | 9.57(3) | 6.47(3) | 7.83(3) | 93.75(12)° | [7] |
| $TmCl_3 \cdot 6H_2O$ | 9.550(15) | 6.450(10) | 7.820(12) | 93°40′ (5′) | [4] |
| $LuCl_3 \cdot 6H_2O$ | 9.545(2) | 6.452(1) | 7.798(2) | 93.65(2)° | [18] |

In the data of [2, 7, 18], a and c are interchanged with respect to the original papers; the constants of [2] are converted into Å. The constants for $PmCl_3 \cdot 6H_2O$ are interpolated. As elucidated by a figure from Ivanov [3], a and c decrease in the series of hexahydrates between Nd and Tm more slowly than linearly and b approximately linearly, while $\beta$ (within the particular measurement sequence [3, 4]) is practically constant. For $YCl_3 \cdot 6H_2O$ a value of $\beta = 92°$ was determined by Singh, from Dieke, Crosswhite [1].

References:

[1] G.H. Dieke, H.M. Crosswhite (J. Opt. Soc. Am. **46** [1956] 885/9, 886). – [2] V.I. Iveronova, V.P. Tarasova, M.M. Umanskii (Izv. Akad. Nauk SSSR Ser. Fiz. **15** [1951] 164/9; Structure Reports, Vol. 15, 1951, p. 248). – [3] V.I. Ivanov (Kristallografiya **13** [1968] 905/7; Soviet Phys.-Cryst. **13** [1968] 786/8). – [4] E.J. Graeber, G.H. Conrad, S.F. Duliere (Acta Cryst. **21** [1966] 1012/3). – [5] N.K. Bel'skii, Yu.T. Struchkov (Kristallografiya **10** [1965] 21/8; Soviet Phys.-Cryst. **10** [1965] 15/20).

[6] M. Marezio, H.A. Plettinger, W.H. Zachariasen (Acta Cryst. **14** [1961] 234/6). – [7] I.V. Ugro (IS-T-168 [1967] 1/86, 38/59; C.A. **69** [1968] No. 55126; Diss. Iowa State Univ. Sci. Technol. 1967, pp. 38/59; Diss. Abstr. B **28** [1967] 1891). – [8] V.V. Bakakin, R.F. Klevstova, L.P. Solov'eva (Zh. Strukt. Khim. **15** [1974] 820/30; J. Struct. Chem. [USSR] **15** [1974/75] 723/32). – [9] G.H. Dieke, L. Leopold (J. Opt. Soc. Am. **47** [1957] 944/54). – [10] H. Lämmermann (Z. Physik **150** [1958] 551/72).

[11] H.G. Kahle (Z. Physik **155** [1959] 157/69). – [12] G. Gramberg (Z. Physik **159** [1960] 125/48). – [13] J. Stöhr, E.R. Seidel, J.B. Gruber (J. Chem. Phys. **61** [1974] 4820/3). – [14] M.B. Schulz, C.D. Jeffries (Phys. Rev. [2] **159** [1967] 277/84). – [15] A. Habenschuss, F.H. Spedding (Cryst. Struct. Commun. **9** [1980] 71/6).

[16] A. Habenschuss, F.H. Spedding (Cryst. Struct. Commun. **9** [1980] 207/11). – [17] A. Habenschuss, F.H. Spedding (Cryst. Struct. Commun. **9** [1980] 213/8). – [18] A. Habenschuss, F.H. Spedding (Cryst. Struct. Commun. **9** [1980] 157/60). – [19] B.W. Berringer (Proc. 14th Rare Earth Res. Conf., Fargo, N. Dak., 1979 [1980], pp. 457/62).

### 11.2.7.4 Lattice Vibrations

Introduction

The vibrations in these lattices include the translational vibrations of all lattice building blocks ($\nu_T$), the stretching vibrations of the $MO_n$ polyhedra ($\nu$(M–O)), for which in the hexahydrates the $M^{3+}$ ion has the site symmetry $C_2$, and the external and internal vibrations of the $H_2O$ molecules. The external, so-called librational modes, include the rocking vibrations ($\varrho$) and the wagging vibrations ($\gamma$). Sometimes these librational modes together with the stretching vibrations $\nu(M{-}OH_2)$ are called the pendulum vibrations around the three inertial axes of the molecule ($\nu_{R(i)}$). The internal vibrations include the symmetric ($\nu_1$) and antisymmetric ($\nu_3$) stretching vibrations, displaced in comparison to the free molecules, as well as the deformation vibration ($\nu_2$).

Considering the ion complexes $[(H_2O)_7MCl_2M(H_2O)_7]^{4+}$ in the heptahydrate (see pp. 187/8) and $[M(H_2O)_6Cl_2]^+$ in the hexahydrate (see p. 191), bands are to be expected which should be assignable to the M–Cl vibrations in the complexes. Such bands have not been observed so far as assignment has taken place. Therefore these Cl atoms (which in comparison with the O-atoms of the $H_2O$ molecules lie further removed from the $M^{3+}$ ion, see pp. 188/9 and 192) probably do not belong to the coordination sphere of the $M^{3+}$.

Lattice vibrations from the optical absorption studies of compounds whose composition is reported as $CeCl_3 \cdot 6H_2O$, $PrCl_3 \cdot 3H_2O$, and $HoCl_3 \cdot 9H_2O$, are listed along with the hexahydrates.

### 11.2.7.4.1 $MCl_3 \cdot 7H_2O$ and $MCl_3 \cdot 7D_2O$

In the reflection spectra of nonoriented $LaCl_3 \cdot 7H_2O$ and $LaCl_3 \cdot 7D_2O$ samples at 80 K, the band maxima corresponding to the characteristic vibrations occur at the following wave numbers in the region from 4000 to 50 $cm^{-1}$, Haage et al. [1]:

| | | | | | translational vibrations $\nu_T$ | | | | | | | |
|---|---|---|---|---|---|---|---|---|---|---|---|---|
| $LaCl_3 \cdot 7H_2O$ . . | 55 | 78 | 85 | 96 | 108 | 129 | 170 | 215 | 237 | 269 | 312 | 368 |
| $LaCl_3 \cdot 7D_2O$ . . | – | – | – | 78 | 93 | 125 | 169 | 203 | 227 | 256 | 303 | 359 |

| | librational vibrations | | |
|---|---|---|---|
| | $\nu_{RIII}$ | $\nu_{RII}$ | $\nu_{RI}$ |
| $LaCl_3 \cdot 7H_2O$ . . | 417, 466 | 596, 619 | 716, 741, 765 |
| $LaCl_3 \cdot 7D_2O$ . . | 397 | 441, 455 | 527, 546 |

| | internal vibrations | | |
|---|---|---|---|
| | $\nu_2$ | $\nu_1$ | $\nu_3$ |
| $LaCl_3 \cdot 7H_2O$ . . | 1578, 1590, 1608 | 3068 | 3274, 3364, 3426 |
| $LaCl_3 \cdot 7D_2O$ . . | 1168 | 2260 | 2300, 2380 |

Vertically paired wave numbers relate to bands of similar intensity and shape. The same wave numbers (within the measurement uncertainty) are observed for the corresponding Ce compounds. Deviations occur for $CeCl_3 \cdot 7H_2O$ in the range from 90 to 110 $cm^{-1}$ and at 146 $cm^{-1}$, and for $CeCl_3 \cdot 7D_2O$ at 107 $cm^{-1}$ and between about 150 and 169 $cm^{-1}$; they are caused by the crystal field splitting of the ground level $^2F_{5/2}$ of $Ce^{3+}$. From the appearance of groups of three for the reflection bands for $\nu_R$ and $\nu_i$, it can be inferred that the structure contains at least three different sites for $H_2O$ molecules with distinctly different potentials. The pendular vibrations $\nu_{RI}$ and $\nu_{RII}$ show on the transition from the light to the heavy isotope a displacement proportional to the reciprocal square root of the moment of inertia. For $\nu_{RIII}$ this displacement is less pronounced, the motion is obviously composed of both libration and translation components. The bands of the pure translational and the internal vibrations show familiar displacements, Haage et al. [1].

In the absorption spectra of $MCl_3 \cdot 7H_2O$ (M = La, Ce, and Pr) in a Vaseline suspension, the vibrations of the $H_2O$ molecules at room temperature from studies at 400 to 1000, 1500 to 1800, and 3100 to 3700 $cm^{-1}$ appear at the following wave numbers:

| | external vibrations | $\nu_2$ | $\nu_1$, $\nu_3$ |
|---|---|---|---|
| $LaCl_3 \cdot 7H_2O$ . . | 462, 595 | 1620 | 3160, 3370 |
| $CeCl_3 \cdot 7H_2O$ . . | 465, 600 | 1613 | 3175, 3387 |
| $PrCl_3 \cdot 7H_2O$ . . | 475, 600 | 1634 | 3175, 3400 |

The bands with the exception of the deformation vibration $\nu_2$ are all broad, Kharzeeva, Serebrennikov [2]. The intense broad band between 2000 and 2400 $cm^{-1}$ observed at 20 K in the absorption spectrum of unoriented single crystal films of $LaCl_3 \cdot 7H_2O$ arises through combination vibrations $\nu_2+\nu_{RII}$ or $\nu_{RI}$. For the Ce compound various 4f−4f transitions lie in this region. Combinations of three phonons (e.g., $\nu_2+\nu_{RIII}+\nu_{RII}$) occur for both compounds as lower intensity bands between about 2450 and 2600 $cm^{-1}$. In the 1900 to 2700 $cm^{-1}$ range for the deuterated La and Ce chlorides an intense absorption of the water of crystallization exists between 2300 and 2700 $cm^{-1}$, which is brought about by the stretching vibrations $\nu_1$ and $\nu_3$ of the $D_2O$. The bands at 2064, 2167, and 2248 $cm^{-1}$ in the spectrum of $CeCl_3 \cdot 7D_2O$ are due to 4f−4f transitions, Haage et al. [1].

References:

[1] N. Haage, K.H. Hellwege, J. Jäger, G. Schaack (Physik Kondensierten Materie **10** [1969] 144/51). – [2] S.E. Kharzeeva, V.V. Serebrennikov (Tr. Tomsk. Gos. Univ. **204** [1971] 350/5; C.A. **77** [1972] No. 54506).

### 11.2.7.4.2 $MCl_3 \cdot 6H_2O$ and $MCl_3 \cdot 6D_2O$

In the region of lower wave numbers only the assignment of bands for $YCl_3 \cdot 6H_2O$ has in some degree been executed. Thus the lines observed between 145 and 208 $cm^{-1}$ in the Raman spectrum excited with the 435.8 nm Hg line correspond to the translational vibrations. At higher wave numbers bands are observed in the IR and Raman (R) spectra of the hydrates and deuterates which are assigned to the vibrations of the $Y(X_2O)_6^{3+}$ octahedra (X=D or H). Wave numbers in $cm^{-1}$ at room temperature:

| | | | | |
|---|---|---|---|---|
| $YCl_3 \cdot 6H_2O$ . . | 430, ? $\nu_3(F_{1u})$ | 305, $\nu_4(F_{1u})$ | 248, $\nu_5(F_{2g})$ | R: 243, $\nu_5(F_{2g})$ |
| $YCl_3 \cdot 6D_2O$ . . | ? | 285, $\nu_4(F_{1u})$ | 234, $\nu_5(F_{2g})$ | – |
| $\nu_H/\nu_D$ . . . . . | | 1.07 | 1.06 | |

The appearance of the $\nu_5(F_{2g})$ vibration in the IR spectrum indicates that the site symmetry of the $Y^{3+}$ ion is lower than $O_h$, in agreement with crystallographic results, see "Introduction", p. 194. The location of the $\nu_3(F_{1u})$ vibration is inexact, since it is superimposed on a broad librational vibration of the crystal water molecules, Sipachev, Grigor'ev [1].

The following intense maxima are found in the Raman spectrum of $ScCl_3 \cdot 6H_2O$ single crystals excited by the 435.8 nm Hg line: 116, 132, 157, 198, 203, 231, 246, 259, 372(m), 406, 484(m) (m=medium intensity). Three bands are observed in the IR spectrum at 250(w), 298(s), and 381(m) $cm^{-1}$ (w=weak, s=strong), which in view of their scarcely displaced locations in the deuterate (248, 299, and 367 $cm^{-1}$; $\hat{=}\nu_H/\nu_D \approx 1$, 1, 1.04) are assigned to vibrations of the Cl atoms in the crystal lattice. From this it at first was concluded that in the $ScCl_3 \cdot 6H_2O$ possibly no $Sc(H_2O)_6^{3+}$ units are present in the lattice [1]. In later studies, however, absorption bands have been observed which are assigned to the $\nu$(Sc-O) vibration, namely at about 480 $cm^{-1}$(s, b) in Nujol suspension by Stotz, Melson [2] or at 455(m), 465(s), and 540 $cm^{-1}$(m) in mineral oil suspension by Arkhangel'skii et al. [3] (b=broad). According to the authors [2] the Cl ions do not participate in the coordination of the $Sc^{3+}$, since a corresponding band is not observed.

The absorption bands observed between 80 and 400 $cm^{-1}$ in the IR spectra of $HoCl_3 \cdot 6H_2O$ and $TmCl_3 \cdot 6H_2O$ (in petrolatum white jelly, coated on a polyethylene plate) have been assigned to the wagging vibrations of the $H_2O$ molecules and to

OH···H interactions; wave numbers in $cm^{-1}$, at room temperature, from Berringer et al. [4]:

| | | | | | | | |
|---|---|---|---|---|---|---|---|
| $HoCl_3 \cdot 6H_2O$ . . . . . | 105 | 115 | 162 | 195 | 222 | 245 | 305 |
| $TmCl_3 \cdot 6H_2O$ . . . . . | 106 | 120 | 165 | 200 | 227 | 250 | 304 |

Additional wave numbers derived from the vibronic spectrum at 4.2 K are 56, 67, and 347 $cm^{-1}$ for $HoCl_3 \cdot 6H_2O$, Stöhr, Gruber [5]. In total the wave numbers 43, 71, 106, 120, 155, 197, 244, 260, 308, 347, 419, 517, 577, and 732 $cm^{-1}$ have been found similarly for $TmCl_3 \cdot 6H_2O$ by Olsen, Gruber [6].

For $EuCl_3 \cdot 6H_2O$ the wave numbers 115 and 127 $cm^{-1}$ have been determined from the absorption spectrum in the visible at room temperature; they have been assigned after comparison with studies at 77 K to thermally activated lattice vibrations, Weissman, Freed [7].

Various bands have been observed at room temperature in the region of the external vibrations of the $H_2O$ molecules (400 to 800 $cm^{-1}$) in $MCl_3 \cdot 6H_2O$ with M=Sc in mineral oil suspension [3] and M=Y, Pr to Lu in kerosene [9] and Nujol suspension [8, 14], and probably in $PrCl_3 \cdot 3H_2O$ and $HoCl_3 \cdot 9H_2O$ in Nujol suspension [8]. Several of these bands after comparison with the spectra of the deuterated compounds have been assigned to the rocking vibrations $\varrho$ in the molecular plane and to wagging vibrations $\gamma$ perpendicular to this plane [10, 14]. In view of their width, these bands overlap probable bands which should be assigned to the vibrations $\nu$(M–O) of the $MO_n$ polyhedra (the twisting mode is IR inactive) [14]; wave numbers in $cm^{-1}$ for $MCl_3 \cdot 6H_2O$:

| | Nujol suspension [14] | | | Nujol suspension [8] | | | | kerosene suspension [9] | | |
|---|---|---|---|---|---|---|---|---|---|---|
| M | $\gamma$ | $\varrho$ | $\varrho$ | ? | ? | librational modes | | external modes | | |
| Sc | | (695 | 730)[a] | | | | | | | |
| Y | (453 | 620)[b] | | | | | | 475 | 596 | 646 |
| Ce | | | | 462s | 578s | 680vs | 740vs | | | |
| Pr | | | | (465vs | 564vs | | 723m)[c] | 470 | 581 | 608 |
| Nd | 493 | 598 | 618 | 469s | 567s | 687vs | 750vs | 475 | 587 | 612 |
| Sm | | | | 476s | 523m | 693s | 760s | 473 | 592 | 623 |
| Eu | | | | 472vs,b | 577s | 697m | 761m | 480 | 595 | 621 |
| Gd | 498 | 602 | 638 | 475w,b | 583vs | 700w | 767w | 475 | 596 | 631 |
| Tb | | | | 467w | 592vs | 700w | 770s | 481 | 600 | 633 |
| Dy | | | | 479s,b | 596s | 693m | 772s | 483 | 604 | 637 |
| Ho | | | | (468s | 605m | 698m | 770m)[d] | 487 | 608 | 645 |
| Er | 495 | 616 | | 477s | 615vs | | 760s[e] | 477 | 610 | 650 |
| Tm | | | | 478m,b | 607w | | 780m,b | 495 | 608 | 654 |
| Yb | | | | 483s | 607m | 710s | 775s | 500 | 612 | 656 |
| Lu | 500 | 620 | | | | | | 500 | 600 | 656 |

w=weak, m=medium, s=strong, vs=very strong, b=broad

[a] Mineral oil suspension [3]; only one $\varrho$ vibration was observed by [10] at 747 $cm^{-1}$. — [b] [10]. — [c] Probably $PrCl_3 \cdot 3H_2O$. — [d] Probably $HoCl_3 \cdot 9H_2O$, additional bands at 714 $cm^{-1}$ (m) and 805 $cm^{-1}$ (w). — [e] An additional band at 815 $cm^{-1}$ (m).

The band at about 770 cm$^{-1}$ is assigned to an OH libration mode around an axis perpendicular to the symmetry axis [8]. Within the series of hexahydrates the maxima of the comparable absorption bands shift with increasing atomic number of the M atom (≙decreasing M-$H_2O$ distances in the compounds) essentially toward higher wave numbers. This corresponds to a strengthening of the hydration of the $M^{3+}$ ions, Kharzeeva, Serebrennikov [9].

Wave numbers for the deuterated compounds $MCl_3 \cdot 6\,D_2O$ in Nujol suspension:

| M | $\gamma$ | $\varrho$ | $\varrho$ | Ref. | ? | ? | libration | Ref. |
|---|---|---|---|---|---|---|---|---|
| Y | 341 | 485 | | [10] | | | | |
| Pr | | | | | (455m | | 550m)$^{a)}$ | [8] |
| Nd | 373 | 455 | 495sh | [14] | 474s | | 568s | [8] |
| Sm | 375 | 420 | 462 | [14] | 460m | | 578m | [8] |
| Dy | 370 | 445 | 498sh | [14] | | | | |
| Ho | | | | | (483vs | 602vs)$^{b)}$ | | [8] |
| Er | 360 | 450 | | [14] | | | | |
| Yb | 370 | 475 | | [14] | | | | |

sh = shoulder

$^{a)}$ Probably $PrCl_3 \cdot 3\,D_2O$. — $^{b)}$ Probably $HoCl_3 \cdot 9\,D_2O$.

The libration modes at 550 to 578 cm$^{-1}$ in the deuterates correspond to the libration modes at 723 to 760 cm$^{-1}$ in the hydrates [8]. In the Y-containing compound the ratio of the wave numbers of the hydrate to those of the deuterate for the $\gamma$ vibration is 1.33 and for the $\varrho$ vibration 1.28 [10].

One band, which is observed at 1083 to 1109 cm$^{-1}$ in $MCl_3 \cdot 6\,H_2O$ with M = Nd to Lu, at 1048 cm$^{-1}$ in $CeCl_3 \cdot 6\,H_2O$, and at 1015 cm$^{-1}$ in the partially deuterated Sm-containing compound in Nujol suspension, is probably a combination of a libration (near 770 cm$^{-1}$) and a $\nu$(M-O) vibration, whereby $\nu$(M-O) ≈ 320 cm$^{-1}$ is estimated [8].

The following absorption bands correspond to the internal vibrations of the $H_2O$ molecules of the hexahydrates in kerosene suspension, Kharzeeva, Serebrennikov [9]:

| M | $\nu_2$ | $\nu_1$ | $\nu_3$ | M | $\nu_2$ | $\nu_1$ | $\nu_3$ |
|---|---|---|---|---|---|---|---|
| Y | 1634 | 3230 | 3350 | Dy | 1625 | 3200 | 3375 |
| Pr | 1625 | 3192 | 3350 | Ho | 1625 | 3212 | 3350 |
| Nd | 1620 | 3225 | 3330 | Er | 1620 | 3212 | 3342 |
| Sm | 1625 | 3225 | 3375 | Tm | 1625 | 3234 | 3350 |
| Eu | 1625 | 3220 | 3345 | Yb | 1625 | 3230 | 3350 |
| Gd | 1610 | 3220 | 3350 | Lu | 1630 | 3234 | 3350 |
| Tb | 1630 | 3225 | 3366 | | | | |

In an analogous assignment Basile et al. [14] report similar wave numbers for $NdCl_3 \cdot 6\,H_2O$ and $ErCl_3 \cdot 6\,H_2O$. The bands corresponding to the stretching vibrations are broad [9, 14]. Differing from this, in earlier studies of $MCl_3 \cdot 6\,H_2O$ with M = Ce and Nd to Yb in Nujol suspension by Morrow, Palm [8] the band at ≈3220 cm$^{-1}$, which was not observed in all compounds, was assigned to the first overtone of the bending mode,

$2\nu_2$. The symmetrical stretching mode $\nu_1$ was observed only in the intense spectra of the Nd, Gd, and Tb compounds [8]:

| $MCl_3 \cdot 6H_2O$ ..... | $\nu_1$ | $\nu_3$ | $2\nu_2$ | $\nu_2+\nu_R$ [a)] | $\nu_2$ | ? |
|---|---|---|---|---|---|---|
| $NdCl_3 \cdot 6H_2O$ .... | 3484vs | 3348vs | 3218vs | 2205m | 1621w | |
| $GdCl_3 \cdot 6H_2O$ .... | 3517vs | 3333vs | 3248vs | 2217w | 1625vs | 2040w |
| $TbCl_3 \cdot 6H_2O$ .... | 3542vs,b | 3347vs,b | 3238vs,b | 2225m | 1626s | |

w = weak, m = medium, s = strong, vs = very strong, b = broad

a) $\nu_R \approx 600\ cm^{-1}$.

Wave numbers of the other hexahydrates can be found in the paper by Morrow, Palm [8] and for $ScCl_3 \cdot 6H_2O$ and $YCl_3 \cdot 6H_2O$ by Sipachev, Grigor'ev [10].

In $ScCl_3 \cdot 6D_2O$, $YCl_3 \cdot 6D_2O$ [10] and in $NdCl_3 \cdot 6(H,D)_2O$ and $SmCl_3 \cdot 6(H,D)_2O$ [8], all in Nujol suspensions, the bands have the following wave numbers:

| M | $\nu_3$(OH) | ? | $\nu_3$(OD) | $2\nu_2$ | ? | $\nu_2$(HOH) | $\nu_2$(HOD) | $\nu_2$(DOD) |
|---|---|---|---|---|---|---|---|---|
| Sc | | | 2493 | | | | | 1196 |
| Y | | | 2477 | | | | | 1197 |
| Nd | 3370vs | 2865w | 2495m | | | 1617m | 1413w | 1195w |
| Sm | 3330vs | | 2455vs,b | 2372vs | 1960w | 1632m | | |

For $NdCl_3 \cdot 6D_2O$ and $ErCl_3 \cdot 6D_2O$, on the contrary Basile et al. [14] report the following results (sh = shoulder):

| | $\nu_3$, $\nu_1$ | $\nu_2$ |
|---|---|---|
| $NdCl_3 \cdot 6D_2O$ | 2670s,b; 2580s,b | 1196s; 1190s |
| $ErCl_3 \cdot 6D_2O$ | 2550s,b; 2480s,b; 2400sh | 1267s; 1195s; 1185s |

Wave numbers derived from the vibronic spectrum for the internal vibrations in the hydrates and deuterates of $NdCl_3$ [11], $GdCl_3$ [12], and $TbCl_3$ [13] can be found in the reference papers.

References:

[1] V.A. Sipachev, A.I. Grigor'ev (Zh. Strukt. Khim. **10** [1969] 820/4; J. Struct. Chem. [USSR] **10** [1969] 710/4). – [2] R.W. Stotz, G.A. Melson (Inorg. Chem. **11** [1972] 1720/1). – [3] I.V. Arkhangel'skii, L.N. Komissarova, V.M. Shatskii, N.P. Shepelev (Zh. Neorgan. Khim. **17** [1972] 310/4; Russ. J. Inorg. Chem. **17** [1972] 163/5). – [4] B.W. Berringer, J.B. Gruber, D.N. Olsen, J. Stöhr (J. Inorg. Nucl. Chem. **34** [1972] 373/5). – [5] J. Stöhr, J.B. Gruber (J. Chem. Phys. **55** [1971] 4457/62).

[6] D.N. Olsen, J.B. Gruber (J. Chem. Phys. **54** [1971] 2077/88, 2082). – [7] S.I. Weissman, S. Freed (J. Chem. Phys. **8** [1940] 227/8). – [8] R.J. Morrow, A. Palm (J. Inorg. Nucl. Chem. **24** [1962] 239/44; UCRL-6564-T [1962] 1/5 from N.S.A. **16** [1962] No. 31609). – [9] S.E. Kharzeeva, V.V. Serebrennikov (Tr. Tomsk. Gos. Univ. **204** [1971] 350/5; C.A. **77** [1972] No. 54506). – [10] V.A. Sipachev, A.I. Grigor'ev (Zh. Neorgan. Khim. **15** [1970] 1762/6; Russ. J. Inorg. Chem. **15** [1970] 905/7).

[11] H. Ewald (Ann. Physik [5] **34** [1939] 209/36, 230, 236). – [12] S. Yatsiv, I. Adato, A. Goren (Phys. Rev. Letters **11** [1963] 108/10). – [13] S. Yatsiv, F. Ehrenfreund, U. El-Hanany (J. Chem. Phys. **42** [1965] 743/9). – [14] L.J. Basile, D. Gronert, J.R. Ferraro (Solvent Extr. Res. Proc. 5th Intern. Conf. Solvent Extr. Chem., Jerusalem 1968 [1969], pp. 299/303).

### 11.2.7.5 Mössbauer Effect

Table 11.2.7-1 contains the measured Mössbauer parameters of the rare earth trichloride hexahydrates $MCl_3 \cdot 6\,H_2O$. The signs of the quadrupole coupling constants are specified, if available. For the g-g nuclei $^{154}Sm$, $^{162}Dy$, $^{164,166,168}Er$, and $^{170,172,174}Yb$, due to the zero nuclear spin, in the nuclear ground state neither electrical nor magnetic hyperfine splitting occurs.

The effective magnetic field at the nucleus, $H_{eff}$, is a result at low temperatures of fluctuating electron spins, whose relaxation rate is small compared to the Larmor

Table 11.2.7-1
Mössbauer Parameters for $MCl_3 \cdot 6\,H_2O$, M = Rare Earth.
$E_M$ denotes the resonance energy, $T_A$ and $T_S$ the temperatures of the absorber and the source, respectively, RT room temperature, $\delta$ the isomer shift referred to the source,

| M | isotope | $E_M$ in keV | source | $T_A$ in K | $T_S$ in K | $\delta$ in mm/s |
|---|---|---|---|---|---|---|
| Sm | $^{149}Sm$ | 22 | $^{149}Eu_2O_3$ | 4.2 | RT | – |
| Sm | $^{154}Sm$ | 82.0 | $^{154}Sm$ in $Sm_2O_3$ | 4.2 | 35 | – |
| Sm | $^{154}Sm$ | 82.0 | $^{154}Sm$ in $Sm_2O_3$ | 30 | 30 | – |
| Eu | $^{151}Eu$ | 21.6 | $^{151}Sm$ in $Sm_2O_3$ | RT | RT | −0.31 ± 0.06 |
| Eu | $^{151}Eu$ | 21.6 | $^{151}Sm$ in $Sm_2O_3$ | RT | RT | −0.50 ± 0.02 |
| Gd | $^{155}Gd$ | 86.52 | $^{155}Eu$ in $^{154}Sm$-Al alloy | 4.2 | 10 | – |
| Gd | $^{155}Gd$ | 86.52 | $^{155}Eu$ in $^{154}Sm_2Sn_2O_7$ | 4.2 | 4.2 | 0.05 ± 0.02 |
| Gd | $^{155}Gd$ | 86.52 | $^{155}Gd$ in $EuF_2$ | 4.2 | 4.2 | 0.38 ± 0.04 |
| Dy | $^{161}Dy$ | 25.7 | $^{161}Tb_2O_3$ | 4.2 | – | – |
| Dy | $^{162}Dy$ | 81 | $^{162}Dy$ in $^{161}DyF_3$ | 5 | 5 | – |
| Er | $^{164}Er$ | 91.40 | $^{164}Ho$ in $HoAl_2$ | 4.2 | ≈35 | 0.33 ± 0.25 |
| Er | $^{166}Er$ | 80.56 | $^{166}Ho$ in $HoAl_2$ | 4.2 | ≈30 | 0.30 ± 0.28 |
| Er | $^{166}Er$ | 80.56 | $^{166}Ho$ in $HoAl_2$ | 14.5 | 31.5 | – |
| Er | $^{168}Er$ | 79.80 | $^{168}Tm$ in $TmAl_2$ | 4.2 | 4.2 | 1.0 ± 0.4 |
| Tm | $^{169}Tm$ | 8.412 | $^{169}Er$ in Al | 0.5 to 4.2 | RT | – |
| Yb | $^{170}Yb$ | 84.26 | $^{170}Tm$ in $TmAl_2$ | 4.2 | 4.2 | 0.11 ± 0.04 |
| Yb | $^{170}Yb$ | 84.26 | $^{170}Tm$ in Tm metal | 4.2 | 4.2 | – |
| Yb | $^{171}Yb$ | 66.7 | $^{171}Tm$ in Er-Al alloy | 4.2 | 4.2 | −0.03 ± 0.03 |
| Yb | $^{171}Yb$ | 75.88 | $^{171}Lu$ in $TmAl_2$ | 4.2 | 4.2 | 0.05 ± 0.12 |
| Yb | $^{172}Yb$ | 78.70 | $^{172}Lu$ in Lu metal | 4.2 | 4.2 | 0.22 ± 0.18 |
| Yb | $^{174}Yb$ | 76.46 | $^{174}Lu$ in Lu metal | 4.2 | 4.2 | −0.01 ± 0.13 |
| Yb | $^{174}Yb$ | 76.46 | Yb metal | 4.2 | 4.2 | 0.12 ± 0.16 |
| Yb | $^{176}Yb$ | 82.2 | Yb metal | 4.2 | 4.2 | 0.02 ± 0.3 |

[a] In [1, 2, 4, 7] the coupling constants eqQ and $\mu H$ are in MHz. – [b] In [6, 14, 15] $\delta$ and eqQ are in eV. – [c), d)] The Mössbauer level in the source is populated

frequency of the nuclear spin, see for example [19]. For $YbCl_3 \cdot 6H_2O$ the magnetic hyperfine structure of the Mössbauer spectra with the lines at 66.7 and 84.3 keV of $^{171}Yb$ yields $H_{eff}=2900 \pm 30$ kG at $T \leqq 20$ K [19, 20] and $H_{eff}=2970 \pm 30$ kG at $T=4.2$ K [21]. These studies allow the measurement of the electron spin-lattice relaxation times in $YbCl_3 \cdot 6H_2O$ for temperatures below 77 K, even in the region of very short relaxation times which is not accessible by electron spin resonance methods [19, 20, 24]. $H_{eff}=1570 \pm 30$ kG is derived from the Mössbauer spectrum of $NdCl_3 \cdot 6H_2O$ with the 72.5 keV radiation of $^{145}Nd$ [22].

In $TmCl_3 \cdot 6H_2O$ a magnetic hyperfine interaction of the second order arises through the interaction of the nuclear spin $I_e={}^3/_2$ of $^{169}Tm$ with the nearly degenerate electronic ground state of $Tm^{3+}$ in the lattice (for $Tm^{3+}$ in $TmCl_3 \cdot 6H_2O$ the two lowest crystal field components of the J-multiplet $^3H_6$ are separated by only about 1 $cm^{-1}$). This second-order effect is called a "pseudo-quadrupole interaction", since it, like the electric quadrupole interaction, produces a displacement of the magnetic sublevels of the nucleus

---

$\mu_g H_{eff}$ and $\mu_e H_{eff}$, or $eqQ_g$ and $eqQ_e$, respectively, the coupling constants of the magnetic interaction ($\mu H$) or quadrupole interaction (eqQ) for the ground state (g) and the excited state (e) of the nucleus concerned.

| $eqQ_g$ in mm/s | $eqQ_e$ in mm/s | $\mu_g H_{eff}$ in $10^{-6}$ eV | $\mu_e H_{eff}$ in $10^{-6}$ eV | notes | Ref. |
|---|---|---|---|---|---|
| $\leqq 1.8$ | $-15.1 \pm 1.8$ | $2.07 \pm 0.06$ | – | a) | [1] |
| – | $13.1 \pm 0.7$ | – | $4.23 \pm 0.08$ | a), c) | [2] |
| – | $11.1 \pm 1.0$ | – | $4.22 \pm 0.09$ | a), c) | [2] |
| – | – | – | – | e) | [25] |
| – | – | – | – | | [3] |
| $-4.01 \pm 0.04$ | – | – | – | a) | [4] |
| $4.14 \pm 0.06$ | – | – | – | | [5] |
| $3.6 \pm 0.4$ | – | – | – | b) | [6] |
| 72.4 | – | 2.79 | – | a) | [7] |
| – | – | – | $5.4 \pm 0.2$ | d) | [8] |
| – | $+0.08 \pm 0.8$ | – | $6.45 \pm 0.15$ | | [9] |
| – | $+1.6 \pm 2.1$ | – | $5.85 \pm 0.12$ | | [9, 10] |
| – | – | – | $5.73 \pm 0.10$ | | [11] |
| – | $-0.8 \pm 1.6$ | – | $6.09 \pm 0.12$ | | [9, 10] |
| – | 140 | – | – | | [12] |
| – | $-0.24 \pm 0.16$ | – | $3.06 \pm 0.04$ | | [9, 13] |
| – | – | – | $3.01 \pm 0.17$ | | [14] |
| – | $-0.22 \pm 0.13$ | $9.0 \pm 0.1$ | – | b) | [15] |
| – | $-0.4 \pm 4.8$ | $8.73 \pm 0.13$ | – | | [16] |
| – | $+0.08 \pm 0.8$ | – | $3.04 \pm 0.06$ | | [9; cf. 17] |
| – | $-0.4 \pm 0.7$ | – | $3.08 \pm 0.06$ | | [9; cf. 18] |
| – | – | – | $3.100 \pm 0.124$ | b), c) | [14] |
| – | – | – | $3.485 \pm 0.146$ | b), c) | [14] |

by c) Coulomb excitation with 3 MeV protons, or d) capture of thermal neutrons in $^{161}Dy$. – e) $Eu_2O_3$ as absorber is reference for $\delta$.

proportional to $m_I^2$. Thereby the shift, the intensities, and the widths of both lines of the quadrupole doublet in the Mössbauer spectrum with the 8.412 keV line of $^{169}Tm$ at $T \leqq 4.2$ K are changed in a characteristic manner as a function of temperature. An asymmetric line broadening is partially attributed to spin-spin relaxation between $Tm^{3+}$ ions [12, 23].

References:

[1] S. Ofer, I. Nowik (Nucl. Phys. A **93** [1967] 689/91). – [2] R.M. Wheeler, U. Atzmony, J.C. Walker (Phys. Rev. [2] **186** [1969] 1280/4). – [3] S. Kumar, G.N. Rao, O.K. Medhi, U. Agarwala (Z. Naturforsch. **29a** [1974] 1778/81). – [4] T.E. Katila, V.K. Typpi, G.K. Shenoy, L. Niinistö (Solid State Commun. **11** [1972] 1147/50). – [5] J.D. Cashion, D.B. Prowse, A. Vas (J. Phys. C **6** [1973] 2611/24).

[6] H. Prange (Z. Physik **212** [1968] 415/34). – [7] H.H. Wickman, I. Nowik (J. Phys. Chem. Solids **28** [1967] 2099/103). – [8] W. Henning, D. Heunemann, W. Weber, P. Kienle, H.J. Körner (Z. Physik **207** [1967] 505/10). – [9] E. Münck (Z. Physik **208** [1968] 184/207). – [10] E. Münck, D. Quitmann, S. Hüfner (Z. Naturforsch. **21a** [1966] 847/8).

[11] H. Dobler, G. Petrich, S. Hüfner, P. Kienle, W. Wiedemann, H. Eicher (Phys. Letters **10** [1964] 319/20). – [12] M.J. Clauser, R.L. Mössbauer (Phys. Rev. [2] **178** [1969] 559/67). – [13] A. Hüller, W. Wiedemann, P. Kienle, S. Hüfner (Phys. Letters **15** [1965] 269/71). – [14] J.S. Eck, Y.K. Lee, J.C. Walker (Phys. Rev. [2] **163** [1967] 1295/8). – [15] W. Henning, P. Kienle, E. Steichele, F. Wagner (Phys. Letters **22** [1966] 446/8).

[16] W. Henning, P. Kienle, H.J. Körner (Z. Physik **199** [1967] 207/10). – [17] E. Münck, D. Quitmann, H. Prange, S. Hüfner (Z. Naturforsch. **21a** [1966] 1318/9). – [18] E. Münck, S. Hüfner, H. Prange, D. Quitmann (Z. Naturforsch. **21a** [1966] 1507/8). – [19] G.M. Kalvius, G.K. Shenoy, B.D. Dunlap (Magn. Resonance Relat. Phenomena Proc. 16th Congr. AMPERE, Bucharest 1970 [1971], pp. 584/8). – [20] B.D. Dunlap, G.K. Shenoy, G.M. Kalvius (Phys. Rev. [3] B **10** [1974] 26/30).

[21] C. Gunther, E. Kankeleit (Phys. Letters **22** [1966] 443/6). – [22] F.T. Parker, J.C. Walker (Hyperfine Interactions Excited Nucl. Proc. Conf., Rehovot-Jerusalem 1970 [1971], pp. 853/9). – [23] M.J. Clauser, E. Kankeleit, R.L. Mössbauer (Phys. Rev. Letters **17** [1966] 5/8; Bull. Am. Phys. Soc. [2] **10** [1965] 1202). – [24] H. Winkler, P. Alwardt, E. Gerdau (J. Phys. [Paris] **40** [1979] Suppl. No. 3, pp. C2-250/C2-252). – [25] D.K. Gupta, D.K. Sharma, K.B. Lal (Indian J. Pure Appl. Phys. **17** [1979] 546/8).

## 11.2.7.6 Electron Paramagnetic Resonance (EPR)

### 11.2.7.6.1 EPR Spectra of Pure $MCl_3 \cdot n\,H_2O$

For $LaCl_3 \cdot 7\,H_2O$, Lancaster, Gordy [1] have observed no EPR signal at room temperature with a frequency $\nu = 46.500$ GHz in magnetic fields up to 15 kG. For $PrCl_3 \cdot 7\,H_2O$, Davis et al. [2] detected two signals of different intensity with a resonance frequency $\nu \approx 9.4$ GHz ($\lambda \approx 3.2$ cm$^{-1}$) at 4 K.

For the chloride hydrates $MCl_3 \cdot 6\,H_2O$ with M = Y, Nd, and Sm no EPR signal was observed in magnetic fields up to 15 kG at room temperature; the following frequencies were used for the individual salts in the search for a resonance: Y 45.602, Nd 46.700, Sm 47.320 GHz. For $GdCl_3 \cdot 6\,H_2O$ at room temperature a signal was observed with $\nu =$ 32.500 GHz at a field strength B = 11.500 kG, and with $\nu = 51.640$ GHz at B = 17.325 kG.

The g-factor and the line width increase with frequency from g=2.02 to g=2.13 and from 2.070 to 3.300 kG, respectively [1]. Without details g=1.99 and a line width of approximately 2 kG were reported by Garif'yanov [3].

References:

[1] F.W. Lancaster, W. Gordy (J. Chem. Phys. **19** [1951] 1181/91). – [2] C.F. Davis, A.F. Kip, R. Malvano (Atti Accad. Nazl. Lincei Rend. Classe Sci. Fis. Mat. Nat. [8] **11** [1951] 77/9). – [3] N.S. Garif'yanov (Dokl. Akad. Nauk SSSR **84** [1952] 923/5).

### 11.2.7.6.2 EPR Spectra of Rare Earth Impurity Ions in $MCl_3 \cdot nH_2O$

EPR spectra of rare earth trichloride hydrates have mostly been obtained for samples in which the ions under study were added in small concentrations to rare earth chloride hydrate hosts, for the same reasons as for the anhydrous rare earth trichlorides, cf. p. 93. Diamagnetic $YCl_3 \cdot 6H_2O$ has been most widely used as a host crystal, whereas systematic EPR studies of rare earth tripositive ions in $LaCl_3 \cdot 7H_2O$ were prohibited by an incomplete knowledge of crystal structure. On the other hand, EPR spectra of $Gd^{3+}$ have been observed and analyzed throughout the whole series of rare earth trichloride hexahydrate hosts of Y and Nd to Yb (excl. Pm and Gd).

Results are presented in Table 11.2.7-2 which is arranged in the same manner as Table 11.2.6-4, pp. 94/99. For symbols and abbreviations see p. 93. The setting of the principal magnetic axes x, y, z is as follows: In the monoclinic rare earth trichloride hexahydrates for symmetry reasons one of the principal magnetic axes must coincide with the crystal b axis. Following the convention adopted in [1], this is identified as the y axis. The angle ∢za is the acute angle between the z axis and the crystal a axis. For the triclinic crystals of $LaCl_3 \cdot 7H_2O$ and $PrCl_3 \cdot 7H_2O$, however, there is no magnetic axis of symmetry, and the axes have been identified according to the following criterion: The direction of magnetic field for which the splitting of the resonance lines attains a maximum is denoted by z. The directions of maximum and minimum splitting in a plane perpendicular to z are then the axes x and y, respectively [5, 6, 8].

Table 11.2.7-2
EPR Results for Trivalent Rare Earth Impurity Ions in Host Lattices of $MCl_3 \cdot 6H_2O$, M=Y and Nd to Yb, and of $MCl_3 \cdot 7H_2O$, M=La, Pr.
For symbols and abbreviations see p. 93. References see p. 206.

| impurity ion | conc. (%) [a] | M of host | T in K | $\nu$ in GHz | observed properties and derived parameters | Ref. |
|---|---|---|---|---|---|---|
| Ce | a: 0.1; 1 | Y | 1.3 to 5 | 9.3 | $g_x=2.67$, $g_y=2.305$, $g_z=0.782$, ∢za=69° (1 at.%); $\tau_{1b}$ [b] | [1] |
| Pr | – | Y | – | 9.3 | no resonance observed for H ≦ 19 kG | [1] |
| Nd | a: 0.1; 1 | Y | 0.2 to 5 | 9.36 | $g_x=0.26$, $g_y=0.33$, $g_z=3.50$, ∢za=63° (0.1 at. %); Hfs constants; $\tau_{1b}$ [b] | [1] |
| Sm | – | Y | – | 9.3 | no resonance observed for H ≦ 19 kG | [1] |
| Gd | 0.2 | Y | 298 | X-band | $g_x=1.994$, $g_z=1.992$; ∢za=63°; CF constants | [2] |

Table 11.2.7-2 [continued]

| impurity ion | conc. (%)[a)] | M of host | T in K | ν in GHz | observed properties and derived parameters | Ref. |
|---|---|---|---|---|---|---|
| Gd | m: 0.2; 1 | Y | 1.77; 77; 295 | 9.45 to 10.37; 34.65 to 35.24 | $g_x=1.992$, $g_y=1.997$, $g_z=1.994$, ∢za = 58° (T = 295 K); CF constants | [3] |
| Gd | m: 0.17 | Y | 1.2 to 330 | 9.4308; 35.144 | $g_x=1.999$, $g_y=1.993$, $g_z=1.988$, ∢xc = 121.6° [c)] (T = 293 K); CF constants | [4] |
| Gd | m: 0.2 | La | 1.59; 4.63; 79.4; 295 | X-band | $g_x=1.996$, $g_y=2.001$, $g_z=1.994$ (T = 295 K); CF constants | [5] |
| Gd | – | La | 77; 298 | 9.4 | $g_x=1.998$, $g_y=2.000$, $g_z=1.989$ (T = 298 K); CF constants | [6] |
| $^{155}$Gd, $^{157}$Gd | m: 0.02 | La, deuterate | 77; 298 | 9.4; 25 | $g_x=g_y=g_z=1.99$ (T = 298 K); Hfs constant; $\mu(^{155,157}Gd)$ | [7] |
| Gd | m: 0.2 | Pr | 1.60; 79; 295 | X-band | $g_x=1.993$, $g_y=1.958$, $g_z=1.993$ (T = 295 K); CF constants | [8] |
| Gd | m: 0.05 | Pr | 298 | X-band | splitting factors, CF constants, referred to unspecified axis system | [9] |
| Gd | 0.2 | Nd | 298 | X-band | $g_x=1.996$, $g_z=1.990$, ∢za = 58.5°; CF constants | [2] |
| Gd | m: 0.1; 0.2; 1 | Nd | 1.65 to 295 | X-band; K-band | $g_x=g_y=1.992$, $g_z=1.993$ (9.4327 GHz, 295 K), ∢za = 58.5°; CF constants | [10] |
| Gd | w: 0.05 | Nd | 297 | 9.4 | $g_x=1.9930$, $g_z=1.9920$; CF constants | [11] |
| Gd | m: 0.05 | Nd | 298 | X-band | splitting factors, CF constants, referred to unspecified axis system | [9] |
| Gd | 0.2 | Sm | 298 | X-band | $g_x=1.997$, $g_z=1.995$, ∢za = 63°; CF constants | [2] |
| Gd | m: 1 | Sm | 1.88; 77; 295 | X-band; 35 | $g_x=1.992$, $g_y=1.997$, $g_z=1.991$ (T = 295 K), ∢za = 62.25°; CF constants | [12] |
| Gd | w: 0.05 | Sm | 297 | 9.4 | $g_x=1.9910$, $g_z=1.9912$; CF constants | [11] |
| Gd | m: 0.05 | Sm | 298 | X-band | splitting factors, CF constants, referred to unspecified axis system | [13] |
| Gd | 0.2 | Eu | 298 | X-band | $g_x=1.989$, $g_z=2.000$, ∢za = 62.5°; CF constants | [2] |

Table 11.2.7-2 [continued]

| impurity ion | conc. (%) [a] | M of host | T in K | ν in GHz | observed properties and derived parameters | Ref. |
|---|---|---|---|---|---|---|
| Gd | m: 0.4 | Eu | 1.60; 79; 295 | X-band | $g_x=2.000$, $g_y=1.994$, $g_z=1.992$, ∢za = 64° (T = 295 K); CF constants | [14] |
| Gd | w: 0.05 | Eu | 297 | 9.4 | $g_x=1.9915$, $g_z=1.9910$; CF constants | [11] |
| Gd | w: 0.2 | Ho | 298 | 9.3 | $g_x=1.9864$, $g_z=1.9880$; CF constants | [15] |
| Gd | 0.2 | Tb to Yb | 298 | X-band | maximum and minimum values are: $g_x=2.002$ (Yb) to 1.976 (Er), $g_z=1.995$ (Dy, Yb) to 1.972 (Er), ∢za = 64.5° (Tm) to 60.5° (Er); CF constants; comparative study for the series of rare earth trichloride hexahydrate hosts [d] | [2] |
| Gd | w: 0.05 | Tb to Yb | 297 | 9.4 | maximum and minimum values are: $g_x=1.9930$ (Tb) to 1.9870 (Yb), $g_z=1.9920$ (Tb) to 1.9881 (Yb); CF constants; comparative study for the series of rare earth trichloride hexahydrate hosts | [11] |
| Dy | a: 0.1; 1; 5 | Y | 1.3 to 5 | 9.36 | $g_x=2.00$, $g_y=1.52$, $g_z=16.52$, ∢za = −23° (0.1 at.%); Hfs constants; $\tau_{1b}$ [b] | [1] |
| Er | a: 0.1; 1 | Y | 0.2 to 5 | 9.3 | $g_x$, $g_y<1$ (sample-dependent), $g_z=13.74$, ∢za = 77° (0.1 at. %); Hfs constants; $\tau_{1b}$ [b] | [1] |
| Tm | m: 0.5 | Y | 1.5 | 24.7; 34.5 to 35.1 | $g_z \cdot \cos\alpha = 13.37$ [e]; Hfs and CF constants; magnetic and electric dipole transitions [f]; line shapes; $\tau_1$ | [17] |
| Tm | 0.5 | Y | 4.2 | 33.7 to 36 | $g_z=14.0$, ∢za = 65°; Hfs and CF constants [f] | [18] |
| Yb | a: 1 | Y | <5 | 9.3 | $g_x$, $g_y<0.5$, $g_z=5.57$, ∢za = 71°; measurements of Hfs and $\tau_{1b}$ [b] attempted but unsuccessful | [1] |

[a] As far as given in the original papers, impurity concentrations are specified as a = at. %, m = mol %, w = wt %. — [b] $\tau_{1b}$ is the time required for the combined system of spins and phonons to relax to the surrounding bath of liquid He. — [c] The setting of the axes x and z in the a-c crystal plane has been changed here from that of the original reference [4] according to x→z, z→x, to conform with the setting adopted in [1 to 3, 10, 14]. — [d] For preliminary results see [16]. — [e] $\alpha$ is a small angle between the z axis and the plane in which the orientation of the magnetic field with respect to the crystal is varied [17]. — [f] Transitions observed for $Tm^{3+}$ occur between the two lowest crystal field states which are both singlets [17, 18].

References:

[1] M.B. Schulz, C.D. Jeffries (Phys. Rev. [2] **159** [1967] 277/84). – [2] S.K. Misra, P. Mikolajczak (J. Magn. Resonance **35** [1979] 87/94). – [3] S.K. Misra, G.R. Sharp (J. Phys. Chem. Solids **37** [1976] 999/1003). – [4] D. Meierling, W. Ullmann (Z. Naturforsch. **23a** [1968] 1971/5). – [5] S.K. Misra, G.R. Sharp (J. Phys. C **9** [1976] 1975/9).

[6] M. Weger, W. Low (Phys. Rev. [2] **111** [1958] 1526/8). – [7] W. Low (Phys. Rev. [2] **103** [1956] 1309). – [8] S.K. Misra, G.R. Sharp (Phys. Status Solidi B **75** [1976] 607/12). – [9] G.B. Singh, G.C. Upreti, P. Venkateswarlu (J. Chem. Phys. **46** [1967] 2885/90). – [10] S.K. Misra, G.R. Sharp (J. Chem. Phys. **64** [1976] 2168/73).

[11] V.M. Malhotra, H.D. Bist, G.C. Upreti (J. Magn. Resonance **27** [1977] 439/53). – [12] S.K. Misra, G.R. Sharp (J. Phys. C **9** [1976] 401/8). – [13] G.B. Singh, P. Venkateswarlu (Proc. Indian Acad. Sci. A **65** [1967] 211/8). – [14] S.K. Misra, G.R. Sharp (Physica B+C **83** [1976] 174/8). – [15] V.M. Malhotra, Y.S. Jain, H.D. Bist (Proc. Nucl. Phys. Solid State Phys. Symp. C **18** [1975] 502/4; C.A. **87** [1977] No. 93175).

[16] S.K. Misra, G.R. Sharp (J. Phys. C **10** [1977] 897/904). – [17] I.E. Rouse, J.B. Gruber (Phys. Rev. [3] B **13** [1976] 3764/73). – [18] J.B. Gruber, E.A. Karlow, D.N. Olsen, U. Ranon (Phys. Rev. [3] B **2** [1970] 49/53).

### 11.2.7.7 Density

Values in $g/cm^3$ for the X-ray densities $D_{calc}$ and the experimentally determined densities $D_{exp}$, which were mostly determined by the flotation method [1, 5 to 11] or on a Berman torsion balance [12] are for the respective $MCl_3 \cdot n\,H_2O$ compounds:

| $MCl_3 \cdot 7\,H_2O$ | $LaCl_3 \cdot 7\,H_2O$ | $CeCl_3 \cdot 7\,H_2O$ | $PrCl_3 \cdot 7\,H_2O$ |
|---|---|---|---|
| $D_{calc}$ . . . . | 2.219 [1], 2.20 [2], 2.223 [3] | 2.244 [4] | 2.280 [6] |
| $D_{exp}$ . . . . | 2.215 [1], 2.23 [2], 2.20 [3] | 2.25 [5] | 2.25 [6] |

| $MCl_3 \cdot 6\,H_2O$ | $NdCl_3 \cdot 6\,H_2O$ | $PmCl_3 \cdot 6\,H_2O$ | $SmCl_3 \cdot 6\,H_2O$ | $EuCl_3 \cdot 6\,H_2O$ |
|---|---|---|---|---|
| $D_{calc}$ . . . . | 2.331 [7], 2.32 [3] | 2.36 [3] | 2.405 [8], 2.407 [12] | 2.423 [12] |
| $D_{exp}$ . . . . | 2.40 [3] a) | – | 2.382 [12] | 2.418 [12] |

| $MCl_3 \cdot 6\,H_2O$ | $GdCl_3 \cdot 6\,H_2O$ | $TbCl_3 \cdot 6\,H_2O$ | $DyCl_3 \cdot 6\,H_2O$ |
|---|---|---|---|
| $D_{calc}$ . . . . | 2.477 [10], 2.478 [12] | 2.513 [12] | 2.554 [12], 2.56 [11] |
| $D_{exp}$ . . . . | 2.468 [12] | 2.492 [12] | 2.525 [12], 2.53 [11] |

| $MCl_3 \cdot 6\,H_2O$ | $HoCl_3 \cdot 6\,H_2O$ | $ErCl_3 \cdot 6\,H_2O$ | $TmCl_3 \cdot 6\,H_2O$ | $LuCl_3 \cdot 6\,H_2O$ |
|---|---|---|---|---|
| $D_{calc}$ . . . . | 2.594 [12] | 2.604 [12], 2.61 [3] | 2.617 [12] | 2.698 [9] |
| $D_{exp}$ . . . . | 2.564 [12] | 2.588 [12], 2.63 [3] b) | 2.608 [12] | – |

a) Probable $NdCl_3$ contamination. – b) Probable $ErCl_3$ contamination.

References:

[1] A. Habenschuss, F.H. Spedding (Cryst. Struct. Commun. **8** [1979] 511/6). – [2] V.V. Bakakin, R.F. Klevstova, L.P. Solov'ev (Zh. Strukt. Khim. **15** [1974] 820/30; J. Struct. Chem. [USSR] **15** [1974] 723/32). – [3] V.I. Ivanov (Kristallografiya **13** [1968] 905/7; Soviet Phys.-Cryst. **13** [1968] 786/8). – [4] E.J. Peterson, E.I. Onstott, R.B. v. Dreele (Acta Cryst. B **35** [1979] 805/9). – [5] K.H. Hellwege, R. v. Klot, G. Weber (Physik Kondensierten Materie **2** [1964] 397/407).

[6] A. Habenschuss, F.H. Spedding (Cryst. Struct. Commun. **7** [1978] 535/41). – [7] A. Habenschuss, F.H. Spedding (Cryst. Struct. Commun. **9** [1980] 71/6). – [8] A.

Habenschuss, F.H. Spedding (Cryst. Struct. Commun. **9** [1980] 207/11). – [9] A. Habenschuss, F.H. Spedding (Cryst. Struct. Commun. **9** [1980] 157/60). – [10] A. Habenschuss, F.H. Spedding (Cryst. Struct. Commun. **9** [1980] 213/8).

[11] I.V. Ugro (IS-T-168 [1967] 1/86, 35; C.A. **69** [1968] No. 55126; Diss. Iowa State Univ. Sci. Technol. 1967, p. 35; Diss. Abstr. B **28** [1967] 1891). – [12] E.J. Graeber, G.H. Conrad, S.F. Duliere (Acta Cryst. **21** [1966] 1012/3).

## 11.2.7.8 Thermal Properties

### 11.2.7.8.1 Melting Point $t_m$ in °C

| | | | | | | | |
|---|---|---|---|---|---|---|---|
| $MCl_3 \cdot nH_2O$ . . | $YCl_3 \cdot 6H_2O$ | | | $LaCl_3 \cdot 7H_2O$ | | $CeCl_3 \cdot 7H_2O$ | |
| $t_m$ . . . . . . . | 163.8 to 165.0 | 160 | ≈162 | 92 | 94 | 90 | 103 |
| Ref. . . . . . | [4] | [1] | [2] | [5] | [1] | [8] | [9] |

| | | | | | | | | | |
|---|---|---|---|---|---|---|---|---|---|
| $MCl_3 \cdot nH_2O$ . . | $CeCl_3 \cdot 6H_2O$ | $PrCl_3 \cdot 7H_2O$ | | $NdCl_3 \cdot 6H_2O$ | | $SmCl_3 \cdot 6H_2O$ | | |
| $t_m$ . . . . . . . | 94 | 109 | 113 | 124 | 127 | ≈142 | 143 | 147 |
| Ref. . . . . . | [1] | [6] | [1] | [2] | [1, 7] | [2] | [1] | [7] |

| | | | | | |
|---|---|---|---|---|---|
| $MCl_3 \cdot nH_2O$ . . | $EuCl_3 \cdot 6H_2O$ | $GdCl_3 \cdot 6H_2O$ | | $TbCl_3 \cdot 6H_2O$ | |
| $t_m$ . . . . . . . | 152 | 148 | 158 | 153 | 158.0 to 158.8 |
| Ref. . . . . . | [1] | [1] | [7] | [1] | [4] |

| | | | | | | | | |
|---|---|---|---|---|---|---|---|---|
| $MCl_3 \cdot nH_2O$ . . | $DyCl_3 \cdot 6H_2O$ | | | $HoCl_3 \cdot 6H_2O$ | | $ErCl_3 \cdot 6H_2O$ | |
| $t_m$ . . . . . . . | 155 | 157 | 161.8 to 163.1 | 160 | 163.4 to 164.5 | 153 | 163.8 to 165.0 |
| Ref. . . . . . | [1] | [7] | [4] | [1] | [4] | [1] | [4] |

| | | | | | | | |
|---|---|---|---|---|---|---|---|
| $MCl_3 \cdot nH_2O$ . . | $TmCl_3 \cdot 6H_2O$ | | $YbCl_3 \cdot 6H_2O$ | | | $LuCl_3 \cdot 6H_2O$ | |
| $t_m$ . . . . . . . | 154 | 162.4 to 163.5 | 153.0 to 155.2 | ≈154 | 155 | 148.0 to 150.5 | 150 |
| Ref. . . . . . | [1] | [4] | [4] | [2] | [1] | [4] | [1] |

For similar data on $t_m$ of $YCl_3 \cdot 6H_2O$, $PrCl_3 \cdot 7H_2O$, and $NdCl_3 \cdot 6H_2O$ see also Matignon [3].

References:

[1] W.W. Wendlandt, R.G. Sewell (Texas J. Sci. **8** [1961] 231/4). – [2] M.D. Williams, H.C. Fogg, C. James (J. Am. Chem. Soc. **47** [1925] 297/301). – [3] C. Matignon (Ann. Chim. Phys. [8] **8** [1906] 243/83, 251, 386/401, 389, 433/9, 434). – [4] Yu.G. Sakharova, T.A. Ezhova (Zh. Neorgan. Khim. **21** [1976] 551/4; Russ. J. Inorg. Chem. **21** [1976] 296/8). – [5] A. Habenschuss, F.H. Spedding (Cryst. Struct. Commun. **8** [1979] 511/6).

[6] A. Habenschuss, F.H. Spedding (Cryst. Struct. Commun. **7** [1978] 535/41). – [7] A. Habenschuss, F.H. Spedding (Cryst. Struct. Commun. **9** [1980] 71/6, 207/11, 213/8, 157/60). – [8] T. Kojima (J. Electrochem. Soc. Japan **19** [1951] 113/7; C.A. **1951** 8726). – [9] M.I. Kadantseva, E.F. Zhuravlev (Zh. Neorgan. Khim. **18** [1973] 537/43; Russ. J. Inorg. Chem. **18** [1973] 281/4).

### 11.2.7.8.2 Thermodynamic Data

**Formation Data**

Heat of hydration $\Delta H_{298}(hyd)$ calculated from the heat of solution at 25 °C, enthalpy of formation $\Delta H^\circ_{298}(f)$, estimated Gibbs free energy $\Delta G^\circ_{298}(f)$, and estimated entropy

$\Delta S^\circ_{298}(f)$ of formation. Values of $\Delta H$ and $\Delta G$ in kcal/mol, and of $\Delta S$ in $cal \cdot mol^{-1} \cdot K^{-1}$ from Flynn, Spedding [1] are:

| $MCl_3 \cdot n\,H_2O$ | $-\Delta H_{298}(hyd)$ | $-\Delta H^\circ_{298}(f)$ | $-\Delta H^\circ_{298}(f)$ [a)] | $-\Delta G^\circ_{298}(f)$ | $-\Delta S^\circ_{298}(f)$ |
|---|---|---|---|---|---|
| $YCl_3 \cdot 6\,H_2O$ | 37.28±0.08 | 681.63±0.12 | 679.87 | 581.1 | 337.2 |
| $LaCl_3 \cdot 7\,H_2O$ | 25.37±0.02 | 759.27±0.20 | 759.50 | 645.0 | 383.5 |
| $PrCl_3 \cdot 7\,H_2O$ | 28.45±0.06 | 759.69±0.16 | 758.76 | 645.4 | 383.5 |
| $PrCl_3 \cdot 6\,H_2O$ | 25.81±0.06 | 688.73±0.16 | 687.80 | 588.2 | 337.2 |
| $SmCl_3 \cdot 6\,H_2O$ | 30.92±0.06 | – | – | – | 337.2 |
| $GdCl_3 \cdot 6\,H_2O$ | 34.38±0.04 | 684.24±0.15 | 684.37 | 583.7 | 337.2 |
| $ErCl_3 \cdot 6\,H_2O$ | 38.22±0.05 | 679.10±0.25 | 677.19 | 578.6 | 337.2 |
| $YbCl_3 \cdot 6\,H_2O$ | 40.67±0.01 | – | – | – | 337.2 |

[a)] Values given by Spedding, Flynn [2].

Values recommended by the National Bureau of Standards for the enthalpies $\Delta H^\circ$ at 0 and 298 K and the Gibbs free energies $\Delta G^\circ_{298}$, for the formation of the $MCl_3 \cdot n\,H_2O$ in kcal/mol (values for $ScCl_3 \cdot 6\,H_2O$ and $YCl_3 \cdot 6\,H_2O$ from [3], for $LaCl_3 \cdot n\,H_2O$ to $LuCl_3 \cdot n\,H_2O$ from [4]):

| $MCl_3 \cdot n\,H_2O$ | $-\Delta H^\circ_0$ | $-\Delta H^\circ_{298}$ | $-\Delta G^\circ_{298}$ |
|---|---|---|---|
| $ScCl_3 \cdot 6\,H_2O$ | – | 671.6 | – |
| $YCl_3 \cdot 6\,H_2O$ | – | 691.3 | 592.1 |
| $LaCl_3 \cdot 7\,H_2O$ | – | 759.7 | 648.5 |
| $CeCl_3 \cdot 7\,H_2O$ | – | 757.5 | – |
| $PrCl_3 \cdot 7\,H_2O$ | – | 759.5 | – |
| $PrCl_3 \cdot 6\,H_2O$ | – | 688.5 | – |
| $NdCl_3 \cdot 6\,H_2O$ | 678.72 | 687.0 | 588.1 |
| $SmCl_3 \cdot 6\,H_2O$ | – | 686.0 | 587.1 |
| $EuCl_3 \cdot 6\,H_2O$ | – | 665.6 | 565.5 |
| $GdCl_3 \cdot 6\,H_2O$ | 675.66 | 685 | 586 |
| $TbCl_3 \cdot 6\,H_2O$ | – | 683.4 | 583.4 |
| $DyCl_3 \cdot 6\,H_2O$ | 676.83 | 686 | 586 |
| $HoCl_3 \cdot 6\,H_2O$ | 678.80 | 687.9 | 588.0 |
| $ErCl_3 \cdot 6\,H_2O$ | 677.92 | 687.0 | 586.6 |
| $YbCl_3 \cdot 6\,H_2O$ | – | 680.2 | 580.6 |
| $LuCl_3 \cdot 6\,H_2O$ | 667.41 | 676.6 | 576.3 |

For older calculated thermodynamic data of formation, see Montgomery [5], Montgomery, Hubert [6].

## Entropy

Recommended values for the entropy $S_{298}$ in $cal \cdot mol^{-1} \cdot K^{-1}$ of $YCl_3 \cdot 6\,H_2O$ from [3] and $LaCl_3 \cdot n\,H_2O$ to $LuCl_3 \cdot n\,H_2O$ from [4], estimated values from [2], and values calculated by Hinchey, Cobble [7] for $S_{298}$ and the electronic component S(e) (with estimated values in parentheses):

| $MCl_3 \cdot nH_2O$ | $S_{298}$ [3, 4] | $S_{298}$ [2] | $S_{298}$ [7] | S(e) [7] |
|---|---|---|---|---|
| $YCl_3 \cdot 6H_2O$ | 92 | 89.1 | (88.0) | 0 |
| $LaCl_3 \cdot 7H_2O$ | 110.6 | 100.3 | 110.6 [a)] | 0 |
| $CeCl_3 \cdot 7H_2O$ | – | – | (112.3) | 3.56 |
| $PrCl_3 \cdot 7H_2O$ | – | 100.3 | 111.2 [a)] | 4.37 |
| $PrCl_3 \cdot 6H_2O$ | – | 90.9 | – | – |
| $NdCl_3 \cdot 6H_2O$ | 99.7 | – | 99.4 [a)] | 4.58 |
| $SmCl_3 \cdot 6H_2O$ | 99 | 91.1 | 98.2 | 3.64 [b)] |
| $EuCl_3 \cdot 6H_2O$ | 97.3 | – | (95.5) | 2.23 [b)] |
| $GdCl_3 \cdot 6H_2O$ | 97.56 | 91.3, 95.79 [c)] | 93.6 [a)] | 4.13 |
| $TbCl_3 \cdot 6H_2O$ | 96.4 | 96.52 [c)] | (95.1) | 5.10 |
| $DyCl_3 \cdot 6H_2O$ | 96.00 | – | 96.0 [a)] | 5.51 |
| $HoCl_3 \cdot 6H_2O$ | 97.08 | 96.44 [c)] | 97.0 [a)] | 5.63 |
| $ErCl_3 \cdot 6H_2O$ | 95.3 | 91.5 | 95.4 [a)] | 5.51 |
| $TmCl_3 \cdot 6H_2O$ | 95.5 | – | – | 5.10 |
| $YbCl_3 \cdot 6H_2O$ | 94.6 | 91.6 | (93.9) | 4.13 |
| $LuCl_3 \cdot 6H_2O$ | 89.9 | 89.51 [c)] | 90.09 [a)] | 0 |

[a)] Extrapolated from the data of Pfeffer [8 to 11]. – [b)] Contains contributions from higher electronic states. – [c)] Values from [13, 14]. They are given in $J \cdot mol^{-1} \cdot K^{-1}$ in the original paper.

See also the data from the publication of Cobble [12] for $S_{298}$ and S(e). The ratio of the lattice entropy to the heat capacity for the heptahydrates is 1.07, for the hexahydrates 1.09 to 1.10 [7]. See [13, 14] also for a comparison of the calorimetrically determined entropies with those calculated from optical data. Numerical values for $S^\circ_T$ between 0 and 300 K can be found in "Rare Earth Elements" C 4b.

**Enthalpy Increment $H^\circ_{298} - H^\circ_0$. Thermodynamic Functions $(H^\circ_{298} - H^\circ_0)/T$ and $(G^\circ_{298} - H^\circ_0)/T$**

Enthalpy increments $H^\circ_{298} - H^\circ_0$ in kcal/mol recommended by [4] and thermodynamic functions $(H^\circ_{298} - H^\circ_0)/T$ and $-(G^\circ_{298} - H^\circ_0)/T$ in $J \cdot mol^{-1} \cdot K^{-1}$ (values in parentheses are in $cal \cdot mol^{-1} \cdot K^{-1}$) from [13, 14] of $LaCl_3 \cdot 7H_2O$ and $MCl_3 \cdot 6H_2O$:

| hydrate | $LaCl_3$ | $NdCl_3$ | $GdCl_3$ | $TbCl_3$ | $DyCl_3$ | $HoCl_3$ | $ErCl_3$ | $LuCl_3$ |
|---|---|---|---|---|---|---|---|---|
| $H^\circ_{298} - H^\circ_0$ | 17.12 | 15.14 | 14.43 | – | 14.42 | 14.49 | 14.39 | 13.96 |
| $(H^\circ_{298} - H^\circ_0)/T$ | – | – | 201.37 | 204.69 | – | 202.14 | – | 195.52 |
| | | | (48.13) | (48.91) | | (48.31) | | (46.73) |
| $-(G^\circ_{298} - H^\circ_0)/T$ | – | – | 199.43 | 199.15 | – | 201.30 | – | 179.06 |
| | | | (47.66) | (47.6) | | (48.11) | | (42.8) |

Numerical values for $(H^\circ_T - H^\circ_0)/T$ and $(G^\circ_T - H^\circ_0)/T$ for these compounds between 0 and 300 K can be found in "Rare Earth Elements" C 4b.

**Heat Capacity**

The molar heat capacities $C_p$ at 25 °C in $cal \cdot mol^{-1} \cdot K^{-1}$ for various hydrated chlorides are:

| $MCl_3 \cdot nH_2O$ | $LaCl_3 \cdot 7H_2O$ | | $PrCl_3 \cdot 7H_2O$ | $NdCl_3 \cdot 6H_2O$ | | $SmCl_3 \cdot 6H_2O$ |
|---|---|---|---|---|---|---|
| $C_{p\,298.15}$ | 103.0 | 103.0 [a)] | 100.2 [a)] | 86.25 | 86.2 [a)] | 86.4 |
| Ref. | [4] | [7] | [7] | [4] | [7] | [4, 7] |

| $MCl_3 \cdot nH_2O$ . . | $EuCl_3 \cdot 6H_2O$ | $GdCl_3 \cdot 6H_2O$ | | $DyCl_3 \cdot 6H_2O$ | | $HoCl_3 \cdot 6H_2O$ | |
|---|---|---|---|---|---|---|---|
| $C_{p\,298.15}$ . . . . | 87.7 | 83.0 | 81.9 [a)] | 82.7 | 81.9 [a)] | 83.0 | 83.0 [a)] |
| Ref. . . . . . | [4, 7] | [4, 7] | [7] | [4] | [7] | [4] | [7] |

| $MCl_3 \cdot nH_2O$ . . | $ErCl_3 \cdot 6H_2O$ | | $YbCl_3 \cdot 6H_2O$ | $LuCl_3 \cdot 6H_2O$ | |
|---|---|---|---|---|---|
| $C_{p\,298.15}$ . . . . | 82.0 | 82.6 [a)] | 81.6 | 82.0 | 81.9 [a)] |
| Ref. . . . . . | [4] | [7] | [4, 7] | [4] | [7] |

[a)] Extrapolated from data in the works of Pfeffer [8 to 11].

For older values see Cobble [14]. $C_p = 354.1$, 354.4, 351.6, and 345.3 $J \cdot mol^{-1} \cdot K^{-1}$ at 298.15 K for the hexahydrates of $GdCl_3$, $TbCl_3$, $HoCl_3$, and $LuCl_3$. Numerical results between 0 and 300 K are given in "Rare Earth Elements" C 4b. The curves for the temperature dependence of the lattice contribution to the heat capacity are similar for the four hydrates, Spedding et al. [13], see also Rulf [14]. Concerning the course of the molar heat capacities of $LaCl_3 \cdot nH_2O$ and $PrCl_3 \cdot nH_2O$ (n probably $\approx 7$) between 4.8 and 260 K, see $LaCl_3 \cdot 7H_2O$ "Rare Earth Elements" C 4b, Section 11.2.11.3.3.1.2.

References:

[1] J.P. Flynn, F.H. Spedding (ISC-379 [1953] 1/82, 65). – [2] F.H. Spedding, J.P. Flynn (J. Am. Chem. Soc. **76** [1954] 1474/7). – [3] D.D. Wagman, W.H. Evans, V.B. Parker et al. (NBS-TN-270-5 [1971] 1/37). – [4] R.H. Schumm, D.D. Wagman, S. Bailey, W.H. Evans, V.B. Parker (NBS-TN-270-7 [1973] 1/75). – [5] R.L. Montgomery (U.S. Bur. Mines Rept. Invest. No. 5468 [1959] 1/23, 19).

[6] R.L. Montgomery, T.D. Hubert (U.S. Bur. Mines Rept. Invest. No. 5525 [1959] 1/8). – [7] R.J. Hinchey, J.W. Cobble (Inorg. Chem. **9** [1970] 917/21). – [8] K.H. Hellwege, U. Johnsen, W. Pfeffer (Z. Physik **154** [1959] 301/9). – [9] K.H. Hellwege, F. Küch, K. Niemann, W. Pfeffer (Z. Physik **162** [1961] 358/62). – [10] W. Pfeffer (Z. Physik **164** [1962] 295/302).

[11] W. Pfeffer (Z. Physik **162** [1961] 413/20). – [12] J.W. Cobble (Ann. Rev. Phys. Chem. **17** [1966] 15/36, 30). – [13] F.H. Spedding, D.C. Rulf, B.C. Gerstein (J. Chem. Phys. **56** [1972] 1498/506). – [14] D.C. Rulf (IS-T-378 [1970] 1/124).

### 11.2.7.9 Magnetic Properties

Of the hydrated rare earth trichlorides so far studied, the following are magnetically ordered at very low temperatures; $T_C$ = transformation temperature for ferromagnetism FM, or $T_N$ for antiferromagnetism AFM:

| $MCl_3 \cdot nH_2O$ . . . | $CeCl_3 \cdot 7H_2O$ | $GdCl_3 \cdot 6H_2O$ | $DyCl_3 \cdot 6H_2O$ | $ErCl_3 \cdot 6H_2O$ |
|---|---|---|---|---|
| $T_C$ or $T_N$ in K . . | $\approx 0.07$ (FM) | 0.182 (AFM) | 0.289 (FM) | 0.356 (FM) |

Whether ferromagnetism or antiferromagnetism occurs is in part still controversial. The origin of the magnetic ordering is (as in the anhydrous trichlorides) the combined effect of dipole-dipole and exchange interactions. $CeCl_3 \cdot 7H_2O$ is notable insofar as it may be one of the few examples of an exchange pair system (like $Cu(NO_3)_2 \cdot 2.5H_2O$): every two $Ce^{3+}$ ions are coupled into an isolated pair by an isotropic antiferromagnetic interaction.

Above $T_C$ or $T_N$, but still at very low temperatures (measurements at higher temperatures apparently do not exist), the paramagnetic susceptibility obeys the Curie-Weiss law. In the monoclinic crystal system to which the investigated hexahydrates belong

($CeCl_3 \cdot 7H_2O$ crystallizes triclinically), only one of the mutually orthogonal principal axes x′, y′, z′ of the susceptibility tensor, namely y′, coincides with one of the crystal axes, the b axis; the other two lie in the monoclinic a-c plane.

The paramagnetic spin-lattice relaxation appears to be derived in almost all hydrates studied from two mechanisms, one direct (at lower temperature, high field strength) and another for which the Raman or Orbach processes are involved. Additional data on magnetic properties are found in the Sections 11.2.7.5, p. 200 and 11.2.7.6, p. 202.

### 11.2.7.10 Optical Properties

#### Color

The following colors are reported for the transparent crystals:

| | |
|---|---|
| Heptahydrates | |
| $LaCl_3 \cdot 7H_2O$ | colorless [1] from [2] |
| $CeCl_3 \cdot 7H_2O$ | colorless [3, 14] |
| $PrCl_3 \cdot 7H_2O$ | green [4]; in thin layers yellow-green [5] |

In the papers [2, 3, 5] these are reported as hexahydrates, see 11.2.7.3.1, p. 185.

| | | | |
|---|---|---|---|
| Hexahydrates | | | |
| $NdCl_3 \cdot 6H_2O$ | yellow [5] | $TbCl_3 \cdot 6H_2O$ | colorless [9, 12] |
| $PmCl_3 \cdot 6H_2O$ | green [6, 7] | $DyCl_3 \cdot 6H_2O$ | yellow [13]; faintly yellow [9] |
| $SmCl_3 \cdot 6H_2O$ | yellow [8, 9]; greenish yellow [10] | $HoCl_3 \cdot 6H_2O$ | honey yellow [9] |
| $EuCl_3 \cdot 6H_2O$ | colorless [9, 11] | $ErCl_3 \cdot 6H_2O$ | light pink [9] |
| $GdCl_3 \cdot 6H_2O$ | colorless [9, 10] | $TmCl_3 \cdot 6H_2O$ | light green [9] |

#### Optical Constants

The compounds $MCl_3 \cdot 6H_2O$ with M=Y, Sm to Tm are optically negative. Values for the refractive indices $n_x$, $n_y$, $n_z$ by the immersion method (±0.002 accuracy) and the optical axis angle $2V_x$ at the 589.3 nm Na-D line are:

| $MCl_3 \cdot 6H_2O$ | $n_x$ | $n_y$ | $n_z$ | $2V_x$ | Ref. |
|---|---|---|---|---|---|
| $YCl_3 \cdot 6H_2O$ | | | | 68.5° | [15] |
| $SmCl_3 \cdot 6H_2O$ | 1.564 | 1.569 | 1.573 | large | [10] |
| $SmCl_3 \cdot 6H_2O$ | 1.570 | 1.578 | 1.582 | ≈60° | [9] a) |
| $EuCl_3 \cdot 6H_2O$ | 1.5702 | 1.5784 | 1.5816 | 69° | [11] |
| $EuCl_3 \cdot 6H_2O$ | 1.570 | 1.578 | 1.582 | ≈70° | [9] |
| $GdCl_3 \cdot 6H_2O$ | 1.565 | 1.570 | 1.575 | ≈75° | [10] |
| $GdCl_3 \cdot 6H_2O$ | 1.572 | 1.580 | 1.584 | ≈75° | [9] |
| $TbCl_3 \cdot 6H_2O$ | 1.572 | 1.580 | 1.584 | ≈75° | [9] |
| $DyCl_3 \cdot 6H_2O$ | 1.574 | 1.580 | 1.586 | ≈75° | [9] a) |
| $HoCl_3 \cdot 6H_2O$ | 1.574 | 1.582 | 1.586 | ≈80° | [9] |
| $ErCl_3 \cdot 6H_2O$ | 1.574 | 1.582 | 1.586 | ≈80° | [9] |
| $TmCl_3 \cdot 6H_2O$ | 1.574 | 1.582 | 1.586 | 80° to 85° | [9] a) |

a) R.E. Thoma, H. Insley, according to Graeber et al. [9].

In all of these crystals $n_y = n_b$; $n_x$ and $n_z$ run in relation to the structure of Fig. 28, p. 191, approximately parallel to the diagonals of the spatial quadrangles defined by the $H_2O$ molecules O(2) and O(3), as shown by Bel'skii, Struchkov [11] for $EuCl_3 \cdot 6H_2O$. Extinction angles for the crystallographic orientation a>c>b at the 589.3 nm Na-D line between $n_x$ and c are 52° and 36° for $YCl_3 \cdot 6H_2O$ [15] and $EuCl_3 \cdot 6H_2O$ [11], respectively. Angles between $n_z$ and c are −33° and −50° for $SmCl_3 \cdot 6H_2O$ and $GdCl_3 \cdot 6H_2O$, respectively [10]. In another crystallographic and optical orientation (c, b, a; $n_y$, $n_z$, $n_x$), the absolute values determined for the extinction angle between $n_x$ and a (corresponding to the angle between $n_z$ and c in the above-chosen orientation) are 33° in $SmCl_3 \cdot 6H_2O$ for yellow light, Lämmermann [16]; 34° according to Kahle [17] or 37° according to Nordmann [18] in $EuCl_3 \cdot 6H_2O$ for green light; 35.8° in $DyCl_3 \cdot 6H_2O$ for blue light, Gramberg [19].

The plane of the optical axes is (010). In the $SmCl_3 \cdot 6H_2O$, $GdCl_3 \cdot 6H_2O$ [10], and $EuCl_3 \cdot 6H_2O$ crystals one of the optical axes is nearly perpendicular to the a axis [11] and also approximately to the cleavage surface (100), Pabst [10].

**Spectra**

Data concerning the IR and Raman spectra of these compounds are found under "Lattice Vibrations", Section 11.2.7.4, p. 194. The spectra in the visible and UV region are determined by electronic transitions in the trivalent rare earth ions and are discussed in another volume of this Handbook.

References:

[1] C. de Marignac (Oeuvres Completes, Vol. 1, Genf - Paris - Berlin 1902, pp. 355/426, 362). – [2] P. Groth (Chemische Krystallographie, Pt. 1, Leipzig 1906, p. 250). – [3] A. Fock (Z. Krist. **22** [1894] 29/42, 32/4). – [4] C.F. Davis, A.F. Kip, R. Malvano (Atti Accad. Nazl. Lincei Rend. Classe Sci. Fis. Mat. Nat. [8] **11** [1951] 77/9). – [5] H.C. Jones, J.A. Anderson (Carnegie Inst. Wash. Publ. No. 110 [1909] 68/99, 72, 94). (For references [6 to 19] see p. 213).

Table 11.2.7-3
Thermal Decomposition of Rare Earth Chloride Hydrates.

| $MCl_3 \cdot nH_2O$ | dehydration | | |
|---|---|---|---|
| | temp. range in °C | $t_{max}$ (DTG) in °C | $t_{max}$ (DTA) in °C |
| $YCl_3 \cdot 6H_2O$ | 160 to 220 | 185, 230 | 190, 240 |
| $LaCl_3 \cdot 7H_2O$ | 95 to 200 | 100, 150, 160, 200 | 100, 150, 160, 200 |
| $CeCl_3 \cdot 7H_2O$ | 100 to 240 | 140, 160, 235 | 100, 160, 235 |
| $PrCl_3 \cdot 7H_2O$ | 110 to 250 | 120, 160, 180, 230 | 120, 160, 180, 230 |
| $NdCl_3 \cdot 6H_2O$ | 115 to 270 | 140, 180, 200, 250 | 140, 180, 200, 250 |
| $SmCl_3 \cdot 6H_2O$ | 140 to 300 | 150, 170, 200, 270 | 150, 170, 200, 270 |
| $EuCl_3 \cdot 6H_2O$ | 150 to 270 | 160, 190, 210, 280 | 160, 190, 210, 280 |
| $GdCl_3 \cdot 6H_2O$ | 160 to 310 | 180, 200, 210, 300 | 160, 180, 200, 280 |
| $DyCl_3 \cdot 6H_2O$ | 160 to 290 | 160, 200, 270 | 160, 200, 270 |
| $HoCl_3 \cdot 6H_2O$ | 150 to 210 | 160, 170, 200 | 160, 170, 200 |
| $ErCl_3 \cdot 6H_2O$ | 150 to 215 | 160, 200 | 160, 200 |
| $TmCl_3 \cdot 6H_2O$ | 150 to 250 | 165, 180, 205 | 165, 180, 205 |
| $YbCl_3 \cdot 6H_2O$ | 150 to 220 | 170, 190, 220 | 170, 190, 220 |
| $LuCl_3 \cdot 6H_2O$ | 140 to 220 | 150, 175, 210 | 150, 175, 210 |

[6] F. Weigel, V. Scherer (Radiochim. Acta **7** [1967] 40/6). – [7] G.W. Parker, P.M. Lantz (ORNL-75 [1948]; AECD-2160 [1948]). – [8] F.H. Spedding, R.S. Bear (Phys. Rev. [2] **42** [1932] 58/75, 59). – [9] E.J. Graeber, G.H. Conrad, S.F. Duliere (Acta Cryst. **21** [1966] 1012/3). – [10] A. Pabst (Am. J. Sci. [5] **22** [1931] 426/30).

[11] N.K. Bel'skii, Yu.T. Struchkov (Kristallografiya **10** [1965] 21/8; Soviet Phys.-Cryst. **10** [1965] 15/20). – [12] G. Urbain, G. Jantsch (Compt. Rend. **146** [1908] 127/9). – [13] J.V. Ugro (Diss. Iowa State Univ. Sci. Technol. 1967, p. 39; Diss. Abstr. B **28** [1967] 1891). – [14] E.J. Peterson, E.I. Onstott, R.B. v. Dreele (Acta Cryst. B **35** [1979] 805/9). – [15] S. Singh from G.H. Dieke, H.M. Crosswhite (J. Opt. Soc. Am. **46** [1956] 885/9).

[16] H. Lämmermann (Z. Physik **150** [1958] 551/72, 563). – [17] H.G. Kahle (Z. Physik **155** [1959] 157/69, 157/8). – [18] H. Nordmann (Phys. Status Solidi A **2** [1970] 475/84, 479). – [19] G. Gramberg (Z. Physik **159** [1960] 125/48, 136).

## 11.2.7.11 Chemical Reactions

### 11.2.7.11.1 Reactions on Heating

The thermal decomposition of $MCl_3 \cdot 7H_2O$ (M = La, Ce, Pr) and $MCl_3 \cdot 6H_2O$ (M = Y, Nd to Lu except Tb) was examined by differential thermographic and thermoanalytic methods (DTG, DTA) from 20 to 1000 °C ± 5 °C with a heating rate of 10 K/min in quartz crucibles in air (?). For the dehydration to $MCl_3$ (contaminated by MOCl), a maximum of 4 steps is found. The temperature of the initial dehydration increases from La to Gd but does not change significantly from Dy to Lu. The final decomposition products are oxidechlorides, sesquioxides, mixtures thereof, or $CeO_2$. The decomposition temperature decreases from La to Lu. Details are given in the Table 11.2.7-3 from Sokolova et al. [1] (see also Abalduev et al. [2]).

These results are at variance with earlier measurements. On heating $MCl_3 \cdot nH_2O$ in a stream of dry air or in air-HCl mixtures (1:1) at a heating rate of 0.5 to 2.5 K/min

| oxidation | | | final product |
|---|---|---|---|
| temp. range in °C | $t_{max}$ (DTG) in °C | $t_{max}$ (DTA) in °C | |
| 360 to 460 | 430 | 430 | YOCl; $Y_3O_4Cl$; $Y_2O_3$ |
| 430 to 560 | 540 | 500 | LaOCl |
| | | | $CeO_2$ |
| 410 to 510 | 510 | 470 | PrOCl |
| 380 to 530 | 490 | 470 | NdOCl |
| 320 to 430 | 380 | 380 | SmOCl |
| 350 to 410 | 380 | 380 | $Eu_2O_3$; EuOCl |
| 360 to 410 | 380 | 380 | $Gd_2O_3$; GdOCl |
| 270 to 410 | 390 | 390 | $Dy_2O_3$ |
| 340 to 390 | 360 | 360 | $Ho_2O_3$ |
| 390 to 415 | 400 | 400 | $Er_2O_3$ |
| 360 to 400 | 380 | 380 | $Tm_2O_3$ |
| 360 to 410 | 390 | 400 | $Yb_2O_3$ |
| 300 to 360 | 330 | 330 | $Lu_2O_3$ |

$MCl_3$ was obtained only for M = La to Nd. For M = Y, Sm to Lu anhydrous $MCl_3$ is found only in air-HCl with partial intermediate formation of $M(OH)Cl_2$, $MOCl \cdot H_2O$, and $MOCl \cdot MCl_3 \cdot 2H_2O$. This is explained by the higher temperature of formation of MOCl, depending on the HCl partial pressure. Final products were always MOCl or $CeO_2$ (cf. "Rare Earth Elements" C 4b, Section 11.2.11.4.4.1). For the heavier M noticeable decomposition of MOCl to $M_2O_3$ takes place, Haeseler, Matthes [3]. Similar data are given for the decomposition of $MCl_3 \cdot n\,H_2O$ (M = Sc to Lu) in a slow stream of air at a heating rate of 5.4 and 10.5 K/min. Under these conditions ScOCl is not observed (see, however, contradictory results given in "Rare Earth Elements" C 4b, Section 11.2.11.1.12), Wendlandt, Bear [4], Wendlandt [5, 6]. See the sections on the individual compounds for details on the various steps of dehydration and decomposition. Thermal decomposition of $MCl_3 \cdot 6H_2O$ (M = Ce, Gd, Tb) in air, Ar, or Ar-$H_2$ at a heating rate of 1 to 5 K/min is also reported by Haire [7]. Dehydration of $MCl_3 \cdot 6H_2O$ (M = Sm, Eu, Yb) in vacuum starts at ≈50, 40, and 35 °C respectively. Final products are SmOCl or mixtures of $MCl_3$, $MCl_2$, and MOCl (M = Eu, Yb). The pure $MCl_3$ is not observed, Il'in et al. [8].

When oxidation is avoided by heating to 280 °C in an $N_2$ stream at 4, 8, and 12 K/min $MCl_3$ is obtained. The following table gives the heats of decomposition ΔH in kcal/mol of $MCl_3 \cdot n\,H_2O$ ($n > 6$ since the compounds contained some occluded water) and of their intermediate dehydration products with the numbers in parentheses designating the numbers of $H_2O$ molecules present in the hydrate after each stage of dehydration to which the specific ΔH values belong:

| | | | | |
|---|---|---|---|---|
| $LaCl_3 \cdot 6.7H_2O$ | 53.9 (3) | 9.1 (2) | 20.0 (1) | 12.1 ± 1.0 (0) |
| $CeCl_3 \cdot 6.4H_2O$ | 44.1 (3) | 18.0 (2) | 14.1 (1) | 11.8 ± 0.2 (0) |
| $PrCl_3 \cdot 6.1H_2O$ | 42.8 (3) | 15.1 (2) | 12.4 (1) | 12.4 ± 0.3 (0) |
| $NdCl_3 \cdot 6.2H_2O$ | 40.3 (3) | 14.4 (2) | 12.8 (1) | 12.2 ± 0.3 (0) |
| $SmCl_3 \cdot 6.4H_2O$ | 36.9 (4) | 18.0 (2) | 14.9 (1) | 10.8 ± 0.7 (0) |
| $EuCl_3 \cdot 6.5H_2O$ | 44.7 (3) | 16.7 (2) | 13.9 (1) | 12.6 ± 1.0 (0) |
| $GdCl_3 \cdot 6.5H_2O$ | 43.1 (3) | 17.8 (2) | 12.7 (1) | 13.0 ± 0.3 (0) |
| $TbCl_3 \cdot 7.1H_2O$ | 37.7 (4) | 24.6 (2) | 14.4 (1) | 10.1 ± 1.0 (0) |
| $DyCl_3 \cdot 7.3H_2O$ | 54.2 (3) | — | 36.3 (1) | 15.0 ± 1.0 (0) |
| $HoCl_3 \cdot 7.0H_2O$ | — | 65.7 (2) | 13.5 (1) | 13.1 ± 0.4 (0) |
| $ErCl_3 \cdot 7.1H_2O$ | 47.7 (4) | 22.6 (2) | 13.2 (1) | 11.3 ± 0.7 (0) |
| $TmCl_3 \cdot 6.8H_2O$ | 41.8 (4) | 23.3 (2) | 9.8 (1) | 9.8 ± 0.6 (0) |
| $YbCl_3 \cdot 7.4H_2O$ | 44.7 (4) | 23.2 (2) | 12.0 (1) | 10.0 ± 0.3 (0) |
| $LuCl_3 \cdot 7.2H_2O$ | 45.4 (4) | 23.3 (2) | 16.2 (1) | 7.0 ± 0.3 (0) |

The temperature at which dehydration begins increases along the series in accord with decreasing radius of the ions, Ashcroft, Mortimer [9]. The dehydration of $MCl_3 \cdot n\,H_2O$ and formation of anhydrous $MCl_3$ in a $CCl_4/N_2$ atmosphere, Revzin [10], in dry HCl, Kleinheksel, Kremers [11], Jantsch et al. [12], in the presence of $NH_4Cl$, Jensen [13], Krestov et al. [14], is described in Section 11.2.6.1.4, p. 62.

Anhydrous chlorides free from oxide chloride and oxides are formed by refluxing chloride hydrates with $SOCl_2$. The presence of air or moisture causes no problems. The approximate times necessary for the dehydration of $MCl_3 \cdot n\,H_2O$ for the different rare earths are:

| M | La | Ce | Pr | Nd | Sm | Gd | Dy | Er |
|---|---|---|---|---|---|---|---|---|
| time in h | 1 | 3 | 4 | 4 | 5 | 15 | 50 | 110 |

Freeman, Smith [5].

References:

[1] L.G. Sokolova, A.V. Lapitskaya, A.F. Bol'shakov, S.B. Pirkes, B.V. Abalduev (Zh. Neorgan. Khim. **26** [1981] 1736/41). – [2] B.V. Abalduev, A.F. Bol'shakov, L.G. Sokolova, A.V. Lapitskaya, S.B. Pirkes (Issled. Obl. Khim. Redkozem. Elem. **1975** 47/9; Ref. Zh. Met. **1975** Abstr. No. 11G279; C.A. **84** [1976] No. 98768). – [3] G. Haeseler, F. Matthes (J. Less-Common Metals **9** [1965] 133/51). – [4] W.W. Wendlandt, J.L. Bear (Anal. Chim. Acta **21** [1959] 439/42). – [5] W.W. Wendlandt (J. Inorg. Nucl. Chem. **9** [1959] 136/9).

[6] W.W. Wendlandt (J. Inorg. Nucl. Chem. **5** [1958] 118/22). – [7] R.G. Haire (in: J.C. Kevane, T. Moeller, Proc. 10th Rare Earth Res. Conf., Carefree, Ariz., 1973 Vol. 2, pp. 882/91; CONF-730402-P2 [1973] 882/91). – [8] V.K. Il'in, V.A. Krenev, V.I. Evdokimov (Zh. Neorgan. Khim. **17** [1972] 2853/4; Russ. J. Inorg. Chem. **17** [1972] 1497/8). – [9] S.J. Ashcroft, C.T. Mortimer (J. Less-Common Metals **14** [1968] 403/6). – [10] G.E. Revzin (Metody Poluch. Khim. Reaktivov Prep. **1967** No. 16, pp. 124/9; C.A. **69** [1968] No. 48783).

[11] J.H. Kleinheksel, H.C. Kremers (J. Am. Chem. Soc. **50** [1928] 959/67, 962). – [12] G. Jantsch, H. Grubitsch, F. Hoffmann, H. Alber (Z. Anorg. Allgem. Chem. **185** [1929] 49/64, 53). – [13] B.S. Jensen (Nucl. Instr. **1** [1957] 323). – [14] G.A. Krestov, V.A. Kobenin, S.V. Semenovskii (Izv. Vysshikh Uchebn. Zavedenii Khim. Khim. Tekhnol. **14** [1971] 462/4; C.A. **75** [1971] No. 29414). – [15] J.H. Freeman, M.L. Smith (J. Inorg. Nucl. Chem. **7** [1958] 224/7).

### 11.2.7.11.2 Reactions with Water, Acids, and Salt Solutions

For the solubility of $MCl_3 \cdot nH_2O$ in water and water-containing solvents see Section 11.2.6.12.4, p. 159. The properties of solutions are treated in Section 11.2.8, p. 216.

Heats of solution were measured by [1 to 8] and extrapolated to infinite dilution by Hinchey, Cobble [6]. Values of $\Delta H^\circ_{298}$ in kcal/mol are given in the table together with the calculated free energies $\Delta G^\circ_{298}$ in kcal/mol and the entropies $\Delta S^\circ_{298}$ in $cal \cdot mol^{-1} \cdot K^{-1}$, from [6]:

| $MCl_3 \cdot nH_2O$ | $-\Delta H^\circ_{298}$ | $-\Delta G^\circ_{298}$ | $-\Delta S^\circ_{298}$ | Ref. |
|---|---|---|---|---|
| $YCl_3 \cdot 6H_2O$ | 11.052 | 7.914 | 10.53 | [2, 3] |
| $LaCl_3 \cdot 7H_2O$ | 7.373 | 5.783 | 5.34 | [3, 4] |
| $CeCl_3 \cdot 7H_2O$ | 6.898 | 5.714 | 3.97 | [2, 5] |
| $PrCl_3 \cdot 7H_2O$ | 7.041 | 5.969 | 3.60 | [3] |
| $NdCl_3 \cdot 6H_2O$ | 9.103 | 6.665 | 8.18 | [2, 5] |
| $SmCl_3 \cdot 6H_2O$ | 9.029 | 6.337 | 9.03 | [6] |
| $EuCl_3 \cdot 6H_2O$ | 8.770 | 6.471 | 7.42 | [2, 5] |
| $GdCl_3 \cdot 6H_2O$ | 8.423 | 6.566 | 6.23 | [7] |
| $TbCl_3 \cdot 6H_2O$ | 9.556 | 6.703 | 9.57 | [8] |
| $DyCl_3 \cdot 6H_2O$ | 10.067 | 6.915 | 10.58 | [7] |
| $HoCl_3 \cdot 6H_2O$ | 10.325 | 7.266 | 10.26 | [2, 3] |
| $ErCl_3 \cdot 6H_2O$ | 11.314 | 7.510 | 12.76 | [2, 3] |
| $YbCl_3 \cdot 6H_2O$ | 10.798 | 7.848 | 9.93 | [6] |
| $LuCl_3 \cdot 6H_2O$ | 11.910 | 7.873 | 13.55 | [1] |

Solution enthalpies were also measured by Pepple [9]. Values (in kcal/mol) of $\bar{L}$ ($\pm$0.4%) $= \phi_L - \Delta H$ (sol.) at infinite dilution and 25 °C are as follows:

| | | | | |
|---|---|---|---|---|
| $MCl_3 \cdot nH_2O$ . . | $LaCl_3 \cdot 7H_2O$ | $PrCl_3 \cdot 7H_2O$ | $NdCl_3 \cdot 6H_2O$ | $SmCl_3 \cdot 6H_2O$ |
| $\bar{L}'$ . . . . . . | 6.684 | 7.000 | 9.134 | 8.611 |
| $MCl_3 \cdot nH_2O$ . . | $EuCl_3 \cdot 6H_2O$ | $GdCl_3 \cdot 6H_2O$ | $DyCl_3 \cdot 6H_2O$ | $ErCl_3 \cdot 6H_2O$ |
| $\bar{L}'$ . . . . . . | 8.707 | 9.112 | 9.959 | 10.735 |
| $MCl_3 \cdot nH_2O$ . . | $TmCl_3 \cdot 6H_2O$ | $YbCl_3 \cdot 6H_2O$ | $LuCl_3 \cdot 6H_2O$ | |
| $\bar{L}'$ . . . . . . | 11.112 | 11.449 | 11.851 | |

$\bar{L}'$ increases regularly from $GdCl_3 \cdot 6H_2O$ to $LuCl_3 \cdot 6H_2O$ including the values for $TbCl_3 \cdot 6H_2O$ and $HoCl_3 \cdot 6H_2O$ by [7], DeKock [12], see the figure in [9]. For the heats of solution of $MCl_3 \cdot 6H_2O$ (M = La, Pr, Nd, Sm, Gd) and $LaCl_3 \cdot 5H_2O$ in aqueous solutions of salting-out agents, e.g., $MgCl_2$ and $NH_4Cl$, at 25 °C see Nosova [10].

Regarding the enthalpy of solution of $LaCl_3 \cdot 5H_2O$, $PrCl_3 \cdot 6H_2O$, $NdCl_3 \cdot 6H_2O$, and $GdCl_3 \cdot 6H_2O$ in aqueous solution of the chlorides of Li, Na, K, Cs, Mg, Ca, and Ba, see Aleshko-Ozhevskii [11].

References:

[1] H.H. Petheram, F.H. Spedding (IS-770 [1963] 1/69; N.S.A. **18** [1964] No. 13722). – [2] F.H. Spedding, J.P. Flynn (J. Am. Chem. Soc. **76** [1954] 1474/7). – [3] F.H. Spedding, A.W. Naumann, R.E. Eberts (J. Am. Chem. Soc. **81** [1959] 23/8). – [4] F.H. Spedding, C.F. Miller (J. Am. Chem. Soc. **74** [1952] 4195/8; ISC-167 [1951] 1/131; N.S.A. **6** [1952] No. 560; ISC-191 [1951] 1/14; N.S.A. **6** [1952] No. 840). – [5] F.H. Spedding, D.A. Csejka, C.W. DeKock (J. Phys. Chem. **70** [1966] 2423/9).

[6] R.J. Hinchey, J.W. Cobble (Inorg. Chem. **9** [1970] 917/21). – [7] F.H. Spedding, C.W. DeKock (IS-1356 [1965] 1/162, 44; C.A. **66** [1967] No. 59520). – [8] D.A. Csejka, F.H. Spedding (IS-417 [1961] 1/143; N.S.A. **16** [1962] No. 31584). – [9] G.W. Pepple (Diss. Iowa State Univ. 1967, pp. 1/124, 87; Diss. Abstr. B **28** [1967/68] 1455). – [10] T.A. Nosova (Radiokhimiya **8** [1966] 477/9; Soviet Radiochem. **8** [1966] 440/2).

[11] Yu.P. Aleshko-Ozhevskii (Zh. Fiz. Khim. **45** [1971] 988/9; Russ. J. Phys. Chem. **45** [1971] 554/5). – [12] C.W. DeKock (Diss. Iowa State Univ. 1967, pp. 1/166, 130; Diss. Abstr. **26** [1965] 112).

## 11.2.8 Comparative Data for $MCl_3$ Solutions

### 11.2.8.1 Aqueous $MCl_3$ Solutions

Aqueous $MCl_3$ solutions are easily obtainable by dissolving the oxides or carbonates in hydrochloric acid as well as by direct dissolution of the anhydrous or hydrated chlorides in water. The solubilities in the $MCl_3$-HCl-$H_2O$ systems are treated in Section 11.2.6.12.4, p. 159, together with the thermodynamics of the dissolution process.

#### 11.2.8.1.1 Structure of the Solutions

Chloro-complexes do not occur in dilute aqueous $MCl_3$ solutions. The numerous measurements conducted over a wide concentration range for various physical properties are explained by a preponderant interaction of the $M^{3+}$ ions with the water, i.e., the inner coordination sphere around the $M^{3+}$ ions is occupied by $H_2O$ molecules. In the presence of high concentrations of $Cl^-$ ions (produced by addition of HCl or alkali chlorides), $MCl^{2+}$ and occasionally also $MCl_2^+$ ions are observed. Whether the $Cl^-$ ions thus partially enter into the inner coordination sphere is not yet completely clarified. In any case, it is certain that the metal ions in pure aqueous $MCl_3$ solutions retain their first hydration sphere up to high $MCl_3$ concentrations. The coordination number shrinks with increasing atomic number of the lanthanide ions in the range between Nd and Tb, namely probably from 9 to 8, as for example is shown by the trend of the molar volumes $V_m$ measured by Spedding et al. [1, 2] as a function of the ionic radius $r_M$. The transition region, in which $V_m$ no longer changes uniformly with $r_M$, is observed both at infinite dilution and also for saturated solutions [1]. The change of the water coordination number in this transition region is also shown by the trend of the viscosity, Spedding et al. [3], the heat capacity, Pepple [4], the activity coefficients, Spedding et al. [5], and the electrical conductance, Spedding et al. [6].

X-ray diffraction measurements on a nearly saturated aqueous $PrCl_3$ solution yield for the inner coordination sphere an $H_2O$ coordination number of 9, on nearly saturated $TbCl_3$ and $ErCl_3$ solutions of 8, Martin, Spedding [7]. A decrease of the inner-sphere water coordination number of the rare earth ions from 9 to 8 between $Nd^{3+}$ and $Tb^{3+}$ has recently also been confirmed by X-ray diffraction measurements on concentrated aqueous $MCl_3$ solutions with M=La, Pr, Nd, Sm, Eu, Tb, Dy, Er, Tm, and Lu, see Habenschuss, Spedding [19 to 21]. An Ar laser ($\lambda$=488.0 nm) has been used to generate the Raman spectra of about 0.5 to 3.5 m aqueous $LaCl_3$, $GdCl_3$, and $LuCl_3$ solutions that contain about 5 mol % $D_2O$. The frequency of the OD or OM bands increases with growing $MCl_3$ concentration with simultaneous line sharpening, up to an $MCl_3$ concentration of about 1.5 m, then the frequency drops and the bands become broader. Analyses of the valence vibration bands can be explained by the assumption that the number of $H_2O$ molecules in the first coordination sphere of $La^{3+}$ is larger than for $Gd^{3+}$ or $Lu^{3+}$, Mundy, Spedding [8]. Comparative absorption spectroscopic studies on 0.028 to 2.24 M $NdCl_3$ solutions and solid hydrated neodymium salts indicate that the $Cl^-$ ions in the $NdCl^{2+}$ species present in the concentrated solutions are arranged in an outer coordination sphere, and thus the $NdCl_2^+$ complexes are actually ionic associates. The aquo-complexes present in dilute solutions possibly occur with two different structures. Since the spectra of 0.025 to 2 M $ErCl_3$ solutions do not change with the concentration, the chloro-complex ions of Er, like the chloro-complexes of Nd, are ionic associates, Davidenko, Lugina [9]. The formation of ion pairs instead of complexes is also supported by a comparison of literature results for measured stability constants of $MCl_3$ solutions (M=La, Ce, Eu, Lu) with theoretically calculated values for 6- and 8-fold hydrated ions, Manning [10]. This is contradictory with the dichloro-complex formulated by Brady [11] with direct bonding [9]. Spectroscopic studies on 0.0054 to 0.452 molar $M(ClO_4)_3$ solutions (M=Pr to Yb), at an ionic strength of 3 M adjusted with LiCl+$LiClO_4$ and 25 °C, show that the observed $MCl^{2+}$ species should be treated as an outer-sphere complex, Kozachenko, Batyaev [12].

The spectroscopic studies of Sayre et al. [13] on 0.01 to 0.5 M aqueous $EuCl_3$ solutions, which yield a $D_{2h}$ symmetry around the $Eu^{3+}$ ion, can be explained by the assumption that the environment of the $Eu^{3+}$ ion is occupied by 7 or 8 water molecules [9]. X-ray studies on 0.95, 1.47, and 3.05 m $ErCl_3$ solutions led Brady [11] to conclude that the $H_2O$ molecules are arranged octahedrally around the $Er^{3+}$ ion. In the middle of two opposite octahedral edges are located two $Cl^-$ ions, in agreement with the field symmetry in $EuCl_3$ solutions found by [13]. The spacings Er↔Cl and $H_2O \leftrightarrow H_2O$ are about 4.6 and 3.2 Å, respectively. In the rest of the solution the $H_2O \leftrightarrow H_2O$ distances are about 2.95 Å. The $Er^{3+}$ field increases the degree of ordering of the solvent, the structure approaches that of ice [11]. A comparison of the absorption spectra of 0.0535 m $NdCl_3$ solutions with additions of 0 to 12 mol/L of HCl, LiCl, and $HClO_4$, with spectra of the solid compounds which have a coordination number for the $Nd^{3+}$ of 9 or 8 (e.g., $Nd(BrO_3)_3 \cdot 9H_2O$ or $Nd_2(SO_4)_3 \cdot 8H_2O$ and $NdCl_3 \cdot 6H_2O$, respectively), shows that the coordination number drops from 9 to 8 at very high concentrations. Supplementary investigations on $HoCl_3$ and $ErCl_3$ solutions lead to comparable results. Raising the temperature has a similar influence, as for example is shown by the spectra of $NdCl_3$ in 11 molar LiCl solutions at 25 and 42 °C, Karraker [14]. Data concerning the distance of closest approach of the ions, given by the Debye-Hückel relationship, i.e., the Debye-Hückel ion-size parameter, are found under the discussion of activity coefficients, Section 11.2.8.1.6, p. 232. For a more comprehensive discussion of the aquo-complexes see "Rare Earth Elements" D 3, 1981, pp. 1/14.

Literature results for the activity coefficients of aqueous $LaCl_3$ solutions and the spectra of $EuCl_3$ solutions are explained by Frank, Evans [15] by assuming a "superstructure" in the solutions. An analysis of literature data on the equilibrium distribution of $MCl_3$ (M especially La, Lu), dissolved in concentrated LiCl solutions, and anion exchangers as well as on extraction equilibria with organic amines, indicates that for LiCl concentrations above about 8 mol/L predominantly $MCl_2^+$ ions exist in the aqueous phase and $MCl_4^-$ ions in the organic phase, Marcus [16]. The nuclear resonance of $^{35}Cl$ in 3 molar NaCl solution after addition of $LaCl_3$, $NdCl_3$, or $SmCl_3$ and an interpretation of the lower than expected resonance absorption in the case of $LaCl_3$ through complex formation are found in Wertz [17].

The difference in the degree of hydration of pairs of $M^{3+}$ ions caused by salting-out agents has been determined by measuring the enthalpies of solution in the salt solutions (LiCl, NaCl, KCl, CsCl, $MgCl_2$, $CaCl_2$, and $BaCl_2$) relative to those in pure $H_2O$. It was found that for all investigated $MCl_3$ pairs, namely La+Pr, La+Nd, Pr+Nd, La+Gd, Pr+Gd, and Nd+Gd, a maximum separation is reached at KCl in the alkali chloride series, and at $CaCl_2$ in the alkaline earth chloride series, Aleshko-Ozhevskii [18].

References:

[1] F.H. Spedding, V.W. Saeger, K.A. Gray et al. (J. Chem. Eng. Data **20** [1975] 72/81). – [2] F.H. Spedding, M.G. Pikal, B.O. Ayers (J. Phys. Chem. **70** [1966] 2440/9). – [3] F.H. Spedding, D.L. Witte, L.E. Shiers, J.A. Rard (J. Chem. Eng. Data **19** [1974] 369/73). – [4] G.W. Pepple (Diss. Iowa State Univ. 1967, pp. 1/124, 105/6; Diss. Abstr. B **28** [1967/68] 1455). – [5] F.H. Spedding, H.O. Weber, V.W. Saeger, H.H. Petheram, J.A. Rard, A. Habenschuss (J. Chem. Eng. Data **21** [1976] 341/60).

[6] F.H. Spedding, J.A. Rard, V.W. Saeger (J. Chem. Eng. Data **19** [1974] 369/73). – [7] L.L. Martin, F.H. Spedding (from [1]). – [8] W.C. Mundy, F.H. Spedding (J. Chem. Phys. **59** [1973] 2183/90). – [9] N.K. Davidenko, L.N. Lugina (Zh. Neorgan. Khim. **13** [1968] 980/7; Russ. J. Inorg. Chem. **13** [1968] 512/7). – [10] P.G. Manning (Can. J. Chem. **43** [1965] 2911/3).

[11] G.W. Brady (J. Chem. Phys. **33** [1960] 1079/82). – [12] N.N. Kozachenko, I.M. Batyaev (Zh. Neorgan. Khim. **16** [1971] 125/7; Russ. J. Inorg. Chem. **16** [1971] 66/7). – [13] E.V. Sayre, D.G. Miller, S. Freed (J. Chem. Phys. **26** [1957] 109/13). – [14] D.G. Karraker (Inorg. Chem. **7** [1968] 473/9). – [15] H.S. Frank, M.W. Evans (J. Chem. Phys. **13** [1945] 507/32, 530/1).

[16] Y. Marcus (J. Inorg. Nucl. Chem. **28** [1966] 209/19). – [17] J.E. Wertz (J. Phys. Chem. **61** [1957] 51/5). – [18] Yu.P. Aleshko-Ozhevskii (Zh. Fiz. Khim. **45** [1971] 988/9; Russ. J. Phys. Chem. **45** [1971] 554/5). – [19] A. Habenschuss, F.H. Spedding

(J. Chem. Phys. **70** [1979] 3758/63). – [20] A. Habenschuss, F.H. Spedding (J. Chem. Phys. **70** [1979] 2797/806).

[21] A. Habenschuss, F.H. Spedding (J. Chem. Phys. **73** [1980] 442/50).

### 11.2.8.1.2 Hydrogen Ion Concentration

The pH value of aqueous trichloride solutions drops with increasing $MCl_3$ concentration, as shown by **Fig. 30** from Spedding et al. [1] for 25 °C. The older values measured by Kleinheksel, Kremers [2] differ in part significantly from these values, as shown in the following table of the pH as a function of the normality N for aqueous solutions of rare earth trichlorides $MCl_3$:

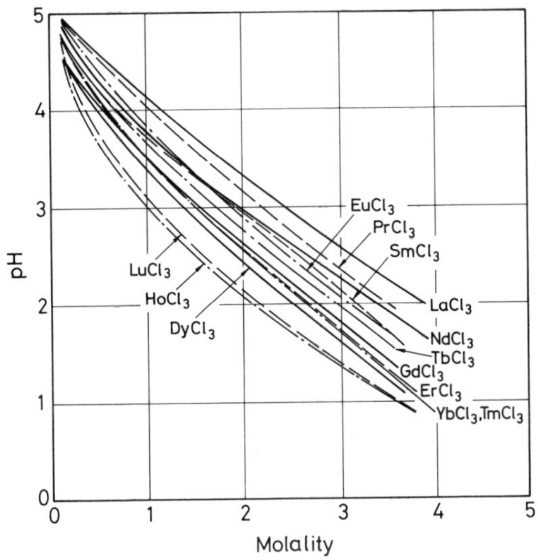


Fig. 30

pH values of aqueous rare earth chloride solutions as a function of molality.

| $MCl_3$ | N | pH | $MCl_3$ | N | pH |
|---|---|---|---|---|---|
| $YCl_3$ | 0.002635 | 1.673 | $SmCl_3$ | 0.004188 | 2.453 |
| | 0.02635 | 1.056 | | 0.04188 | 1.582 |
| | 0.2635 | 0.9971 | | 0.4188 | 1.053 |
| $LaCl_3$ | 0.00495 | 5.497 | $DyCl_3$ | 0.003204 | 5.180 |
| | 0.0495 | 4.279 | | 0.03204 | 4.580 |
| | 0.4950 | 3.375 | | 0.3204 | 4.140 |
| $CeCl_3$ | 0.00578 | 2.659 | $HoCl_3$ | 0.000209 | 2.253 |
| | 0.0578 | 1.876 | | 0.00209 | 1.386 |
| | 0.5780 | 1.293 | | 0.0209 | 0.8957 |
| $PrCl_3$ | 0.00333 | 5.417 | $TmCl_3$ | 0.001108 | 2.679 |
| | 0.0333 | 3.880 | | 0.01108 | 1.638 |
| | 0.3330 | 2.662 | | 0.1108 | 1.095 |
| $NdCl_3$ | 0.004333 | 3.724 | | | |
| | 0.04333 | 2.835 | | | |
| | 0.4333 | 2.319 | | | |

The pH values of the $MCl_3$ solutions indicate that hydrolysis occurs to a lesser extent than for the analogous perchlorate solutions [1]. Complete hydrolysis at infinite dilution is deemed probable by [2].

References:

[1] F.H. Spedding, J.A. Rard, V.W. Saeger (J. Chem. Eng. Data **19** [1974] 373/8). – [2] J.H. Kleinheksel, H.C. Kremers (J. Am. Chem. Soc. **50** [1928] 959/67, 965).

### 11.2.8.1.3 Density and Molar Volume of Aqueous $MCl_3$ Solutions

#### Individual Values at Room Temperature

At 25 °C the pycnometrically determined densities D in $g/cm^3$ for aqueous rare earth trichloride $MCl_3$ solutions are, as a function of the molality m:

| $MCl_3$ | molality m | | | | | | |
|---|---|---|---|---|---|---|---|
| | 0.05 | 0.1 | 0.5 | 1.0 | 1.5 | 2.0 | 2.2 |
| $ScCl_3$ | 1.0039 | 1.0101 | 1.0613 | 1.1209 | 1.1763 | 1.2280 | – |
| $YCl_3$ | 1.0056 | 1.0141 | 1.0820 | 1.1622 | 1.2357 | 1.3090 | – |
| $LaCl_3$ | 1.0082 | 1.0194 | 1.1057 | 1.2075 | 1.3036 | 1.3923 | 1.4266 |
| $CeCl_3$ | 1.0084 | 1.0195 | 1.1074 | 1.2120 | 1.3100 | 1.4020 | 1.4380 |
| $PrCl_3$ | 1.0089 | 1.0204 | 1.1085 | 1.2129 | 1.3115 | 1.4047 | 1.4407 |
| $NdCl_3$ | 1.0090 | 1.0203 | 1.1105 | 1.2190 | 1.3214 | 1.4176 | 1.4553 |

For additional intermediate values, see the paper, Mason [1]. Table 11.2.8-1 presents densities and apparent molar volumes $\Phi_V$ of aqueous $MCl_3$ solutions at 25 °C as a function of the molality m up to the saturation concentration, from pycnometric measurements by Spedding et al. [2]. Individual values in the table at $m \approx 0.02$ to 0.2 mol/kg have also been determined by the magnetically controlled float method, namely for M = La, Pr, Nd, Sm, Gd, Tb, Dy, Ho, Er, and Yb by Spedding et al. [3] and for M = Eu, Tm, and Lu by Spedding et al. [4]. The highest concentration data correspond to the respective saturation concentrations.

Table 11.2.8-1
Density and Apparent Molar Volume of Aqueous $MCl_3$ Solutions at Room Temperature.

| | | | | | | |
|---|---|---|---|---|---|---|
| $YCl_3$ | | | | | | |
| m . . . . . . . | 0.019698 | 0.050454 | 0.11449 | 0.98630 | 2.5070 | 3.9478 |
| D in $g/cm^3$ . . | 1.00059 | 1.00603 | 1.01724 | 1.16016 | 1.37464 | 1.54379 |
| $\Phi_V$ in mL/mol . | 16.29 | 17.15 | 18.30 | 25.37 | 32.17 | 36.52 |
| $LaCl_3$ | | | | | | |
| m . . . . . . . | 0.020724 | 0.055827 | 0.11094 | 0.97118 | 2.5285 | 3.8959 |
| D in $g/cm^3$ . . | 1.00177 | 1.00966 | 1.02191 | 1.20228 | 1.48478 | 1.69433 |
| $\Phi_V$ in mL/mol . | 18.02 | 18.99 | 20.30 | 27.74 | 34.90 | 38.82 |

Table 11.2.8-1 [continued]

| | | | | | | |
|---|---|---|---|---|---|---|
| $PrCl_3$ | | | | | | |
| m | 0.017303 | 0.063399 | 0.11366 | 1.0564 | 2.6858 | 3.8910 |
| D in g/cm³ | 1.00110 | 1.01172 | 1.02318 | 1.22580 | 1.52786 | 1.71626 |
| $\Phi_V$ in mL/mol | 13.96 | 15.41 | 16.53 | 24.57 | 32.11 | 36.06 |
| $NdCl_3$ | | | | | | |
| m | 0.025985 | 0.048954 | 0.11599 | 1.0224 | 2.5811 | 3.9292 |
| D in g/cm³ | 1.00322 | 1.00861 | 1.02421 | 1.22345 | 1.52157 | 1.73794 |
| $\Phi_V$ in mL/mol | 13.40 | 14.11 | 15.60 | 23.33 | 30.71 | 35.38 |
| $SmCl_3$ | | | | | | |
| m | 0.022158 | 0.044960 | 0.11417 | 0.96565 | 2.6945 | 3.6401 |
| D in g/cm³ | 1.00242 | 1.00790 | 1.02435 | 1.21672 | 1.55670 | 1.71441 |
| $\Phi_V$ in mL/mol | 14.52 | 15.07 | 16.67 | 23.49 | 31.10 | 34.45 |
| $EuCl_3$ | | | | | | |
| m | 0.019377 | 0.040986 | 0.10009 | 1.0130 | 2.5873 | 3.5889 |
| D in g/cm³ | 1.00176 | 1.00697 | 1.02105 | 1.22727 | 1.53892 | 1.70875 |
| $\Phi_V$ in mL/mol | 15.81 | 16.08 | 17.71 | 24.78 | 31.37 | 34.78 |
| $GdCl_3$ | | | | | | |
| m | 0.023050 | 0.051437 | 0.12354 | 1.0011 | 2.6051 | 3.5906 |
| D in g/cm³ | 1.00276 | 1.00970 | 1.02717 | 1.22892 | 1.55221 | 1.72291 |
| $\Phi_V$ in mL/mol | 16.27 | 17.27 | 18.76 | 25.50 | 32.14 | 35.33 |
| $TbCl_3$ | | | | | | |
| m | 0.021266 | 0.057714 | 0.11505 | 0.98552 | 2.5095 | 3.5713 |
| D in g/cm³ | 1.00235 | 1.01132 | 1.02533 | 1.22678 | 1.53661 | 1.72349 |
| $\Phi_V$ in mL/mol | 16.47 | 17.54 | 18.51 | 25.69 | 32.32 | 35.56 |
| $DyCl_3$ | | | | | | |
| m | 0.025036 | 0.049460 | 0.12118 | 1.0551 | 2.5897 | 3.3610 |
| D in g/cm³ | 1.00339 | 1.00950 | 1.02730 | 1.24627 | 1.56110 | 1.74509 |
| $\Phi_V$ in mL/mol | 15.69 | 16.78 | 18.20 | 25.67 | 32.30 | 35.67 |
| $HoCl_3$ | | | | | | |
| m | 0.020900 | 0.056431 | 0.11311 | 1.1536 | 2.6690 | 3.6942 |
| D in g/cm³ | 1.00241 | 1.01142 | 1.02571 | 1.27204 | 1.58466 | 1.76694 |
| $\Phi_V$ in mL/mol | 15.25 | 16.16 | 16.95 | 25.34 | 31.86 | 35.25 |
| $ErCl_3$ | | | | | | |
| m | 0.023577 | 0.046163 | 0.11280 | 0.97087 | 2.4989 | 3.7821 |
| D in g/cm³ | 1.00318 | 1.00898 | 1.02596 | 1.23379 | 1.55991 | 1.79389 |
| $\Phi_V$ in mL/mol | 13.97 | 14.80 | 16.41 | 23.57 | 30.59 | 34.74 |
| $TmCl_3$ | | | | | | |
| m | 0.020380 | 0.041670 | 0.10263 | 1.0036 | 2.4182 | 3.8773 |
| D in g/cm³ | 1.00242 | 1.00794 | 1.02368 | 1.24454 | 1.55202 | 1.82148 |
| $\Phi_V$ in mL/mol | 12.23 | 13.68 | 14.95 | 22.49 | 29.08 | 34.06 |
| $YbCl_3$ | | | | | | |
| m | 0.025889 | 0.046279 | 0.11559 | 1.0857 | 2.2400 | 4.0028 |
| D in g/cm³ | 1.00398 | 1.00937 | 1.02758 | 1.26886 | 1.52637 | 1.86137 |
| $\Phi_V$ in mL/mol | 12.01 | 12.73 | 14.33 | 22.33 | 27.79 | 33.76 |

Table 11.2.8-1 [continued]

| | | | | | | |
|---|---|---|---|---|---|---|
| $LuCl_3$ | | | | | | |
| m . . . . . . . | 0.021500 | 0.045520 | 0.10334 | 1.2032 | 2.4095 | 4.1165 |
| D in $g/cm^3$ . . | 1.00286 | 1.00930 | 1.02466 | 1.30015 | 1.56913 | 1.89291 |
| $\Phi_V$ in mL/mol . | 11.44 | 11.87 | 13.28 | 22.07 | 27.54 | 33.32 |

Numerous intermediate values can be found in the papers [2 to 4]. Similar densities of saturated aqueous $MCl_3$ solutions with M=Y and La to Lu at 25 °C are found by Nikolaev et al. [5], cf. p. 225. Additional results for the density and molar volume of aqueous $MCl_3$ solutions at 25 °C for M=La, Ce, Pr, Nd, Sm, Eu, Er, and Yb are given by Spedding et al. [6], for M=La, Nd, Sm, Gd, Dy, Er, and Yb by Saeger, Spedding [7], Saeger [8], for M=Pr, Sm, Gd, Tb, Dy, Ho, and Er by Pikal, Spedding [9], Pikal [10], for M=Eu, Tm, and Lu by Cullen [11, 12]. Data for the apparent molar volume of dilute aqueous $MCl_3$ solutions with M=La, Nd, Er, and Yb are also given by Spedding, Ayers [13].

**Approximation Relationships for the Concentration Dependence of the Density at Room Temperature**

The change of the density D with the molality m at 25 °C can be described by the empirical relationship $D^{25}=D_0^{25}+A_1m+A_2m^{3/2}+A_3m^2+A_4m^{5/2}+A_5m^3$ with an accuracy of $\pm2\times10^{-5}$ $g/cm^3$, where for $D_0^{25}$ (the density of pure water) the literature value of 0.997075 $g/cm^3$ is used. The following constants have been determined by Spedding et al. [2]:

| $MCl_3$ | $A_1$ | $A_2$ | $A_3$ | $A_4$ | $A_5$ |
|---|---|---|---|---|---|
| $YCl_3$ | 0.178986 | −0.005717 | −0.008864 | 0.000827 | −0.000063 |
| $LaCl_3$ | 0.224751 | 0.002423 | −0.021341 | 0.005968 | −0.000893 |
| $PrCl_3$ | 0.233521 | −0.010274 | −0.002719 | −0.004376 | 0.001039 |
| $NdCl_3$ | 0.238275 | −0.012536 | 0.001171 | −0.006628 | 0.001440 |
| $SmCl_3$ | 0.242300 | −0.007779 | −0.006516 | −0.000962 | −0.000005 |
| $EuCl_3$ [a] | 0.241825 | −0.004160 | −0.011170 | 0.001704 | −0.000534 |
| $EuCl_3$ [a] | 0.242397 | −0.006135 | −0.009472 | 0.000993 | −0.000408 |
| $GdCl_3$ | 0.246545 | −0.004738 | −0.011443 | 0.001628 | −0.000419 |
| $TbCl_3$ | 0.249022 | −0.008117 | −0.006771 | −0.001698 | 0.000409 |
| $DyCl_3$ | 0.252882 | −0.007005 | −0.008178 | −0.001154 | 0.000324 |
| $HoCl_3$ | 0.255492 | −0.003264 | −0.014093 | 0.002695 | −0.000583 |
| $ErCl_3$ | 0.259748 | −0.009487 | −0.003585 | −0.004234 | 0.000984 |
| $TmCl_3$ | 0.262170 | −0.006955 | −0.007359 | −0.001589 | 0.000325 |
| $YbCl_3$ | 0.268177 | −0.012089 | 0.000125 | −0.006067 | 0.001249 |
| $LuCl_3$ | 0.270202 | −0.008345 | −0.005401 | −0.002486 | 0.000425 |

[a] Different experimental series.

Constants for a polynomial of the form $D^{25}=A_0+A_1m+A_2m^2+A_3m^3+A_4m^4+A_5m^5$ are given by Spedding et al. [14]. Older approximations for the calculation of the density at 25 °C of aqueous $MCl_3$ solutions with M=La, Ce, Pr, Nd, Sm, Eu, Er, and Yb can be found in Spedding et al. [6], and for M=La, Ce, Pr, and Nd in Mason, Leland [15].

The trend of the apparent molar volume with the molality can according to [2] be described by the Redlich-Meyer equation of the following type:

$$\Phi_V = \Phi_V^0 + S_V \sqrt{D_0} m^{1/2} + B_2 m^{3/4} + B_3 m + B_4 m^{5/4} + B_5 m^{3/2},$$

where $\Phi_V^0$ is the apparent molar volume at infinite dilution, $D_0$ is the density of $H_2O$, and $S_V$ is the Debye-Hückel limiting slope, which for a 3:1 salt has a value of 27.45. The following constants were determined for 25 °C:

| $MCl_3$ | $\Phi_V^0$ | $B_2$ | $B_3$ | $B_4$ | $B_5$ |
|---|---|---|---|---|---|
| $YCl_3$ | 13.354 | −29.465 | 8.250 | 11.995 | −6.130 |
| $LaCl_3$ | 14.688 | −19.407 | −10.901 | 25.934 | −9.841 |
| $PrCl_3$ | 11.237 | −25.617 | 3.867 | 13.352 | −6.035 |
| $NdCl_3$ | 10.367 | −23.051 | − 3.401 | 19.336 | −7.529 |
| $SmCl_3$ | 11.374 | −29.818 | 15.392 | 0.993 | −1.681 |
| $EuCl_3$ | 12.343 | −24.301 | 2.733 | 10.524 | −4.120 |
| $GdCl_3$ | 13.319 | −30.402 | 17.026 | − 0.437 | −1.416 |
| $TbCl_3$ | 13.520 | −30.677 | 14.561 | 4.664 | −3.699 |
| $DyCl_3$ | 12.903 | −29.136 | 12.018 | 5.924 | −3.790 |
| $HoCl_3$ | 12.027 | −27.269 | 5.570 | 12.647 | −5.945 |
| $ErCl_3$ | 10.827 | −21.848 | − 4.539 | 19.385 | −7.511 |
| $TmCl_3$ | 9.479 | −20.321 | − 6.437 | 19.394 | −7.054 |
| $YbCl_3$ | 9.433 | −30.718 | 14.121 | 5.353 | −3.801 |
| $LuCl_3$ | 8.063 | −27.507 | 10.187 | 6.313 | −3.570 |

Fig. 31

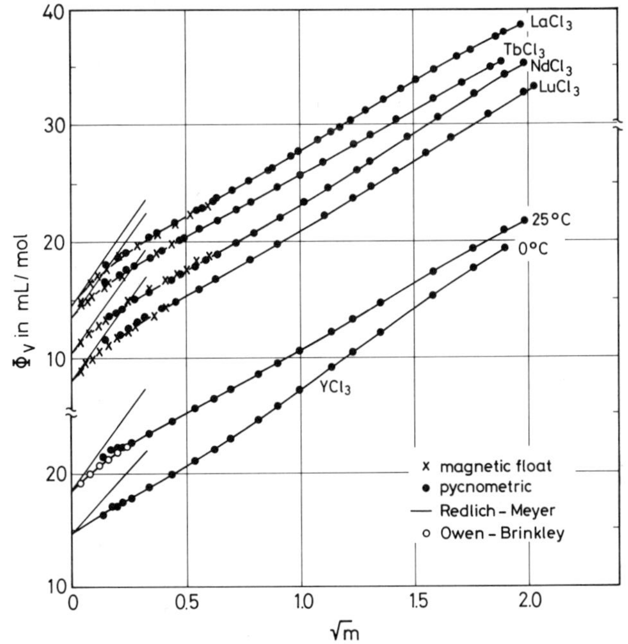


Apparent molar volumes $\Phi_V$ of aqueous rare earth chloride solutions, obtained by various methods. Straight lines represent Debye-Hückel limiting slopes.

In the determination of the constants, the densities were also taken into consideration which had been obtained by the magnetically controlled float method for dilute solutions, by [3] for M = La, Pr, Nd, Sm, Gd, Tb, Dy, Ho, Er, and Yb, and by [4] for M = Eu, Tm, and Lu. The accuracy varies between ±0.5 mL/mol for 0.02 m solutions and ±0.1 mL/mol at the saturation limit. A more precise determination of $\Phi_V^0$ is achieved by extrapolation of the densities of [3, 4] determined by means of the float method with use of a modified Owen-Brinkley equation which takes into account the effect of the ion size [2]. Parameters for the Owen-Brinkley equation and for the approximation relationship for calculation of $\Phi_V$ of dilute aqueous $MCl_3$ solutions are given by [3, 4].

The trend of $\Phi_V$ with $\sqrt{m}$ is portrayed for several $MCl_3$ solutions in **Fig. 31**, p. 223, from [2]. This figure shows both pycnometric and float method values in comparison with those which have been calculated by means of the modified Redlich-Meyer or Owen-Brinkley equations. The Debye-Hückel limiting slope is also shown.

**Fig. 32** from [2] plots the partial molar volume $\bar{V}_2$ of the listed $MCl_3$ compounds for various molal concentrations as a function of the ionic radius of the rare earth elements (for partial molar volumes of the solvent see a figure in the paper). The values for infinite dilution $\bar{V}_2^0$ are identical with $\Phi_V^0$ and are based on the measurements of [3, 4]. The trend of $\bar{V}_2$ with the ionic radius is explained by a reduction of the water coordination number in the inner coordination sphere of the $M^{3+}$ ions in the transition range approximately from M = Nd to Tb, cf. p. 217 [2]. Molar volumes of aqueous $MCl_3$ solutions at 25 °C with M = La, Nd, Sm, Gd, Dy, Er, and Yb for concentrations of about 0.02 mol/kg up to saturation are given in [7, 8]. Results for dilute solutions with M = La, Nd, Er, and Yb can be found in [13], for M = Pr, Sm, Gd, Tb, Dy, Ho, and Er in [9, 10], and for M = Eu, Tm, and Lu in [11, 12].

Fig. 32

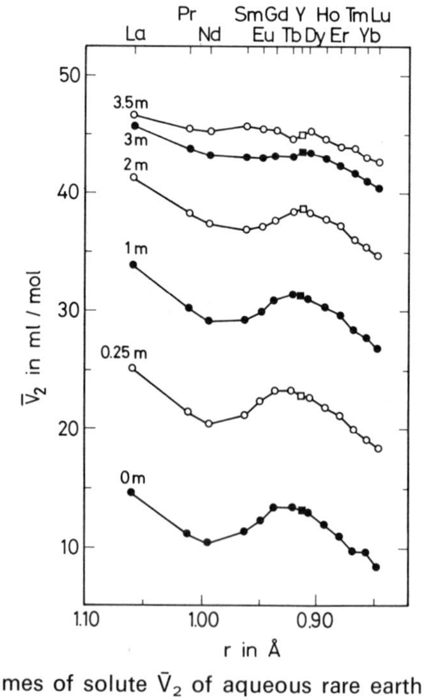


Partial molar volumes of solute $\bar{V}_2$ of aqueous rare earth chloride solutions.

**Densities and Molar Volumes at Various Temperatures and Temperature Functions**

The densities D of saturated aqueous $MCl_3$ solutions at t=10, 25, and 50 °C determined by [5] are given in the following table. The saturation concentration $c_s$ is in wt %. The saturation molality $m_s$ is calculated from the $MCl_3:H_2O$ mole ratio given in [5].

| $MCl_3$ . . . | $YCl_3$ | $LaCl_3$ | $CeCl_3$ | $PrCl_3$ | $NdCl_3$ | $SmCl_3$ | $EuCl_3$ | $GdCl_3$ |
|---|---|---|---|---|---|---|---|---|
| $c_s$ (10 °C) . | 43.4 | 48.2 | 48.1 | 48.2 | 49.3 | 47.8 | 47.6 | 48.3 |
| $m_s$ (10 °C) . | 3.937 | 3.805 | 3.764 | 3.766 | 3.882 | 3.568 | 3.518 | 3.548 |
| D (10 °C) . | 1.545 | 1.677 | 1.693 | 1.700 | 1.731 | 1.711 | 1.703 | 1.717 |
| $c_s$ (25 °C) . | 43.5 | 48.8 | 48.6 | 49.0 | 49.7 | 48.3 | 48.3 | 48.7 |
| $m_s$ (25 °C) . | 3.893 | 3.885 | 3.256 | 3.885 | 3.940 | 3.636 | 3.615 | 3.598 |
| D (25 °C) . | 1.544 | 1.696 | 1.706 | 1.716 | 1.739 | 1.714 | 1.709 | 1.724 |
| $c_s$ (50 °C) . | 45.2 | 51.6 | 52.2 | 52.2 | 52.2 | 50.5 | 50.4 | 50.8 |
| $m_s$ (50 °C) . | 4.222 | 4.347 | 4.395 | 4.420 | 4.357 | 3.971 | 3.926 | 3.907 |
| D (50 °C) . | 1.559 | 1.749 | 1.768 | 1.769 | 1.770 | 1.742 | 1.738 | 1.748 |

| $MCl_3$ . . . | $TbCl_3$ | $DyCl_3$ | $HoCl_3$ | $ErCl_3$ | $TmCl_3$ | $YbCl_3$ | $LuCl_3$ |
|---|---|---|---|---|---|---|---|
| $c_s$ (10 °C) . | 48.8 | 49.4 | 49.9 | 51.0 | 51.6 | 53.3 | 54.2 |
| $m_s$ (10 °C) . | 3.593 | 3.633 | 3.674 | 3.808 | 3.874 | 4.088 | 4.209 |
| D (10 °C) . | 1.718 | 1.738 | 1.764 | 1.792 | 1.821 | 1.862 | 1.897 |
| $c_s$ (25 °C) . | 49.1 | 49.9 | 50.7 | 51.3 | 52.2 | 53.3 | 54.2 |
| $m_s$ (25 °C) . | 3.636 | 3.704 | 3.792 | 3.850 | 3.965 | 4.085 | 4.206 |
| D (25 °C) . | 1.729 | 1.744 | 1.767 | 1.795 | 1.823 | 1.860 | 1.897 |
| $c_s$ (50 °C) . | 50.7 | 51.1 | 51.8 | 52.5 | 53.1 | 54.8 | 55.1 |
| $m_s$ (50 °C) . | 3.877 | 3.885 | 3.951 | 4.040 | 4.112 | 4.330 | 4.361 |
| D (50 °C) . | 1.752 | 1.773 | 1.796 | 1.817 | 1.843 | 1.880 | 1.912 |

The accuracy is $\pm 0.003$ g/cm$^3$. Data regarding molar volumes are found in the paper, [5]. Densities and molar volumes of aqueous $YCl_3$ solutions at 0 °C are given in [2]; see Fig. 31, p. 223, for apparent molar volumes. See Novikov et al. [22] for molar volumes of aqueous $ScCl_3$, $YCl_3$, and $LaCl_3$ at 25 °C.

The following table shows densities of saturated aqueous hydrochloric acid solutions of $YCl_3$, $NdCl_3$, and $SmCl_3$ at 0 °C as a function of the HCl content:

| wt % HCl . . | 0 | 5 | 10 | 15 | 20 | 25 | 30 |
|---|---|---|---|---|---|---|---|
| $YCl_3$ . . . . | 1.48 | 1.42 | 1.36 | 1.30 | 1.24 | 1.20 | 1.18 |
| $NdCl_3$ . . . | – | 1.57 | 1.44 | 1.34 | 1.26 | 1.20 | 1.17 |
| $SmCl_3$ . . . | 1.69 | 1.55 | 1.43 | 1.33 | 1.25 | 1.20 | 1.18 |

Above 35% HCl the densities of the solutions practically coincide with those of pure hydrochloric acid, Fischer et al. [16].

A dilatometric method was used to determine the densities of about 0.1 to 3.5 m $MCl_3$ solutions in intervals of 5 °C between 5 and 80 °C for M=Pr and Nd, as well as between 20 and 80 °C for M=La, Gildseth et al. [17], and between 5 and 80 °C for M=Sm, Gd, Dy, Er, and Yb, Habenschuss, Spedding [18]. Compare also the studies of aqueous $MCl_3$ solutions with M=La and Nd by Gildseth, Spedding [19] and with M=La, Pr, Nd, Sm, Gd, Dy, Er, and Yb by Habenschuss [25].

The densities measured by [17 to 20] have been used to calculate the apparent and partial molar volumes as well as the thermal expansion coefficients and to describe through approximation relationships the numerous individual values reported in the original papers. The temperature dependence of the density in the temperature ranges named can be described by the equation $D=C_0+C_1t+C_2t^2+C_3t^3+C_4t^4+C_5t^5$. An analogous equation results for the apparent molar volume $\Phi_V$. Constants $C_i$ with i from 1 to 5 for description of D and $A_i$ with i from 0 to 5 for description of $\Phi_V$ for the named concentration ranges are compiled in the papers, namely for M=La, Pr, and Nd by [17], for the saturated $MCl_3$ solutions by [18].

The concentration dependence of the apparent molar volume between 5 and 80 °C can be only qualitatively described by the Masson equation $\Phi_V=\Phi_V^*+S_V^*\sqrt{m}$. Numerical values for the constants can be found in the papers. A significantly better fit is achieved with equations of the Redlich-Meyer type, of the form $\Phi_V=\sum_i\sum_j B_{ij}\,m^{j/2}t^i$. An accuracy of ±0.05 mL/mol at lower molalities m and 0.01 mL/mol at higher concentrations can be attained for $\Phi_V$ with the constants $B_{ij}$, for M=La, Pr, and Nd from [17], and the others from [18], given in Table 11.2.8-2.

Table 11.2.8-2
Constants $B_{ij}$ for the Redlich-Meyer Type Equation of the Form $\Phi_V=\sum_i\sum_j B_{ij}\,m^{j/2}\,t^i$.

| i | $B_{0j}$ | $B_{1j}\times10$ | $B_{2j}\times10^3$ | $B_{3j}\times10^5$ | $B_{4j}\times10^7$ | $B_{5j}\times10^9$ |
|---|---|---|---|---|---|---|
| $LaCl_3$ | | | | | | |
| 0 | 13.16045 | 3.281874 | −6.366580 | 2.365504 | 1.899787 | −1.967494 |
| 1 | 3.00208 | 2.204681 | −7.569032 | 20.830004 | −25.848852 | 12.190078 |
| 2 | 13.05398 | −5.179800 | 16.111709 | −34.762909 | 39.303245 | −17.461539 |
| 3 | −5.66861 | 2.588094 | −8.691832 | 19.366320 | −22.046183 | 9.764624 |
| 4 | 0.72472 | −0.420563 | 1.551038 | −3.658856 | 4.271842 | −1.913151 |
| $PrCl_3$ | | | | | | |
| 0 | 7.33653 | 4.211035 | −10.416601 | 12.621156 | −10.918243 | 3.969154 |
| 1 | 9.26902 | 0.632370 | 0.030499 | −0.586084 | 3.977471 | −2.590980 |
| 2 | 4.96866 | −3.731125 | 9.007725 | −12.711765 | 6.158433 | −0.397690 |
| 3 | −1.24015 | 1.915427 | −5.327945 | 8.159326 | −4.384721 | 0.510897 |
| 4 | −0.08435 | −0.299908 | 0.946806 | −1.503887 | 0.750997 | −0.051895 |
| $NdCl_3$ | | | | | | |
| 0 | 7.01944 | 4.049902 | −9.946627 | 11.914813 | −10.041062 | 3.485562 |
| 1 | 6.47009 | 1.582629 | −2.931885 | 4.765540 | −1.908215 | 0.078915 |
| 2 | 8.58704 | −5.389763 | 15.245779 | −25.819582 | 20.774818 | −6.738754 |
| 3 | −3.41336 | 3.051321 | −9.973919 | 18.594805 | −16.146067 | 5.557504 |
| 4 | 0.41501 | −0.561968 | 2.067342 | −4.107507 | 3.699134 | −1.310753 |
| $SmCl_3$ | | | | | | |
| 0 | 8.79272 | 3.633290 | −9.207126 | 10.581492 | −8.498357 | 2.845232 |
| 1 | 5.68926 | 1.719658 | −1.688217 | 3.257708 | −2.437776 | 0.820360 |
| 2 | 8.84064 | −4.750366 | 9.697633 | −15.967735 | 13.899013 | −4.856224 |
| 3 | −4.45278 | 2.530876 | −5.316871 | 9.771461 | −9.019395 | 3.217975 |
| 4 | 0.84321 | −0.439665 | 0.905332 | −1.873181 | 1.810152 | −0.656795 |

Table 11.2.8-2 [continued]

| i | $B_{0j}$ | $B_{1j} \times 10$ | $B_{2j} \times 10^3$ | $B_{3j} \times 10^5$ | $B_{4j} \times 10^7$ | $B_{5j} \times 10^9$ |
|---|---|---|---|---|---|---|
| $GdCl_3$ | | | | | | |
| 0 | 12.51134 | 3.214004 | − 9.104461 | 10.601625 | − 8.002207 | 2.302221 |
| 1 | 0.82659 | 2.356689 | − 2.465442 | 5.645571 | − 7.683728 | 4.432602 |
| 2 | 16.15518 | −5.954562 | 12.111485 | −22.385991 | 24.727501 | −11.411894 |
| 3 | − 9.12490 | 3.436574 | − 7.368226 | 14.857264 | −16.845767 | 7.689722 |
| 4 | 1.80696 | −0.654865 | 1.420572 | − 3.130848 | 3.672479 | − 1.686999 |
| $DyCl_3$ | | | | | | |
| 0 | 11.69037 | 3.265827 | − 8.568245 | 8.328155 | − 5.039354 | 1.027088 |
| 1 | 1.18948 | 2.261705 | − 3.387633 | 9.019313 | −11.661217 | 5.867764 |
| 2 | 16.07697 | −5.629731 | 11.661891 | −21.430774 | 22.926090 | −10.074390 |
| 3 | − 8.41597 | 3.074596 | − 6.361837 | 11.992867 | −12.856155 | 5.589461 |
| 4 | 1.51815 | −0.561024 | 1.106200 | − 2.187188 | 2.402396 | − 1.053455 |
| $ErCl_3$ | | | | | | |
| 0 | 9.03266 | 3.466844 | − 8.904948 | 9.300320 | − 6.557997 | 1.867643 |
| 1 | 4.79945 | 1.954430 | − 3.187235 | 7.658626 | − 8.668782 | 3.906149 |
| 2 | 10.60908 | −5.287963 | 11.810209 | −20.743802 | 20.423988 | − 8.170145 |
| 3 | − 4.90249 | 2.887697 | − 6.497133 | 11.772823 | −11.746071 | 4.684687 |
| 4 | 0.75779 | −0.526137 | 1.130281 | − 2.138067 | 2.192527 | − 0.883361 |
| $YbCl_3$ | | | | | | |
| 0 | 5.63238 | 3.712794 | − 9.978995 | 11.994744 | − 9.905657 | 3.459956 |
| 1 | 10.11309 | 1.354181 | 0.537387 | − 2.032094 | 3.282440 | − 1.700918 |
| 2 | 3.12236 | −4.540127 | 6.816343 | − 7.660794 | 4.416068 | − 0.734692 |
| 3 | − 0.50469 | 2.463077 | − 3.600014 | 4.181916 | − 2.521691 | 0.429468 |
| 4 | − 0.13014 | −0.439087 | 0.528115 | − 0.560368 | 0.284882 | − 0.006570 |

Partial molar volumes $\bar{V}_2$ of the $MCl_3$ in 2 m solutions at 5 to 80 °C as a function of ionic radius are shown in **Fig. 33**, p. 228; for other concentrations see the figures in [18]. For infinite dilution, where the partial and apparent molar volumes are equal, the following values for $\Phi_V(0) = \bar{V}_2(0)$ in mL/mol are found [18]:

| t in °C | $LaCl_3$ | $PrCl_3$ | $NdCl_3$ | $SmCl_3$ | $GdCl_3$ | $DyCl_3$ | $ErCl_3$ | $YbCl_3$ |
|---|---|---|---|---|---|---|---|---|
| 5 | 11.74 | 7.30 | 6.52 | 7.94 | 10.60 | 9.84 | 7.90 | 5.55 |
| 25 | 14.54 | 10.40 | 9.66 | 10.68 | 12.82 | 12.16 | 10.37 | 8.16 |
| 50 | 13.69 | 9.33 | 8.69 | 9.48 | 11.07 | 10.37 | 9.35 | 6.61 |
| 75 | 8.39 | 4.77 | 4.27 | 4.92 | 6.02 | 5.27 | 3.71 | 1.65 |
| 80 | 7.07 | 3.50 | 3.03 | 3.66 | 4.64 | 3.91 | 2.35 | 0.32 |

Pitzer et al. [21] make an attempt to fit volumetric data from the literature to a general equation, the constants of which are listed for most of the rare earth chlorides.

In the plot of $\bar{V}_2$ vs. temperature a maximum is observed for dilute solutions, which in the case of infinite dilution lies at the following temperatures $t_{max}$ (in °C) [18]:

| | | | | | | | | |
|---|---|---|---|---|---|---|---|---|
| $MCl_3$ . . . . | $LaCl_3$ | $PrCl_3$ | $NdCl_3$ | $SmCl_3$ | $GdCl_3$ | $DyCl_3$ | $ErCl_3$ | $YbCl_3$ |
| $t_{max}$ . . . . | 28.8 | 30.6 | 30.9 | 29.8 | 27.1 | 26.7 | 27.7 | 28.2 |

Fig. 33

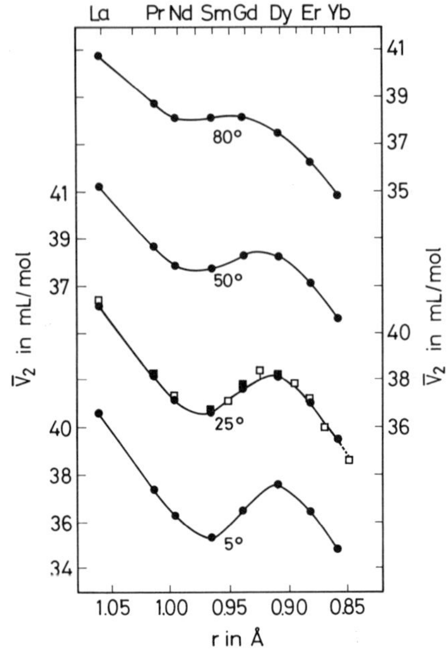


Partial molar volumes $\bar{V}_2$ of solute in 2 m aqueous $MCl_3$ solutions at 5 to 80 °C as a function of the ionic radius r.

Fig. 34

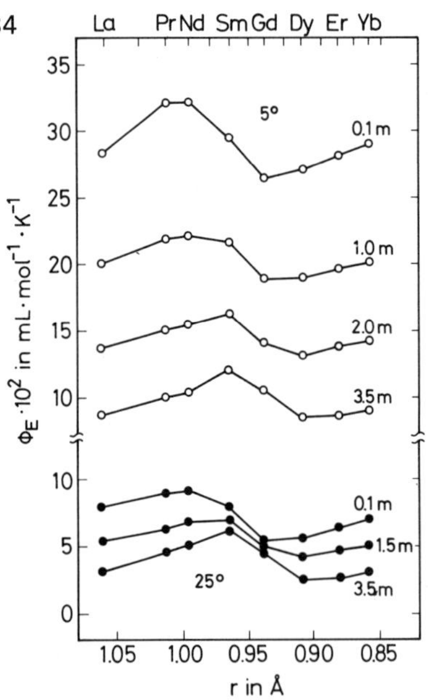


Fig. 35

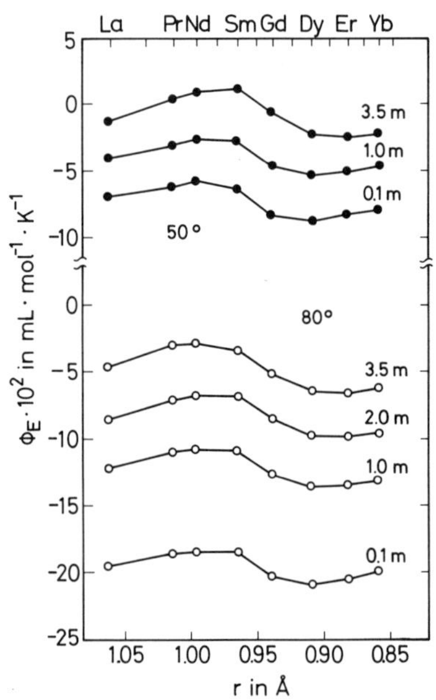


Apparent molar expansibility $\Phi_E$ of aqueous $MCl_3$ solutions at various temperatures (in °C) and concentrations (molality m) as a function of the ionic radius r.

The apparent molar expansibility $\Phi_E$ of aqueous $MCl_3$ solutions at various temperatures and concentrations is shown in **Fig. 34** and **35**, respectively. At infinite dilution, where $\Phi_E$ equals the partial molar thermal expansibility $\bar{E}_2(0)$ of the $MCl_3$, the following values have been determined for $\Phi_E(0) = \bar{E}_2(0)$ (in $mL \cdot mol^{-1} \cdot K^{-1}$):

| t in °C | $LaCl_3$ | $PrCl_3$ | $NdCl_3$ | $SmCl_3$ | $GdCl_3$ | $DyCl_3$ | $ErCl_3$ | $YbCl_3$ |
|---|---|---|---|---|---|---|---|---|
| 5 | 0.255 | 0.292 | 0.291 | 0.263 | 0.238 | 0.246 | 0.252 | 0.260 |
| 25 | 0.036 | 0.046 | 0.049 | 0.036 | 0.011 | 0.013 | 0.020 | 0.026 |
| 50 | −0.107 | −0.118 | −0.114 | −0.120 | −0.139 | −0.143 | −0.113 | −0.137 |
| 75 | −0.248 | −0.243 | −0.237 | −0.241 | −0.264 | −0.262 | −0.260 | −0.256 |
| 80 | −0.281 | −0.265 | −0.259 | −0.262 | −0.289 | −0.282 | −0.282 | −0.276 |

Additional results for the apparent and partial expansibility of aqueous $MCl_3$ solutions as a function of composition and temperature are compiled in figures in the paper [18]; cf. also [20]. Results for solutions of $LaCl_3$, $PrCl_3$, and $NdCl_3$ can also be found in [17].

References:

[1] C.M. Mason (J. Am. Chem. Soc. **60** [1938] 1638/47). – [2] F.H. Spedding, V.W. Saeger, K.A. Gray et al. (J. Chem. Eng. Data **20** [1975] 72/81). – [3] F.H. Spedding, M.J. Pikal, B.O. Ayers (J. Phys. Chem. **70** [1966] 2440/9). – [4] F.H. Spedding, P.F. Cullen, A. Habenschuss (J. Phys. Chem. **78** [1974] 1106/10). – [5] A.V. Nikolaev, A.A. Sorokina, N.P. Sokolova et al. (Izv. Sibirsk. Otd. Akad. Nauk SSSR Ser. Khim. Nauk **1977** No. 4, pp. 84/90; C.A. **87** [1977] No. 173507).

[6] F.H. Spedding, P.E. Porter, J.M. Wright (J. Am. Chem. Soc. **74** [1952] 2778/81; ISC-169 [1952] from N.S.A. **6** [1952] No. 361). – [7] V.W. Saeger, F.H. Spedding (IS-338 [1960] 1/202, 55/62; N.S.A. **16** [1962] No. 88). – [8] V.W. Saeger (Diss. Iowa State Univ. 1960, pp. 1/209, 55/62; C.A. **1961** 15086). – [9] M.J. Pikal, F.H. Spedding (IS-1344 [1966] 1/167, 64/9; C.A. **65** [1966] 17781). – [10] M. Pikal (Diss. Iowa State Univ. 1966 from Diss. Abstr. B **27** [1966] 134).

[11] P.F. Cullen (Diss. Iowa State Univ. 1969 from Diss. Abstr. Intern. B **31** [1970/71] 610). – [12] P.F. Cullen (IS-T-342 [1969] 1/120, 61/2, 70, 72; N.S.A. **24** [1970] No. 1367). – [13] F.H. Spedding, B.D. Ayers (from F.H. Spedding, G. Atkinson in: W.J. Hamer, The Structure of Electrolytic Solutions, London 1959, pp. 319/39, 330/1). – [14] F.H. Spedding, J.A. Rard, V.W. Saeger (J. Chem. Eng. Data **19** [1974] 373/8). – [15] C.M. Mason, H.L. Leland (J. Am. Chem. Soc. **57** [1935] 1507/8).

[16] W. Fischer, H. Bauer, I. Dillo, T. Molaug, J. Nier, H. Rohrer, K. Trovaag (Z. Anorg. Allgem. Chem. **357** [1968] 177/83). – [17] W.M. Gildseth, A. Habenschuss, F.H. Spedding (J. Chem. Eng. Data **20** [1975] 292/309). – [18] A. Habenschuss, F.H. Spedding (J. Chem. Eng. Data **21** [1976] 95/113). – [19] W.M. Gildseth, F.H. Spedding (IS-992 [1964] 1/181, 61/2, 80/3, 114/23, 148/57; C.A. **62** [1965] 9823). – [20] A. Habenschuss (IS-T-600 [1973] 1/318, 202/3, 210, 229/30, 240/1, 260/7, 281/307; C.A. **80** [1974] No. 74779).

[21] K.S. Pitzer, J.R. Peterson, L.F. Silvester (J. Solution Chem. **7** [1978] 45/56, 50). – [22] S.N. Novikov, V.A. Vasilev, M.K. Karapet'yants (Deposited Doc. VINITI – 3433-77 [1977] 79/83; C.A. **90** [1979] No. 110931).

#### 11.2.8.1.4 Ultrasonic Absorption and Compressibility

Aqueous solutions of lanthanide chlorides display no significant relaxation of the ultrasonic absorption, Grecsek [1], as shown also by earlier measurements on 0.2 and 0.4 molar $NdCl_3$ and $GdCl_3$ solutions, Garnsey, Ebdon [2]. The absorption coefficient $\alpha$ of a 1 molar $LaCl_3$ solution at 25 °C is as a function of the frequency $\nu$:

| | | | | | | | |
|---|---|---|---|---|---|---|---|
| $\nu$ in MHz . . . . . . | 30.1 | 50.5 | 70.4 | 89.8 | 110.2 | 130.2 | 150.5 |
| $\alpha$ in neper/cm . . . . | 0.219 | 0.600 | 1.14 | 1.98 | 2.87 | 3.98 | 5.35 |

The average quotient $\alpha/\nu^2$ is about $23.6\times10^{-17}$ neper·s²·cm⁻¹ [1]. This ratio is about $22\times10^{-17}$ and $24\times10^{-17}$ $neper\cdot s^2\cdot cm^{-1}$ for 0.2 and 0.4 molar $NdCl_3$ solutions, respectively. $GdCl_3$ solutions of the same concentrations yield $24.5\times10^{-17}$ and $27.5\times10^{-17}$, and pure water gives $21.7\times10^{-17}$ $neper\cdot s^2\cdot cm^{-1}$ [2].

The apparent molar compressibility $\Phi_K$ of aqueous $MCl_3$ solutions at 25 °C has been determined from ultrasonic measurements at constant frequency. Values for $10^2\,\Phi_K$ as a function of the square root of the molality, are in $cm^3\cdot bar^{-1}\cdot mol^{-1}$:

| $\sqrt{m}$ . . . . | 0 | 0.05 | 0.10 | 0.20 | 0.30 | 0.40 | 0.50 | 0.60 |
|---|---|---|---|---|---|---|---|---|
| $LaCl_3$ . . . | 1.536 | 1.516 | 1.494 | 1.452 | 1.410 | 1.369 | 1.327 | 1.285 |
| $NdCl_3$ . . . | 1.651 | 1.627 | 1.604 | 1.557 | 1.510 | 1.462 | 1.415 | 1.368 |
| $ErCl_3$ . . . . | 1.631 | 1.608 | 1.585 | 1.539 | 1.494 | 1.448 | 1.402 | 1.357 |
| $YbCl_3$ . . . | 1.755 | 1.725 | 1.695 | 1.635 | 1.575 | 1.514 | 1.454 | 1.394 |

An approximation relationship for description of the concentration dependence of $\Phi_K$ can be found in the paper, Spedding, Atkinson [3].

References:

[1] J.J. Grecsek (Diss. Univ. Maryland 1970, pp. 1/103, 54, 63; Diss. Abstr. Intern. B **31** [1970] 2599/60). – [2] R. Garnsey, D.W. Ebdon (J. Am. Chem. Soc. **91** [1969] 50/6). – [3] F.H. Spedding, G. Atkinson (in: W.J. Hamer, The Structure of Electrolytic Solutions, London 1959, pp. 319/39, 333/4).

#### 11.2.8.1.5 Viscosity

The relative viscosity $\eta_r=\eta(\text{soln})/\eta(H_2O)$ of aqueous $MCl_3$ solutions has been measured at 25 °C with an Ubbelohde viscosimeter in the concentration range from about 0.05 m up to the saturation limit, on solutions with M = La, Pr, Nd, Sm, Tb, Dy, Ho, and Er by Spedding, Pikal [1], Pikal [2], with M = Pr, Eu, Gd, Tm, Yb, and Lu by Spedding et al. [3], and with M = Pr, Sm, Gd, Tb, Dy, Ho, and Er by Pikal, Spedding [4]. **Fig. 36** displays $\eta_r$ for aqueous $LaCl_3$ and $LuCl_3$ solutions as a function of the molality. The ratio of the relative viscosities of various $MCl_3$ solutions to the relative viscosity of aqueous $LaCl_3$ solutions, $\eta_r(MCl_3):\eta_r(LaCl_3)$, is presented in **Fig. 37** [3]. The change of $\eta_r$ with the molar concentration can be described by the Jones-Dole equation $\eta_r = 1+A\sqrt{c}+Bc$, with c in mol/L, for which the following constants A are calculated and B are measured:

| $MCl_3$ | $LaCl_3$ | $PrCl_3$ | $NdCl_3$ | $SmCl_3$ | $EuCl_3$ | $GdCl_3$ | $TbCl_3$ |
|---|---|---|---|---|---|---|---|
| A | 0.0285 | 0.0285 | 0.0286 | 0.0288 | 0.0290 | 0.0293 | 0.0293 |
| B | 0.554 | 0.567 | 0.557 | 0.584 | 0.602 | 0.626 | 0.633 |
| Ref. | [1] | [3] | [1] | [1] | [3] | [3] | [1] |

| $MCl_3$ | $DyCl_3$ | $HoCl_3$ | $ErCl_3$ | $TmCl_3$ | $YbCl_3$ | $LuCl_3$ |
|---|---|---|---|---|---|---|
| A | 0.0297 | 0.0295 | 0.0296 | 0.0298 | 0.0299 | 0.0300 |
| B | 0.639 | 0.650 | 0.646 | 0.655 | 0.652 | 0.659 |
| Ref. | [1] | [1] | [1] | [3] | [3] | [3] |

The B coefficients were determined from measurements on <0.1 m solutions, with a probable error of ±0.005 L/mol. The concentration dependence of $\eta_r$ can also be described by a combination of the equations of Pitts and Vand. Constants for this equation as well as a comparison of the values calculated with it with the experimental $\eta_r$ values, are given in the cited papers for $MCl_3$ with M = La, Pr, Nd, Sm, Tb, Dy, Ho, and Er [1] (cf. also [2]) and for M = Pr, Eu, Gd, Tm, Yb, and Lu [3].

Fig. 36

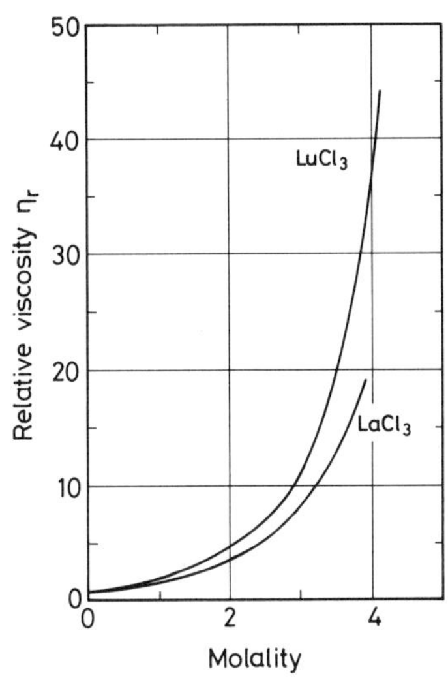


Fig. 37

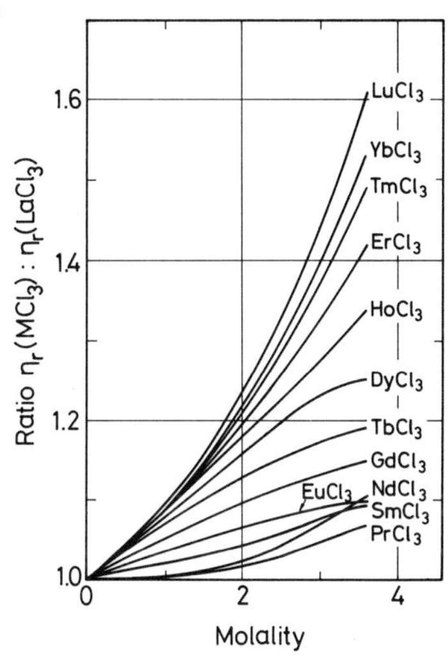


Relative viscosity $\eta_r = \eta(soln)/\eta(H_2O)$ of aqueous $LaCl_3$ and $LuCl_3$ solutions at 25 °C as a function of molality, Fig. 36. Ratio of $\eta_r$ of various $MCl_3$ solutions to $\eta_r$ of aqueous $LaCl_3$ solutions at 25 °C as a function of molality, Fig. 37.

Isomolal sequences of $\eta_r$ as a function of the ionic radius $r_M$ form an S-shaped curve similar to that for the molar volume (cf. p. 224, 228), which likewise is explained by the existence of two $MCl_3$ series (M = La to Nd and Eu to Lu), in which the $M^{3+}$ ions are differently hydrated, with $\eta_r$ of the $SmCl_3$ solutions lying in the transition zone [3]. Viscosities of aqueous $MCl_3$ solutions with M = Ho, Er, Dy, and Tb as a function of the densities of the solutions are given in a figure in Andres [5].

References:

[1] F.H. Spedding, M.J. Pikal (J. Phys. Chem. **70** [1966] 2430/40). – [2] M. Pikal (Diss. Iowa State Univ. 1966 from Diss. Abstr. B **27** [1966] 134). – [3] F.H. Spedding, D.L. Witte, L.E. Shiers, J.A. Rard (J. Chem. Eng. Data **19** [1974] 369/73). – [4] M.J.

Pikal, F.H. Spedding (IS-1344 [1966] 1/167, 138/41; C.A. **65** [1966] 17781). – [5] U.T. Andres (Mater. Sci. Eng. **26** [1976] 269/75, 273).

### 11.2.8.1.6 Vapor Pressure, Activity, and Osmotic Coefficients

Introductory Remarks

The osmotic coefficient is the ratio of the actually measured osmotic pressure of the solution to that which is given theoretically from the number of particles present in the solution. The osmotic coefficient $\Phi$ is defined by $\Phi = -(1000 \ln a_1)/\nu m M_1$, for which $a_1$ is the activity of the solvent $H_2O$ and $M_1$ its molecular weight, $\nu$ is the total number of ions formed on complete dissociation of the solute, and m is the molality. The vapor pressure p and $\Phi$ can for example be determined by isopiestic equilibration with solutions for which $\Phi$ is known. By use of the Gibbs-Duhem equation the mean activity coefficient of the salt $\gamma_\pm$ can be determined from $\Phi$, see for example Spedding et al. [1].

**Vapor Pressure**

Selected values for the vapor pressure p in Torr and the activity of the water $a(H_2O)$ of aqueous $MCl_3$ solutions at 25 °C as a function of the molality are given in the following table:

| $a(H_2O)$ | p in Torr | molality of the solution | | | | | | |
|---|---|---|---|---|---|---|---|---|
| | | $YCl_3$ | $LaCl_3$ | $CeCl_3$ | $NdCl_3$ | $SmCl_3$ | $EuCl_3$ | $GdCl_3$ |
| 0.8898 | 21.14 | 1.225 | 1.257 | 1.279 | 1.279 | 1.247 | 1.253 | 1.257 |
| 0.8750 | 20.79 | 1.332 | 1.368 | 1.397 | 1.397 | 1.358 | 1.364 | 1.367 |
| 0.8400 | 19.96 | 1.559 | 1.591 | 1.639 | 1.612 | 1.575 | 1.593 | 1.607 |
| 0.8110 | 19.27 | 1.716 | 1.758 | 1.782 | 1.770 | 1.731 | 1.754 | 1.785 |
| 0.7860 | 18.68 | 1.885 | 1.938 | 2.002 | 1.989 | 1.927 | 1.938 | 1.951 |
| 0.7436 | 17.68 | 2.116 | 2.209 | 2.259 | 2.237 | 2.172 | 2.175 | 2.187 |
| 0.6906 | 16.41 | 2.405 | 2.511 | 2.584 | 2.553 | 2.472 | 2.485 | 2.499 |
| 0.6456 | 15.34 | 2.646 | 2.855 | 2.863 | 2.824 | 2.728 | 2.753 | 2.767 |
| 0.5960 | 14.16 | 2.985 | 3.139 | 3.189 | 3.128 | 3.069 | 3.073 | 3.089 |
| 0.5574 | 13.24 | 3.126 | 3.359 | 3.430 | 3.363 | 3.245 | 3.272 | 3.292 |
| 0.5245 | 12.46 | 3.320 | 3.567 | 3.664 | 3.573 | 3.462 | 3.498 | 3.518 |
| 0.5020 | 11.93 | 3.517 | 3.754 | 3.814 | 3.714 | – | – | – |

| $a(H_2O)$ | p in Torr | molality of the solution | | | | | |
|---|---|---|---|---|---|---|---|
| | | $TbCl_3$ | $DyCl_3$ | $HoCl_3$ | $ErCl_3$ | $YbCl_3$ | $LuCl_3$ |
| 0.8898 | 21.14 | 1.255 | 1.235 | 1.236 | 1.213 | 1.239 | 1.240 |
| 0.8750 | 20.79 | 1.358 | 1.347 | 1.344 | 1.317 | 1.347 | 1.349 |
| 0.8400 | 19.96 | 1.584 | 1.566 | 1.578 | 1.564 | 1.579 | 1.572 |
| 0.8110 | 19.27 | 1.750 | 1.726 | 1.733 | 1.715 | 1.730 | 1.729 |
| 0.7860 | 18.68 | 1.945 | 1.908 | 1.972 | 1.901 | 1.904 | 1.949 |
| 0.7436 | 17.68 | 2.195 | 2.148 | – | 2.151 | 2.135 | 2.193 |
| 0.6906 | 16.41 | 2.518 | 2.444 | 2.230 | 2.448 | 2.445 | 2.494 |
| 0.6456 | 15.34 | 2.785 | 2.697 | 2.781 | 2.718 | 2.676 | 2.745 |
| 0.5960 | 14.16 | 3.102 | 3.035 | 3.036 | 3.012 | 3.014 | 3.026 |
| 0.5574 | 13.24 | 3.315 | 3.206 | 3.305 | 3.220 | 3.227 | 3.252 |
| 0.5245 | 12.46 | 3.500 | 3.422 | 3.482 | 3.418 | 3.387 | 3.462 |
| 0.5020 | 11.93 | – | 3.608 | 3.595 | 3.559 | 3.550 | 3.573 |

The values were determined isopiestically, using as the standard for $a(H_2O) \geqq 0.75$ aqueous NaCl solutions and for <0.75 aqueous $CaCl_2$ or $H_2SO_4$ solutions, Kotlyar-Shapirov et al. [2]. The same method yields the following vapor pressures for the saturated aqueous $MCl_3$ solutions at 25 °C (saturation concentration $c_s$ in wt % and the saturation molality $m_s$ calculated from the mole ratios given in [3]), Nikolaev et al. [3]:

| $MCl_3$ . . . | $YCl_3$ | $LaCl_3$ | $CeCl_3$ | $PrCl_3$ | $NdCl_3$ | $SmCl_3$ | $EuCl_3$ |
|---|---|---|---|---|---|---|---|
| p in Torr . . | 10.02 | 11.21 | 11.83 | 11.52 | 11.07 | 11.78 | 12.02 |
| $c_s$ . . . . | 43.5 | 48.8 | 48.6 | 49.0 | 49.7 | 48.3 | 48.3 |
| $m_s$ . . . . | 3.893 | 3.885 | 3.836 | 3.885 | 3.940 | 3.636 | 3.615 |

| $MCl_3$ . . . | $GdCl_3$ | $TbCl_3$ | $DyCl_3$ | $HoCl_3$ | $ErCl_3$ | $YbCl_3$ | $LuCl_3$ |
|---|---|---|---|---|---|---|---|
| p in Torr . . | 12.11 | 11.92 | 11.40 | 11.12 | 10.69 | 9.98 | 9.50 |
| $c_s$ . . . . | 48.7 | 49.1 | 49.9 | 50.7 | 51.3 | 53.3 | 54.2 |
| $m_s$ . . . . | 3.598 | 3.636 | 3.704 | 3.792 | 3.850 | 4.085 | 4.206 |

**Individual Values for Activities**

Mean molar activity coefficients $y_\pm$ of dilute aqueous $MCl_3$ solutions at 25 °C are shown below as a function of the concentration c in mol/L, plus the mean distance of closest approach of ions å used in the Debye-Hückel equation (cf. pp. 235/6):

| $MCl_3$ | c in mol/L | | | | | | | | å in Å |
|---|---|---|---|---|---|---|---|---|---|
| | 0.0005 | 0.001 | 0.002 | 0.003 | 0.005 | 0.01 | 0.02 | 0.03 | |
| $LaCl_3$ | 0.8348 | 0.7758 | 0.7257 | 0.6869 | 0.6311 | 0.5557 | 0.4798 | 0.4356 | 5.75 |
| $CeCl_3$ | 0.8353 | 0.7753 | 0.7241 | 0.6858 | 0.6303 | 0.5546 | 0.4794 | 0.4364 | 5.75 |
| $PrCl_3$ | 0.8350 | 0.7762 | 0.7263 | 0.6871 | 0.6314 | 0.5553 | 0.4792 | 0.4354 | 5.75 |
| $NdCl_3$ | 0.8354 | 0.7765 | 0.7250 | 0.6863 | 0.6300 | 0.5488 | 0.4720 | 0.4282 | 5.70 |
| $SmCl_3$ | 0.8346 | 0.7756 | 0.7243 | 0.6858 | 0.6301 | 0.5536 | 0.4769 | 0.4330 | 5.63 |
| $EuCl_3$ | 0.8345 | 0.7756 | 0.7243 | 0.6859 | 0.6297 | 0.5523 | 0.4759 | 0.4319 | 5.60 |
| $GdCl_3$ | 0.8346 | 0.7757 | 0.7245 | 0.6857 | 0.6306 | 0.5527 | 0.4759 | 0.4337 | 5.63 |
| $DyCl_3$ | 0.8336 | 0.7740 | 0.7236 | 0.6832 | 0.6263 | 0.5432 | 0.4661 | 0.4232 | 5.32 |
| $HoCl_3$ | 0.8361 | 0.7788 | 0.7289 | 0.6891 | 0.6367 | 0.5618 | 0.4854 | 0.4431 | 6.04 |
| $ErCl_3$ | 0.8357 | 0.7783 | 0.7284 | 0.6894 | 0.6350 | 0.5597 | 0.4832 | 0.4408 | 5.92 |
| $TmCl_3$ | 0.8329 | 0.7802 | 0.7285 | 0.6886 | 0.6345 | 0.5554 | 0.4798 | 0.4378 | 5.81 |
| $YbCl_3$ | 0.8331 | 0.7793 | 0.7281 | 0.6899 | 0.6318 | 0.5591 | 0.4834 | 0.4410 | 5.90 |
| $LuCl_3$ | 0.8453 | 0.7808 | 0.7301 | 0.6897 | 0.6389 | 0.5647 | 0.4896 | 0.4469 | 6.07 |

With the above given å values the results follow the Debye-Hückel theory with an accuracy better than 0.1% over this concentration range. The values were compiled by Spedding, Atkinson [4] from emf measurements on $MCl_3$ solutions with M = La, Ce, Pr, Nd, Sm, Eu, Er, and Yb of Spedding et al. [5], with M = Gd of Spedding, Jaffe [6], with M = Dy of Spedding, Dye [7], and in addition from unpublished investigations of $LuCl_3$ solutions. Selected values for the mean molal activity coefficients $\gamma_\pm$ and osmotic coefficients $\Phi$ of concentrated aqueous $MCl_3$ solutions at 25 °C as a function of the molality m from Spedding et al. [1] and Mason [8], are given in Table 11.2.8-3, pp. 234/5.

Table 11.2.8-3
Mean Molal Activity Coefficients $\gamma_\pm$ and Osmotic Coefficients $\Phi$ of Aqueous $MCl_3$ Solutions at 25 °C.

| $MCl_3$ | m | 0.05 | 0.1 | 0.5 | 1.0 | 1.6 | 2.0 | 3.0 | 3.6 | Ref. |
|---|---|---|---|---|---|---|---|---|---|---|
| $ScCl_3$ | $\gamma_\pm$ | 0.412 | 0.357 | 0.334 | 0.492 | 0.961 | – | – | – | [8] |
| | $\Phi$ | 0.799 | 0.797 | 0.970 | 1.285 | 1.727 | – | – | – | [8] |
| $YCl_3$ | $\gamma_\pm$ | – | 0.3383 | 0.3020 | 0.4188 | 0.7643 | 1.221 | 4.197 | 8.441 | [1] |
| | $\gamma_\pm$ | 0.417 | 0.357 | 0.318 | 0.436 | 0.804 | 1.330 | – | – | [8] |
| | $\Phi$ | – | 0.7838 | 0.9405 | 1.2251 | 1.6345 | 1.9301 | 2.6484 | 3.0136 | [1] |
| | $\Phi$ | 0.801 | 0.789 | 0.940 | 1.217 | 1.638 | 1.965 | – | – | [8] |
| $LaCl_3$ | $\gamma_\pm$ | – | 0.3312 | 0.2794 | 0.3598 | 0.5921 | 0.8679 | 2.265 | 3.786 | [1] |
| | $\gamma_\pm$ | 0.417 | 0.356 | 0.303 | 0.387 | 0.639 | 0.954 | – | – | [8] |
| | $\Phi$ | – | 0.7760 | 0.9046 | 1.1514 | 1.5035 | 1.7473 | 2.2953 | 2.5498 | [1] |
| | $\Phi$ | 0.806 | 0.788 | 0.912 | 1.149 | 1.506 | 1.766 | – | – | [8] |
| $CeCl_3$ | $\gamma_\pm$ | 0.394 | 0.338 | 0.290 | 0.372 | 0.614 | 0.946 | – | – | [8] |
| | $\Phi$ | 0.783 | 0.782 | 0.915 | 1.152 | 1.508 | 1.796 | – | – | [8] |
| $PrCl_3$ | $\gamma_\pm$ | – | 0.3298 | 0.2809 | 0.3630 | 0.6026 | 0.8954 | 2.489 | 4.355 | [1] |
| | $\gamma_\pm$ | 0.402 | 0.343 | 0.290 | 0.371 | 0.618 | 0.929 | – | – | [8] |
| | $\Phi$ | – | 0.7773 | 0.9099 | 1.1566 | 1.5142 | 1.7683 | 2.3631 | 2.6478 | [1] |
| | $\Phi$ | 0.778 | 0.784 | 0.906 | 1.148 | 1.512 | 1.777 | – | – | [8] |
| $NdCl_3$ | $\gamma_\pm$ | – | 0.3238 | 0.2753 | 0.3566 | 0.5978 | 0.8958 | 2.563 | 4.602 | [1] |
| | $\gamma_\pm$ | 0.403 | 0.344 | 0.294 | 0.380 | 0.642 | 0.984 | – | – | [8] |
| | $\Phi$ | – | 0.7767 | 0.9086 | 1.1579 | 1.5232 | 1.7829 | 2.3971 | 2.7003 | [1] |
| | $\Phi$ | 0.794 | 0.783 | 0.914 | 1.159 | 1.531 | 1.809 | – | – | [8] |
| $SmCl_3$ | $\gamma_\pm$ | – | 0.3271 | 0.2811 | 0.3688 | 0.6258 | 0.9447 | 2.735 | 4.922 | [1] |
| | $\gamma_\pm$ | 0.408 | 0.351 | 0.304 | 0.403 | 0.695 | 1.079 | – | – | [8] |
| | $\Phi$ | – | 0.7800 | 0.9155 | 1.1709 | 1.5411 | 1.8040 | 2.4207 | 2.7212 | [1] |
| | $\Phi$ | 0.796 | 0.789 | 0.922 | 1.180 | 1.559 | 1.845 | – | – | [8] |
| $EuCl_3$ | $\gamma_\pm$ | – | 0.3264 | 0.2839 | 0.3774 | 0.6507 | 0.9913 | 2.916 | 5.208 | [1] |
| | $\gamma_\pm$ | 0.412 | 0.355 | 0.308 | 0.413 | 0.720 | 1.140 | – | – | [8] |
| | $\Phi$ | – | 0.7814 | 0.9228 | 1.1843 | 1.5624 | 1.8293 | 2.4505 | 2.7464 | [1] |
| | $\Phi$ | 0.800 | 0.794 | 0.927 | 1.187 | 1.575 | 1.875 | – | – | [8] |
| $GdCl_3$ | $\gamma_\pm$ | – | 0.3279 | 0.2871 | 0.3867 | 0.6789 | 1.047 | 3.172 | 5.789 | [1] |
| | $\Phi$ | – | 0.7833 | 0.9272 | 1.1968 | 1.5848 | 1.8580 | 2.4943 | 2.8004 | [1] |
| $TbCl_3$ | $\gamma_\pm$ | – | 0.3353 | 0.2948 | 0.4023 | 0.7195 | 1.124 | 3.521 | 6.384 | [1] |
| | $\Phi$ | – | 0.7857 | 0.9309 | 1.2085 | 1.6072 | 1.8880 | 2.5415 | 2.8409 | [1] |
| $DyCl_3$ | $\gamma_\pm$ | – | 0.3399 | 0.3007 | 0.4118 | 0.7435 | 1.174 | 3.784 | 7.188 | [1] |
| | $\Phi$ | – | 0.7866 | 0.9346 | 1.2130 | 1.6181 | 1.9064 | 2.5775 | 2.9020 | [1] |
| $HoCl_3$ | $\gamma_\pm$ | – | 0.3430 | 0.3068 | 0.4219 | 0.7673 | 1.220 | 4.059 | 7.912 | [1] |
| | $\Phi$ | – | 0.7878 | 0.9405 | 1.2190 | 1.6278 | 1.9201 | 2.6144 | 2.9561 | [1] |
| $ErCl_3$ | $\gamma_\pm$ | – | 0.3386 | 0.3006 | 0.4138 | 0.7555 | 1.208 | 4.114 | 8.180 | [1] |
| | $\Phi$ | – | 0.7868 | 0.9366 | 1.2180 | 1.6305 | 1.9273 | 2.6388 | 2.9949 | [1] |

Table 11.2.8-3 [continued]

| $MCl_3$ | m | 0.05 | 0.1 | 0.5 | 1.0 | 1.6 | 2.0 | 3.0 | 3.6 | Ref. |
|---|---|---|---|---|---|---|---|---|---|---|
| $TmCl_3$ | $\gamma_\pm$ | – | 0.3342 | 0.2974 | 0.4102 | 0.7513 | 1.208 | 4.186 | 8.501 | [1] |
| | $\Phi$ | – | 0.7780 | 0.9378 | 1.2198 | 1.6345 | 1.9354 | 2.6589 | 3.0313 | [1] |
| $YbCl_3$ | $\gamma_\pm$ | – | 0.3334 | 0.2967 | 0.4089 | 0.7514 | 1.211 | 4.248 | 8.747 | [1] |
| | $\Phi$ | – | 0.7822 | 0.9374 | 1.2193 | 1.6368 | 1.9399 | 2.6720 | 3.0549 | [1] |
| $LuCl_3$ | $\gamma_\pm$ | – | 0.3358 | 0.2992 | 0.4118 | 0.7571 | 1.221 | 4.301 | 8.921 | [1] |
| | $\Phi$ | – | 0.7848 | 0.9380 | 1.2185 | 1.6368 | 1.9406 | 2.6760 | 3.0650 | [1] |

The following values have been obtained for aqueous solutions saturated at 25 °C ($m_s$= saturation molality) [1]:

| $MCl_3$ | $YCl_3$ | $LaCl_3$ | $PrCl_3$ | $NdCl_3$ | $SmCl_3$ | $EuCl_3$ | $GdCl_3$ |
|---|---|---|---|---|---|---|---|
| $\gamma_\pm$ | 12.47 | 4.794 | 5.633 | 6.220 | 5.119 | 5.208 | 5.789 |
| $\Phi$ | 3.2092 | 2.6597 | 2.7697 | 2.8456 | 2.7411 | 2.7464 | 2.8004 |
| $m_s$ | 3.9478 | 3.8944 | 3.8969 | 3.9307 | 3.6414 | 3.5839 | 3.5898 |
| $MCl_3$ | $TbCl_3$ | $DyCl_3$ | $HoCl_3$ | $ErCl_3$ | $TmCl_3$ | $YbCl_3$ | $LuCl_3$ |
| $\gamma_\pm$ | 6.384 | 7.414 | 8.767 | 10.02 | 11.73 | 13.98 | 16.66 |
| $\Phi$ | 2.8409 | 2.9170 | 3.0052 | 3.0962 | 3.1946 | 3.2937 | 3.3875 |
| $m_s$ | 3.5733 | 3.630 | 3.6987 | 3.7840 | 3.8814 | 4.0018 | 4.1239 |

The coefficients given by Mason [8] in Table 11.2.8-3 were determined isopiestically, with use of KCl and NaCl solutions as standards; the results for $MCl_3$ with M=Sc, Y, La, Ce, Pr, and Nd were obtained by recalculation of the values of Mason [9], and those for $SmCl_3$ and $EuCl_3$ solutions were newly determined. The data of [1] rest on isopiestic studies using KCl and $CaCl_2$ solutions as standards. For the conversion of the osmotic into the mean activity coefficients the older emf measurements mentioned on p. 233, for dilute solutions by [4 to 7] and for aqueous $TbCl_3$ solutions by Spedding et al. [10], were also taken into consideration. See [1] for the activities of the water in the $MCl_3$ solutions. Similar values for the activity coefficients were determined by Spedding and coworkers by emf measurements on $MCl_3$ solutions with M=La, Sm, Eu, and Yb [11] and M=Nd, Dy, Ho, Er, Tm, and Yb [12]. For data concerning isopiestically determined osmotic and activity coefficients for $MCl_3$ solutions with M=Y and Er, see Petheram, Spedding [13], with M=La, Nd, Sm, Gd, and Dy, see Saeger, Spedding [14]. A compilation of recalculated osmotic and activity coefficients based on measurements by [9] and complemented by their own results for $LaCl_3$, $SmCl_3$, and $EuCl_3$ solutions, is found in Robinson, Stokes [15], older compilations of activity coefficients are given in Robinson, Harned [16]. A plot of the mean activity coefficients compiled from literature data against the ionic radius at various molalities is given in a figure in Aksel'rud, Spivakovskii [17]. Numerical values for 0.30 to 1.8 m $ErCl_3$ and $YbCl_3$ solutions are given by Aksel'rud, Spivakovskii [18].

### Debye-Hückel Equation

Systematic deviations from the Debye-Hückel law even occur in the mean activity coefficients measured on dilute aqueous $MCl_3$ solutions by Spedding and coworkers [5 to 7, 10]. These deviations can be corrected by an additional linear term Dm. With a deviation of <0.6% the concentration dependence of the mean molal activity constants

$\gamma_{\pm}$ of $\leqq$0.02 to 0.04 m solutions can be described by the modified form

$$\ln \gamma_{\pm} = \ln(\gamma_{\pm}/\gamma'_{\pm}) + \ln \gamma'_{\pm} = -A\sqrt{m}(1+B\sqrt{m})^{-1} + D\,m$$

This gives for the osmotic coefficients

$$\Phi = 1 - AB^{-3}m^{-1}[(1+B\sqrt{m}) - (1+B\sqrt{m})^{-1} - 2\ln(1+B\sqrt{m})] + 0.5\,D\,m,$$

where A=8.6430 and B=0.80515 å with å=Debye-Hückel ion-size parameter in Å; $\gamma'_{\pm}$ is the activity coefficient of a reference solution. The following table shows the parameters for the modified Debye-Hückel relationship and the upper concentration limits of validity m(max) in mol/kg as well as the parameters for the unmodified equation (D=0). The calculations are based on the earlier measurements on dilute aqueous $MCl_3$ solutions (cf. p. 233) of Spedding et al. [5 to 7], which are designated in the following table as Ref. (Orig.):

| $MCl_3$ | å | å(D=0) | $\gamma'_{\pm}$ | $\gamma'_{\pm}$(D=0) | D | m(max.) | Ref. (Orig.) |
|---|---|---|---|---|---|---|---|
| $LaCl_3$ | 6.662 | 5.798 | 0.4346 | 0.4320 | −1.60 | 0.03119 | [5] |
| $CeCl_3$ | 6.213 | 5.830 | 0.4098 | 0.4089 | −0.70 | 0.04009 | [5] |
| $PrCl_3$ | 6.175 | 5.731 | 0.4284 | 0.4271 | −0.85 | 0.03235 | [5] |
| $NdCl_3$ | 6.087 | 5.573 | 0.4201 | 0.4189 | −1.05 | 0.03352 | [7] |
| $SmCl_3$ | 5.950 | 5.659 | 0.4119 | 0.4110 | −0.55 | 0.03746 | [5] |
| $EuCl_3$ | 5.696 | 5.616 | 0.4156 | 0.4153 | −0.15 | 0.03532 | [5] |
| $GdCl_3$ | 6.044 | 5.679 | 0.4183 | 0.4172 | −0.70 | 0.03515 | [6] |
| $HoCl_3$ | 6.446 | 6.032 | 0.4314 | 0.4302 | −0.75 | 0.03423 | [7] |
| $ErCl_3$ | 6.520 | 5.982 | 0.4235 | 0.4221 | −1.00 | 0.03640 | [7] |
| $TmCl_3$ | 6.649 | 5.895 | 0.4731 | 0.4714 | −1.65 | 0.02178 | [7] |
| $YbCl_3$ | 6.021 | 5.748 | 0.4076 | 0.4068 | −0.50 | 0.04017 | [5] |
| $YbCl_3$ | 6.800 | 5.903 | 0.4276 | 0.4254 | −1.65 | 0.03417 | [7] |

Corresponding parameters for aqueous $TbCl_3$, $DyCl_3$, and $LuCl_3$ solutions are not included in the table, since the values scatter widely, Spedding et al. [1]. Constants of the Debye-Hückel equation for $MCl_3$ solutions with M=Sc, Y, La, Ce, Pr, and Nd are given by [9]. Debye-Hückel ion-size parameters from [4] are found on p. 233. For calculation of $\Phi$ and $\gamma_{\pm}$ at higher concentrations the measured individual values (cf. Table 11.2.8-3, pp. 234/5) were fitted to a polynomial of the form

$$\Phi = 1 - (A/3)\sqrt{m} + \sum_i A_i m^{r_i} \quad \text{or} \quad \ln\gamma_{\pm} = -A\sqrt{m} + \sum_i A_i[(r_i+1)r_i^{-1}]m^{r_i}$$

Numerical values for the parameters $A_i$ and $r_i$ with i=1 to 7 are given in the paper [1]. A representation of the literature data by means of a modified Debye-Hückel equation and data for the parameters determined can also be found in Pitzer et al. [19].

### Structural Interpretation of the Concentration Dependence

The passage through a minimum of $\Phi$ or $\gamma_{\pm}$ in very dilute solutions, see for example **Fig. 38** for the osmotic coefficients as well as the numerical values for $\Phi$ and $\gamma_{\pm}$ on pp. 233 to 235, can be explained by a slight formation of ion pairs. The S-shaped curve of $\gamma_{\pm}$ vs. the ionic radius, see **Fig. 39**, is attributed to a change of the water coordination number in the inner coordination sphere of the lanthanide ions between Nd and Tb. Figures for $\Phi$ and $H_2O$ activities at various concentrations as a function of the ionic radius are given in the paper [1]. On the basis of literature data for the

Fig. 38

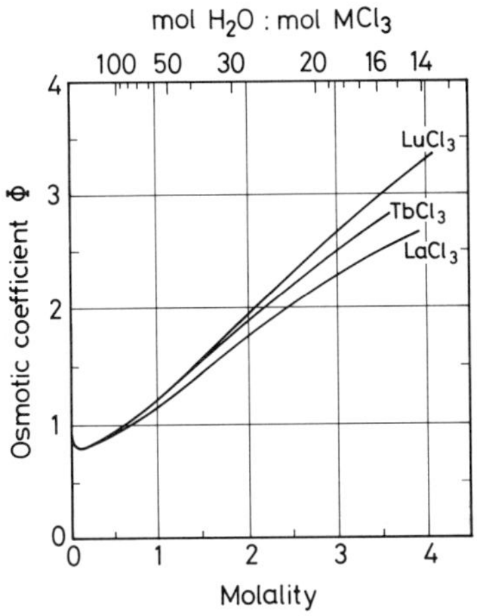


Osmotic coefficient of aqueous $LaCl_3$, $TbCl_3$, and $LuCl_3$ solutions at 25 °C as a function of molality.

Fig. 39

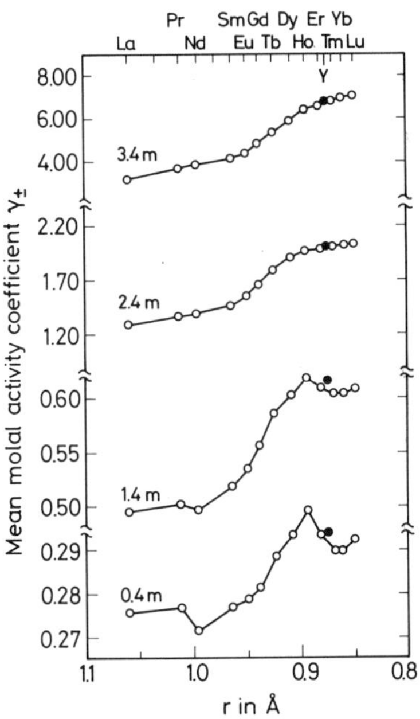


Mean molal activity coefficient $\gamma_{\pm}$ of aqueous $MCl_3$ solutions at various molalities as a function of the ionic radius r.

$H_2O$ activities of aqueous $MCl_3$ solutions with M = La to Eu at 25 °C and for solutions with different lanthanide ions, a relationship between the differences of the ionic radii $r_M - r_{M'}$ and the corresponding differences of the activities $a(MCl_3) - a(M'Cl_3)$ of isomolal solutions was derived and the influence of the hydration of the ions on the activity of the water discussed, Novikov et al. [20].

**Additional Results**

Comparative considerations of the concentration dependence of $\gamma_{\pm}$ of aqueous $MCl_3$ solutions with M = Sc, Y, and La through Eu have been published by Moriyama [21]; for a classification scheme, see Moriyama [22], for a semiempirical theoretical treatment of the pecularities of $\gamma_{\pm}$, see Moriyama [23]. Theoretical calculations of the activity constants of aqueous $MCl_3$ solutions with M = Sc, Y, and La through Eu have been conducted by Ikeda [24]. For calculation of individual activity coefficients of ions in water, among others Sc, Y, Ce, Pr, Nd, and Sm as cations and Cl as anion, see Kieland [25].

References:

[1] F.H. Spedding, H.O. Weber, V.W. Saeger, H.H. Petheram, J.A. Rard, A. Habenschuss (J. Chem. Eng. Data **21** [1976] 341/60). – [2] G.S. Kotlyar-Shapirov, N.P. Sokolova, A.A. Sorokina (Izv. Sibirsk. Otd. Akad. Nauk SSSR Ser. Khim. Nauk **1976** No. 4, pp. 44/7; C.A. **85** [1976] No. 113300). – [3] A.V. Nikolaev, A.A. Sorokina, N.P. Sokolova, G.S. Kotlyar-Shapirov, N.P. Anoshina, L.I. Bagryantseva, A.P. Olovyanishnikova (Izv. Sibirsk. Otd. Akad. Nauk SSSR Ser. Khim. Nauk **1977** No. 4, pp. 84/90; C.A. **87** [1977] No. 173507). – [4] F.H. Spedding, G. Atkinson (in: W.J. Hamer, The Structure of Electrolytic Solutions, London 1959, pp. 319/39, 328). – [5] F.H. Spedding, P.E. Porter, J.M. Wright (J. Am. Chem. Soc. **74** [1952] 2781/3).

[6] F.H. Spedding, I.S. Yaffe (J. Am. Chem. Soc. **74** [1952] 4751/5). – [7] F.H. Spedding, J.L. Dye (J. Am. Chem. Soc. **76** [1954] 879/81). – [8] C.M. Mason (J. Am. Chem. Soc. **63** [1941] 220/3). – [9] C.M. Mason (J. Am. Chem. Soc. **60** [1938] 1638/47). – [10] F.H. Spedding, R.A. Nelson, J.A. Rard (J. Chem. Eng. Data **19** [1974] 379/81).

[11] F.H. Spedding, J.M. Wright (ISC-143 [1951] 1/93, 73/9; N.S.A. **5** [1951] No. 5601). – [12] J.L. Dye, F.H. Spedding (ISC-355 [1953] 1/119, 86/92; C.A. **1956** 6880). – [13] H.H. Petheram, F.H. Spedding (IS-770 [1963] 1/69, 45/8, 56/8; N.S.A. **18** [1964] No. 13722). – [14] V.W. Saeger, F.H. Spedding (IS-338 [1960] 1/202, 165/74, 178/82, 184/5; N.S.A. **16** [1962] No. 88). – [15] R.A. Robinson, R.H. Stokes (Trans. Faraday Soc. **45** [1949] 612/24, 618/9, 622).

[16] R.A. Robinson, H.S. Harned (Chem. Rev. **28** [1941] 419/76, 448). – [17] N.V. Aksel'rud, V.B. Spivakovskii (Zh. Neorgan. Khim. **5** [1960] 348/55; Russ. J. Inorg. Chem. **5** [1960] 168/71). – [18] N.V. Aksel'rud, V.B. Spivakovskii (Zh. Neorgan. Khim. **5** [1960] 547/57; Russ. J. Inorg. Chem. **5** [1960] 262/7). – [19] K.S. Pitzer, J.R. Peterson, L.F. Silvester (J. Solution Chem. **7** [1978] 45/56, 49). – [20] S.N. Novikov, M.K. Karapet'yants, V.A. Vasilev (Izv. Vysshikh Uchebn. Zavedenii Khim. Khim. Tekhnol. **19** [1976] 1047/50; C.A. **85** [1976] No. 1675222).

[21] M. Moriyama (Naturwissenschaften **43** [1956] 515). – [22] M. Moriyama (Z. Physik. Chem. [Frankfurt] **27** [1961] 34/41). – [23] M. Moriyama (Z. Physik. Chem. [Frankfurt] **25** [1960] 310/20). – [24] T. Ikeda (Bull. Soc. Chem. Japan **24** [1951] 101/5). – [25] J. Kieland (J. Am. Chem. Soc. **59** [1937] 1675/8).

### 11.2.8.1.7 Thermodynamic Data

Values for the enthalpy of formation $\Delta H^{\circ}_{298}$, Gibbs free energy of formation $\Delta G^{\circ}_{298}$, entropy $S^{\circ}_{298}$, and heat capacity $C^{\circ}_{p,298}$ of aqueous $MCl_3$ solutions in the standard state at 25 °C from the critical tables of Schumm et al. [1] are, in kcal/mol or cal · mol$^{-1}$ · K$^{-1}$:

| $MCl_3$ | $LaCl_3$ | $CeCl_3$ | $PrCl_3$ | $NdCl_3$ | $SmCl_3$ | $EuCl_3$ | $GdCl_3$ |
|---|---|---|---|---|---|---|---|
| $-\Delta H^\circ_{298}$ | 288.9 | 286.2 | 288.3 | 286.3 | 285.2 | 264.4 | 284 |
| $-\Delta G^\circ_{298}$ | 257.5 | 254.7 | 256.4 | 254.7 | 253.4 | 231.3 | 253 |
| $-S^\circ_{298}$ | 12 | 9 | 10 | 9 | 10.2 | 13 | 8.8 |
| $-C^\circ_{p,298}$ | 101 | – | 105 | 103 | 103 | 96 | 98 |

| $MCl_3$ | $TbCl_3$ | $DyCl_3$ | $HoCl_3$ | $ErCl_3$ | $TmCl_3$ | $YbCl_3$ | $LuCl_3$ |
|---|---|---|---|---|---|---|---|
| $-\Delta H^\circ_{298}$ | 283.0 | 286 | 288.4 | 288.5 | 286.6 | 281.1 | 279 |
| $-\Delta G^\circ_{298}$ | 249.9 | 253 | 255.1 | 254.0 | 252.3 | 248.0 | 244 |
| $-S^\circ_{298}$ | 14 | 14.8 | 13.8 | 18.0 | 18 | 17 | 23 |
| $-C^\circ_{p,298}$ | 94 | 93 | 94 | 93 | 92 | 92 | 92 |

With increasing molar ratios $H_2O:MCl_3$, the enthalpies of formation at 25 °C decrease as follows (selected values in kcal/mol with negative sign) [1]:

| $H_2O:MCl_3$ | $LaCl_3$ | $CeCl_3$ | $PrCl_3$ | $NdCl_3$ | $SmCl_3$ | $GdCl_3$ |
|---|---|---|---|---|---|---|
| 300 | 287.495 | 284.80 | 286.912 | 284.903 | 283.781 | 282.528 |
| 500 | 287.681 | 284.99 | 287.099 | 285.105 | 283.976 | 282.744 |
| 600 | 287.736 | 285.05 | 287.162 | 285.170 | 284.041 | 282.813 |
| 800 | 287.826 | 285.14 | 287.256 | 285.267 | 284.134 | 282.917 |
| 1000 | 287.892 | 285.21 | 287.325 | 285.333 | 284.202 | 282.990 |
| 2000 | 288.087 | 285.41 | 287.523 | 285.516 | 284.403 | 283.190 |
| 4000 | 288.271 | 285.58 | 287.692 | 285.683 | 284.581 | 283.371 |
| 5000 | 288.324 | 285.63 | 287.741 | 285.730 | 284.632 | 283.422 |
| 10000 | 288.467 | 285.77 | 287.873 | 285.865 | 284.771 | 283.572 |
| 20000 | 288.578 | 285.88 | 287.980 | 285.975 | 284.879 | 283.674 |
| 50000 | 288.686 | 285.99 | 288.087 | 286.085 | 284.987 | 283.785 |
| 100000 | 288.745 | 286.05 | 288.146 | 286.145 | 285.046 | 283.845 |

| $H_2O:MCl_3$ | $TbCl_3$ | $DyCl_3$ | $HoCl_3$ | $ErCl_3$ | $TmCl_3$ | $YbCl_3$ |
|---|---|---|---|---|---|---|
| 300 | 281.512 | 284.517 | 286.924 | 287.07 | 285.145 | 279.640 |
| 500 | 281.732 | 284.725 | 287.139 | 287.27 | 285.358 | 279.854 |
| 600 | 281.800 | 284.794 | 287.210 | 287.34 | 285.428 | 279.923 |
| 800 | 281.902 | 284.898 | 287.314 | 287.44 | 285.530 | 280.023 |
| 1000 | 281.976 | 284.974 | 287.389 | 287.51 | 285.598 | 280.093 |
| 2000 | 282.182 | 285.193 | 287.598 | – | 285.798 | 280.296 |
| 4000 | 282.367 | 285.380 | 287.767 | – | 285.982 | 280.477 |
| 5000 | 282.420 | 285.432 | 287.816 | – | 286.032 | 280.528 |
| 10000 | 282.563 | 285.573 | 287.956 | – | 286.172 | 280.667 |
| 20000 | 282.676 | 285.680 | 288.071 | – | 286.279 | 280.777 |
| 50000 | 282.785 | 285.787 | 288.184 | – | 286.387 | 280.886 |
| 100000 | 282.845 | 285.846 | 288.244 | – | 286.446 | 280.945 |

Enthalpies of formation and $H_2O:MCl_3$ ratios of saturated solutions are [1]:

| $MCl_3$ | $LaCl_3$ | $PrCl_3$ | $TbCl_3$ | $HoCl_3$ | $YbCl_3$ |
|---|---|---|---|---|---|
| $H_2O:MCl_3$ | 14.18 | 13.96 | 15.50 | 14.84 | 13.86 |
| $-\Delta H^\circ_{298}$ | 281.28 | 280.63 | 275.12 | 280.23 | 272.720 |

The enthalpies of formation of infinite dilute solutions from gaseous ions $M^{3+}$ and $Cl^-$ have been found by cycle process calculations starting from experimental heats of solution. Average values for these "enthalpies of hydration" in kJ/mol are:

| $MCl_3$ . . | $CeCl_3$ | $PrCl_3$ | $NdCl_3$ | $SmCl_3$ | $GdCl_3$ | $DyCl_3$ | $HoCl_3$ | $ErCl_3$ | $YbCl_3$ |
|---|---|---|---|---|---|---|---|---|---|
| $-\Delta H^\circ_{298}$. . | 4455.7 | 4500.8 | 4542.9 | 4582.4 | 4645.6 | 4703.2 | 4746.2 | 4778.1 | 4836.3 |

The calculated values reveal an effect very similar to the tetrad effect observed in extraction equilibria of rare earth (III) compounds, Blachnik, Selle [2].

Heats of dilution of aqueous $MCl_3$ were determined at 25 °C in the concentration ranges $\leqq$0.2 m for M=La, Nd, Er, Yb [3], 0.25 m to saturation for M=La, Ho [4], 0.10 to 0.25 m for M=La, Yb [5, 6], 0 m to saturation for M=Pr, Tb, Yb [4, 7], $\leqq$0.25 m for M=Pr, Sm, Gd, Tb, Dy, Ho, Tm, Yb [8], and 0.02 to 0.25 m for M=Sm, Gd, Dy, Ho, Tm [9]. The experimental results were used to obtain values for the relative apparent molar heat content, which were represented by empirical least squares equations of the form

$$\Phi_L = a\,m^{1/2} - b\,m + c\,m^{3/2} - d\,m^2$$

with the following constants:

| $MCl_3$ | a | b | c | d | Ref. |
|---|---|---|---|---|---|
| $LaCl_3$ | 6925 | 14575 | 14178 | – | [3, 6] |
| $PrCl_3$ | 6945 | 18118 | 31512 | 21862 | [8] |
| $NdCl_3$ | 6925 | 16056 | 17523 | – | [3] |
| $SmCl_3$ | 6850 | 16004 | 24210 | 14374 | [8] |
| | 6788 | 15926 | 25037 | 15902 | [9] |
| $GdCl_3$ | 7026 | 17507 | 29802 | 19897 | [8] |
| | 7234 | 19171 | 34487 | 24139 | [9] |
| $TbCl_3$ | 6872 | 15042 | 20995 | 10611 | [8] |
| $DyCl_3$ | 6766 | 14641 | 21234 | 11941 | [8] |
| | 6654 | 15816 | 30889 | 26415 | [9] |
| $HoCl_3$ | 7128 | 18831 | 34662 | 24974 | [8] |
| | 6907 | 15222 | 19588 | 7447 | [9] |
| $ErCl_3$ | 4952 | 5077 | 1926 | – | [3] |
| $TmCl_3$ | 6917 | 16663 | 26693 | 16438 | [8] |
| | 6670 | 14951 | 22708 | 13465 | [9] |
| $YbCl_3$ | 4783 | 5122 | 2644 | – | [3, 6] |
| | 6859 | 16098 | 25379 | 15585 | [8] |

Relationships derived therefrom for the relative partial heat contents $\bar{L}_1$ and $\bar{L}_2$ of water and $MCl_3$, respectively, are

$$\bar{L}_1 = c_1 m^{3/2} + d_1 m^2 - e_1 m^{5/2} + f_1 m^3$$

$$\bar{L}_2 = a_2 m^{1/2} - b_2 m + c_2 m^{3/2} - d_2 m^2$$

with the following constants:

| $MCl_3$ | $c_1$ | $d_1$ | $e_1$ | $f_1$ | $a_2$ | $b_2$ | $c_2$ | $d_2$ | Ref. |
|---|---|---|---|---|---|---|---|---|---|
| $LaCl_3$ | −62.36 | 262.47 | 382.98 | – | 10388 | 29150 | 35445 | – | [6] |
| $SmCl_3$ | −61.146 | 286.92 | 676.61 | 572.98 | 10182 | 31852 | 62593 | 47706 | [9] |
| $GdCl_3$ | −65.164 | 345.38 | 931.99 | 869.78 | 10851 | 38342 | 86218 | 72417 | [9] |
| $DyCl_3$ | −59.939 | 284.94 | 834.74 | 951.88 | 9981 | 31632 | 77223 | 79245 | [9] |
| $HoCl_3$ | −62.227 | 274.24 | 529.35 | 268.33 | 10361 | 30444 | 48970 | 22341 | [9] |
| $TmCl_3$ | −60.083 | 269.36 | 613.66 | 485.17 | 10005 | 29902 | 56770 | 40395 | [9] |
| $YbCl_3$ | −43.09 | 92.278 | 71.451 | – | 7175 | 10244 | 6610 | – | [6] |

The "short chord" method of treating the heat of dilution data was used for the very dilute concentration range. This treatment yielded different equations for the concentration dependence of $\bar{P}_i$, which is the slope of the $\Phi_L$ vs. $m^{1/2}$ plots. Integration of the equation deemed most suitable for a given salt resulted in the expression

$$\Phi_L \text{ (in cal/mol)} = a'm^{1/2} - b'm - c'm^{3/2}$$

valid for the short chord range. Parameters a', b', and c' are the following:

| $MCl_3$ | a' | b' | c' | Ref. | $MCl_3$ | a' | b' | c' | Ref. |
|---|---|---|---|---|---|---|---|---|---|
| $LaCl_3$ | 6925 | 15413 | – | [3] | $DyCl_3$ | 6925 | 16272 | – | [8] |
| $PrCl_3$ | 6925 | 16011 | – | [8] | $HoCl_3$ | 6925 | 12935 | – | [8] |
| $NdCl_3$ | 6925 | 14248 | – | [3] | $ErCl_3$ | 465 | −121100 | 955700 | [3] |
| $SmCl_3$ | 6925 | 15968 | – | [8] | $TmCl_3$ | 6925 | 16061 | – | [8] |
| $GdCl_3$ | 6925 | 14068 | – | [8] | $YbCl_3$ | 6925 | 15175 | – | [8] |
| $TbCl_3$ | 6925 | 14667 | – | [8] | | 247 | −104500 | 738100 | [3] |

Pepple [10] measured anew heats of dilution of aqueous $MCl_3$ (M = Nd, Sm, Eu, Gd, Dy, Er, Tm, and Lu) from zero concentration to saturation at 25 °C and determined short chord data by diluting solutions of concentrations higher than m = 0.25, striving to obtain dilution heats large enough to be measured with a minimum of relative error. Evaluating the experimental results by an approach different from that of the above authors he developed $\Phi_L(m)$ equations for three concentration ranges:

(1) $\Phi_L = Am^{1/2} - Bm$ (m = 0 to 0.007 = limiting law region)
(2) $\Phi_L = A'm^{1/2} + B'm + C'm^{3/2} + D'm^2$ (m = 0 to 0.25 = dilute region)
(3) $\Phi_L = A''m^{1/2} + B''m + C''m^{3/2} + D''m^2 + I$ (m > 0.25 = concentrated region)

The parameters are for equation (1):

| $MCl_3$ . | $NdCl_3$ | $SmCl_3$ | $EuCl_3$ | $GdCl_3$ | $DyCl_3$ | $ErCl_3$ | $TmCl_3$ | $LuCl_3$ |
|---|---|---|---|---|---|---|---|---|
| A . . | 6925 | 6925 | 6925 | 6925 | 6925 | 6925 | 6925 | 6925 |
| B . . | 15330 | 15150 | 14070 | 13990 | 14560 | 13580 | 16060 | 14810 |

and for equations (2) and (3):

| $MCl_3$ | A' | −B' | C' | −D' | A'' | −B'' | C'' | D'' | I |
|---|---|---|---|---|---|---|---|---|---|
| $NdCl_3$ | 7007.0 | 18844.0 | 33186.9 | 22740.0 | 2101.60 | 596.02 | 141.34 | 256.28 | 588.70 |
| $SmCl_3$ | 6915.7 | 16603.5 | 25786.5 | 15437.8 | 3788.14 | 2815.90 | 1508.91 | − 39.06 | 182.07 |

| $MCl_3$ | A′ | −B′ | C′ | −D′ | A″ | −B″ | C″ | D″ | I |
|---|---|---|---|---|---|---|---|---|---|
| $EuCl_3$ | 7015.5 | 17566.1 | 29615.3 | 19318.3 | 4180.22 | 3484.37 | 2146.88 | −223.66 | 125.82 |
| $GdCl_3$ | 6945.3 | 15856.5 | 23471.1 | 12939.0 | 3538.88 | 2440.69 | 1557.25 | − 97.08 | 283.66 |
| $DyCl_3$ | 6927.1 | 15864.4 | 24874.3 | 15283.3 | 3768.19 | 2965.70 | 2066.18 | −228.32 | 262.01 |
| $ErCl_3$ | 7050.0 | 17413.7 | 29170.4 | 18882.7 | 3844.14 | 2959.29 | 1957.54 | −193.32 | 215.97 |
| $TmCl_3$ | 6851.2 | 16191.8 | 26079.7 | 16255.1 | 3101.74 | 2002.86 | 1414.68 | − 88.49 | 398.20 |
| $LuCl_3$ | 7029.4 | 18450.1 | 33269.8 | 23136.6 | 3021.44 | 1879.21 | 1333.97 | − 84.84 | 410.82 |

Values for the relative partial molar heat contents were represented as least square polynomials

$$\bar{L}_1 = C_1 m^{3/2} + D_1 m^2 + E_1 m^{5/2} + F_1 m^3$$
$$\bar{L}_2 = A_2 m^{1/2} + B_2 m + C_2 m^{3/2} + D_2 m^2 + U$$

with the following constants:

m = 0 to 0.25

| $MCl_3$ | $C_1$ | $-D_1$ | $E_1$ | $-F_1$ | $A_2$ | $-B_2$ | $C_2$ | $-D_2$ |
|---|---|---|---|---|---|---|---|---|
| $NdCl_3$ | 63.142 | 339.74 | 897.57 | 820.05 | 10510.0 | 37688.0 | 82964.0 | 68217.0 |
| $SmCl_3$ | 62.334 | 299.41 | 697.51 | 556.75 | 10374.0 | 33207.0 | 64468.0 | 46314.0 |
| $EuCl_3$ | 63.223 | 316.67 | 800.85 | 696.40 | 10523.0 | 35132.0 | 74042.0 | 57954.0 |
| $GdCl_3$ | 62.619 | 286.12 | 635.48 | 467.25 | 10418.0 | 31713.0 | 58677.0 | 38817.0 |
| $DyCl_3$ | 62.452 | 286.23 | 673.30 | 551.63 | 10391.0 | 31729.0 | 62186.0 | 45850.0 |
| $ErCl_3$ | 63.501 | 313.70 | 788.26 | 680.38 | 10534.0 | 34459.0 | 71881.0 | 55704.0 |
| $TmCl_3$ | 54.646 | 211.88 | 420.20 | 262.69 | 10277.0 | 32383.0 | 65200.0 | 48767.0 |
| $LuCl_3$ | 63.301 | 332.21 | 898.56 | 833.20 | 10544.0 | 36903.0 | 83183.0 | 69419.0 |

m > 0.25

| $MCl_3$ | $C_1$ | $-D_1$ | $E_1$ | $F_1$ | $A_2$ | $-B_2$ | $C_2$ | $D_2$ | U |
|---|---|---|---|---|---|---|---|---|---|
| $NdCl_3$ | 18.882 | 10.631 | 3.746 | 9.2502 | 3152.9 | 1192.7 | 353.7 | 768.77 | 588.58 |
| $SmCl_3$ | 34.181 | 50.875 | 38.188 | −1.4366 | 5681.6 | 5631.0 | 3621.8 | −117.09 | 182.22 |
| $EuCl_3$ | 37.671 | 62.812 | 58.045 | −8.0657 | 6268.9 | 6966.6 | 5365.9 | −670.68 | 126.17 |
| $GdCl_3$ | 31.898 | 44.020 | 42.120 | −3.4620 | 5308.4 | 4881.4 | 3893.1 | −288.70 | 283.61 |
| $DyCl_3$ | 32.132 | 49.727 | 53.370 | −7.6903 | 5550.9 | 5805.1 | 5098.5 | −672.14 | 291.05 |
| $ErCl_3$ | 34.636 | 53.339 | 52.923 | −6.9719 | 5767.6 | 5920.6 | 4895.0 | −580.20 | 215.63 |
| $TmCl_3$ | 14.453 | 1.8518 | 12.218 | 2.9937 | 4582.4 | 3918.5 | 3490.6 | −256.67 | 418.37 |
| $LuCl_3$ | 28.135 | 36.228 | 37.946 | −3.5288 | 4532.7 | 3759.2 | 3335.4 | −254.62 | 410.69 |

See Pitzer et al. [28] for a general equation for the dependence of $\Phi_L$ on temperature and pressure.

Values for $\Phi_L$ were also derived from integral heats of solution at various concentrations. Data for aqueous $CeCl_3$ and $NdCl_3$ from Spedding, Miller [11] are presented in "Rare Earth Elements" C4b, Sections 11.2.11.4.5.1.2 and 11.2.11.6.5, respectively; see also [12 to 14]. Measurements of the heat of solution of $NdCl_3 \cdot 6H_2O$ by Eberts, Spedding [6] yielded the equation $\Phi_L = 6925\ m^{1/2} - 17987\ m + 24509\ m^{3/2}$ which is at variance with the results of [11]. Values of $\Phi_L$ obtained from the heats of solution of the hydrated chlorides generally agree with those derived from the heats of dilution, but not with those obtained from the heats of solution of anhydrous salts, as is demonstrated for $NdCl_3$, $ErCl_3$ [3], $LaCl_3$, $SmCl_3$, $GdCl_3$, $DyCl_3$, $HoCl_3$, $TmCl_3$ [9]. At a concentration of 0.05 m, the $\Phi_L$ results from heat of solution measurements on $PrCl_3$ and $YbCl_3$ by [15] are approximately 300 cal/mol larger than results from heats of dilution. This discrepancy increases with increasing concentration [8]. It is attributed either to a slow last reaction step of the dissolution process, whose reaction heat cannot be detected in the isothermally jacketed calorimeters used, or to the occurrence of hydrolysis during dissolution; see the discussions in the cited papers.

The data given above indicate that the investigated chlorides, with the possible exception of $ErCl_3$ and $YbCl_3$, obey the Debye-Hückel limiting law in dilute aqueous solutions. The average of the experimental limiting slopes of the $\Phi_L$ function of 11 chlorides from [3, 8] is only 4.7% below the theoretical value of 6925. If the strongly deviating value for $ErCl_3$ is omitted, the average is 6770 which is only 2.3% below 6925. The apparent anomalies of $ErCl_3$ and $YbCl_3$ are probably due to slight hydrolysis with formation of polymeric species at the most dilute molalities.

Fig. 40

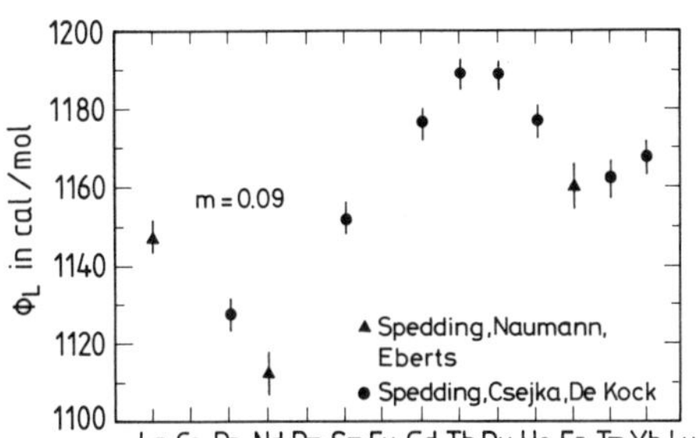


Fig. 41

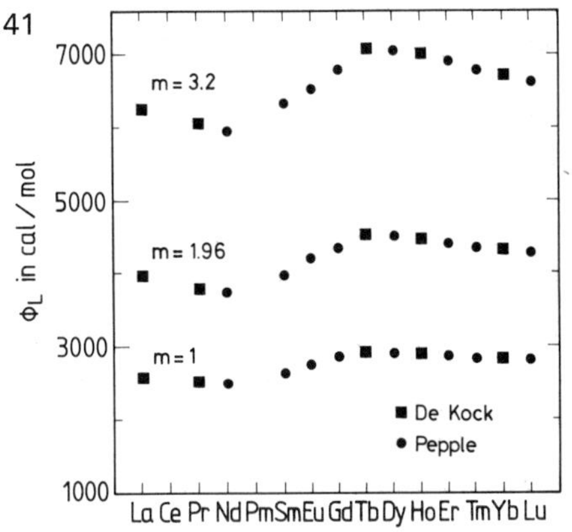


Relative apparent molar heat contents $\Phi_L$ of aqueous rare earth chloride solutions at 25 °C and constant molalities m=0.09 (Fig. 40), 1, 1.96, and 3.2 (Fig. 41) from different publications.

Plots of $\Phi_L$ across the rare earth series at constant molalities are shown in **Fig. 40** and **41** from [8] and [10], respectively; data from [3] and [4] are also entered into the figures. The trend of the $\Phi_L$ data, which is similar to those of data for the apparent molar volumes and heat capacities and other solution properties, is interpreted in terms of a change in water coordination number of the rare earth ions across the series, see p. 217.

Relative partial molar entropies of the water, $\bar{S}_1-\bar{S}_1^0$, and of the dissolved chlorides, $\bar{S}_2-\bar{S}_2^0$, were determined from the $\bar{L}_1$ and $\bar{L}_2$ values and from the activity coefficient data for the $MCl_3$ solutions; see [7] and [10] for numerical values for $T(\bar{S}_1-\bar{S}_1^0)$ and $T(\bar{S}_2-\bar{S}_2^0)$ at 25 °C and rounded concentrations. The partial molar entropy of the $H_2O$ in aqueous $LaCl_3$ at 100 °C is presented in "Rare Earth Elements" C4b, Section 11.2.11.3.4.1.4, along with additional data for $\Phi_L$, $\bar{L}_1$, $\bar{L}_2$, and other solution properties.

The heat capacities $C_p$ of aqueous $MCl_3$ solutions at 25 °C and 0.1 m to saturation were determined calorimetrically for M = La, Nd, Dy, Er, and Yb by [16, 17] for M = Pr, Sm, Eu, Gd, Tb, Ho, Tm, and Lu by [18 to 20]. The experimental results were used to obtain apparent molar heat capacities, $\Phi_C$, partial molar heat capacities of the solutes, $\bar{C}_2$, and partial molar heat capacities of water, $\bar{C}_1$, taking the value $c_p = 0.9989\ cal \cdot g^{-1} \cdot K^{-1}$ ($= 4.1793\ J \cdot g^{-1} \cdot K^{-1}$) for the specific heat capacity of pure water. The following values of $C_p$, $\Phi_C$, $\bar{C}_2$, and $\bar{C}_1$ in $cal \cdot mol^{-1} \cdot K^{-1}$ are given by [16, 18] for saturated solutions (c(sat.) = $MCl_3$ concentration at saturation in mol/kg $H_2O$):

| $MCl_3$ | $LaCl_3$ | $PrCl_3$ | $NdCl_3$ | $SmCl_3$ | $EuCl_3$ | $GdCl_3$ | $TbCl_3$ |
|---|---|---|---|---|---|---|---|
| c(sat.) | 3.8959 | 3.8912 | 3.9291 | 3.6412 | 3.5888 | 3.5899 | 3.5713 |
| $C_p$ | 0.5092 | 0.5212 | 0.5214 | 0.5242 | 0.5232 | 0.5084 | 0.5061 |
| $-\Phi_C$(exp.) | 0.8 | − 6.1 | − 9.1 | − 4.2 | − 2.6 | 2.6 | 3.7 |
| $-\Phi_C$(calc.) | 0.9 | − 6.1 | − 9.1 | − 4.2 | − 2.5 | 2.6 | 3.8 |
| $\bar{C}_2$ | 41.3 | 66.7 | 77.4 | 62.8 | 54.0 | 43.0 | 32.0 |
| $\bar{C}_1$ | 15.033 | 13.75 | 13.161 | 14.15 | 14.66 | 15.05 | 15.70 |

| $MCl_3$ | $DyCl_3$ | $HoCl_3$ | $ErCl_3$ | $TmCl_3$ | $YbCl_3$ | $LuCl_3$ |
|---|---|---|---|---|---|---|
| c(sat.) | 3.6309 | 3.6941 | 3.7822 | 3.8774 | 4.0028 | 4.1164 |
| $C_p$ | 0.4999 | 0.4941 | 0.4868 | 0.4818 | 0.4702 | 0.4601 |
| $-\Phi_C$(exp.) | 3.0 | 2.6 | 2.2 | 0.7 | 0.7 | 1.4 |
| $-\Phi_C$(calc.) | 3.0 | 2.6 | 2.2 | 0.9 | 0.8 | 1.5 |
| $\bar{C}_2$ | 32.4 | 29.5 | 26.8 | 26.9 | 24.5 | 19.6 |
| $\bar{C}_1$ | 15.676 | 15.86 | 16.016 | 16.06 | 16.173 | 16.43 |

The concentration dependence of $\Phi_C$ was expressed by polynomial equations of the form

$$\Phi_C = A + B\,m^{1/2} + C\,m + D\,m^{3/2} + E\,m^2$$

The following parameters of these equations were derived by a weighted least-squares treatment of the experimental $\Phi_C$ values:

| $MCl_3$ | −A | B | −C | D | −E | Ref. |
|---|---|---|---|---|---|---|
| $LaCl_3$ | 111.35 | 71.22 | 52.453 | 44.768 | 11.204 | [16] |
| $PrCl_3$ | 121.36 | 100.730 | 85.391 | 59.5617 | 12.9533 | [18] |
| $NdCl_3$ | 115.07 | 80.14 | 59.044 | 44.626 | 9.732 | [16] |
| $SmCl_3$ | 124.14 | 126.700 | 129.140 | 87.9856 | 19.1966 | [18] |
| $EuCl_3$ | 115.63 | 121.103 | 131.466 | 91.8997 | 20.5198 | [18] |
| $GdCl_3$ | 117.80 | 119.362 | 121.432 | 82.7956 | 18.4807 | [18] |
| $TbCl_3$ | 113.74 | 120.535 | 126.387 | 87.0900 | 19.9319 | [18] |
| $DyCl_3$ | 107.70 | 97.03 | 94.920 | 69.738 | 16.539 | [16] |
| $HoCl_3$ | 114.79 | 123.367 | 128.620 | 87.8105 | 20.0228 | [18] |
| $ErCl_3$ | 110.93 | 112.86 | 118.806 | 83.889 | 19.466 | [16] |
| $TmCl_3$ | 106.52 | 93.985 | 90.081 | 66.8670 | 16.0079 | [18] |
| $YbCl_3$ | 104.84 | 85.81 | 79.761 | 61.075 | 14.822 | [16] |
| $LuCl_3$ | 103.65 | 79.672 | 71.556 | 56.3495 | 13.9014 | [18] |

The values $\Phi_C$ (calc.) in the first table were computed using these parameters.

Measurements by Spitzer et al. [21] on $LaCl_3$ solutions in a flow calorimeter at 25 °C resulted in $\Phi_C$ values ranging from $-419.7$ J·mol$^{-1}$·K$^{-1}$ ($-100.31$ cal·mol$^{-1}$·K$^{-1}$) at 0.013967 m to $-386.8$ J·mol$^{-1}$·K$^{-1}$ ($-92.45$ cal·mol$^{-1}$·K$^{-1}$) at 0.089901 m. For the limiting value of $\Phi_C$ at infinite dilution $\Phi_C^0 = -456$ to $-475$ J·mol$^{-1}$·K$^{-1}$ is derived depending on the value chosen for the limiting slope and on the concentration range of the experimental results used for evaluation. Calorimetric $\Phi_C$ data of Novikov et al. [22] for aqueous solutions of $ScCl_3$, $YCl_3$, and $LaCl_3$ at 25 °C are shown in **Fig. 42**. $\Phi_C^0$ values for these solutions are $-90$, $-102$, and $-110$ cal·mol$^{-1}$·K$^{-1}$ ($-376.6$, $-426.8$, and $-460.2$ J·mol$^{-1}$·K$^{-1}$), respectively; see also Vasil'ev, Novikov [23] for a discussion of the concentration dependence of $\Phi_C$. Measurements by Karapet'yants, Vasilev et al. [24, 25] led to $\Phi_C^0 = -100$ cal·mol$^{-1}$·K$^{-1}$ ($-418$ J·mol$^{-1}$·K$^{-1}$) for aqueous $YCl_3$ and to $\Phi_C^0 = -111.4$ cal·mol$^{-1}$·K$^{-1}$ ($-466$ J·mol$^{-1}$·K$^{-1}$) for aqueous $CeCl_3$. Least squares treatments of the experimental $C_p$ results at 25 °C yielded the equations ($C_p$ in cal·mol$^{-1}$·K$^{-1}$):

$C_p(YCl_3) = 0.9979 - 0.26855\ m + 0.06695\ m^2 + 0.0071089\ m^3$ (m = 0.1 to 3.3) [24],

$C_p(CeCl_3) = 0.9979 - 0.3295\ m + 0.093625\ m^2 - 0.011171\ m^3$ (m = 0.040 to 3.058) [25].

The concentration dependence of $\Phi_C$ for aqueous $CeCl_3$ is represented by

$\Phi_C(CeCl_3) = -111.38 + 46.339\ m^{1/2} + 4.8107\ m + 0.75498\ m^{3/2}$ [25].

Jekel et al. [26, 27] used enthalpies of solution at several temperatures from 0.8 to 95 °C to derive $\Phi_C^0 = -104$ cal·mol$^{-1}$·K$^{-1}$ ($-435$ J·mol$^{-1}$·K$^{-1}$) for aqueous $GdCl_3$ at 25 °C.

Fig. 42

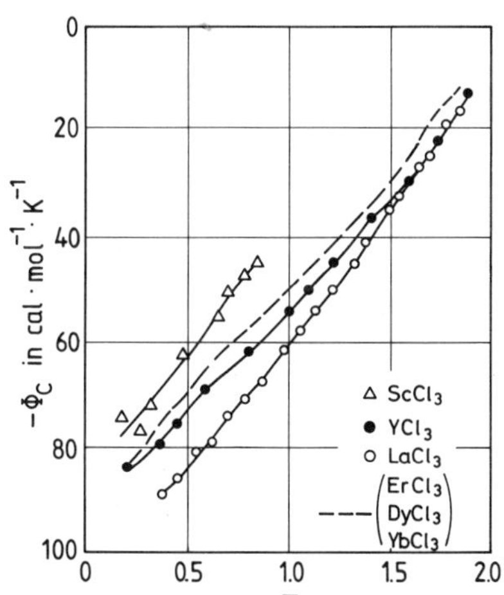


Apparent molar heat capacity $\Phi_C$ of aqueous $ScCl_3$, $YCl_3$, and $LaCl_3$ as a function of the square root of the molality m from Novikov et al. [22]. The dashed curve represents data for $ErCl_3$, $DyCl_3$, and $YbCl_3$ solutions from Spedding et al. [16]

Smoothed values for the relative apparent and partial molar heat capacities of aqueous $CeCl_3$ and $NdCl_3$ at 25 °C and m = 0.0001 to 0.36 are given in Sections 11.2.11.4.5.1.2 and 11.2.11.6.5 of "Rare Earth Elements" C4b.

Fig. 43

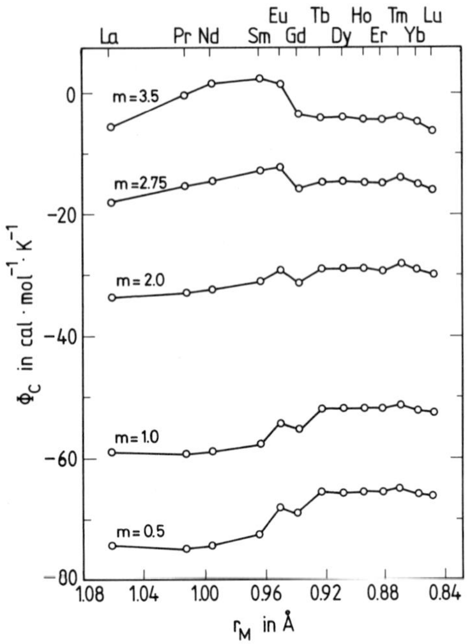


Apparent molar heat capacity $\Phi_C$ of aqueous $MCl_3$ solutions at given molalities m as a function of the ionic radius $r_M$ at 25 °C.

Fig. 44

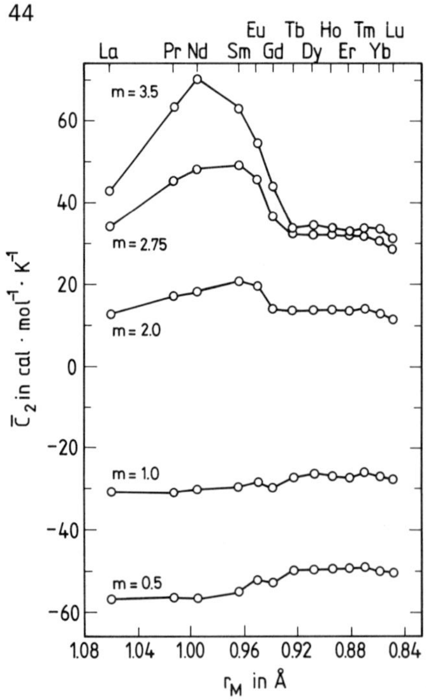


Fig. 45

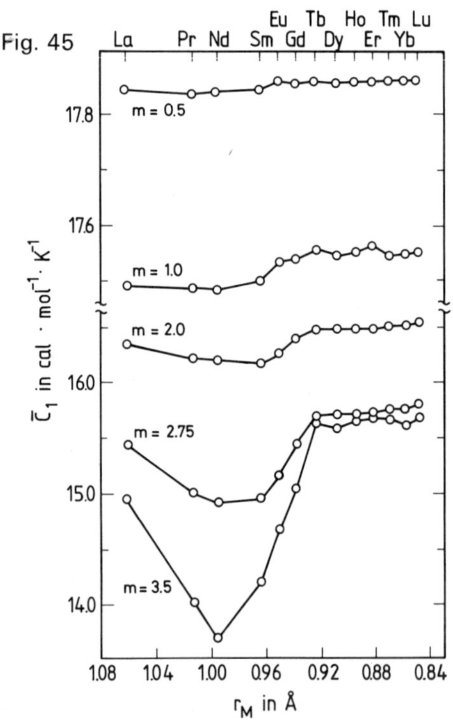


Partial molar heat capacities of solute $\bar{C}_2$ (Fig. 44) and of water $\bar{C}_1$ (Fig. 45) of aqueous $MCl_3$ solutions at given molalities m as a function of the ionic radius $r_M$ at 25 °C.

**Fig. 43** to **45** illustrate the trends of $\Phi_C$, $\bar{C}_2$, and $\bar{C}_1$ across the rare earth series, at constant molalities, as a function of ionic radius. Values for $\Phi_C$ and $\bar{C}_2$ for rare earth chloride solutions, like those for the rare earth perchlorate solutions, are negative at low concentrations because of the ions' disruptive effect on the hydrogen bonding of water and the strong orienting effect exerted by the ions on the surrounding water dipoles. Only between 1.55 and 1.70 m does the heat capacity gained in formation of ion hydrates predominate and $\bar{C}_2$ become positive. Values for $\bar{C}_1$ across the rare earth series do not show a monotonic trend with rare earth ionic radius but rather a two series effect caused by different water coordination numbers of the light and heavy rare earth cations. The differences in the concentration dependence of the heat capacity properties between rare earth chloride and rare earth perchlorate solutions are discussed in terms of ion-ion and ion-solvent interactions [16, 18]. At high concentrations, the $\Phi_C$ vs. $\sqrt{m}$ curves for light and heavy lanthanide chlorides are much closer to one another than at low concentrations. It is thus extremely likely that the hydrate coordination number of the rare earth ions changes with concentration. While in dilute solutions the coordination number is chiefly determined by the structure of water and the ionic radius, in concentrated solutions the interaction of the ions with water, i.e., the strength of the bond between the water molecule and the ions, is apparently of decisive importance in the formation of aquo-complexes [24].

References:

[1] R.H. Schumm, D.D. Wagman, S. Bailey et al. (NBS-TN-270-7 [1973] 1/75). – [2] R. Blachnik, D. Selle (Thermochim. Acta **33** [1979] 301/10, 308). – [3] F.H. Spedding, A.W. Naumann, R.E. Eberts (J. Am. Chem. Soc. **81** [1959] 23/8). – [4] C.W. DeKock (Diss. Iowa State Univ. 1965, pp. 1/166 from Diss. Abstr. **26** [1965] 112). – [5] R.E. Eberts (Diss. Iowa State College 1957, pp. 1/154 from Diss. Abstr. **17** [1957] 2834).

[6] R.E. Eberts, F.H. Spedding (ISC-941 [1957] 1/147 from C.A. **1959** 6745). – [7] C.W. DeKock, F.H. Spedding (IS-1356 [1965] 1/162 from C.A. **66** [1967] No. 59520). – [8] F.H. Spedding, D.A. Csejka, C.W. DeKock (J. Phys. Chem. **70** [1966] 2423/9). – [9] D.A. Csejka, F.H. Spedding (IS-417 [1961] 1/138; N.S.A. **16** [1962] No. 31584). – [10] G.W. Pepple (Diss. Iowa State Univ. Sci. Technol. 1967, pp. 1/124; Diss. Abstr. B **28** [1967] 1455).

[11] F.H. Spedding, C.F. Miller (J. Am. Chem. Soc. **74** [1952] 3158/62). – [12] F.H. Spedding, C.F. Miller (ISC-167 [1951] 1/131 from N.S.A. **6** [1952] No. 560). – [13] F.H. Spedding, C.F. Miller (ISC-190 [1951] 1/17 from N.S.A. **6** [1952] No. 839). – [14] F.H. Spedding, C.F. Miller (ISC-191 [1951] 1/14 from N.S.A. **6** [1952] No. 840). – [15] F.H. Spedding, J.P. Flynn (J. Am. Chem. Soc. **76** [1954] 1474/7, 1477/80).

[16] F.H. Spedding, K.C. Jones (J. Phys. Chem. **70** [1966] 2450/5). – [17] K.C. Jones (Diss. Iowa State Univ. 1965, pp. 1/84 from Diss. Abstr. **26** [1966] 5743). – [18] F.H. Spedding, J.P. Walters, J.L. Baker (J. Chem. Eng. Data **20** [1975] 438/43). – [19] J.P. Walters, F.H. Spedding (IS-1988 [1968] 1/96 from N.S.A. **23** [1969] No. 13689). – [20] J.P. Walters (Diss. Iowa State Univ. 1968, pp. 1/98 from Diss. Abstr. B **29** [1968] 1327).

[21] J.J. Spitzer, I.V. Olofsson, P.P. Singh, L.G. Hepler (Can. J. Chem. **57** [1979] 2798/803). – [22] S.N. Novikov, V.A. Vasil'ev, M.K. Karapet'yants (Deposited Doc. VINITI-3433-77 [1977] 79/83; C.A. **90** [1979] No. 110931). – [23] V.A. Vasil'ev, S.N. Novikov (7th Vses. Konf. Kalorim. Rasshir. Tezisy Dokl., Moscow 1977, Vol. 1, pp. 227/31;

C.A. **91** [1979] No. 217897). – [24] M.K. Karapet'yants, V.A. Vasilev, S.N. Novikov (Zh. Fiz. Khim. **50** [1976] 1031/3; Russ. J. Phys. Chem. **50** [1976] 622/3). – [25] V.A. Vasilev, S.N. Novikov, M.K. Karapet'yants (Zh. Fiz. Khim. **49** [1975] 1936/7; Russ. J. Phys. Chem. **49** [1975] 1137/8).

[26] E.C. Jekel (Diss. Purdue Univ. 1964 from Diss. Abstr. **26** [1965] 1907). – [27] E.C. Jekel, C.M. Criss, J.W. Cobble (J. Am. Chem. Soc. **86** [1964] 5404/7). – [28] K.S. Pitzer, J.R. Peterson, L.F. Silvester (J. Solution Chem. **7** [1978] 45/56).

### 11.2.8.1.8 Magnetic Properties

The magnetic susceptibility of aqueous $MCl_3$ solutions with M = Tb, Dy, Ho, and Er as a function of the density of the solutions is given in **Fig. 46**, Andres [1]. Older susceptibility measurements on aqueous solutions with M = Y, Ce, Pr, Nd, Sm, Gd, Er, and Yb at 18 °C are found in Du Bois, Liebknecht [2]. For magnetooptical data see p. 253.

Fig. 46

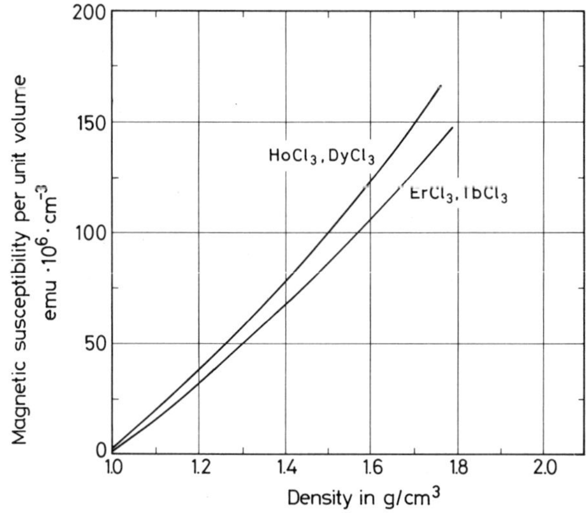


Magnetic susceptibility of aqueous $MCl_3$ solutions as a function of the density of the solution.

References:

[1] U.T. Andres (Mater. Sci. Eng. **26** [1976] 269/75). – [2] H. Du Bois, O. Liebknecht (Ann. Physik [4] **1** [1900] 189/98).

### 11.2.8.1.9 Electrical Properties

#### Conductance

Conductance measurements on about 0.0002 to 0.1 N aqueous $DyCl_3$ and $TbCl_3$ solutions at 0.5, 1, and 2 kHz and 25 °C show no noticeable frequency dependence, Spedding et al. [1]. The following table presents a compilation of values for the equivalent conductance $\Lambda$ in $cm^2 \cdot \Omega^{-1} \cdot equiv^{-1}$ of dilute aqueous $MCl_3$ solutions at 25 °C, from Spedding, Atkinson [2]. The data are based on studies of Spedding et al. [3 to 5], designated in the following as Ref. (Orig.), which are supplemented by measurements of $LuCl_3$ solutions.

| $MCl_3$ | c in mol/L | | | | | | | | | | Ref. |
|---|---|---|---|---|---|---|---|---|---|---|---|
| | 0 | 0.0001 | 0.0005 | 0.001 | 0.002 | 0.003 | 0.005 | 0.01 | 0.02 | 0.03 | (Orig.) |
| $LaCl_3$ | 146.1 | 141.5 | 135.5 | 130.3 | 126.1 | 122.8 | 118.2 | 111.5 | 104.5 | 100.2 | [3] |
| $CeCl_3$ | 146.2 | 141.8 | 135.9 | 130.5 | 126.3 | 122.9 | 118.3 | 111.6 | 104.6 | 100.3 | [3] |
| $PrCl_3$ | 146.0 | 141.3 | 135.3 | 130.1 | 125.9 | 122.6 | 118.0 | 111.3 | 104.2 | 100.0 | [3] |
| $NdCl_3$ | 145.7 | 141.0 | 135.1 | 130.0 | 126.0 | 122.5 | 117.9 | 111.4 | 104.3 | 100.2 | [4] |
| $SmCl_3$ | 144.9 | 139.8 | 134.2 | 129.4 | 125.3 | 122.0 | 117.5 | 110.9 | 103.9 | 99.7 | [3] |
| $EuCl_3$ | 144.2 | 139.6 | 133.6 | 128.7 | 124.9 | 121.4 | 116.7 | 110.3 | 103.5 | 99.3 | [3] |
| $GdCl_3$ | 143.7 | 138.9 | 132.6 | 127.8 | 124.1 | 120.8 | 116.4 | 110.1 | 103.4 | 99.4 | [5] |
| $DyCl_3$ | 142.7 [a)] | 137.6 | 132.0 | 127.3 | 123.2 | 120.0 | 115.7 | 109.1 | 102.3 | 98.3 | [4] |
| $HoCl_3$ | 142.6 | 138.1 | 132.3 | 127.5 | 123.3 | 120.0 | 115.7 | 109.1 | 102.4 | 98.4 | [4] |
| $ErCl_3$ | 142.3 | 137.7 | 132.1 | 127.4 | 123.3 | 120.0 | 115.7 | 109.1 | 102.3 | 98.2 | [4] |
| $TmCl_3$ | 141.8 | 137.0 | 131.4 | 127.1 | 123.2 | 119.9 | 115.4 | 108.8 | 102.1 | 97.7 | [4] |
| $YbCl_3$ | 141.5 | 137.2 | 131.6 | 127.2 | 123.2 | 120.0 | 115.5 | 109.0 | 102.4 | 98.1 | [4] |
| $LuCl_3$ | 141.1 | 136.3 | 130.7 | 126.4 | 122.6 | 119.1 | 114.8 | 108.6 | 101.8 | 97.5 | [2] |

a) Value corrected by [1] for $\Lambda_\infty$ (=extrapolated conductance at infinite dilution).

The reported values for $LaCl_3$ solutions are in good agreement with the older measurements of Jones, Bickford [6] and of Longworth, MacInnes [7], cf. "Rare Earth Elements" C 4 b. In aqueous $TbCl_3$ solutions $\Lambda_\infty = 143.3$ at 25 °C; $\Lambda$ of about 0.003 to 0.1 N $TbCl_3$ and $DyCl_3$ solutions can be found in the paper [1]. Also the measurements of Jantsch et al. [8] on about 0.0004882 to 0.0625 molar $MCl_3$ solutions (reported as dilutions from 2048 to 16 L/mol) at 25 °C (and 0 °C) show no important deviations from the values given in the table. For the 0.0004882 and 0.03125 molar solutions, for example, the following molar conductances $\Lambda(1)$ and $\Lambda(2)$ in $cm^2 \cdot \Omega^{-1} \cdot mol^{-1}$ have been determined:

| $MCl_3$ . . . . | $YCl_3$ | | $LaCl_3$ | $NdCl_3$ | $SmCl_3$ | |
|---|---|---|---|---|---|---|
| t in °C . . . | 0 | 25 | 0 | 0 | 0 | 25 |
| $\Lambda(1)$ . . . . | 194.5 | 378.3 | 206.9 | 197.7 | 196.4 | 393.1 |
| $\Lambda(2)$ . . . . | 158.2 | 299.5 | 161.1 | 161.0 | 162.0 | 304.1 |

| $MCl_3$ . . . . | $GdCl_3$ | | $ErCl_3$ | | $YbCl_3$ | |
|---|---|---|---|---|---|---|
| t in °C . . . | 0 | 25 | 0 | 25 | 0 | 25 |
| $\Lambda(1)$ . . . . | 207.8 | 397.9 | 194.2 | 384.5 | 198.8 | 391.3 |
| $\Lambda(2)$ . . . . | 160.1 | 301.9 | 158.6 | 299.6 | 158.7 | 299.3 |

Conductance measurements on aqueous $MCl_3$ solutions with M = La and Pr to Lu have been carried out in the concentration range from about 0.02 m up to the saturation limit at 25 °C by Spedding et al. [9]. From about 0.1 m to saturation the following relationship describes the decrease of the equivalent conductance with increasing molality m:

$$\Lambda = B_0 - B_1 m^{1/2} + B_2 m - B_3 m^{3/2} + B_4 m^2 - B_5 m^{5/2} + B_6 m^3 - B_7 m^{7/2}$$

Coefficients for this polynomial are compiled in [9]. The difference of the equivalent conductance relative to that of $LuCl_3$ solutions, which increases with growing $MCl_3$ concentration, according to a figure in [9] is nearly the same for the light rare earth elements with M = La, Pr, Nd, and Sm (≈10 and 25% at m = 1 and 3, respectively).

Fig. 47

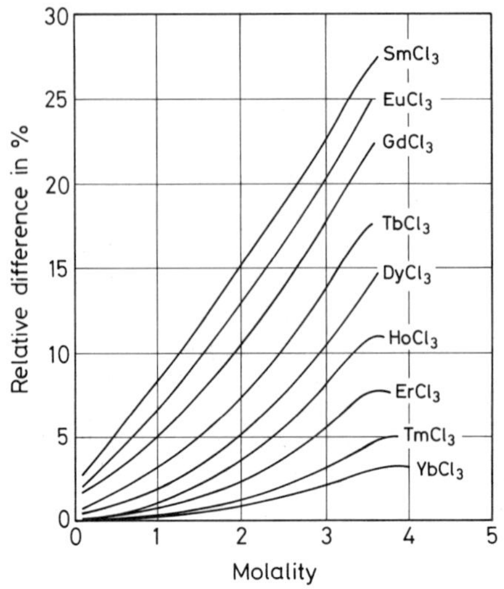


Percent differences in equivalent conductances relative to $LuCl_3$ for aqueous $MCl_3$ solutions as a function of the molality.

The relative difference for $MCl_3$ solutions with M = Sm to Yb is shown in **Fig. 47**. The equivalent conductance of the lanthanide chlorides declines at concentrations above about 0.16 m to somewhat lower values than that of the perchlorates. This observation is discussed from the viewpoint of different ion hydration, anion-cation interaction, cation hydrolysis, and different effects of the electrolyte on the hydrogen bonds of the water structure [9]. The description by the Onsager theory of the concentration dependence of the electrical conductance of dilute solutions of unsymmetrical electrolytes, especially $NdCl_3$ and $ErCl_3$, can be extended to higher concentrations, when a graphic evaluation of the integrals is employed. Comparison of calculated and measured values can be found in the paper, Dye, Spedding [10]. Experimental data on the concentration dependence of the electrical conductance of aqueous $MCl_3$ solutions with M = La, Sm, Eu, and Yb are found in Spedding, Wright [11], for M = La, Nd, Sm, Gd, Dy, Er, and Yb in Saeger, Spedding [12], Saeger [13], and for M = Nd, Dy, Ho, Er, Tm, and Yb in Dye, Spedding [14].

Limiting values of the ionic conductance of the $M^{3+}$ ions with M = La to Lu, measured on aqueous solutions of $MCl_3$ and other rare earth salts, are found in [1]. With increasing concentration the ionic equivalent conductance shrinks for the $M^{3+}$ ions in $MCl_3$ solutions. The values for M = La to Sm lie practically on one curve, those for the ions $Eu^{3+}$, $Er^{3+}$, and $Yb^{3+}$ are lower. The equivalent conductance values for the $Cl^-$ ion in all $MCl_3$ solutions lie on one curve, see the figure in the paper, Spedding et al. [15].

## Transference Numbers

The following table presents cation transference numbers $t_+$ of aqueous $MCl_3$ solutions at 25 °C as a function of concentration, from a compilation by [2] based on investigations of Spedding et al. [4, 5, 15] and the authors' own measurements on $LuCl_3$ solutions:

| $MCl_3$ | c in mol/L | | | | | | Ref. |
|---|---|---|---|---|---|---|---|
| | 0.005 | 0.010 | 0.015 | 0.020 | 0.025 | 0.030 | |
| $LaCl_3$ | 0.4598 | 0.4543 | 0.4501 | 0.4467 | 0.4434 | 0.4406 | [15] |
| $CeCl_3$ | 0.4608 | 0.4551 | 0.4507 | 0.4473 | 0.4438 | 0.4409 | [15] |
| $PrCl_3$ | 0.4610 | 0.4554 | 0.4511 | 0.4476 | 0.4442 | 0.4413 | [15] |
| $NdCl_3$ | 0.4607 | 0.4550 | 0.4506 | 0.4472 | 0.4437 | 0.4408 | [4] |
| $SmCl_3$ | 0.4606 | 0.4550 | 0.4507 | 0.4472 | 0.4438 | 0.4409 | [15] |
| $EuCl_3$ | 0.4586 | 0.4527 | 0.4482 | 0.4446 | 0.4411 | 0.4381 | [15] |
| $GdCl_3$ | 0.4576 | 0.4510 | 0.4459 | 0.4418 | 0.4378 | 0.4344 | [5] |
| $DyCl_3$ | 0.4508 | 0.4447 | 0.4400 | 0.4363 | 0.4326 | 0.4295 | [4] |
| $HoCl_3$ | 0.4459 | 0.4395 | 0.4346 | 0.4307 | 0.4268 | 0.4235 | [4] |
| $ErCl_3$ | 0.4456 | 0.4396 | 0.4350 | 0.4314 | 0.4277 | 0.4246 | [4] |
| $TmCl_3$ | 0.4460 | 0.4400 | 0.4354 | 0.4318 | 0.4281 | 0.4250 | [4] |
| $YbCl_3$ | 0.4451 | 0.4388 | 0.4339 | 0.4301 | 0.4263 | 0.4230 | [4] |
| $LuCl_3$ | 0.4426 | 0.4362 | 0.4313 | 0.4274 | 0.4235 | 0.4202 | [2] |

These values were determined by the moving phase boundary method. $t_+$ is a linear function of $\sqrt{c}$ in the concentration region from c=0.0025 to 0.04 mol/L, albeit except at very low concentrations larger than is calculated with the Onsager theory. The constants of the empirical relationship $t_+=a-b\sqrt{c}$ are:

| $MCl_3$ . . . | $LaCl_3$ | $CeCl_3$ | $PrCl_3$ | $NdCl_3$ | $SmCl_3$ | $EuCl_3$ | $GdCl_3$ |
|---|---|---|---|---|---|---|---|
| a . . . . . . | 0.4730 | 0.4745 | 0.4746 | 0.4744 | 0.4742 | 0.4727 | 0.4737 |
| b . . . . . . | 0.187 | 0.194 | 0.192 | 0.194 | 0.192 | 0.200 | 0.227 |

| $MCl_3$ . . . | $DyCl_3$ | $HoCl_3$ | $ErCl_3$ | $TmCl_3$ | $YbCl_3$ | $LuCl_3$ |
|---|---|---|---|---|---|---|
| a . . . . . . | 0.4655 | 0.4613 | 0.4600 | 0.4604 | 0.4604 | 0.4581 |
| b . . . . . . | 0.218 | 0.208 | 0.204 | 0.204 | 0.216 | 0.219 |

The values for $t_+$ at infinite dilution according to a figure in the paper change only slightly between $LaCl_3$ and $GdCl_3$ and then decrease with increasing atomic number with formation of a distinct break in the curve [2]. Cation transport numbers of aqueous $MCl_3$ solutions with M=La, Sm, Eu, and Yb are also found in Spedding, Wright [11], with M=Nd, Dy, Ho, Er, Tm, and Yb in Dye, Spedding [14]. For $TbCl_3$ solutions [1] gives the relation $t_+=0.4624-0.1235\sqrt{N}$ (N=normality); individual values for 0.012532 to 0.12532 N solutions are found in the paper. Like the description of the concentration dependence of the electrical conductance by the Onsager theory (cf. p. 250), the range of validity for the transference numbers can be extended by graphical evaluation of the integrals [10].

References:

[1] F.H. Spedding, R.A. Nelson, J.A. Rard (J. Chem. Eng. Data **19** [1974] 379/81). – [2] F.H. Spedding, B. Atkinson (in: W.J. Hamer, The Structure of Electrolytic Solutions, London 1959, pp. 319/39, 325/9). – [3] F.H. Spedding, P.E. Porter, J.M. Wright (J. Am. Chem. Soc. **74** [1952] 2055/8; AECU-1374 [1951] from N.S.A. **5** [1951] No. 4706). – [4] F.H. Spedding, J.L. Dye (J. Am. Chem. Soc. **76** [1954] 879/81). – [5] F.H. Spedding, I.S. Yaffe (J. Am. Chem. Soc. **74** [1952] 4751/5).

[6] G. Jones, C.F. Bickford (J. Am. Chem. Soc. **56** [1934] 602/11). – [7] L.G. Longworth, D.A. MacInnes (J. Am. Chem. Soc. **60** [1938] 3070/4). – [8] G. Jantsch,

H. Grubitsch, E. Lischka (Z. Elektrochem. **43** [1937] 293/6). – [9] F.H. Spedding, J.A. Rard, V.W. Saeger (J. Chem. Eng. Data **19** [1974] 337/8). – [10] J.L. Dye, F.H. Spedding (J. Am. Chem. Soc. **76** [1954] 888/92).

[11] F.H. Spedding, J.M. Wright (ISC-143 [1951] 1/93, 20/2, 24/5, 54/9; N.S.A. **5** [1951] No. 5601). – [12] V.W. Saeger, F.H. Spedding (IS-338 [1960] 1/202, 118/30; N.S.A. **16** [1962] No. 88). – [13] V.W. Saeger (Diss. Iowa State Univ. 1960, pp. 1/209, 118/24; Diss. Abstr. **21** [1961] 2510). – [14] J.L. Dye, F.H. Spedding (ISC-355 [1953] 1/119, 36/45, 67/71; C.A. **1956** 6880). – [15] F.H. Spedding, P.E. Porter, J.M. Wright (J. Am. Chem. Soc. **74** [1952] 2778/81; ISC-169 [1952] from N.S.A. **6** [1952] No. 563).

### 11.2.8.1.10 Optical Properties

#### Refractive Index and Molar Refraction

The following table presents the refractive index n and the molar refraction $R_{mol}$ of saturated aqueous $MCl_3$ solutions at 10, 25, and 50 °C. Saturation concentrations $c_s$ are given in wt%, the saturation molality $m_s$ is calculated from the mole ratios reported in the paper. The wave length used is not given in the paper.

| $MCl_3$ . . . | $YCl_3$ | $LaCl_3$ | $CeCl_3$ | $PrCl_3$ | $NdCl_3$ | $SmCl_3$ | $EuCl_3$ | $GdCl_3$ |
|---|---|---|---|---|---|---|---|---|
| $c_s$ (10 °C) . | 43.4 | 48.2 | 48.1 | 48.2 | 49.3 | 47.8 | 47.6 | 48.3 |
| $m_s$ (10 °C) . | 3.937 | 3.805 | 3.764 | 3.766 | 3.882 | 3.568 | 3.518 | 3.548 |
| n (10 °C) . | 1.4618 | 1.4678 | 1.4702 | 1.4720 | 1.4740 | 1.4649 | 1.4615 | 1.4590 |
| $R_{mol}$ (10 °C) | 79.96 | 84.33 | 84.78 | 84.50 | 82.56 | 86.79 | 87.57 | 86.90 |
| $c_s$ (25 °C) . | 43.5 | 48.8 | 48.6 | 49.0 | 49.7 | 48.3 | 48.3 | 48.7 |
| $m_s$ (25 °C) . | 3.893 | 3.885 | 3.836 | 3.885 | 3.940 | 3.636 | 3.615 | 3.598 |
| n (25 °C) . | 1.4612 | 1.4695 | 1.4723 | 1.4739 | 1.4742 | 1.4650 | 1.4610 | 1.4590 |
| $R_{mol}$ (25 °C) | 79.84 | 82.65 | 83.26 | 82.63 | 81.54 | 85.75 | 85.93 | 85.87 |
| $c_s$ (50 °C) . | 45.2 | 51.6 | 52.2 | 52.2 | 52.2 | 50.5 | 50.4 | 50.8 |
| $m_s$ (50 °C) . | 4.222 | 4.347 | 4.395 | 4.420 | 4.357 | 3.971 | 3.926 | 3.907 |
| n (50 °C) . | 1.4605 | 1.4775 | 1.4795 | 1.4790 | 1.4760 | 1.4655 | 1.4620 | 1.4605 |
| $R_{mol}$ (50 °C) | 75.98 | 76.84 | 75.81 | 75.95 | 76.52 | 80.77 | 81.22 | 81.52 |

| $MCl_3$ . . . | $TbCl_3$ | $DyCl_3$ | $HoCl_3$ | $ErCl_3$ | $TmCl_3$ | $YbCl_3$ | $LuCl_3$ |
|---|---|---|---|---|---|---|---|
| $c_s$ (10 °C) . | 48.8 | 49.4 | 49.9 | 51.0 | 51.6 | 53.3 | 54.2 |
| $m_s$ (10 °C) . | 3.593 | 3.633 | 3.674 | 3.808 | 3.874 | 4.088 | 4.209 |
| n (10 °C) . | 1.4570 | 1.4582 | 1.4601 | 1.4621 | 1.4650 | 1.4690 | 1.4720 |
| $R_{mol}$ (10 °C) | 86.16 | 85.46 | 84.45 | 82.33 | 81.00 | 78.40 | 76.64 |
| $c_s$ (25 °C) . | 49.1 | 49.9 | 50.7 | 51.3 | 52.2 | 53.3 | 54.2 |
| $m_s$ (25 °C) . | 3.636 | 3.704 | 3.792 | 3.850 | 3.965 | 4.085 | 4.206 |
| n (25 °C) . | 1.4574 | 1.4580 | 1.4594 | 1.4621 | 1.4660 | 1.4682 | 1.4690 |
| $R_{mol}$ (25 °C) | 85.14 | 84.27 | 82.82 | 81.66 | 80.12 | 78.35 | 76.22 |
| $c_s$ (50 °C) . | 50.7 | 51.1 | 51.8 | 52.5 | 53.1 | 54.8 | 55.1 |
| $m_s$ (50 °C) . | 3.877 | 3.885 | 3.951 | 4.040 | 4.112 | 4.330 | 4.361 |
| n (50 °C) . | 1.4585 | 1.4587 | 1.4600 | 1.4615 | 1.4650 | 1.4670 | 1.4700 |
| $R_{mol}$ (50 °C) | 81.52 | 81.10 | 79.96 | 78.80 | 77.77 | 75.32 | 74.57 |

Additional results for 30 °C can be found in the paper, Nikolaev et al. [1]. The concentration dependence of the refractive index n of aqueous $PrCl_3$, $SmCl_3$, and $EuCl_3$ solutions at 25 °C can be described ($\pm$0.0002) by the relationship $n = n(H_2O) + am - bm^2$, where $n(H_2O)$ is the refractive index of water, m is the molality, and a and b are constants. The following values of a and b were determined as a function of the wavelength $\lambda$:

| $\lambda$ in nm | | $PrCl_3$ | | $SmCl_3$ | | $EuCl_3$ | |
|---|---|---|---|---|---|---|---|
| | $n(H_2O)$ | a | b | a | b | a | b |
| 481.1 | 1.33678 | – | – | – | – | 0.04661 | 0.00276 |
| 508.6 | 1.33545 | 0.04658 | 0.00261 | 0.04628 | 0.00266 | 0.04642 | 0.00278 |
| 546.1 | 1.33386 | 0.04704 | 0.00289 | 0.04614 | 0.00269 | 0.04586 | 0.00269 |
| 589.3 | 1.33238 | – | – | 0.04578 | 0.00267 | 0.04516 | 0.00259 |
| 636.2 | 1.33112 | 0.04638 | 0.00285 | – | – | 0.04527 | 0.00268 |
| 643.8 | 1.33071 | 0.04644 | 0.00287 | 0.04572 | 0.00276 | 0.04538 | 0.00272 |

Mason et al. [2].

The refractive indices measured by Slack et al. [3] on aqueous $CeCl_3$ solutions (cf. "Rare Earth Elements" C4b, Section 11.2.11.4.5.1.3) and by [2] on $PrCl_3$, $SmCl_3$, and $EuCl_3$ solutions yield for infinite dilution the following values for the molar refraction of the dissolved $MCl_3$ species, Okazaki [4]:

| $\lambda$ in nm | $CeCl_3$ | $PrCl_3$ | $SmCl_3$ | $EuCl_3$ |
|---|---|---|---|---|
| 643.8 | – | 30.50 | 29.45 | 29.50 |
| 636.2 | – | 30.60 | – | 29.52 |
| 589.3 | 31.9 | (30.60) | 29.50 | 29.56 |
| 546.1 | – | 30.70 | 29.65 | 29.90 |
| 508.6 | – | 30.60 | 29.70 | 30.15 |
| 481.1 | – | – | – | 30.22 |

A relationship of the Kramers-Kronig type between the refractive index and the absorption coefficient in the region of the electronic transitions has been experimentally determined by Canit et al. [5] for aqueous solutions of $NdCl_3$, $ErCl_3$, and $EuCl_3$.

## Extinction

The extinction coefficients of a 0.169 molar $PrCl_3$ solution in the range from 9200 to 10400 $cm^{-1}$ and of a 0.283 molar $SmCl_3$ solution from 8800 to 9600 and 10300 to 10900 $cm^{-1}$ are somewhat higher than those of the corresponding nitrate solutions of the same concentration. In the case of 0.884 molar $YbCl_3$ and $Yb(NO_3)_3$ solutions between 9700 and 11300 $cm^{-1}$, they are equal, Hoogschagen [6].

## Magnetooptical Properties

The Faraday effect has been studied at various wavelengths on aqueous $MCl_3$ solutions as a function of the concentration, namely for M=La and Nd by Mason et al. [7], M=Ce by Slack et al. [3], see "Rare Earth Elements" C4b, Section 11.2.11.4.5.1.3, and M=Pr, Sm, and Eu by Mason [2]. For the Verdet constant $\omega$ of $LaCl_3$ and $NdCl_3$ solutions the following selected values for $10^3\ \omega$ in $min \cdot cm^{-1} \cdot Oe^{-1}$ ($=290\ rad \cdot m \cdot Wb^{-1}$) as a function of the molality m have been determined at temperatures t between 10 and 50 °C and the wavelengths $\lambda$:

$LaCl_3$

| t in °C | λ in nm | molality m 0.00 | 0.10 | 0.40 | 0.80 | 1.40 | 2.20 | 2.50 | 2.80 | 3.40 | 3.964 |
|---|---|---|---|---|---|---|---|---|---|---|---|
| 10 | 546.1 | 15.40 | 15.73 | 17.07 | 18.35 | 20.42 | 22.43 | 23.37 | 24.10 | 25.44 | 26.18 |
| 10 | 589.3 | 13.19 | 13.46 | 14.40 | 15.67 | 17.21 | 19.08 | 19.89 | 20.56 | 21.56 | 21.96 |
| 25 | 546.1 | 15.40 | 15.75 | 16.85 | 18.17 | 20.35 | 22.50 | 23.28 | 24.04 | 25.43 | 25.88 |
| 25 | 589.3 | 13.07 | 13.47 | 14.37 | 15.55 | 17.22 | 19.14 | 19.83 | 20.35 | 21.52 | 22.03 |
| 50 | 546.1 | 15.40 | 15.80 | 16.87 | 18.41 | 20.22 | 22.43 | 23.30 | 23.90 | 25.11 | 25.71 |
| 50 | 589.3 | 13.10 | 13.53 | 14.33 | 15.33 | 17.14 | 19.10 | 19.69 | 20.15 | 21.23 | 21.69 |

$NdCl_3$ (λ=546.1 nm)

| t in °C | molality m 0.00 | 0.05 | 0.10 | 0.20 | 0.30 | 0.40 | 0.50 |
|---|---|---|---|---|---|---|---|
| 10 | 15.40 | 15.40 | 14.60 | 13.84 | 13.03 | 12.30 | 11.45 |
| 25 | 15.40 | 15.40 | 14.66 | 13.99 | 13.32 | 12.47 | 11.72 |
| 50 | 15.33 | 15.40 | 14.73 | 14.16 | 13.59 | 12.81 | 12.14 |

The temperature dependence of the Verdet constant is significantly larger for the $NdCl_3$ solutions than for the $LaCl_3$ solutions. It can be described in the case of $NdCl_3$ by an equation of the form $\omega=-a/T+b$; however, the constants a and b presented in the paper do not yield good agreement with the experimental data tabulated above [7]. The following selected values are reported by [2] for the Verdet constants $10^3\ \omega$ in $min \cdot cm^{-1} \cdot Oe^{-1}$ of aqueous $PrCl_3$, $SmCl_3$, and $EuCl_3$ solutions at 25 °C:

| $MCl_3$ | λ in nm | molality m 0.00 | 0.25 | 0.50 | 1.00 | 1.50 | 2.00 | 2.50 | 3.00 |
|---|---|---|---|---|---|---|---|---|---|
| $PrCl_3$ | 508.6 | 18.00 | 14.03 | 10.17 | 2.25 | −5.84 | −13.71 | −21.53 | −29.28 |
| | 546.1 | 15.50 | 12.32 | 8.92 | 2.20 | −4.39 | −10.77 | −17.41 | −23.78 |
| | 643.8 | 10.81 | 8.52 | 6.32 | 1.71 | −2.89 | − 7.50 | −12.13 | −16.66 |
| $SmCl_3$ | 508.6 | 18.00 | 18.88 | 20.25 | 21.69 a) | 23.46 | 24.96 | 26.24 | 27.69 |
| | 546.1 | 15.50 | 16.29 | 17.25 | 18.64 a) | 20.17 | 21.37 | 22.55 | 23.52 |
| | 643.8 | 10.82 | 11.50 | 12.43 | 13.34 a) | 14.08 | 15.00 | 15.96 | 16.55 |
| $EuCl_3$ | 481.1 | 20.57 | 21.37 | 22.23 | 23.94 | 25.06 | 26.20 | 26.84 | 27.16 b) |
| | 508.6 | 18.00 | 18.73 | 19.44 | 20.73 | 21.99 | 22.98 | 23.73 | 23.94 b) |
| | 546.1 | 15.50 | 16.12 | 16.82 | 17.94 | 18.85 | 19.77 | 20.41 | 20.71 b) |
| | 643.8 | 10.81 | 11.47 | 11.62 | 12.85 | 13.71 | 14.35 | 14.92 | 15.06 b) |

a) 1.003 m. — b) 2.847 m

The magnetic molecular rotation $[M]_\omega$ of 0.1 to 0.5 m aqueous $NdCl_3$ solutions increases insignificantly with increasing concentration. The following average values of $[M]_\omega$ (in $min \cdot cm^2 \cdot Oe^{-1}$, λ=546.1 nm) were determined for these solutions as a function of the temperature t:

| t in °C . . . . . | 10 | 20 | 25 | 30 | 40 | 50 |
|---|---|---|---|---|---|---|
| $-[M]_\omega$ . . . . | 3.16 | 3.04 | 2.98 | 2.93 | 2.82 | 2.72 |

The temperature dependence can be described by a relationship analogous to the Curie-Weiss law, $[M]_\omega = C/(T-\Theta)$, with $C=-790.1$ and $\Theta=32.6$ [7]. The following average values for $[\bar{M}]_\omega$ of the dissolved $MCl_3$ at 25 °C and various wavelengths $\lambda$ have been determined, with consideration of the measurements of [7] on aqueous $LaCl_3$ and $NdCl_3$ solutions [2]:

| $MCl_3$ | $\lambda$ in nm 481.1 | 508.6 | 546.1 | 589.3 | 636.2 | 643.8 |
|---|---|---|---|---|---|---|
| $LaCl_3$ | – | – | 4.05 | 3.45 | – | – |
| $PrCl_3$ | – | −16.20 | −13.44 | – | −9.66 | −9.35 |
| $NdCl_3$ | – | – | − 6.99 | – | – | – |
| $SmCl_3$ | – | 3.96 | 3.54 | 3.01 | – | 2.60 |
| $EuCl_3$ | 3.60 | 3.16 | 2.75 | 2.45 | 2.37 | 2.24 |

For the specific magnetic rotation, see the paper [2]. The molecular rotation constants at infinite dilution calculated from these data for $PrCl_3$, $SmCl_3$, and $EuCl_3$, as well as for $CeCl_3$ from the Verdet constants and refractive indices measured by [3], can be found in Okazaki [4]. The magnetooptical rotation spectra of weak hydrochloric acid (pH=1) $MCl_3$ solutions at room temperature in the range of about 250 to 600 nm for M=Ce, Sm, Eu, Gd, Tb, Ho, Er, and Tm are reported by Briat [8], for M=Eu and Er by Briat et al. [9], and for M=Pr, Nd, and Dy at pH=4 to 5 by Briat [10].

References:

[1] A.V. Nikolaev, A.A. Sorokina, N.P. Sokolova, G.S. Kotlyar-Shapirov, N.P. Anoshina, L.I. Bagryantseva, A.P. Olovyanishnikova (Izv. Sibirsk. Otd. Akad. Nauk SSSR Ser. Khim. Nauk **1977** No. 4, pp. 84/90; C.A. **87** [1977] No. 173507). – [2] C.M. Mason, J.W. Hickery, W.K. Wilson (J. Am. Chem. Soc. **62** [1940] 95/9). – [3] F.G. Slack, R.L. Reeves, J.A. Peoples (Phys. Rev. [2] **46** [1934] 724/7). – [4] A. Okazaki (J. Phys. Soc. Japan **1** [1946] 1/3). – [5] J.C. Canit, M. Billardon, J. Badoz (J. Phys. [Paris] **32** [1971] 691/7).

[6] J. Hoogschagen (Physica **11** [1946] 513/7). – [7] C.M. Mason, R.D. Gray, G.L. Ernst (J. Am. Chem. Soc. **58** [1936] 1600/4). – [8] B. Briat (Compt. Rend. **260** [1965] 3335/8). – [9] B. Briat, M. Billardon, J. Badoz (Physica **33** [1967] 230/7). – [10] B. Briat (Compt. Rend. **260** [1965] 853/6).

### 11.2.8.2 Nonaqueous Solutions

#### $MCl_3$ Solutions in Alcohols and Alcohol-Water Mixtures

Structure of the Solutions

UV spectra of $MCl_3$ and $MCl_3 \cdot 6H_2O$ (with M being Pr and Nd in particular), dissolved in methanol, ethanol, and n-propanol, have been taken in the range from 10000 to 50000 $cm^{-1}$ and the results interpreted from the viewpoint of the solvate structure. For anhydrous alcoholic solutions it is assumed that ethanol behaves as a ligand similar to $H_2O$, so that in $MCl_3 \cdot 6H_2O$ solutions the $C_2H_5OH$ and $H_2O$ molecules compete. $CH_3OH$ has the highest dielectric constant of the three solvent alcohols, but it hardly competes in the $MCl_3 \cdot 6H_2O$ solutions with the $H_2O$ as a ligand, Bukietyńska et al. [1]. Nevertheless, Johnston et al. [2] conclude from electrical conductance measurements on alcoholic $MCl_3$ solutions with M=La and Pr to Yb that methanol is a stronger complex-forming ligand than is ethanol. Water in the solvate shell is displaced by n-propanol. Charge-transfer bands due to $Cl^-$ ions are observed on $EuCl_3$ dissolved in alcohols (and amides), but in contrast not on aqueous solutions. In anhydrous alcoholic solutions of $PrCl_3$ the $Cl^-$ ions exist in the first coordination sphere. The dominant species in the ethanolic $PrCl_3$ solution is the solvate $PrCl_3 \cdot 6C_2H_5OH$ with the coordination number 9 and probably $D_{3h}$ symmetry. Species with lower symmetry also occur in small quantities, which arise through partial dissociation. In methanol two or more species exist, among them probably the solvate $[PrCl_2 \cdot 7CH_3OH]^+$ with $C_{2v}$ symmetry predominates. In n-propanol the steric factor is probably more important than dissociation, the preponderant species is a solvate with the coordination number 8 and probably lower symmetry than $D_{3h}$. Spectra for $MCl_3$ and $MCl_3 \cdot 6H_2O$, dissolved in the named alcohols, are compared in [1] with the corresponding spectra of the aqueous solutions.

In continuation of the UV spectroscopic investigations on alcoholic $MCl_3$ and $MCl_3 \cdot 6H_2O$ solutions with M = Pr, Sm, Eu, and Yb, it was determined that in methanolic and ethanolic solutions of $PrCl_3$ and $PrCl_3 \cdot 6H_2O$ the $Cl^-$ ions and alcohol molecules are located in the inner coordination sphere of $M^{3+}$. In the case of $PrCl_3 \cdot 6H_2O$ dissolved in n-propanol, this sphere also still contains water molecules. The alcoholic solutions of $SmCl_3$, $EuCl_3$, and $YbCl_3$ show charge-transfer bands which are not observed for aqueous solutions. They are assigned to transitions of $Cl^-$ ions and alcohol molecules around the $M^{3+}$ ion, Keller et al. [3]. The absorption spectra of $EuCl_3$ in absolute methanol or ethanol at room temperature and the fluorescence spectra at low temperatures can be explained by $C_{2v}$ symmetry on the $Eu^{3+}$ ion. On addition of 2 to 3% $H_2O$ the spectrum begins to change. At $\geqq$15% $H_2O$ a spectrum occurs from which $D_{2h}$ symmetry is deduced, Sayre et al. [4]. See Miller [5] for a model representation for the configuration of the solvate based on these results.

A comparison of the absorption spectra of $MCl_3 \cdot 6H_2O$ with M = Nd, Eu, and Er, dissolved in methanol, methanol-water mixtures, or water, with the spectra of solutions of the corresponding bromates and perchlorates shows that the influence of the anion decreases in the sequence Nd > Eu > Er, while at the same time the spectra of the methanolic and aqueous solutions become more similar. On addition of >10% $H_2O$ the spectra match those of the pure aqueous solutions. The charge-transfer bands of the $EuCl_3$ solution, which correspond to transitions of $Cl^-$ ions, disappear on addition of 15 to 20% $H_2O$. Bands based on transitions of the $CH_3OH$ are still identifiable on addition of 70% $H_2O$. In methanol solutions with 5 and 10% $H_2O$ as well as 2 mol/L LiCl + $LiClO_4$, the stability constants for $NdCl^{2+}$ are 0.9 and ≈0.3 and for $ErCl^{2+}$ 0.5 and ≈0.2, respectively, Davidenko et al. [6]. The tendency to form $MCl_6^{3-}$ ions in hydrochloric acid-ethanol solutions increases with decreasing ionic radius according to spectroscopic investigations. In 75% ethanol $NdCl_6^{3-}$ exists at an HCl concentration of 8 mol/L and $EuCl_6^{3-}$ at 5 mol/L HCl. $ErCl_6^{3-}$ is observed in ethanol solutions at nearly all investigated HCl contents, at very low concentrations deviations from the octahedral structure occur, Barbanel [7]. Spectroscopic studies of complex formation in aqueous hydrochloric acid $MCl_3$ solutions with M = Ce, Pr, Nd, and Sm and in hydrochloric acid-ethanol-$CeCl_3$ solutions are reported by Jørgensen [8].

## Physical Properties

The following table summarizes values of the molar conductance $\Lambda$ (in $cm^2 \cdot \Omega^{-1} \cdot mol^{-1}$) of methanol and ethanol $MCl_3$ solutions at the temperature t as a function of the concentration c:

$MCl_3$ in Methanol

| $MCl_3$ | t in °C | c in mmol/L 0.5 | 0.7 | 1.0 | 2.5 | 5.0 | 10.0 | 12.5 | 15.0 | 20.0 |
|---|---|---|---|---|---|---|---|---|---|---|
| $LaCl_3$ | 20 | 115.0 | 106.8 | 99.3 | 82.9 | 71.9 | 63.3 | 60.6 | 58.7 | 56.1 |
| | 45 | 133.1 | 125.5 | 118.0 | 100.4 | 90.3 | 80.2 | 76.8 | 74.5 | 71.1 |
| $GdCl_3$ | 20 | 124.7 | 117.2 | 109.5 | 91.4 | 78.9 | 67.3 | 64.7 | 62.7 | 60.0 |
| | 45 | 132.2 | 126.5 | 120.3 | 105.0 | 94.4 | 82.5 | 78.9 | 77.1 | 75.1 |
| $YbCl_3$ | 20 | 86.5 | 82.8 | 79.2 | 70.0 | 64.2 | 59.8 | 58.1 | 56.5 | 55.1 |
| | 45 | 110.7 | 106.6 | 102.7 | 91.4 | 84.5 | 79.1 | 77.9 | 77.1 | 75.6 |

$MCl_3$ in Ethanol

| $MCl_3$ | t in °C | c in mmol/L 1.0 | 1.5 | 2.0 | 2.5 | 3.0 | 3.5 | 4.0 | 5.0 | 6.0 |
|---|---|---|---|---|---|---|---|---|---|---|
| $LaCl_3$ | 20 | 21.4 | 19.3 | 17.4 | 16.1 | 15.2 | 14.5 | 14.0 | 13.0 | 12.3 |
| | 45 | 28.6 | 23.1 | 21.3 | 19.5 | 18.2 | 17.3 | 16.6 | 15.4 | 14.6 |
| $PrCl_3$ | 20 | 22.7 | 20.9 | 19.7 | 18.7 | 17.9 | 17.2 | 16.5 | 15.4 | 14.5 |
| | 45 | 29.7 | 26.7 | 24.4 | 22.8 | 21.3 | 20.2 | 19.4 | 18.0 | 17.0 |
| $NdCl_3$ | 20 | 24.3 | 22.3 | 21.1 | 20.1 | 19.3 | 18.7 | 18.1 | 17.0 | 15.9 |
| | 45 | 32.1 | 28.6 | 26.7 | 25.3 | 24.0 | 23.0 | 22.0 | 20.3 | 18.9 |
| $SmCl_3$ | 20 | 26.0 | 24.6 | 23.3 | 22.3 | 21.5 | 21.0 | 20.8 | 19.8 | 19.0 |
| | 45 | 35.5 | 32.4 | 30.0 | 27.9 | 27.1 | 26.1 | 25.2 | 23.8 | 22.7 |
| $GdCl_3$ | 20 | 28.9 | 27.1 | 25.9 | 25.0 | 23.7 | 23.4 | 23.0 | 22.3 | 21.8 |
| | 45 | 38.9 | 36.1 | 34.5 | 33.0 | 31.3 | 30.1 | 29.2 | 28.2 | 26.2 |
| $DyCl_3$ | 20 | 27.3 | 26.4 | 25.5 | 24.5 | 23.8 | 23.2 | 22.8 | 22.1 | 21.7 |
| | 45 | 38.8 | 35.5 | 33.4 | 31.6 | 30.3 | 29.4 | 28.5 | 27.2 | 26.0 |
| $HoCl_3$ | 20 | 26.6 | 25.9 | 25.2 | 24.4 | 23.6 | 23.0 | 22.6 | 21.9 | 21.6 |
| | 45 | 37.3 | 34.2 | 32.5 | 31.2 | 30.2 | 29.3 | 28.1 | 26.0 | 24.5 |
| $ErCl_3$ | 20 | 27.0 | 25.7 | 24.5 | 23.8 | 23.2 | 22.8 | 22.4 | 21.4 | 21.0 |
| | 45 | 36.3 | 33.4 | 31.4 | 29.8 | 28.4 | 27.3 | 26.2 | 24.4 | 22.9 |
| $YbCl_3$ | 20 | 26.3 | 24.1 | 22.8 | 21.7 | 20.8 | 20.1 | 19.5 | 18.7 | 18.2 |
| | 45 | 33.3 | 30.2 | 27.8 | 26.1 | 24.3 | 23.1 | 22.2 | 20.8 | 19.9 |

The molar conductance of alcoholic $MCl_3$ and $MCl_3 \cdot nH_2O$ solutions of equal concentrations goes through a maximum at M = Gd, see **Fig. 48**, p. 258; the plot for $MCl_3 \cdot nH_2O$ is based on measurements of Johnston [9]. The concentration dependence of $\Lambda$ of the $MCl_3$ solutions is typical for weak electrolytes. Plotting of $\Lambda$ against $\sqrt{c}$ produces a nearly hyperbolic curve, Johnston et al. [2]. The viscosities, electrical conductances, and spectra of 0.0004 to 0.4 molar methanolic $MCl_3 \cdot nH_2O$ solutions at 5, 20, and 45 °C as well as the conductance and spectra of $CeCl_3 \cdot 7H_2O$ dissolved in methanol/$CCl_4$ mixtures, 1-propanol, 2-propanol, and isobutanol, are reported by [9].

On multiple passes of water-containing alcoholic $MCl_3$ solutions with M = Pr, Nd, Eu, and Er through a magnetic field perpendicular to the stream of liquid, the intensity

Fig. 48

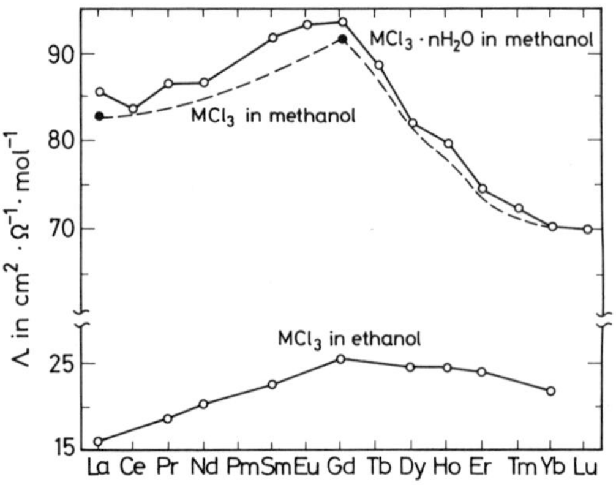


Molar conductances of 0.0025 M solutions at 20 °C of anhydrous $MCl_3$ in methanol and ethanol and of $MCl_3 \cdot nH_2O$ in methanol.

of the optical absorption bands increases. For a 0.04 molar $ErCl_3$ solution in a mixture of 1.5 parts water and 3.5 parts ethanol, the optical density of the band at $\lambda = 379.2$ nm increases within 15 min, for example, around 0.04, when the solution flows through a magnetic field of 2.5 kOe at 3 mL/min. The effect is explained by a change of the solvate structure, Shashkin [10].

## $MCl_3$ Solutions in Additional Organic Solvents

In aprotic solvents complexes of the type $[MCl_6]^{3-}$ can occur, which are observed for example in a solution of $MCl_3$ with M = Ce, Sm, Eu, and Tb in acetonitrile which is saturated with $(C_2H_5)_4NCl$. An extensive discussion of the optical spectra can be found in the paper, Ryan, Jørgensen [11].

$CeCl_3$ and $YbCl_3$ are relatively weak Lewis acids in acetone solutions, as shown by conductometric titrations with trimethylphosphine oxide, trimethylarsine oxide, hexamethylphosphoramide, dimethylsulfoxide, and N,N-dimethylacetamide, Du Preez et al. [12]. Potentiometric titrations of the acetone solutions show that the degree of ionization of $CeCl_3$ and $YbCl_3$ solutions corresponds to that of $ThCl_4$ solutions, Du Preez et al. [13].

Spectroscopic studies in the range from 8000 to 42000 $cm^{-1}$ of $MCl_3$ and $MCl_3 \cdot nH_2O$ with M = Pr, Nd, Ho, and Er, dissolved in formamide, methylformamide, and dimethylformamide, and interpretation of the intensity analysis as well as the nephelauxetic effects from the point of view of the solvate structure are reported by Legendziewicz et al. [14].

Densities, viscosities, and refractive indices at 25, 50, and 75 °C of the saturated solutions of $MCl_3$ hydrates (M = La, Ce, Pr, Nd, Sm) in hexachloro-1,3-butadiene have been published by Shevtsova et al. [16].

## $MCl_3$ Solutions in Molten Salts

Ions of the type $MCl_2^+$ and $MCl^{2+}$ are present in $(K,Na)NO_3$ melts, as shown by polarographic studies on $(1.83 \text{ to } 3.93) \times 10^{-3}$ m $LaCl_3$ and $(1.51 \text{ to } 2.86) \times 10^{-3}$ m $YbCl_3$ solutions; the $KNO_3:NaNO_3$ ratio is in each case 1:1. The chloride complexes of La have stability constants $\beta_1=90$ and $\beta_2=3935$; Yb has $\beta_1=28$, $\beta_2=2184$, Skurupafis, Sevryukov [15].

References:

[1] K. Bukietyńska, B. Ježowska-Trzebiatowska, B. Keller, J. Legendziewicz (Roczniki Chem. **42** [1968] 421/31; C.A. **69** [1968] No. 31739). – [2] D.O. Johnston, J.R. Boone, R.F. Kimberlin (J. Inorg. Nucl. Chem. **32** [1970] 1501/7). – [3] B. Keller, K. Bukietyńska, B. Ježowska-Trzebiatowska (Bull. Acad. Polon. Sci. Ser. Sci. Chim. **24** [1976] 736/9; C.A. **86** [1977] No. 81258). – [4] E.V. Sayre, D.G. Miller, S. Freed (J. Chem. Phys. **26** [1957] 109/13). – [5] D.G. Miller (J. Am. Chem. Soc. **80** [1958] 3576/9).

[6] N.K. Davidenko, L.N. Lugina, K.B. Yatsimirskii (Zh. Neorgan. Khim. **17** [1972] 104/8; Russ. J. Inorg. Chem. **17** [1972] 54/7). – [7] Yu. A. Barbanel (Proc. Moscow Symp. Chem. Transuranium Elem., Moscow 1972 [1976]; J. Inorg. Nucl. Chem. Suppl. **1976** 79/84; C.A. **85** [1976] No. 69040). – [8] C.K. Jørgensen (Mat. Fys. Medd. Danske Videnskab. Selskab. **30** No. 22 [1956] 1/38, 4, 8, 11, 17/21). – [9] D.O. Johnston (Diss. Univ. Mississippi 1963, pp. 1/262 from Diss. Abstr. **25** [1964] 127/8). – [10] M.A. Shashkin (Zh. Obshch. Khim. **45** [1975] 937/8; J. Gen. Chem. [USSR] **45** [1975] 919).

[11] J.L. Ryan, C.K. Jørgensen (J. Phys. Chem. **70** [1966] 2845/57). – [12] J.G.H. Du Preez, R.A. Edge, M.L. Gibson, H.E. Rohwer, C.P.J. van Vuuren (J. S. African Chem. Inst. [2] **29** [1976] 15/23). – [13] J.G.H. Du Preez, B.J. Gellatky, M.L. Gibson, D.R. Groot, H.E. Rohwer (J. S. African Chem. Inst. [2] **29** [1976] 2/3). – [14] J. Legendziewicz, K. Bukietynska, B. Jezowska-Trzebiatowska (Acta Phys. Acad. Sci. Hung. **35** [1974] 167/74). – [15] K.A. Skurupafis, N.N. Sevryukov (Elektrokhimiya **9** [1973] 207/11; Soviet Electrochem. **9** [1973] 198/201).

[16] Z.N. Shevtsova, B.G. Korshunov, V.V. Safonov, L.M. Kogan, V.I. Gudkova (Zh. Neorgan. Khim. **13** [1968] 3096/9; Russ. J. Inorg. Chem. **13** [1968] 1596/8).

## 11.2.9 Comparative Data for Chloride Systems and Compounds Containing Two Rare Earths

### 11.2.9.1 $MCl_3$-$M'Cl_3$ Systems

The equilibrium diagrams of the isomorphous Ce subgroup $MCl_3$-$M'Cl_3$ systems $LaCl_3$-$PrCl_3$, $LaCl_3$-$NdCl_3$, $LaCl_3$-$GdCl_3$, $LaCl_3$-$SmCl_3$, $CeCl_3$-$PrCl_3$, $PrCl_3$-$NdCl_3$, $NdCl_3$-$SmCl_3$, and $SmCl_3$-$GdCl_3$ are shown in **Fig. 49**, p. 260. They are characterized by the formation of continuous series of solid solutions without extrema. Only the system $PrCl_3$-$NdCl_3$ has a "diffuse" maximum at 23 mol % $NdCl_3$ and 801 °C, Nisel'son, Lyzlov [1]. Earlier Korshunov et al. [2] had found a slight maximum in the $LaCl_3$-$SmCl_3$ system, too, but [1] finds this to be unreliable. The isomorphous Y subgroup $MCl_3$-$M'Cl_3$ systems $TbCl_3$-$DyCl_3$, $DyCl_3$-$HoCl_3$, $TbCl_3$-$LuCl_3$, $YbCl_3$-$LuCl_3$, and $HoCl_3$-$LuCl_3$ also form continuous series of solid solutions without extrema, as shown in **Fig. 50**, p. 261, Lyzlov, Nisel'son [3]. On the other hand, the diagrams of the nonisomorphous $MCl_3$-$M'Cl_3$ systems $LaCl_3$-$LuCl_3$, $LaCl_3$-$DyCl_3$, $LaCl_3$-$TbCl_3$, $NdCl_3$-$ErCl_3$, $GdCl_3$-$LuCl_3$, $GdCl_3$-$DyCl_3$, and $GdCl_3$-$TbCl_3$, see **Fig. 51**, p. 262, are of the eutectic type with restricted mutual solid solubility. The only exception is $GdCl_3$-$TbCl_3$ with complete miscibility and a diffuse minimum at ≈58 mol % $TbCl_3$ and ≈575 °C. This is explained by the extremely small relative difference in the radii (1.6%) and lattice energies (0.5%) of Gd and Tb. For the other systems the relative difference in the M radii and lattice energies is 3 to 20% and 1 to 8%, respectively, Lyzlov, Nisel'son [4].

EPR spectra of doped rare earth trichlorides see Section 11.2.6.6.3, p. 93.

Fig. 49

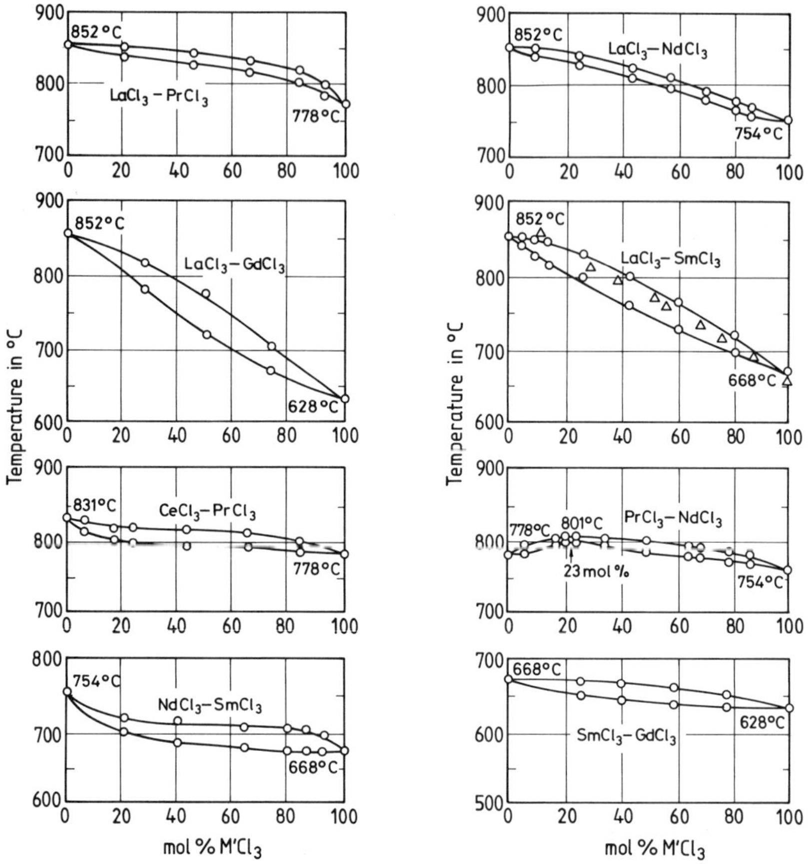


Equilibrium diagrams of the cerium subgroup $MCl_3$-$M'Cl_3$ systems; circles: results of Nisel'son, Lyzlov [1], triangles: results of Korshunov et al. [2].

References:

[1] L.A. Nisel'son, Yu.N. Lyzlov (Zh. Neorgan. Khim. **21** [1976] 3344/7; Russ. J. Inorg. Chem. **21** [1976] 1842/4). – [2] B.G. Korshunov, D.V. Drobot, L.V. Durinina (Zh. Neorgan. Khim. **10** [1965] 2120/3; Russ. J. Inorg. Chem. **10** [1965] 1154/6). – [3] Yu.N. Lyzlov, L.A. Nisel'son (Zh. Neorgan. Khim. **22** [1977] 2245/7; Russ. J. Inorg. Chem. **22** [1977] 1215/6). – [4] Yu.N. Lyzlov, L.A. Nisel'son (Zh. Neorgan. Khim. **23** [1978] 857/9; Russ. J. Inorg. Chem. **23** [1978] 476/7).

### 11.2.9.2 $MCl_3$-$M'Cl_3$-$H_2O$ Systems

In the systems $MCl_3$-$M'Cl_3$-$H_2O$ solid solutions are formed by the $MCl_3$ hydrates. If both $MCl_3$ are heptahydrates (M = La, Ce, Pr, and Nd?), continuous series of mixed crystals of the Roozeboom type I are formed with cationic radii difference $\Delta r \leqq 5.7\%$

Fig. 50

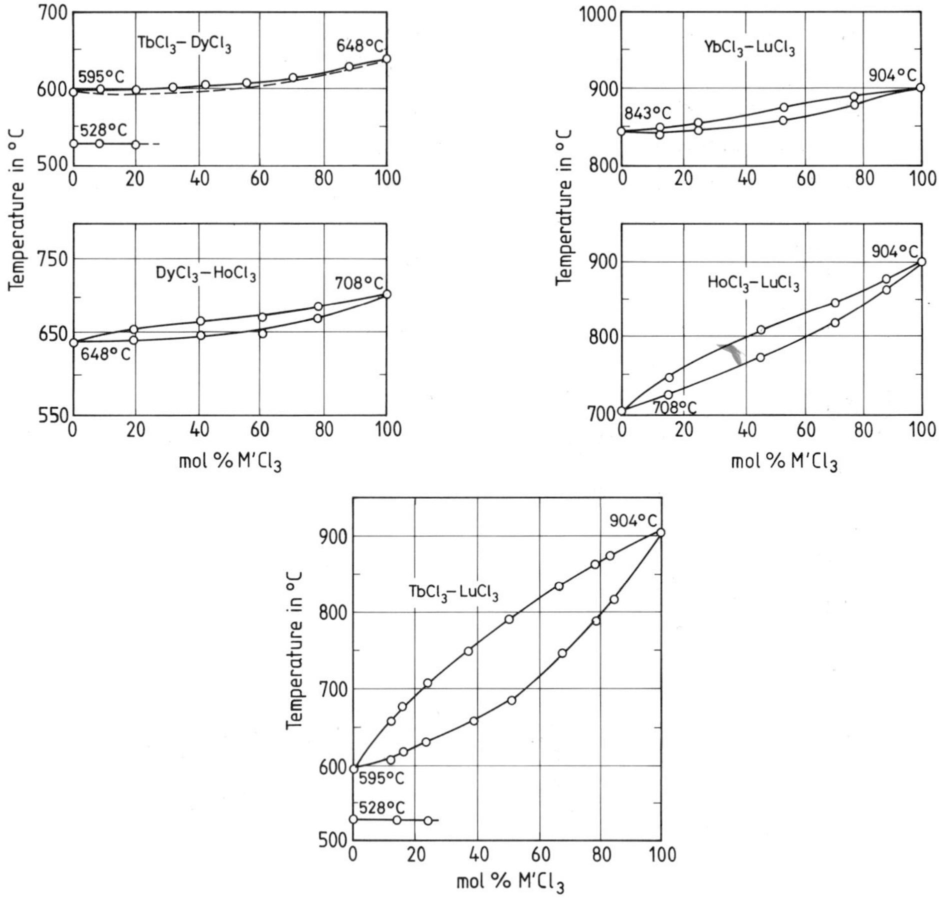


Equilibrium diagrams of the yttrium subgroup $MCl_3$-$M'Cl_3$ systems (in systems with $TbCl_3$ the polymorphic transformation temperature is shown at 528 °C).

from La to Nd (a heptahydrate of $NdCl_3$ seems to be stable only at cristallization at lower temperatures). If both $MCl_3$ are hexahydrates (M = Y, Nd to Lu) or mixed hydrated, solid solutions of type I, III, IV, and V are found depending on $\Delta r$, Nikolaev et al. [1]. For instance, in the $CeCl_3$-$MCl_3$-$H_2O$ systems with M = La to Nd and $\Delta r \leqq 3.4\%$ type I is found, and with M = Y, Sm to Lu and $\Delta r \leqq 16.7\%$ type IV. The solubility of hydrated $MCl_3$, M = Y, Dy to Ho, in $CeCl_3$ hydrate is practically zero, Nikolaev, Sorokina [2], see also Nikolaev et al. [3]. $EuCl_3 \cdot 6H_2O$ forms type I solid solutions with $MCl_3 \cdot 6H_2O$, M = Y, Nd to Ho, where $\Delta r$ is smaller than $\approx 7\%$; type V with $MCl_3 \cdot 6H_2O$, M = Er to Lu, where $\Delta r = 7.9$ to $12.4\%$; and type IV with $MCl_3 \cdot 7H_2O$, M = La to Pr, with $\Delta r = 2.5$ to 8%. The solubility of Eu in La and of Y and Lu in Eu trichloride hydrates is close to zero. It is shown in general that for solid solutions of type IV and V the solubility of the second $MCl_3$ hydrate in a solid salt with a large cationic radius is very often low [2], see also Nikolaev et al. [4]. While in the $YbCl_3$-$LaCl_3$-$H_2O$ system no solid solutions are formed, the mixed crystals in $YbCl_3$-$CeCl_3$ (or $PrCl_3$)-$H_2O$ belong to type IV,

Fig. 51

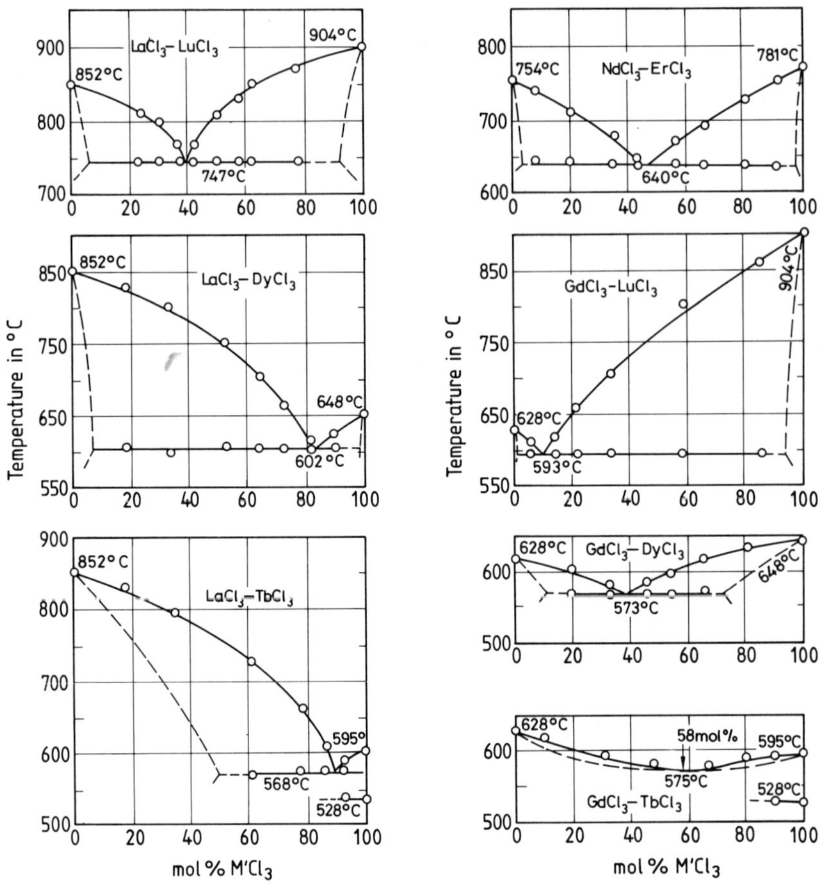


Equilibrium diagrams of $MCl_3$-$M'Cl_3$ systems of nonisomorphous rare earth trichlorides (in systems with $TbCl_3$ the polymorphic transformation temperature is shown at 528 °C).

in $YbCl_3$-$MCl_3$-$H_2O$ with M = Eu, Gd, or Dy to type V, and in $YbCl_3$-$ErCl_3$-$H_2O$ to type I, Nikolaev et al. [5, 6].

Out of 40 systems investigated (and 120 considered), 16 (39 expected) belong to type I, 13 (32) to type IV, 8 (39) to type V, 2 (2) to type III ($NdCl_3 \cdot 6H_2O$ with $HoCl_3 \cdot 6H_2O$ and $ErCl_3 \cdot 6H_2O$ only) and 1 (4) to a eutonic type ($LaCl_3 \cdot 7H_2O$ with Yb subgroup) [1].

The following miscibility gaps in Roozeboom type IV systems were found by Brunisholz, Nozari [7]:

| system | gap in at. % Nd | system | gap in at. % Sm |
|---|---|---|---|
| $NdCl_3$-$LaCl_3$-$H_2O$ | 87.0 to 99.3 | $SmCl_3$-$LaCl_3$-$H_2O$ | 12.5 to 95.0 |
| $NdCl_3$-$PrCl_3$-$H_2O$ | 81.3 to 90.0 | $SmCl_3$-$CeCl_3$-$H_2O$ | 12.3 to 80.5 |
| | | $SmCl_3$-$PrCl_3$-$H_2O$ | 9.5 to 57.0 |

Due to the easy transition $NdCl_3 \cdot 6H_2O \rightleftharpoons NdCl_3 \cdot 7H_2O$, type I is also found for $NdCl_3$–$MCl_3$–$H_2O$, M = Ce, Pr, at lower temperatures; see for instance Sokolova et al. [8] and "Rare Earth Elements" C4b, Section 11.2.11.6.8.

The assignment is partially inconsistent with respect to $YCl_3$ and $NdCl_3$, Nikolaev, Sorokina [9, 10]. For a more detailed discussion of the systems see the sections on individual compounds in "Rare Earth Elements" C4b.

References:

[1] A.V. Nikolaev, A.A. Sorokina, A.A. Mazurova, V.E. Fedorov (Izv. Sibirsk. Otd. Akad. Nauk SSSR Ser. Khim. Nauk **1976** No. 2, pp. 3/11; C.A. **85** [1976] No. 99933). – [2] A.V. Nikolaev, A.A. Sorokina (Izv. Sibirsk. Otd. Akad. Nauk SSSR Ser. Khim. Nauk **1970** No. 3, pp. 89/91; Sib. Chem. J. **1970** 376/7). – [3] A.V. Nikolaev, A.A. Sorokina, A.Ya. Vilenskaya (Izv. Sibirsk. Otd. Akad. Nauk SSSR Ser. Khim. Nauk **1967** No. 6, pp. 5/21; C.A. **69** [1968] No. 90501). – [4] A.V. Nikolaev, A.A. Sorokina, A.Ya. Vilenskaya (Dokl. Akad. Nauk SSSR **185** [1969] 608/9; Dokl. Chem. Proc. Acad. Sci. USSR **184/189** [1969] 251/2). – [5] A.V. Nikolaev, A.A. Sorokina, N.L. Kozyreva (Zh. Strukt. Khim. **15** [1974] 1108/11; J. Struct. Chem. [USSR] **15** [1974] 983/6).

[6] A.V. Nikolaev, A.A. Sorokina, N.L. Kozyreva (Izv. Sibirsk. Otd. Akad. Nauk SSSR Ser. Khim. Nauk **1974** No. 6, pp. 76/92; C.A. **82** [1974] No. 65075). – [7] G. Brunisholz, M. Nozari (Helv. Chim. Acta **52** [1969] 2303/14, 2304, 2307). – [8] N.P. Sokolova, L.I. Bagryantseva, O.G. Potapova (Izv. Sibirsk. Otd. Akad. Nauk SSSR Ser. Khim. Nauk **1979** No. 9, pp. 79/83; C.A. **91** [1979] No. 163775). – [9] A.V. Nikolaev, A.A. Sorokina (Izv. Sibirsk. Otd. Akad. Nauk SSSR Ser. Khim. Nauk **1971** No. 4, pp. 61/73; C.A. **77** [1972] No. 10305). – [10] A.V. Nikolaev, A.A. Sorokina (Redkozemel. Metal. Splavy Soedin. Mater. 7th Soveshch., Moscow 1972 [1973], pp. 27/31; C.A. **81** [1974] No. 96823).

### 11.2.9.3 $(M,M')Cl_3 \cdot nH_2O$

The crystallographic studies of mixed crystals are limited to individual reports. Mixed crystals of $LaCl_3 \cdot 7H_2O$ with 9% $GdCl_3 \cdot 6H_2O$ are triclinic, Dieke, Leopold [1], those with ≧50% $NdCl_3 \cdot 6H_2O$ monoclinic, Donnay, according to Dieke, Heroux [2]. Mixed crystals with only monoclinic components crystallize tabularly like the pure compounds, for example $(Ho, Dy)Cl_3 \cdot 6H_2O$ with about 20% Dy ($\beta \approx 95°$), $(Ho,Y)Cl_3 \cdot 6H_2O$ with Ho:Y ≈ 1:10, which is strongly contaminated with Er, and $(Er,Y,Ho)Cl_3 \cdot 6H_2O$ with an undetermined content of the individual components, Kahle [3].

In view of the rather extensive miscibility of the monoclinic and triclinic components, reasons have been developed concerning whether and how a topotactical transition $(M,M')Cl_3 \cdot 7H_2O \rightleftharpoons (M',M)Cl_3 \cdot 6H_2O$ can take place. The following fundamental relationships exist between the pseudomonoclinic cell of the heptahydrate (index "7") and the monoclinic cell of the hexahydrate (index "6"): $\vec{a}_6 = \vec{b}_7$, $2\vec{b}_6 = \vec{a}_7 + \vec{c}_7$, and $\vec{c}_6 = (3\vec{a}_7 - \vec{c}_7)/4$; see a figure in [4]. The crystallochemical regions with similar thematic configurations of the dimer complex in the heptahydrate structure and the monomer complex in the hexahydrate structure are discussed with the aid of the projections along $[010]_6$ and on the $(104)_7$ plane as well as along $[100]_6$ and $[010]_7$ (see in this respect also the Fig. 28, 27, 29, and 26, pp. 191, 189, 192, 188). In addition the atomic displacements are explained which on the basis of this comparison can occur initially in a topotactical transition, Bakakin et al. [4].

In mixed crystals of the approximate composition $Ho_{0.8}Dy_{0.2}Cl_3 \cdot 6H_2O$ the absolute value of extinction angle between the x axis of the optical indicatrix and the elongated crystal edge (c axis) is 47°, in those of $(Y,Ho)Cl_3 \cdot 6H_2O$ with Y:Ho≈10:1, 49°, Kahle [3].

References:

[1] G.H. Dieke, L. Leopold (J. Opt. Soc. Am. **47** [1957] 944/54, 946). – [2] G.H. Dieke, L. Heroux (Phys. Rev. [2] **103** [1956] 1227/39, 1229). – [3] H.G. Kahle (Z. Physik **145** [1956] 347/60, 348/9). – [4] V.V. Bakakin, R.F. Klevstova, L.P. Solov'eva (Zh. Strukt. Khim. **15** [1974] 820/30; J. Struct. Chem. [USSR] **15** [1974/75] 723/32).

### 11.2.10 Comparative Data on Ions in Solutions and Solid Compounds

For rare earth chloro ions in the gaseous state and their properties see Section 11.1, p. 1. The existence of the species $MCl^+$, $MCl_2^+$, and/or $MCl_3^+$ for M=La, Tb, Ho, and Er in the vapor over ACl-$MCl_3$ melts (A=alkali metal) has also been mentioned in "Seltenerdelemente" C 5, 1977, pp. 151/2, 238, 250 which in the following will be cited as "C 5".

#### Occurrence in Solutions and Melts

Rare earth chloro ions which form in aqueous and nonaqueous solutions will be described in the last Volume of "Rare Earth Elements" Series D together with other complexes of the rare earth elements. Therefore, in this section, the ions are just briefly referred to, and a complete literature coverage is not intended.

**$MCl_3^-$, $MCl_4^{2-}$** for M=Sm(II) and Yb(II) are present in KCl-$MCl_2$ melts with M=Sm or Yb (C 5, pp. 188, 199).

**$MCl^{2+}$, $MCl_2^+$.** Complexes of these types for M=La and Yb form in a molten equimolar $KNO_3$-$NaNO_3$ mixture at 290 °C in the presence of chloride ions as evidenced by polarographic studies. With an $La^{3+}$ or $Yb^{3+}$ concentration in the melt of $(1.83 \text{ to } 3.93) \times 10^{-3}$ or $(1.51 \text{ to } 2.86) \times 10^{-3}$ mol/kg and a $Cl^-$ concentration <0.1 or ≦0.3 mol/kg, the overall formation constants $\beta$ and the stepwise stability constants $k_j$ are: $\beta=90$, $k_1=90$ for $LaCl^{2+}$; $\beta=3935$, $k_1=90$, $k_2=43.7$ for $LaCl_2^+$; $\beta=28$, $k_1=28$ for $YbCl^{2+}$; and $\beta=2184$, $k_1=28$, $k_2=78$ for $YbCl_2^+$, Skurupafis, Sevryukov [1]. The existence of the $MCl_2^+$ ion for M=Y and La in ACl-$MCl_3$ melts with A=alkali metal has been reported in C 5, pp. 126, 170, 239/40. This $MCl_2^+$ ion has also been assumed by Ryan, Jørgensen [2] to exist in ethanol. From conductivity measurements the existence of $LaCl^{2+}$ and $LaCl_2^+$ in ethanol is suggested by El-Aggan et al. [51]. This assumption is also based on the charge transfer spectra of halide complexes for M=Sm, Eu, and Yb, which have been readily observed by Barnes [3] in ethanol at 18 °C. The $MCl_2^+$ ion seems not to exist in water [3].

**$MCl_4^-$, $MCl_5^{2-}$.** The presence of both or only one of these ions in mixed ACl-$MCl_3$ melts and in aqueous ACl-$MCl_3$ solutions (A=alkali metal) has been reported in C 5, Section 17, pp. 120ff.

**$MCl_6^{3-}$.** This ion exists in mixed ACl-$MCl_3$ melts and in aqueous ACl-$MCl_3$ solutions (A=alkali metal) as already reported in C 5, Section 17, pp. 120ff. Thus for instance, Raman [4, 5, 6], electromotive force [7 to 10], and calorimetric [7, 9, 11, 12] studies of molten mixtures have indicated that for compositions rich in ACl (75 to 80 mol % ACl) maximum concentrations of octahedral configurations are formed in the melt with

M=Y [6, 9], La [4, 5, 7], Ce [8, 10, 11], and Gd [12]. The content of the $MCl_6^{3-}$ ion formed in the melts increases in the series from LiCl to CsCl as shown by the electronic absorption spectra of ACl-$NdCl_3$ melts at 830 °C, Barbanel' et al. [13]. Continuous changes of the KCl-$YCl_3$ (vibrational) Raman spectra with compositions $\leq$75 mol % KCl have been interpreted by assuming a continuously increasing distortion of the $YCl_6^{3-}$ octahedra in the melt with increasing $YCl_3$ concentrations, from $O_h$ to possibly $D_3$ symmetry. Moreover, for a melt of fixed $YCl_3$ composition, the substitution of an alkali metal cation by another of smaller radius weakens the Y–Cl attraction and facilitates the Cl–Cl repulsion, Papatheodorou [6]. For octahedral ($O_h$) $MCl_6^{3-}$ the minimum of the enthalpy interaction parameters in the molten mixtures should be situated at 25 mol % $YCl_3$. The shift of the minimum to lower $YCl_3$ concentrations as found in mixtures containing RbCl, KCl, and NaCl probably may be interpreted as due to the existence of distorted octahedra even at these low $YCl_3$ concentrations, produced by sharing of $Cl^-$ by the $YCl_6^{3-}$ ions, Papatheodorou et al. [9]. – The melts of stoichiometric $MCl_3$ do not contain free $MCl_6^{3-}$ species, see [6, 13].

In solutions of HCl in absolute ethanol, the tendency to form octahedral $MCl_6^{3-}$ complexes increases with the decrease of the $M^{3+}$ radius as evidenced by the electronic absorption spectra. Complex formation is achieved at an HCl concentration of about 8 mol/L with M=Nd and at about 5 mol/L with M=Eu. For M=Er, the $MCl_6^{3-}$ complex is present at any concentration of HCl, but at low HCl concentrations the complex is a distorted octahedron, Barbanel [14].

$MCl_6^{3-}$ complexes are stable in aprotic solvents such as nitriles. For preparation, for example, the triphenylphosphonium salts $[(C_6H_5)_3PH]_3MCl_6$ are dissolved in 85% succinonitrile-15% acetonitrile with a small excess of $(C_2H_5)_4NCl$ or even $(C_2H_5)_3NHCl$, or hydrated $MCl_3$ is dissolved in a saturated solution of $(C_2H_5)_4NCl$ in acetonitrile. The $Cl^-$ ions have been added to prevent dissociation of the $MCl_6^{3-}$ in the nitriles (but addition of a large excess of HCl to the $MCl_6^{3-}$ solutions actually destroys rather than stabilizes the $MCl_6^{3-}$ complexes). The f-f absorption spectra (including hypersensitive pseudoquadrupolar transitions) of $MCl_6^{3-}$ with M=Pr, Nd, Sm, Dy, Ho, Er, and Tm, the electron-transfer spectra of $MCl_6^{3-}$ with M=Sm, Eu, and Yb, and the f$\rightarrow$d absorption spectra for M=Ce and Tb in these solutions were studied and extensively discussed. Molar extinction coefficients and optical electronegativities have been given too. Data have been interpreted on the basis that the complexes have $O_h$ symmetry, but some deviation from $O_h$ symmetry might be possible, Ryan, Jørgensen [2]. Oscillator strengths of f-f transitions in $NdCl_6^{3-}$ in the mixed nitrile solution (including the hypersensitive transitions at about 17000 $cm^{-1}$) have been derived from the absorption spectra (10640 to 28990 $cm^{-1}$) by Henrie, Henrie [15] and later analyzed. It appears that the symmetry of the $NdCl_6^{3-}$ complex deviates more from $O_h$ symmetry than that of the $NdBr_6^{3-}$ complex, Henrie, Henrie [16].

$MCl_6^{3-}$ complexes exist also in fused pyridinium chloride at 160 °C [14].

**Further Ions.** The existence of $M_2Cl_7^-$ for M=Sm and Gd and of $M_3Cl_{10}^-$ for M=Dy and Gd in the systems ACl-$MCl_3$ with A= alkali metal has already been reported (C5, pp. 133, 192, 249). Data for the $Sc_2Cl_9^{3-}$ ion in ACl-$ScCl_3$ melts are given in C5, pp. 166, 225, 226, 239.

The $CeCl_6^{2-}$ ion can be obtained by dissolving $[(C_2H_5)_4N]_2CeCl_6$ in acetonitrile. Excess $Cl^-$ was not required (in contrast to $CeCl_6^{3-}$, see above) to maintain the $CeCl_6^{2-}$ complex in this solvent. Electron transfer spectra and f$\rightarrow$d absorption spectra have been studied. If a quaternary ammonium bromide and HBr are added to an acetonitrile or nitromethane solution of $CeCl_6^{2-}$, the $CeCl_6^{2-}$ is converted immediately to $CeBr_6^{2-}$, Ryan, Jørgensen [2].

## Occurrence in Solid Compounds

### $MCl_4^-$

The $LaCl_4^-$ ion in $CsLaCl_4 \cdot 7H_2O$ has tetrahedral ($T_d$) symmetry, for estimated force constants see C 5, p. 241. Presumably, the $CeCl_4^-$ ion exists in $2CeCl_3 \cdot [(CH_2)_6(NH_3)_2]Cl_2$, see C 5, p. 222.

### $MCl_6^{3-}$

This ion is assumed to be present, for instance, in the ammoniumchlorometallates of Pr, Eu, and Tb (C 5, pp. 211; 212) and in the trimethylammonium compounds $[(CH_3)_3NH]_3MCl_6$ (C 5, pp. 213/24). In the pyridinium compounds $[C_5H_5NH]_3MCl_6$, M = Nd, Eu, Gd, Dy, Er, and Yb (C 5, p. 217) the ion is assumed to have octahedral ($O_h$) symmetry, from Raman and IR spectra.

**In triphenylphosphonium compounds** $[(C_6H_5)_3PH]_3MCl_6$. From earlier Raman and IR investigations on the compounds with M = Nd, Eu, Dy, Er, and Yb, the respective ions are said to have $O_h$ symmetry. However, from a recent structure analysis of $[(C_6H_5)_3PH]_3PrCl_6$ based on X-ray data, the $PrCl_6^{3-}$ ion very closely approximates $D_{2d}$ symmetry, the Pr–Cl distances are from 2.726 to 2.767 Å and on the average 2.745 Å, Majeste et al. [17]. The previous observation of the $^3H_4 \rightarrow {}^3P_0$ transition in the absorption spectrum of this compound at 300 K (at 20600 $cm^{-1}$) and at 80 K by Ryan, Jørgensen [2] is consistent with this symmetry [17]. A slight distortion from octahedral symmetry was also determined for $NdCl_6^{3-}$ from the absorption spectra at 77 K, Gruber et al. [18], and for $EuCl_6^{3-}$ in $[C_5H_5NH]_3EuCl_6$ from the emission spectra at 77 K, Butter, Seifert [19]. On the assumption that both these ions have $D_{2d}$ symmetry as well, ligand polarizability contributions can account quantitatively for the hypersensitive intensity seen in the spectra of these ions; for values of the intensity parameters see the original paper, Peacock, Stewart [20].

**In caesiumsodiumchlorometallates** $Cs_2NaMCl_6$. In these elpasolite-type compounds $MCl_6^{3-}$ has $O_h$ symmetry at room temperature according to X-ray structure analysis. Data consistent with this symmetry of the $MCl_6^{3-}$ ion are obtained from Raman spectra, electron paramagnetic resonance (EPR) and ENDOR as well as from absorption and magnetic circular dichroism (MCD) spectra already cited in C 5, pp. 251/7. In the case of $PrCl_6^{3-}$ some doubt exists about the assignment of the transition at about 20631 $cm^{-1}$, which is presumed to be due to $^3H_4 \leftrightarrow {}^3P_1$, Schwartz [21], Amberger [22], instead of $^3H_4 \leftrightarrow {}^3P_0$ (see above at $[(C_6H_5)_3PH]_3PrCl_6$) which is $O_h$ forbidden [21]. At low temperatures the $YCl_6^{3-}$ ion has $O_h$ symmetry, too, as shown by the EPR spectra at 77 K [23] and at 4.2 K [23, 24] and the absorption and MCD spectra at liquid helium temperature [25] of doped $Cs_2NaYCl_6$. However, $LaCl_6^{3-}$ (C 5, p. 255) and $PrCl_6^{3-}$ from the EPR spectra at 77 K of Gd-doped $Cs_2NaMCl_6$ (M = La, Pr) seem below the phase transition of the substance to have lower symmetry (but still an inversion center), presumably $C_{4h}$, $C_{2h}$, or $C_i$ [23]. For $PrCl_6^{3-}$ this was confirmed by means of the electronic Raman effect at 70 K [22] and at 10 K (C 5, p. 256). Though $CeCl_6^{3-}$ incorporated into $Cs_2NaYCl_6$ has $O_h$ symmetry at 4.2 K from an EPR spectrum [24] and from a study of the f→d transitions using absorption and MCD techniques [25], its symmetry seems to be lower when in $Cs_2NaCeCl_6$ as evidenced by the five electronic Raman bands (instead of four predicted for the crystal-field-ground state $\Gamma_7(^2F_{5/2})$ of $Ce^{3+}$)

which were observed at about 35 K, Amberger et al. [26]. Temporarily, a symmetry less than $O_h$ has also been presumed for $EuCl_6^{3-}$ by Schwartz et al. [23] because $Cs_2NaEuCl_6$ (like $Cs_2NaPrCl_6$ [21]) due to severe strain gave no MCD spectra at 4.2 K, Schwartz [27].

However, presumably except for M = Tm [28], the low-temperature electronic Raman bands for M = Nd, Sm [28], Eu [29], Dy, and presumably Er [28] at about 35 K and for M = Yb at about 56 K [30] could be identified on the basis of crystal field theory for exact $O_h$ symmetry. A list of the bands (for M = Ce and Pr too) is given by Amberger et al. [28]. In the systems with M = Tb (when excited with 647.1 nm), and Ho no electronic Raman bands were observed at about 35 K [28]. – The Raman shifts of the four predicted purely vibrational bands in the majority of these compounds show a regular dependence on the M atomic number and on the temperature between 30 and 300 K. A striking exception is, however, provided by the $\nu_2(E_g)$ band of $YbCl_6^{3-}$. Both the Raman shift and the shape of this band undergo unusual changes with temperature and suggest exceptionally strong vibronic coupling, Amberger et al. [31]. IR spectra (30 to 300 $cm^{-1}$) for all the $MCl_6^{3-}$ (M = Pr to Lu except Pm and Tm) are portrayed by Barbanel' et al. [32] with explicit values for $\nu_3$ and $\nu_4$. Both these show an irregularly increasing trend on going from $PrCl_6^{3-}$ to $LuCl_6^{3-}$. For totally assigned IR spectra, see for M = Pr [22], Eu [33], and Tm [34].

From analysis of the M–Cl force constants [35] it is concluded that the M–Cl bond strength increases from the start to the end of the 4f series, whereas the degree of f covalence of these bonds is significantly reduced in the first half of the series and increases slightly in the second, Barbanel' et al. [32, 35]. Remarkable covalent contributions to the M–Cl bonding have also been predicted by Kanellakopulos et al. [30] on considering the distances M–Cl in the ions and the ratios of the crystal field parameters $B_4^\circ(Pr)/B_4^\circ(Yb)$, and $B_6^\circ(Pr)/B_6^\circ(Yb)$. A list of the crystal field parameters is given by Amberger et al. [28], but in which the value of $B_4^\circ$ for M = Pr should be corrected to 271 $cm^{-1}$ (see Amberger et al. [36]). According to Amberger et al. [29] the overlapping integrals between the 4f functions and the 3s or 3p functions of the Cl ion decrease monotonically on going from $CeCl_6^{3-}$ to $YbCl_6^{3-}$. – In the optical f-f spectra of these (quasi) octahedral ions magnetic-dipole-allowed transitions (sometimes associated with a more or less extensive vibrational structure) and extensive vibrational structure around electric-dipole-forbidden transitions have been observed; see, for example, for M = Ce the absorption spectra (700 to 3200 $cm^{-1}$) at 300 and 88 K by Barbanel' et al. [32]; for M = Pr the emission spectra (12500 to 20950 $cm^{-1}$) at 300 and 12 K and the absorption spectra (4000 to 25000 $cm^{-1}$) at 290 and 4.2 K by Amberger [22] as well as the absorption and MCD spectra (15000 to 23000 $cm^{-1}$) at 300 K and (except MCD) at 4.2 K by Schwartz [21]; for M = Eu the luminescence Raman spectra (15850 to 16575 $cm^{-1}$, 19350 to 19700 $cm^{-1}$) at 300 K by Amberger [33], the emission spectra (490 to 650 nm) at 300 and 77 K by Serra, Thompson [37], the absorption and MCD spectra (16700 to 21800) at 300 K and (except MCD) at 4.2 K by Schwartz [27], and the absorption spectra (700 to 1250 $cm^{-1}$) and the emission spectrum (at about 16000 $cm^{-1}$) at 300 K by Barbanel' et al. [32]; for M = Tb the absorption, MCD, emission, and magnetic circularly polarized emission spectra (16000 to 20900 $cm^{-1}$) at 20 and 4.2 K by Schwartz [38] and the emission spectra at 300 and 77 K by Thompson et al. [39]; for M = Tm the absorption and MCD spectra (12500 to 28200 $cm^{-1}$) at 77, 20 and 4.2 K by Schwartz et al. [34]; and for M = Yb the absorption spectra at about 10000 $cm^{-1}$ at 300 and 77 K by Schwartz [40]. Assignment of the energy levels, the crystal-field states, and the vibrational structure associated with the octahedral $MCl_6^{3-}$ ions has been made in most cases, see also Serra et al. [41] and Amberger [42] for M = Eu. The vibrational structure is assigned to vibronically induced electric-dipole transitions. Intensity analysis of parts of the spectra (see for M=Eu [43], Tb [44], Tm [34], and Yb [40]) were based on an intensity model which included explicit consideration of pure magnetic-dipole transitions and vibronically induced transitions. It was assumed that the major features of the spectra could be accounted for in terms of localized $M^{3+}$ f-f electronic transitions and vibronic interactions involving localized (ungerade) vibrational modes of the $MCl_6^{3-}$ ion for M=Eu and Tb. The vibronic intensity model included both multipole ($M^{3+}$)-charge (ligand) and multipole ($M^{3+}$)-dipole (ligand) interactions between the electronic charge distributions on the metal ion and the ligands [43, 44]. In the case of M=Tm and Yb the bands could not be assigned based solely on vibrations of the $MCl_6^{3-}$ moiety [34, 40]. Two ligand-to-metal charge-transfer transitions (at 38315 and 47410 $cm^{-1}$) were measured in Yb doped $Cs_2NaYCl_6$ at about 10 K using absorption and MCD spectra [40].

### $MCl_6^{2-}$

The ions $PrCl_6^{2-}$ and $CeCl_6^{2-}$ are present in the compounds of type $R_2MCl_6$ (C5, pp. 223/4). From X-ray structure analysis, the $CeCl_6^{2-}$ ion has octahedral ($O_h$) symmetry in $[(CH_3)_4N]_2CeCl_6$, Cromer, Kline [45], Kiselev et al. [46], but a symmetry less than $O_h$ in the non-cubic compounds (their space group is given in parantheses) with $R=(C_2H_5)_4N^+$(Fmmm$-D_{2h}^{23}$), $R=(C_3H_7)_4N^+$($P2_122_1-D_2^3$), $R=(C_2H_5)_3C_7H_7N^+$ and $(C_2H_5)_2NH_2^+$($P2_12_12_1-D_2^4$), $R=C_5H_5NH^+$($P2_12_12-D_2^3$), Kiselev et al. [46], $R=(C_6H_5)_2I^+$, Adams, Morris [47], $R=(C_6H_5)_3PH^+$ [46], and with R=Cs($P\bar{3}m-D_{3d}^2$) [46], Kaatz, Marcovich [48]. In the vibrational IR and Raman spectra of these compounds the reduced symmetry is clearly evidenced only for $CeCl_6^{2-}$ in $[(C_4H_9)_4N]_2CeCl_6$ at −140 °C, Brandt et al. [49], and in $[(C_2H_5)_2NH_2]_2CeCl_6$ [49], $[(C_6H_5)_2I]_2CeCl_6$ [47], $[(C_6H_5)_3PH]_2CeCl_6$ and $Cs_2CeCl_6$ [49] at room temperature. Except for $CeCl_6^{2-}$ in $[(C_6H_5)_2I]_2CeCl_6$, the Raman-active vibration $\nu_2$ has not been observed in either of these compounds but Raman-active overtones and/or Raman- and IR-active combination bands were seen in the spectra. A list of all observed and assigned vibrations is given by Brandt et al. [49]. According to these authors [49], the assignment by Adams, Morris [47] of the vibrations at 265 and 277 $cm^{-1}$ of $CeCl_6^{2-}$ in $[(C_6H_5)_2I]_2CeCl_6$ should be corrected from $\nu_2$ into $\nu_3$. Based on the data by [47], a normal coordinate analysis was carried out using Wilson's GF matrix method; force constants, mean amplitudes of vibration, and Bastiansen-Morino shrinkages are given in the paper by Avasthi, Mehta [50].

### Further Ions

The ions $CeCl_4^-$ in $2CeCl_3\cdot[(CH_2)_6(NH_3)_2]Cl_2$, $CeCl_7^{4-}$ in $CeCl_3\cdot4[(CH_3)_2NH_2]Cl$, $LaCl_8^{5-}$ in $LaCl_3\cdot5[(CH_3)_3NH]Cl$ have been assumed to exist (C5, pp. 220/2). The complex $[CeCl_4(H_2O)_2]^-$ is present in $CeCl_3\cdot[C_5H_5NH]Cl\cdot2H_2O$ (C5, p. 223). The trichloride heptahydrates are said to have the complexes $[(H_2O)_7MCl_2M(H_2O)_7]^{4+}$ and the trichloride hexahydrates are said to be built up of the $[M(H_2O)_6Cl_2]^+$ complexes, see pp. 187/8, 191, and 194.

References:

[1] K.A. Skurupafis, N.N. Sevryukov (Elektrokhimiya **9** [1973] 207/11; Soviet Electrochem. **9** [1973] 198/201). – [2] J.L. Ryan, C.K. Jørgensen (J. Phys. Chem. **70** [1966] 2845/57). – [3] J.C. Barnes (J. Chem. Soc. **1964** 3880/5). – [4] V.A. Maroni, E.J. Hathaway, G.N. Papatheodorou (J. Phys. Chem. **78** [1974] 1134/5). – [5] G.N. Papatheodorou (Inorg. Nucl. Chem. Letters **11** [1975] 483/90).

[6] G.N. Papatheodorou (J. Chem. Phys. **66** [1977] 2893/900). – [7] G.N. Papatheodorou, T. Østvold (J. Phys. Chem. **78** [1974] 181/5). – [8] J.J. Egan, J. Bracker (J. Chem. Thermodyn. **6** [1974] 9/16). – [9] G.N. Papatheodorou, O. Waernes, T. Østvold (Acta Chem. Scand. A **33** [1979] 173/8). – [10] S. Senderoff, G.W. Mellors, R.I. Bretz (Ann. N.Y. Acad. Sci. **79** [1959] 878/96).

[11] G.N. Papatheodorou, O.J. Kleppa (J. Phys. Chem. **78** [1974] 178/81), – [12] F. Dienstbach, R. Blachnik (Z. Anorg. Allgem. Chem. **412** [1975] 97/109). – [13] Yu.A. Barbanel', V.R. Klokman, V.P. Kotlin, V.V. Kolin (Radiokhimiya **21** [1979] 694/705; Soviet Radiochem. **21** [1979] 600/10). – [14] Yu.A. Barbanel (Proc. Moscow Symp. Chem. Transuranium Elem., Moscow 1972 [1976], pp. 79/84). – [15] D.E. Henrie, B.K. Henrie (J. Inorg. Nucl. Chem. **36** [1974] 2125/8).

[16] D.E. Henrie, B.K. Henrie (J. Inorg. Nucl. Chem. **39** [1977] 1583/6). – [17] R.J. Majeste, D. Chriss, L.M. Trefonas (Inorg. Chem. **16** [1977] 188/91). – [18] J.B. Gruber, E.R. Menzel, J.L. Ryan (J. Chem. Phys. **51** [1969] 3816/9). – [19] E. Butter, W. Seifert (Z. Anorg. Allgem. Chem. **380** [1971] 189/95). – [20] R.D. Peacock, B. Stewart (Chem. Phys. Letters **55** [1978] 295/6).

[21] R.W. Schwartz (Mol. Phys. **31** [1976] 1909/21). – [22] H.-D. Amberger (Z. Anorg. Allgem. Chem. **439** [1978] 48/56). – [23] R.W. Schwartz, S.F. Watkins, C.F. O'Connor, R.L. Carlin (J. Chem. Soc. Faraday Trans. II **72** [1976] 565/70). – [24] R.W. Schwartz, J.N. Hill (J. Chem. Soc. Faraday Trans. II **70** [1974] 124/31). – [25] R.W. Schwartz, P.N. Schatz (Phys. Rev. [3] B **8** [1973] 3229/36).

[26] H.-D. Amberger, G.G. Rosenbauer, R.D. Fischer (Mol. Phys. **32** [1976] 1291/8). – [27] R.W. Schwartz (Mol. Phys. **30** [1975] 81/95). – [28] H.-D. Amberger, G.G. Rosenbauer, R.D. Fischer (J. Phys. Chem. Solids **38** [1977] 379/85). – [29] H.-D. Amberger, R.D. Fischer, G.G. Rosenbauer (Z. Physik. Chem. [Frankfurt] **102** [1976] 279/82). – [30] B. Kanellakopulos, H.-D. Amberger, G.G. Rosenbauer, R.D. Fischer (J. Inorg. Nucl. Chem. **39** [1977] 607/11).

[31] H.-D. Amberger, R.D. Fischer, G.G. Rosenbauer (Transition Metal Chem. [Weinheim] **1** [1976] 242/6; C.A. **86** [1977] No. 130392). – [32] Yu.A. Barbanel', R.B. Dushin, N.K. Mikhailova, G.P. Chudnovskaya (Radiokhimiya **21** [1979] 706/13; Soviet Radiochem. **21** [1979] 610/6). – [33] H.-D. Amberger (Spectrochim. Acta A **34** [1978] 627/31). – [34] R.W. Schwartz, T.R. Faulkner, F.S. Richardson (Mol. Phys. **38** [1979] 1767/80). – [35] Yu.A. Barbanel', R.B. Dushin, V.V. Kolin, N.K. Mikhailova, G.P. Chudnovskaya (Koord. Khim. **1** [1975] 411/5 from C.A. **83** [1975] No. 35106).

[36] H.-D. Amberger, R.D. Fischer, G.G. Rosenbauer (Ber. Bunsenges. Physik. Chem. **79** [1975] 1226/30). – [37] O.A. Serra, L.C. Thompson (Inorg. Chem. **15** [1976] 504/7). – [38] R.W. Schwartz, H.B. Brittain, J.P. Riehl, W. Yeakel, F.S. Richardson (Mol. Phys. **34** [1977] 361/79). – [39] L.C. Thompson, O.A. Serra, J.P. Riehl, F.S. Richardson, R.W. Schwartz (Chem. Phys. **26** [1977] 393/401). – [40] R.W. Schwartz (Inorg. Chem. **16** [1977] 1694/8).

[41] O.A. Serra, L.C. Thompson, J.A. Koningstein (Chem. Phys. Letters **46** [1977] 253/5). – [42] H.-D. Amberger (Z. Physik. Chem. [Wiesbaden] **109** [1978] 125/8). – [43] T.R. Faulkner, F.S. Richardson (Mol. Phys. **35** [1978] 1141/61). – [44] T.R. Faulkner, F.S. Richardson (Mol. Phys. **36** [1978] 193/214). – [45] D.T. Cromer, R.J. Kline (J. Am. Chem. Soc. **76** [1954] 5282/3).

[46] Yu.M. Kiselev, A. Brandt, L.I. Martynenko, V.I. Spitsyn (Dokl. Akad. Nauk SSSR **246** [1979] 879/82; Dokl. Chem. Proc. Acad. Sci. USSR **244/249** [1979] 271/3). –

[47] D.M. Adams, D.M. Morris (J. Chem. Soc. A **1968** 694/5). – [48] T. Kaatz, M. Marcovich (Acta Cryst. **21** [1966] 1011). – [49] A. Brandt, Yu.M. Kiselev, L.I. Martynenko (Z. Anorg. Allgem. Chem. **474** [1981] 233/40). – [50] M.N. Avasthi, M.L. Mehta (Spectrosc. Letters **2** [1969] 383/7).

[51] A.M. El-Aggan, D.C. Bradley, W. Wardlaw (J. Chem. Soc. **1958** 2092/9).

## Table of Conversion Factors

Following the notation in Landolt-Börnstein [7], values which have been fixed by convention are indicated by a bold-face last digit. The conversion factor between calorie and Joule that is given here is based on the thermochemical calorie, $cal_{th\,ch}$, and is defined as 4.184**0** J/cal. However, for the conversion of the ,,Internationale Tafelkalorie'', $cal_{IT}$, into Joule, the factor 4.1868 J/cal is to be used [1, p. 147]. For the conversion factor for the British thermal unit, the Steam Table Btu, $BTU_{ST}$, is used [1, p. 95].

| **Force** | N | dyn | kp |
|---|---|---|---|
| 1 N (Newton) | 1 | $10^5$ | 0.1019716 |
| 1 dyn | $10^{-5}$ | 1 | $1.019716 \times 10^{-6}$ |
| 1 kp | 9.8066**5** | 9.8066**5** $\times 10^5$ | 1 |

| **Pressure** | Pa | bar | kp/m$^2$ | at | atm | Torr | lb/in$^2$ |
|---|---|---|---|---|---|---|---|
| 1 Pa (Pascal) = 1 N/m$^2$ | 1 | $10^{-5}$ | $1.019716 \times 10^{-1}$ | $1.019716 \times 10^{-5}$ | $0.986923 \times 10^{-5}$ | $0.750062 \times 10^{-2}$ | $145.0378 \times 10^{-6}$ |
| 1 bar = $10^6$ dyn/cm$^2$ | $10^5$ | 1 | $10.19716 \times 10^3$ | 1.019716 | 0.986923 | 750.062 | 14.50378 |
| 1 kp/m$^2$ = 1 mm $H_2O$ | 9.8066**5** | 0.98066**5** $\times 10^{-4}$ | 1 | $10^{-4}$ | $0.967841 \times 10^{-4}$ | $0.735559 \times 10^{-1}$ | $1.422335 \times 10^{-3}$ |
| 1 at = 1 kp/cm$^2$ | 0.98066**5** $\times 10^5$ | 0.98066**5** | $10^4$ | 1 | 0.967841 | 735.559 | 14.22335 |
| 1 atm = 76**0** Torr | 1.0132**5** $\times 10^5$ | 1.0132**5** | $1.033227 \times 10^4$ | 1.033227 | 1 | 760 | 14.69595 |
| 1 Torr = 1 mm Hg | 133.3224 | $1.333224 \times 10^{-3}$ | 13.59510 | $1.359510 \times 10^{-3}$ | $1.315789 \times 10^{-3}$ | 1 | $19.33678 \times 10^{-3}$ |
| 1 lb/in$^2$ = 1 psi | $6.89476 \times 10^3$ | $68.9476 \times 10^{-3}$ | 703.069 | $70.3069 \times 10^{-3}$ | $68.0460 \times 10^{-3}$ | 51.7149 | 1 |

| **Work, Energy, Heat** | J | kWh | kcal | Btu | MeV |
|---|---|---|---|---|---|
| 1 J (Joule) = 1 Ws = 1 Nm = $10^7$ erg | 1 | $2.778 \times 10^{-7}$ | $2.39006 \times 10^{-4}$ | $9.4781 \times 10^{-4}$ | $6.242 \times 10^{12}$ |
| 1 kWh | $3.\mathbf{6} \times 10^{6}$ | 1 | 860.4 | 3412.14 | $2.247 \times 10^{19}$ |
| 1 kcal | 4184.**0** | $1.1622 \times 10^{-3}$ | 1 | 3.96566 | $2.6117 \times 10^{16}$ |
| 1 Btu (British thermal unit) | 1055.06 | $2.93071 \times 10^{-4}$ | 0.25164 | 1 | $6.5858 \times 10^{15}$ |
| 1 MeV | $1.602 \times 10^{-13}$ | $4.450 \times 10^{-20}$ | $3.8289 \times 10^{-17}$ | $1.51840 \times 10^{-16}$ | 1 |

| **Power** | kW | h. p. (PS) | kp m/s | kcal/s |
|---|---|---|---|---|
| 1 kW = $10^{10}$ erg/s | 1 | 1.35962 | 101.972 | 0.239006 |
| 1 h. p. (PS) | 0.73550 | 1 | **75** | 0.17579 |
| 1 kp m/s | $9.8066\mathbf{5} \times 10^{-3}$ | 0.01333 | 1 | $2.34384 \times 10^{-3}$ |
| 1 kcal/s | 4.184**0** | 5.6886 | 426.650 | 1 |


References:

[1] A. Sacklowski, Die neuen SI-Einheiten, Goldmann, München 1979. (Conversion tables in an appendix.)

[2] International Union of Pure and Applied Chemistry, Manual of Symbols and Terminology for Physicochemical Quantities and Units, Pergamon, London 1979; Pure Appl. Chem. **51** [1979] 1/41.

[3] The International System of Units (SI), National Bureau of Standards Specl. Publ. 330, [1972].

[4] H. Ebert, Physikalisches Taschenbuch, 5th Ed., Vieweg, Wiesbaden 1976.

[5] Kraftwerk Union Information, Technical and Economic Data on Power Engineering, Mülheim/Ruhr 1978.

[6] E. Padelt, H. Laporte, Einheiten und Größenarten der Naturwissenschaften, 3rd Ed., VEB Fachbuchverlag, Leipzig 1976.

[7] Landolt-Börnstein, 6. Aufl., II. Bd., 1. Tl., 1971, S. 1/14.